Professor Boivin (1955–2014) completed his Ph.D. at the Université de Montréal in 1984 under the direction of Paul Gauthier. He joined the University of Western Ontario as an Assistant Professor in 1986, after holding postdoctoral fellowships at UCLA and University College, London. He was promoted to Associate Professor in 1991, and then to Professor in 2004. He was appointed as Chair of Western's Department of Mathematics in 2011. His research specialties were complex analysis and approximation theory, and he was the author of multiple papers in these areas. He gave tireless service to granting agencies and selection committees in Québec and Ontario, and was a frequent conference organizer.

Professor Boivin participated in the conference on *Invariant Subspaces of the Shift Operator* and even contributed to this proceedings. His sudden death as a result of a heart failure was a great shock for all of us. With great affection for André and profound regret for his passing, we devote the present volume to his memory.

CONTEMPORARY MATHEMATICS

638

Centre de Recherches Mathématiques Proceedings

Invariant Subspaces of the Shift Operator

CRM Workshop
Invariant Subspaces of the Shift Operator
August 26–30, 2013
Centre de Recherches Mathématiques,
Université de Montréal, Montréal

Javad Mashreghi
Emmanuel Fricain
William Ross
Editors

American Mathematical Society
Providence, Rhode Island

Centre de Recherches Mathématiques
Montréal, Québec, Canada

2010 *Mathematics Subject Classification.* Primary 47-XX, 30-XX, 31-XX, 32-XX.

Library of Congress Cataloging-in-Publication Data

Invariant subspaces of the shift operator : CRM Workshop on Invariant Subspaces of the Shift Operator, August 26–30, 2013, Centre de Recherches Mathématiques, Université de Montréal, Montréal : Centre de Recherches Mathématiques proceedings / Javad Mashreghi, Emmanuel Fricain, William Ross, editors.

pages cm. – (Contemporary mathematics ; volume 638)

Includes bibliographical references.

ISBN 978-1-4704-1045-2 (alk. paper)

1. Shift operators (Operator theory)—Congresses. 2. Hilbert space—Congresses. 3. Banach spaces—Congresses. I. Mashreghi, Javad, editor. II. Fricain, Emmanuel, 1971– editor. III. Ross, William T., 1964– editor.

QA329.2.I545 2015

515′.39–dc23 2014038149

Contemporary Mathematics ISSN: 0271-4132 (print); ISSN: 1098-3627 (online)

DOI: `http://dx.doi.org/10.1090/conm/638`

---

10 9 8 7 6 5 4 3 2 1 20 19 18 17 16 15

Contents

Preface

The main theme of this proceedings volume is the invariant subspaces of the shift operator S, or its adjoint S^*, on certain reproducing kernel Hilbert spaces of analytic functions on the open unit disk. Such spaces include the Hardy spaces H^2, the Dirichlet space $\mathcal{D}$, the de Branges-Rovnyak spaces $\mathcal{H}(b)$, and the model spaces K_Θ.

Model spaces have many fascinating aspects. For one, they represent, via Beurling's theorem, which characterizes the invariant subspaces of the shift operator $Sf = zf$ on H^2, the complete set of invariant subspaces of the backward shift operator $S^*f = (f - f(0))/z$. Using the theory of pseudo continuations developed by H. S. Shapiro, the description of these backward shift invariant subspaces was described in a seminal paper of Douglas, Shapiro, and Shields. The concept of pseudocontinuation continues to have an uncanny way of appearing in many unexpected areas of analysis.

These backward shift subspaces K_Θ of H^2 became to be known as model spaces via the Nagy-Foiaş theory when the compression S_Θ of the shift S to K_Θ was shown to represent a wide class of contraction operators on Hilbert space. Though these spaces make connections to operator theory and some areas of mathematical physics, they are a fascinating Hilbert space of analytic functions in the own right. They have interesting reproducing kernels and boundary behavior. The rank-one unitary perturbations of S_Θ (the Clark unitary operators) are an interesting class of operators with an even more interesting and useful set of spectral measures (Clark measures).

Parallel to the theory of Sz.-Nagy and Foias, de Branges and Rovnyak developed another model based on $\mathcal{H}(b)$ spaces, where b is an analytic function in the closed unit ball of H^∞ (bounded analytic functions on the open unit disk). When b is inner, $\mathcal{H}(b)$ coincides with the model space K_b. In the general case, $\mathcal{H}(b)$ spaces are not a closed subspace of H^2, but they are equipped with a norm to become a Hilbert space contractively embedded in H^2. Their structure is fascinating and depends on whether or not $\log(1 - |b|)$ is integrable on the unit circle, equivalently, non-extreme or extreme points in the closed unit ball of H^∞. Since the foundation developed by de Branges and Rovnyak, $\mathcal{H}(b)$ spaces continue to be a precious tool in various questions in analysis such as function theory (resolution of the Bieberbach conjecture by de Branges, rigid functions, Schwarz–Pick inequalities), operator theory (invariant subspace problem, composition operators, kernel of the Toeplitz operators), systems and control theory.

Related to the Hardy space, model spaces, and de Branges-Rovnyak spaces, is the Dirichlet space. This space connects to many areas and tools of analysis. For example, the definition of the Dirichlet space (analytic functions on the disk whose image has finite area measure) certainly connects to geometry. As the name Dirichlet suggests, the Dirichlet space connects to Dirichlet's method for solving Laplace's equation. This naturally leads to another connection, logarithmic potential theory. Though much is known about this space, there remain several important open problems, most notably the characterization of its zero sets and of its shift-invariant subspaces.

This proceedings is the outcome of a conference that was held at CRM (Centre de Recherches Mathématiques) in Montreal from 26 to 30 August 2013. Numerous international experts in this area presented their ongoing research. Moreover, for graduate students and the newcomers to the field, there were several mini-courses on the basics of the H^2, K_Θ, $\mathcal{H}(b)$, and $\mathcal{D}$. The Editors would like to thank CRM for hosting this event and for its generous financial support for the invited speakers.

Contemporary Mathematics
Volume **638**, 2015
http://dx.doi.org/10.1090/conm/638/12802

Approximation numbers of composition operators on a Hilbert space of Dirichlet series

Hervé Queffélec

ABSTRACT. By a theorem of Gordon and Hedenmalm, φ generates a bounded composition operator on the Hilbert space $\mathscr{H}^2$ of Dirichlet series $\sum_n b_n n^{-s}$ with square-summable coefficients b_n if and only if $\varphi(s) = c_0 s + \psi(s)$, where c_0 is a nonnegative integer and ψ a Dirichlet series with the following mapping properties: ψ maps the right half-plane into the half-plane $\mathrm{Re}\, s > 1/2$ if $c_0 = 0$ and is either identically zero or maps the right half-plane into itself if c_0 is positive. It is shown that the nth approximation number of a bounded composition operator on $\mathscr{H}^2$ is bounded below by a constant times r^n for some $0 < r < 1$ when $c_0 = 0$ and by a constant times n^{-A} for some $A > 0$ when c_0 is positive. Both results are best possible. The case $\varphi(s) = c_1 + \sum_{j=1}^{d} c_{q_j} q_j^{-s}$ is mentioned. In that case, the nth approximation number behaves as $n^{-(d-1)/2}$, possibly up to a factor $(\log n)^{(d-1)/2}$. Finally, a general transference principle and recent results in the usual Hardy space allow us to exhibit compact composition operators on $\mathscr{H}^2$ whose approximation numbers decay arbitrarily slowly. Estimates rely mainly on a general Hilbert space method involving reproducing kernels. A key role is played by a recently developed interpolation method for $\mathscr{H}^2$ using estimates of solutions of the $\overline{\partial}$ equation.

1. Introduction and statement of main results

This survey is mainly based on the paper [**25**]. Let Ω be an open subset of $\mathbb{C}$, $H(\Omega)$ the space of analytic functions on Ω with its natural topology, and $H \subset H(\Omega)$ be a separable Hilbert space with a specified orthonormal basis (e_n). We assume that H is continuously embedded in $H(\Omega)$, which amounts to say that the point evaluations δ_a, $a \in \Omega$ are continuous on H and therefore given by a scalar product:

$$\delta_a(f) = f(a) = \langle f, K_a \rangle, \text{ where } K_a \in H.$$

The function K_a is called the reproducing kernel of H at a and classically given by

$$K_a(z) = \sum e_n(z)\overline{e_n(a)}$$

with convergence of the series in H. Let now $\varphi : \Omega \to \Omega$ be an analytic self-map and $C_\varphi : H \to H(\Omega)$ the associated composition operator defined by

$$C_\varphi(f) = f \circ \varphi.$$

We want to know when C_φ actually maps H to itself (then, φ is called a "symbol"), and to compare the properties of the function $\varphi : \Omega \to \Omega$ and the operator C_φ :

2010 *Mathematics Subject Classification.* Primary 47B33, 30B50, 30H10.

$H \to H$, in particular its spectrum, compactness, membership in a Schatten class, and more precisely the decay rate of its approximation numbers to be defined. One basic example is that of the Hardy space

$$H^2(\mathbb{D}) = \Big\{ f(z) = \sum_{n=0}^{\infty} b_n z^n : \|f\|^2 := \sum_{n=0}^{\infty} |b_n|^2 < \infty \Big\},$$

a Hilbert space of analytic functions on the open unit disk $\mathbb{D}$, with orthonormal basis $e_n(z) = z^n$, $n = 0, \dots$ and hence reproducing kernel $K_a(z) = \frac{1}{1-\overline{a}z}$. In that case, all analytic self-maps $\varphi : \mathbb{D} \to \mathbb{D}$ are symbols ([**31**]).

By definition, the subject is border-line between soft analysis (here operator theory) and hard analysis (Carleson measures, interpolation sequences, d-bar correction). Moreover, some proofs are rather delicate and long. And accordingly this survey will only detail the involved tools and one proof. It is roughly divided into five general parts:

(1) The specific Hilbert space involved and its composition operators
(2) Definitions and tools from operator theory (Weyl's inequalities, ...)
(3) Definitions and tools from function theory (Carleson measures,...)
(4) General estimates for approximation numbers
(5) Statement of the main results and proof of one.

2. The space $\mathscr{H}^2$ of Dirichlet series

We will denote by $\mathbb{C}_\theta$ the half-plane $\mathbb{C}_\theta = \{s : \mathrm{Re}\, s > \theta\}$, $\theta \in \mathbb{R}$. The Hilbert space $\mathscr{H}^2$ consists of all ordinary Dirichlet series $f(s) = \sum_{n=1}^{\infty} b_n n^{-s}$ such that

$$\|f\|^2_{\mathscr{H}^2} := \sum_{n=1}^{\infty} |b_n|^2 < \infty.$$

This is clearly a Hilbert space of analytic functions on $\mathbb{C}_{1/2}$ (by the Cauchy-Schwarz inequality) with orthonormal basis e_n, $e_n(s) = n^{-s}$, $n = 1, 2, \dots$ and therefore reproducing kernel

$$K_a(s) = \sum_{n=1}^{\infty} e_n(s)\overline{e_n(a)} = \zeta(s + \overline{a})$$

where ζ is the Riemann zeta function.

This space was (re)-introduced in 1997 by Hedenmalm, Lindqvist, Seip ([**13**]) to characterize those functions $f(t) = \sum_{n=1}^{\infty} \sqrt{2} a_n \sin 2\pi n t$ whose dilates $f_k(t) = f(kt)$ form a Riesz sequence (an essential issue in the present work) in $L^2(0,1)$. The authors showed (see also [**6**]) that the space of multipliers of $\mathscr{H}^2$ is (isometrically) the space $\mathcal{H}^\infty$ of Dirichlet series which are bounded in $\mathbb{C}_0$, and hence convergent in this half-plane thanks to a result of Bohr ([**4**]). Their main theorem is as follows

THEOREM 2.1 ([**13**]). *Let $G(s) = \sum_{n=1}^{\infty} a_n n^{-s}$ be the Dirichlet generating function of $f(t) = \sum_{n=1}^{\infty} \sqrt{2} a_n \sin 2\pi n t$. Then, the following are equivalent:*

1. (f_k) is a Riesz sequence in $L^2(0,1)$.
2. G and $1/G$ belong to $\mathcal{H}^\infty$.

We now describe the admissible composition operators on $\mathscr{H}^2$.

3. Bounded composition operators on $\mathscr{H}^2$

The Gordon–Hedenmalm theorem ([**12**]) gives a full description of those analytic maps $\varphi : \mathbb{C}_{1/2} \to \mathbb{C}_{1/2}$ which generate a composition operator on $\mathscr{H}^2$ and reads as follows, under a slightly reinforced form ([**25**]):

THEOREM 3.1 ([**12**]). *The function φ determines a bounded composition operator C_φ on $\mathscr{H}^2$ if and only if*

$$\varphi(s) = c_0 s + \sum_{n=1}^{\infty} c_n n^{-s} =: c_0 s + \psi(s),$$

where c_0 is a nonnegative integer and ψ is a Dirichlet series that converges uniformly in $\mathbb{C}_\varepsilon$ for every $\varepsilon > 0$ and has the following mapping properties:

(a) *If $c_0 = 0$, then $\psi(\mathbb{C}_0) \subset \mathbb{C}_{1/2}$.*
(b) *If $c_0 \geq 1$, then either $\psi \equiv 0$ or $\psi(\mathbb{C}_0) \subset \mathbb{C}_0$.*

We thus see that there are few admissible symbols φ and that, depending on the value of a weird parameter c_0, they have more or less restrictive mapping properties (observe that $c_0 \geq 1$ corresponds to $\varphi(\infty) = \infty$). Also note that this reformulation of the initial result is slightly stronger, in that we claim the uniform convergence of ψ in each half-plane $\mathbb{C}_\varepsilon$, not only in some half-plane $\mathbb{C}_{\sigma_0}$, and that the mapping properties are stated in terms of ψ rather than in terms of φ, which is stronger and was also proved, in a different way, in the paper [**12**].

4. Definitions and tools from operator theory

4.1. Approximation numbers. Let $T : H \to H$ and $n \geq 1$. The n-th approximation number of T is: $a_n(T) = \inf\{\|T - R\|;\ \text{rank } R < n\}$. Then (short list, see [**5**, p. 155]):

1. $\|T\| = a_1(T) \geq a_2(T) \geq \cdots \geq a_n(T) \geq \cdots$

2. T is compact iff $a_n(T) \downarrow 0$. Otherwise (take $T =$ Diag ε_n), the sequence $(a_n(T))$ can be any arbitrary non-increasing sequence.

3. $a_n(T) = a_n(T^*)$.

4. $a_n(ATB) \leq \|A\| a_n(T) \|B\|$ (Ideal property).

5. $a_n(T) = s_n(T)$ (singular number, Allakhverdiev, 1957).

6. $S_p = \{T \in \mathcal{L}(H)\ ;\ \left[\sum_{n=1}^{\infty} (a_n(T))^p\right]^{1/p} < \infty\}$ (Schatten class S_p, $p > 0$). Note that $p < q \Rightarrow S_p \subset S_q$.

The decay rate of $a_n(T)$ (non-commutative approximation theory) measures the degree of compactness of T.

We recall that (Schmidt decomposition, see [**5**], p.46) a compact operator $T : H \to H$ can be written under the form $T(x) = \sum_{n=1}^{\infty} s_n(T) \langle x, v_n \rangle u_n$ where (u_n) and (v_n) are two orthonormal sequences, and that we have the equations

$$T(v_n) = s_n(T) u_n, \quad s_n(T) = a_n(T) = \langle T(v_n), u_n \rangle. \tag{1}$$

But we know neither the v_n's nor the u_n's! We must proceed differently to estimate the decay rate of a_n. For symbolic operators, estimating this decay rate is a delicate issue...

Here are three questions of increasing difficulty which we will consider for $T = C_\varphi$:

(1) Compactness of T?
(2) Schattenness of T?
(3) Decay rate of $a_n(T)$?

Let us emphasize that question 3, the main object of this paper, is more demanding than question 2. For example, we trivially have

$$T \in \cap_{p>0} S_p \Leftrightarrow a_n(T) = O(n^{-A}) \text{ for all } A > 0.$$

For $T = C_\varphi$ viewed on the Hardy space $H^2(\mathbb{D})$ and associated with a lens map φ, such an estimate was obtained in [**30**], under the form $T \in \cap_{p>0} S_p$. But this is less precise than the estimate $a_n(T) \ll e^{-b\sqrt{n}}$, later obtained in [**16**], in fact as a two-sided estimate.
A word on the symbol $\ll$: here and in what follows the notation $f(u) \ll g(u)$ or equivalently $g(u) \gg f(u)$ means that there is a constant C such that $f(u) \leq Cg(u)$ for all parameters u in question.

4.2. Tools for estimating $a_n(T)$.

(1) Lower bounds: we must *erase* R of rank $< n$ and use

$$a_n(T) = \sup_{\dim E = n} \left[\inf_{f \in S_E} \|Tf\| \right] =: n^{th} \text{ Bernstein number of } T,$$

where S_E denotes the unit sphere of E. This works as well with *two* Hilbert spaces H_1, H_2 and $T : H_1 \to H_2$. We refer to [**24**] for more on these lines.
Indeed, if the rank of R is $< n$, let $f_0 \in S_E$ such that $Rf_0 = 0$. Then:

$$\|T - R\| \geq \|Tf_0 - Rf_0\| = \|Tf_0\| \geq \inf_{f \in S_E} \|Tf\|.$$

The issue is a good choice of E, which will later be taken as a *model space,* generated by a Riesz system of reproducing kernels.

(2) Upper bounds: we must *specialize* R of rank $< n$ and use

$$a_n(T) = \inf_{\text{rank} P < n} \|T - TP\| = n^{th} \text{ Gelfand number of } T$$

where P runs over orthoprojections of rank $< n$ (see [**5**]).
Indeed, this follows from $a_n = s_n$ and the choice $P_{n-1} = \sum_{k=1}^{n-1} v_k \otimes v_k$, the orthoprojection on the span of $(v_1, \ldots, v_{n-1})$, for which obviously $\|T - TP_{n-1}\| = s_n = a_n$. Once more, this marvellous P_{n-1} is beyond reach... The issue here is a good choice of P, which will be taken as an orthoprojection on some *model space.*

We shall here essentially dwell on *lower bounds* for the $a_n(T)$. The tools are roughly the same for upper bounds.

4.3. Use of the spectrum. We do not know the approximation numbers a_n of C_φ (this is what we wish to estimate!), but we will have access to its eigenvalues λ_n, rearranged in decreasing order, thanks to a result of F.Bayart ([**3**]). We can exploit this through the Weyl inequalities ([**5**], p.157):

LEMMA 4.1 (H.Weyl). *For $T : H \to H$ with approximation numbers (a_n) and eigenvalues (λ_n) arranged in descending order, we always have:*

$$a_1 \cdots a_n \geq |\lambda_1 \cdots \lambda_n|, \quad n = 1, 2, \ldots .$$

Note that the inequality $a_n \geq |\lambda_n|$ is completely wrong (see [**14**] p.133)! But using that a_n decreases, we can derive a pointwise estimate of a_n by a small tauberian argument: let N be an integer ≥ 2. Applying Weyl with Nn gives

$$|\lambda_1 \cdots \lambda_{Nn}| \leq a_1 \cdots a_{Nn} \leq a_1^n a_n^{(N-1)n}.$$

Taking $(N-1)n$-th roots, we get

$$a_n \geq a_1^{-1/(N-1)} \left(|\lambda_1 \ldots \lambda_{Nn}|\right)^{1/(N-1)n}. \tag{2}$$

4.4. Riesz systems. We now pave the way to the use of Bernstein numbers and model spaces, which requires definitions. Let $(H,\ K_a)$ be a functional Hilbert space on the set Ω. In this general setting, no analyticity is required.

A sequence (x_j) of vectors of H is called:

(1) An upper Riesz system if $\|\sum_j \lambda_j x_j\|^2 \leq C \sum_j |\lambda_j|^2 \|x_j\|^2$. The best constant C is the upper Riesz constant of (x_j).

(2) A lower Riesz system if $\|\sum_j \lambda_j x_j\|^2 \geq c \sum_j |\lambda_j|^2 \|x_j\|^2$. The best constant c is the lower Riesz constant of (x_j).

(3) A Riesz system if it is both an upper and lower Riesz system.

5. Definitions and tools from function theory

5.1. Carleson measures and Carleson sequences. We will use the two following definitions.

1. A probability measure μ on Ω is called a Carleson measure for H if there is a constant C such that

$$\int_\Omega |f(z)|^2 d\mu(z) \leq C\|f\|_H^2, \ \forall f \in H.$$

The best possible C is called the Carleson norm of μ, and is denoted by $C = \|\mu\|_{\mathcal{C},H}$.

Let now $Z = (z_j)$ be a sequence of Ω.

2. Z is a Carleson sequence if the discrete measure

$$\mu_{Z,H} := \sum \|K_{z_j}\|^{-2} \delta_{z_j}$$

is Carleson for H. The Carleson constant of (z_j) is by definition $\|\mu_{Z,H}\|_{\mathcal{C},H}$.

5.2. Interpolation sequences. The sequence $Z = (z_j)$ is an interpolation sequence if for all (w_j) such that $\sum |w_j|^2 \|K_{z_j}\|^{-2} < \infty$, there exists $f \in H$ such that

$$f(z_j) = w_j \text{ and } \|f\|_H \leq C \left(\sum |w_j|^2 \|K_{z_j}\|^{-2}\right)^{1/2}.$$

The best possible C is called the interpolation constant of (z_j), and it is denoted by $C = M_H(Z)$.

5.3. Criteria for Rieszness. The space E generated by the K_{z_j} is called *a model space.* The following proposition gives NSC conditions for a sequence (K_{z_j}) of reproducing kernels to be an upper or lower Riesz basis of E, with control of constants.

LEMMA 5.1 (Boas). *Let $Z = (z_j)$ be a finite sequence in Ω. Then*

(1) $\|\sum_j \lambda_j K_{z_j}\|^2 \leq \|\mu_{Z,H}\|_{\mathcal{C},H} \sum_j |\lambda_j|^2 \|K_{z_j}\|^2$, *and moreover* $\|\mu_{Z,H}\|_{\mathcal{C},H}$ *is the upper Riesz constant of* (K_{z_j}).

(2) $\|\sum_j \lambda_j K_{z_j}\|^2 \geq [M_H(Z)]^{-2} \sum_j |\lambda_j|^2 \|K_{z_j}\|^2$, *and moreover* $[M_H(Z)]^{-2}$ *is the lower Riesz constant of* (K_{z_j}).

We give the proof (simple, but scattered in the literature) for sake of completeness.

(1). Suppose $\mu = \mu_{Z,H}$ Carleson. Let $f = \sum \lambda_j K_{z_j}$ be a finite sum. Observe that $\|f\|^2 = \langle f, \sum \lambda_j K_{z_j}\rangle = \sum \overline{\lambda_j} f(z_j)$. Using Cauchy-Schwarz, we can write

$$\|f\|^4 = \big|\sum \overline{\lambda_j} f(z_j)\big|^2 \leq \big(\sum_j |\lambda_j|^2 \|K_{z_j}\|^2\big)\big(\sum_l |f(z_l)|^2 \|K_{z_l}\|^{-2}\big)$$

$$\leq \|\mu\|_{\mathcal{C},H}\big(\sum_j |\lambda_j|^2 \|K_{z_j}\|^2\big)\|f\|^2$$

and we get the claimed inequality after simplification. Conversely, if (K_{z_j}) is upper Riesz with constant C, let $f \in H$ with $\|f\|_H = 1$. For some (w_j) such that $\sum |w_j|^2 = 1$, we can write

$$\left(\int |f|^2 d\mu\right)^{1/2} = \left(\sum_j \frac{|f(z_j|^2}{\|K_{z_j}\|^2}\right)^{1/2} = \sum_j \overline{w_j} \frac{f(z_j)}{\|K_{z_j}\|}$$

$$= \langle f, \sum_j \frac{w_j}{\|K_{z_j}\|} K_{z_j}\rangle \leq \Big\|\sum_j \frac{w_j}{\|K_{z_j}\|} K_{z_j}\Big\| \leq \sqrt{C}\left(\sum_j \frac{|w_j|^2}{\|K_{z_j}\|^2}\|K_{z_j}\|^2\right)^{1/2} = \sqrt{C}$$

so that Z is Carleson with constant $\leq C$.

(2). Suppose that Z is interpolating and let (λ_j) be a finite sequence. Then, by linearization, we can write

$$S := \big(\sum_j |\lambda_j|^2 \|K_{z_j}\|^2\big)^{1/2} = \sum_j c_j \overline{\lambda_j} \|K_{z_j}\| \text{ with } \sum |c_j|^2 = 1.$$

Next, observe that $\sum_j (|c_j| \|K_{z_j}\|)^2 (\|K_{z_j}\|)^{-2}) = 1$, so that $c_j \|K_{z_j}\| = f(z_j)$ for some $f \in H$ verifying $\|f\|_H \leq M_H(Z)$. We then obtain that

$$S = \sum_j \overline{\lambda_j} f(z_j) = \langle f, \sum_j \lambda_j K_{z_j} \rangle \leq M_H(Z) \| \sum_j \lambda_j K_{z_j} \|.$$

Conversely, suppose that $(f_j) := (K_{z_j})$ is lower Riesz with constant c. Let (w_j) such that $\sum |w_j|^2 \|f_j\|^{-2} = 1$. Define L on the span E of the f_j by $L(\sum_j \lambda_j f_j) = \sum \lambda_j \overline{w_j}$. We have

$$|L(\sum_j \lambda_j f_j)| \leq \big(\sum |\lambda_j|^2 \|f_j\|^2\big)^{1/2} \big(\sum |w_j|^2 \|f_j\|^{-2}\big)^{1/2} \leq 1/\sqrt{c} \| \sum \lambda_j f_j \|.$$

Therefore, $\|L\| \leq 1/\sqrt{c}$ and there exists $f \in H$ with

$$\|f\| \leq 1/\sqrt{c} \text{ and } L(g) = \langle g, f \rangle \quad \forall g \in E.$$

Testing with $g = f_j$, we get $f(z_j) = w_j$ and see that $M_H(Z) \leq 1/\sqrt{c}$, that is $c \leq [M_H(Z)]^{-2}$.

This "Carleson language" could be dropped (sampling,...), but has a geometric flavor, to be exploited later.

6. General estimates for approximation numbers

6.1. The mapping equation. We give this equation in the context of two Hilbert spaces, implicitly used in [**25**], but we will later be content with the same space. Let H_1, H_2 be functional Hilbert spaces on sets Ω_1, Ω_2 with respective reproducing kernels $K_a^{(1)}$, $K_b^{(2)}$ and let $\Phi : \Omega_1 \to \Omega_2$.

The composition operator $C_\Phi : H_2 \to H_1$ is (formally) defined by

$$C_\Phi(f) = f \circ \Phi.$$

The following makes the transition between *operator* theory and *function* theory.

THEOREM 6.1. *If $\Phi : \Omega_1 \to \Omega_2$ generates a bounded operator $C_\Phi : H_2 \to H_1$, we have the mapping equation*

$$C_\Phi^*(K_a^{(1)}) = K_{\Phi(a)}^{(2)} \quad \forall a \in \Omega_1.$$

The proof is purely formal and is omitted. But Theorem 6.1 turns out to be a useful triviality, strong use of which has already been made in [**31**], [**18**], [**16**], [**17**], [**26**] for example.

6.2. A general lower bound. We give an application of the previous equation in the general framework of the preceding subsection.

THEOREM 6.2 ([**25**]). *Let $\Phi : \Omega_1 \to \Omega_2$ generate a composition operator $C_\Phi : H_2 \to H_1$. Let $S' = \{s'_1, \ldots, s'_n\} \subset \Omega_1$ and $S = \{s_1, \ldots, s_n\} \subset \Omega_2$ be two sets of n distinct points such that $\Phi(s'_j) = s_j$, where $1 \leq j \leq n$. Then:*

$$a_n(C_\Phi) \geq [M_{H_2}(S)]^{-1} \|\mu_{S',H_1}\|_{\mathcal{C},H_1}^{-1/2} \inf_{1 \leq j \leq n} \left(\frac{\|K_{s_j}^{(2)}\|}{\|K_{s'_j}^{(1)}\|} \right). \tag{3}$$

Proof: Let $T = C_\Phi^*$ and $E = \langle K^{(1)}_{s'_1}, \dots, K^{(1)}_{s'_n}\rangle$. We have $\dim E = n$ since the s'_j are distinct. The mapping equation gives $T(K^{(1)}_{s'_j}) = K^{(2)}_{s_j}$ so that if $f = \sum_{j=1}^n \lambda_j K^{(1)}_{s'_j} \in S_E$, the unit sphere of E, we have $Tf = \sum_{j=1}^n \lambda_j K^{(2)}_{s_j}$. We now use Lemma 5.1 to get the two inequalities

$$1 = \|f\|^2 \le \|\mu_{S',H_1}\|_{\mathcal{C},H_1} \sum_{j=1}^n |\lambda_j|^2 \|K^{(1)}_{s'_j}\|^2.$$

$$\|Tf\|^2 = \|\sum_{j=1}^n \lambda_j K^{(2)}_{s_j}\|^2 \ge [M_{H_2}(S)]^{-2} \sum_{j=1}^n |\lambda_j|^2 \|K^{(2)}_{s_j}\|^2.$$

That gives, setting $\mu_n = \inf_{1\le j\le n}\left(\|K^{(2)}_{s_j}\|/\|K^{(1)}_{s'_j}\|\right)$:

$$\|Tf\|^2 \ge \mu_n^2 [M_{H_2}(S)]^{-2} \sum_{j=1}^n |\lambda_j|^2 \|K^{(1)}_{s'_j}\|^2 \ge \mu_n^2\, [M_{H_2}(S)]^{-2} \|\mu_{S',H_1}\|^{-1}_{\mathcal{C},H_1},$$

terminating the proof in view of the definition of the a_n as Bernstein numbers and of the relation $a_n(T) = a_n(T^*)$. One will observe that this theorem (a special case of which appears in [**17**]) enters the category of usual suspects: its proof is nearly as short as its statement! Indeed, this is more a working program. All the job remains to do, by choosing properly the related sets S and S' and then by estimating as sharply as possible each of the three terms appearing in the RHS of (3).

Now, as an application of the Gelfand definition of the approximation numbers, we give a general upper bound in the case $c_0 = 0$.

6.3. A general upper bound. We first give a definition. Let K be a compact subset of $\mathbb{C}_{1/2}$, and $\theta = \inf_{s\in K} \operatorname{Re} s > 1/2$. With any sequence $(s_1, \dots, s_{n-1})$ of points in $\mathbb{C}_\theta$, we associate the finite Blaschke product

$$B(s) = \prod_{j=1}^{n-1} \frac{s - s_j}{s - (1/2 + \theta) + \overline{s_j}} \tag{4}$$

which has modulus < 1 on K and modulus 1 on the vertical line $L = \{s : \operatorname{Re} s = \frac{1/2+\theta}{2}\}$, with $\frac{1/2+\theta}{2} < \theta$. Such a function B will be said to be a *Blaschke product adapted to K*. The reason for reflecting through this "middle-line" L is that each of the constituting factors of B has modulus significantly < 1 on K.

THEOREM 6.3 ([**25**]). *Suppose that $c_0 = 0$ and that $\varphi(s) = \sum_{n=1}^\infty c_n n^{-s}$ with $\varphi(\mathbb{C}_0)$ bounded.*

Let K be a compact subset of $\mathbb{C}_{1/2}$, $\theta = \inf_{s\in K} \operatorname{Re} s > 1/2$ and B be a Blaschke product of degree $n-1$ adapted to K. Then, there is a Carleson measure $\mu_{\varphi,K}$ for $\mathscr{H}^2$, carried by $\mathbb{C}_{1/2}\backslash K$, such that:

$$a_n(C_\varphi) \le \left(\sup_{s\in K} |B(s)|^2 \zeta(1/2+\theta) + \|\mu_{\varphi,K}\|_{\mathcal{C},\mathscr{H}^2}\right)^{1/2}.$$

This general upper bound implies

COROLLARY 6.1 ([**25**]). *If $c_0 = 0$ and the closure K of $\varphi(\mathbb{C}_0)$ is a compact subset of $\mathbb{C}_{1/2}$, one has*

$$a_n(C_\varphi) \le C r^n \text{ with } r < 1.$$

Setting $K = \overline{\varphi(\mathbb{C}_0)}$ and $\theta = \inf_{s\in K} \operatorname{Re} s > 1/2$, we have $\mu_{\varphi,K} = 0$ (by definition of $\mu_{\varphi,K}$, see [**25**], this measure is carried both by $\overline{\varphi(\mathbb{C}_0)}$ and by the complement of K in $\mathbb{C}_{1/2}$, so that its support is empty when one is allowed to take $K = \overline{\varphi(\mathbb{C}_0)}$). It now suffices to take any point s_0 in K and the adapted Blaschke product

$$B(s) = \left(\frac{s - s_0}{s - (1/2+\theta) + \overline{s_0}}\right)^{n-1}.$$

This means that $|B(s)| \leq r^{n-1}$, where

$$r := \sup_{s\in K} \left|\frac{s - s_0}{s - (1/2+\theta) + \overline{s_0}}\right| < 1,$$

and hence $a_n(C_\varphi) \ll r^n$ by Theorem 6.3. This example shows that part (a) of Theorem 7.5 to come is best possible.

7. Statement of the main results

7.1. A hard part of interpolation theory.

7.1.1. *The usual Hardy space of the half-plane.* The usual Hardy space $H^2(\mathbb{C}_{1/2})$, denoted in short by H^2, is defined as the set of functions h analytic in $\mathbb{C}_{1/2}$ for which

$$\|h\|_{H^2}^2 := \sup_{\sigma>1/2} \frac{1}{2\pi} \int_{-\infty}^{\infty} |h(\sigma+it)|^2 dt < \infty. \tag{5}$$

Every h in H^2 has a nontangential boundary limit at almost every point of the vertical line $\sigma = 1/2$, and the corresponding limit function $h \mapsto h(1/2+it)$ is in $L^2(\mathbb{R})$; the L^2-norm of this function coincides with the H^2-norm defined by (5). With the chosen normalization, the reproducing kernel of H^2 is

$$K_a(s) = \frac{1}{s + \overline{a} - 1}.$$

We refer to [**23**] for more details. In particular, we will make use of the following resemblance between H^2 and our new space $\mathscr{H}^2$.

LEMMA 7.1 ([**23**]). *If $f \in \mathscr{H}^2$, then $F(s) = \frac{f(s)}{s} \in H^2$ and*

$$\|F\|_{H^2} \leq C_0 \|f\|_{\mathscr{H}^2} \tag{6}$$

where C_0 is an absolute constant.

We refer to [**23**] for the proof, a consequence of the generalized Hilbert inequality ([**19**], [**20**]). An alternative proof can be found in [**22**].
In this good old space H^2, Carleson and interpolation sequences are well-known thanks to the following two theorems, which require two definitions :

(1) The pseudo-hyperbolic distance d in $\mathbb{C}_{1/2}$ is defined by

$$d(a,b) = \left|\frac{a-b}{a+\overline{b}-1}\right|, \quad a,b \in \mathbb{C}_{1/2}.$$

Note that $d < 1$.

(2) The Carleson square Q of side-length $l = l(Q)$ and center $it_0 \in i\mathbb{R}$ is the set

$$Q = Q(t_0, l) = \{s = \sigma + it : 1/2 \leq \sigma \leq 1/2 + l \text{ and } |t - t_0| \leq l/2\}.$$

We can now state (see [**21**, pp.156–158] or [**8**], and [**29**])

THEOREM 7.1 (Carleson embedding theorem, 1962). *Let μ be a positive measure in $\mathbb{C}_{1/2}$. Then, Q denoting a Carleson square, we have:*

$$\|\mu\|_{C,H^2} \ll \sup_Q \frac{\mu(Q)}{l(Q)}.$$

This theorem is also referred to as the RKT (Reproducing Kernel Thesis).

THEOREM 7.2 (Shapiro-Shields's criterion, 1961). *Let $S = (s_j)$ be a sequence in $\mathbb{C}_{1/2}$. Then, S is interpolating for $H = H^2$ if and only if S is both Carleson and separated, namely:* $\inf_{a,b\in S,\, a\neq b} \left|\frac{a-b}{a+\bar{b}-1}\right| = \inf_{a,b\in S,\, a\neq b} d(a,b) := \eta(S) > 0$. *Moreover,*

$$M_H(S) \leq \|\mu_{S,H^2}\|_{C,H^2}^{1/2} \exp[2\pi(1 + 2\log(1/\eta(S))\|\mu_{S,H^2}\|_{C,H^2}].$$

The latter upper bound can e.g. be obtained from a duality argument that can be found in [**27**], and from [**11**] p.279.

7.1.2. *A transference principle.* In the sequel, we will need to export our knowledge of Carleson and interpolation constants from H^2 to $\mathscr{H}^2$. This will be done through the following two lemmas:

LEMMA 7.2 ([**25**]). *If μ is a Carleson measure for H^2, that is supported on the rectangle $1/2 \leq \operatorname{Re} s \leq \theta$, $|\operatorname{Im} s| \leq R$, where $\theta > 1/2$, then*

$$\|\mu\|_{C,\mathcal{H}^2} \leq C(R^2 + \theta^2)\|\mu\|_{C,H^2}$$

where C is an absolute constant.

Proof: This is an easy consequence of Lemma 7.1. Indeed, if $f \in \mathscr{H}^2$, let $F(s) = \frac{f(s)}{s}$. We see that $|f(s)|^2 = |sF(s)|^2 \leq (R^2 + \theta^2)|F(s)|^2$ on the support of μ, whence

$$\int |f|^2 d\mu \leq \int (R^2+\theta^2)|F|^2 d\mu \leq (R^2+\theta^2)\|\mu\|_{C,H^2}\|F\|_{H^2}^2 \leq C_0^2\,(R^2+\theta^2)\|\mu\|_{C,H^2}\|f\|_{H^2}^2$$

where C_0 is as in Lemma 7.1.

LEMMA 7.3 ([**28**], [**25**]). *Suppose $S = (s_j = \sigma_j + it_j)$ is an interpolating sequence for H^2 and that $1/2 < \sigma_j \leq \theta$ for every j. Then, for some constant $C = C_\theta$, we have*

$$M_{\mathcal{H}^2}(S_R) \leq C[M_{H^2}(S)]^{2\theta+6} R^{2\theta+7/2} \tag{7}$$

whenever $R \geq \theta + 1$, where $S_R = \{s \in S \ :\ |\operatorname{Im} s| \leq R\}$.

Proof: This proof is highly more elaborate and is due to K.Seip ([**28**]), even if the dependence on the constants is made more precise in [**25**]. Note that when S is bounded, a simpler version of Lemma 7.3 has been proved in [**23**]. But this turns out to be unsufficient here, since our (finite!) sequences are not *uniformly bounded.*

The idea is roughly the following: if $F \in H^2$ and if S is a "good" sequence in $\mathbb{C}_{1/2}$, there exists $f \in \mathscr{H}^2$ with $f = F$ on S_R and $\|f\|_{\mathscr{H}^2}$ somehow controlled in terms of $\|F\|_{H^2}$. Indeed, by Paley-Wiener's theorem, we may represent F in H^2 as[1]

$$F(s) = \int_0^\infty \varphi(\xi)e^{-(s-1/2)\xi}d\xi \quad \text{with} \quad \varphi \in L^2(\mathbb{R}^+)$$

so that by the Plancherel identity $\|F\|_{H^2} = \|\varphi\|_2$. Next, we can write

$$F(s) = \sum_{n=1}^\infty \int_{\log n}^{\log(n+1)} \varphi(\xi)e^{-(s-1/2)\xi}d\xi \sim \sum_{n=1}^\infty \int_{\log n}^{\log(n+1)} \varphi(\xi)e^{-(s-1/2)\log n}d\xi,$$

that is

$$F(s) \sim f(s) := \sum_{n=1}^\infty a_n n^{-s} \text{ with } a_n = \sqrt{n}\int_{\log n}^{\log(n+1)} \varphi(\xi)d\xi.$$

One easily checks, using Cauchy-Schwarz, that

$$\|f\|_{\mathscr{H}^2} = (\sum_{n=1}^\infty |a_n|^2)^{1/2} \leq \|\varphi\|_2 = \|F\|_{H^2}.$$

But if the norm of f is quite-well controlled, we don't have $f = F$ on S_R and the approximation of F by f is not sufficient, unless the function φ associated with F has far remote support, say in $(\log N, \infty)$, that is

$$F(s) = \int_{\log N}^\infty \varphi(\xi)e^{-(s-1/2)\xi}d\xi$$

or $F \in E_N H^2$ where $E_N(s) = N^{-s+1/2}$. One thus has to define an operator $\tilde{T}_N : H^2 \to E_N H^2$ which preserves the values of functions on S_R, with help of a correction involving a localization and a d-bar argument. This first correction is norm-consuming, but under control. Once we are in $E_N H^2$, the previous discretization of the integral and a second correction of the same type finally provides you with a function $f \in \mathscr{H}^2$ with controlled norm, coinciding with F on S_R. We refer to the recent papers [**25**], [**28**] for more details.

7.2. Specific statements.

7.2.1. *The initial question.* The study of compact composition operators on $\mathscr{H}^2$ was initiated in [**2, 3, 10**]. In [**3**, Theorem 4], F.Bayart succeeded in describing the spectrum of such operators as follows:

THEOREM 7.3 ([**3**]). *Let $\varphi(s) = c_0 s + \sum_{n=1}^\infty c_n n^{-s}$ be a symbol such that C_φ is a compact composition operator on $\mathscr{H}^2$.*

(a) *If $c_0 = 0$, then* $\operatorname{Spec} C_\varphi = \{0, 1\} \bigcup \{[\varphi'(\alpha)]^k : k \geq 1\}$, *where α is the fixed point of φ in $\mathbb{C}_{1/2}$.*
(b) *If $c_0 = 1$, then* $\operatorname{Spec}(C_\varphi) = \{0, 1\} \bigcup \{k^{-c_1} : k \geq 1\}$.
(c) *If $c_0 > 1$, then* $\operatorname{Spec}(C_\varphi) = \{0, 1\}$.

The first point in Theorem 7.3 needs some justification.

LEMMA 7.4. *Let $C_\varphi : \mathscr{H}^2 \to \mathscr{H}^2$ with $c_0 = 0$. Then, φ has a fixed point in $\mathbb{C}_{1/2}$.*

[1]We allow φ to have a different meaning in this section than elsewhere in this paper.

Proof of the lemma: We rely on the following fact ([**2**], Lemma 11):

$$\varphi(s) \neq s + i\tau \Rightarrow \varphi(\mathbb{C}_{1/2}) \subset \mathbb{C}_{1/2+\varepsilon} \text{ for some } \varepsilon > 0. \tag{8}$$

Now, observe that φ cannot be a vertical translation $s + i\tau$ when $c_0 = 0$. Let $T : \mathbb{D} \to \mathbb{C}_{1/2}$ be the conformal mapping given by

$$T(z) = \frac{1}{2} + \frac{1-z}{1+z}.$$

Let $\omega = T^{-1}\varphi T : \mathbb{D} \to \mathbb{D}$. Assume that ω has no fixed point in $\mathbb{D}$. Then, the Denjoy-Wolff theorem ([**31**], p.78) implies the existence of a point $u \in \partial\mathbb{D}$ such that $\omega_n \to u$, where the arrow denotes uniform convergence on compact subsets of $\mathbb{D}$ and ω_n is the n-th iterate of ω. If $u \neq -1$, $\varphi_n = T\omega_n T^{-1} \to T(u) \in \partial\mathbb{C}_{1/2}$. But this is impossible from (8) since $\varphi_n(\mathbb{C}_{1/2}) \subset \mathbb{C}_{1/2+\varepsilon}$. If $u = -1$, we have $\varphi_n \to \infty$. This is impossible again since, by Theorem 3.1, the Dirichlet series $\varphi = \psi$ converges uniformly on $\mathbb{C}_{1/2}$ so that there is a bounded subset B of $\mathbb{C}_{1/2}$ for which $\varphi(\mathbb{C}_{1/2}) \subset B$ as well as $\varphi_n(\mathbb{C}_{1/2}) \subset B$. This shows that ω has a fixed point $u \in \mathbb{D}$ and that φ has the fixed point $\alpha = T(u) \in \mathbb{C}_{1/2}$. □

Let us go back to our approximation numbers in the $\mathscr{H}^2$ setting. Results are rather incomplete compared to what is known in the classical case of $H^2(\mathbb{D})$ [**31**]. The emphasis in [**3, 10**] was on membership in the Hilbert–Schmidt class. Our topic—the rate of decay of the approximation numbers $a_n(C_\varphi)$—is a more delicate issue.

Here is the starting point of this work: we know ([**18**]) that $a_n(C_\varphi) \geq cr^n$ on $H^2(\mathbb{D})$, and this is optimal: if $\varphi(z) = rz$, $a_n(C_\varphi) = r^{n-1}$ since C_φ is a diagonal operator with diagonal entries r^{n-1}. A change of scale ($z^n \to n^{-s}$) occurs when we pass to Dirichlet series. We thus expect $a_n(C_\varphi) \geq cn^{-A}$ for $\mathscr{H}^2$, which would be optimal as well: indeed, if $\varphi(s) = s + A$, $a_n(C_\varphi) = n^{-A}$, C_φ being the diagonal operator with diagonal entries n^{-A}.
The answer to that "conjecture" is NO.

Set $\varphi(s) = c_1 + c_2 2^{-s}$ and $\Delta(\varphi) = \operatorname{Re} c_1 - |c_2| - 1/2 \geq 0$. Then, we can state:

THEOREM 7.4. *If $\Delta(\varphi) > 0$, one has*

$$a_n(C_\varphi) \leq \frac{1}{\sqrt{2\Delta(\varphi)}} r^n \quad \textit{with} \quad r = \frac{|c_2|}{|c_2| + \Delta(\varphi)} < 1.$$

Proof: Write $f(s) = \sum_{n=1}^{\infty} b_n n^{-s}$, then

$$C_\varphi f(s) = \sum_{n=1}^{\infty} b_n n^{-c_1} \exp(-c_2 2^{-s} \log n)$$

expand the exponential, permute sums, approximate by the partial sums of the new Dirichlet series obtained, and perform elementary computations. This seminal result can now be seen as an immediate application of Theorem 6.3 and its corollary.

We have $c_0 = 0$ in that case. But the "revised conjecture" holds true if $c_0 \geq 1$ (that is if $\varphi(\infty) = \infty$). Proving that fact was our starting point. In the wake, we shall see that the geometric rate of decay above cannot be superseded.

7.2.2. *A first statement.* Our first theorem gives general *lower bounds* for $a_n(C_\varphi)$, which is the main theme of this survey.

THEOREM 7.5 ([**25**]). *Suppose that c_0 is a nonnegative integer and that $\varphi(s) = c_0 s + \sum_{n=1}^{\infty} c_n n^{-s}$ generates a* compact *composition operator C_φ on $\mathscr{H}^2$.*

(a) *If $c_0 = 0$, then $a_n(C_\varphi) \gg r^n$ for some $0 < r < 1$.*
(b) *If $c_0 = 1$, then $a_n(C_\varphi) \gg n^{-\operatorname{Re} c_1 - \varepsilon}$ for every $\varepsilon > 0$.*
(c) *If $c_0 > 1$, then $a_n(C_\varphi) \gg n^{-A}$ for some $A > 0$.*

We note that similar estimates and other results regarding approximation numbers in the $H^2(\mathbb{D})$ setting were obtained in [**17**], but the contrast between (a) and (b) + (c) has no parallel in the theory of composition operators on $H^2(\mathbb{D})$. We know from the latter example that the lower bound of (a) is optimal. Theorem 6.3 has shown that this is the case as soon as $c_0 = 0$ and the closure of $\varphi(\mathbb{C}_0)$ is a compact subset of $\mathbb{C}_{1/2}$.

7.2.3. *A second statement and a class of examples.* Our main concern here will be to reveal the relevance of the complex "dimension" of a symbol φ. We will make sense of this by restricting to Dirichlet series of the form

$$\varphi(s) = c_1 + \sum_{j=1}^{d} c_{q_j} q_j^{-s},$$

where d can be any positive integer or $d = \infty$ and the positive integers $q_j \geq 2$ are independent. We also assume that all the coefficients c_{q_j} are nonzero. If we set $q = (q_j)$ and use multi-index notation, then this means that any integer n can be written as $n = q^\alpha$ for at most one multi-index α. For example, $q_1 = 2$ and $q_2 = 6$ are independent. The canonical example of a collection of independent integers is any set, finite or not, of prime numbers.
An essential characteristic of a symbol of this kind, is the number

$$\kappa(\varphi) = \operatorname{Re} c_1 - \sum_{j=1}^{d} |c_{q_j}|.$$

In view of the Gordon–Hedenmalm theorem (Theorem 3.1), the independence of the q_j, and Kronecker's theorem (see [**10**]), C_φ is bounded if and only if $\kappa(\varphi) \geq 1/2$.

If $\kappa(\varphi) = 1/2$, then C_φ is compact if and only $d > 1$, as was proved independently in [**3**] and [**10**]. Moreover, under the same assumption that $\kappa(\varphi) = 1/2$, it was shown in [**10**] that C_φ is in the Schatten class S_4 when $d = 2$ and that C_φ belongs to S_2 if and only if $d > 2$. Thus we have $a_n(C_\varphi) \ll n^{-1/4}$ for $d = 2$ and $a_n(C_\varphi) \ll n^{-1/2}$ for $d > 2$.

Our next result improves these estimates and gives best possible d-dependent upper and lower bounds, up to a factor $(\log n)^{(d-1)/2}$. I suspect that this extra factor cannot be dropped and that the upper bound gives the right order of growth. In any case, one obtains the exact values of p for which C_φ belongs to the Schatten class S_p.

THEOREM 7.6 ([**25**]). *Let $\varphi(s) = c_1 + \sum_{j=1}^{d} c_{q_j} q_j^{-s}$ be a symbol such that the positive integers $q_j \geq 2$ are* independent, *$c_{q_j} \neq 0$ for $1 \leq j \leq d$, and $\kappa(\varphi) = 1/2$.*

(a) *If* $1 \le d < \infty$, *then* $(\frac{1}{n})^{(d-1)/2} \ll a_n(C_\varphi) \ll (\frac{\log n}{n})^{(d-1)/2}$.
(b) *If* $d = \infty$, *then* C_φ *belongs to* $\bigcap_{p>0} S_p$.

In particular, C_φ *belongs to* $\displaystyle\bigcap_{p>2/(d-1)} S_p$ *but not to* $S_{2/(d-1)}$ *for* $1 < d < \infty$.

7.2.4. *A third statement and another transference principle.* Most of our estimates and, in particular, the entire proof of Theorem 7.6 rely on a general method, applicable in the context of Hilbert spaces of analytic functions. These techniques for estimating approximation numbers are not new; they can be found in the proofs of Proposition 6.3 in [**17**] (lower bounds) and in Theorem 2.3 and Theorem 3.2 in [**18**] (respectively upper and lower bounds).We refer to the closely related paper [**26**] for more along these lines.

The aforementioned techniques give a few other results as well. Indeed, it follows from a general transference principle (see [**25**]) that symbols of composition operators on $H^2(\mathbb{D})$, via left and right composition with two fixed analytic maps, give rise to composition operators on $\mathscr{H}^2$, such that estimates for the approximation numbers carry over from $H^2(\mathbb{D})$. Consequently, using results from the recent paper [**26**], one may construct explicit examples of composition operators on $\mathscr{H}^2$ with approximation numbers with essentially any prescribed sub-exponential decay. The most precise result in this direction is obtained in the case of slow decay. We will make use of the two functions:

1. $T(z) := \frac{1}{2} + \frac{1-z}{1+z}$, which maps $\mathbb{D}$ conformally onto $\mathbb{C}_{1/2}$. Recall that $\mathbb{D}$ denotes the open unit disk.

2. $I(s) := 2^{-s}$ which maps $\mathbb{C}_0$ onto $\mathbb{D}\setminus\{0\}$.

Now, if ω is an analytic self-map of $\mathbb{D}$ such that $\omega(\mathbb{D})$ has a positive distance to -1, that is $\inf_{z\in\mathbb{D}} |1+\omega(z)| > 0$, we set (*transferring* ω *to* φ)

$$\varphi = T \circ \omega \circ I : \mathbb{C}_{1/2} \to \mathbb{C}_{1/2}. \tag{9}$$

It is shown in [**25**], using in particular the embedding Lemma 7.2, that the Dirichlet series φ is then the symbol of a bounded composition operator C_φ on $\mathscr{H}^2$ with $c_0 = 0$. Indeed, the composition operators

$$C_T : \mathscr{H}^2 \to H^2(\mathbb{D}) \quad \text{and} \quad C_I : H^2(\mathbb{D}) \to \mathscr{H}^2$$

are bounded and $C_\varphi = C_I\, C_\omega\, C_T$. We can now state, with the notations of Subsections 5.1 and 5.2 (in particular, $\mu_{Z,H^2(\mathbb{D})} = \sum_{j=1}^\infty (1-|z_j|^2)\delta_{z_j}$):

THEOREM 7.7 ([**25**]). *Let* ω *and* φ *be as in (9). There exist positive constants* c *and* A *such that if* $Z = (z_j)$ *is a finite sequence of points in* $\mathbb{D}$ *such that* Z *and* $\omega(Z)$ *both consist of* n *distinct points, then*

$$\begin{aligned} c\,[M_{H^2(\mathbb{D})}(\omega(Z))]^{-A} \|\mu_{Z,H^2(\mathbb{D})}\|_{\mathcal{C},H^2(\mathbb{D})}^{-1/2} \inf_{1\le j\le n} \left(\frac{1-|z_j|^2}{1-|\omega(z_j)|^2}\right)^{1/2} \\ \le a_n(C_\varphi) \le \|C_T\| a_n(C_\omega). \end{aligned} \tag{10}$$

In particular, $C_\varphi : \mathscr{H}^2 \to \mathscr{H}^2$ is compact as soon as $C_\omega : H^2(\mathbb{D}) \to H^2(\mathbb{D})$ is compact.
As a consequence, if (ε_n) is any positive sequence with limit 0, there exists a compact *composition operator C_φ on $\mathscr{H}^2$ such that*

$$a_n(C_\varphi) \geq \delta \varepsilon_n, \; n = 1, 2, \ldots$$

Finally, we can also have two-sided estimates, under the following form: let g be a positive function on $\mathbb{R}^+$ such that $g(x) \searrow 0$ when $x \to \infty$ and $g(x^2)/g(x)$ is bounded below. Then there exists a compact composition operator C_φ on $\mathscr{H}^2$ with

$$a_n(C_\varphi) = e^{O(1)} g(n)$$

when $n \to \infty$.

We sketch a proof of the second item, taking (10) for granted, since this part is not detailed in [**25**]. We can assume that (ε_n) is non-increasing, replacing it by $\varepsilon'_n = \sup_{k \geq n} \varepsilon_k \geq \varepsilon_n$. In [**17**], to a given non-increasing sequence (ε_n) of positive numbers with limit 0, we have associated a self-map $\omega_0 : \mathbb{D} \to \mathbb{D}$, continuous on $\overline{\mathbb{D}}$ and whose image touches $\partial\mathbb{D}$ only at ± 1, as well as a sequence $(z_j)_{j \geq 1}$ of distinct points of $\mathbb{D}$ such that $0 < z_j < 1$, verifying:

$$C_{\omega_0} \text{ is compact and } \omega_0(z_{j+1}) = z_j \quad \text{for all } j \geq 1. \tag{11}$$

$$\frac{1 - z_{j+1}}{1 - z_j} \to 0 \quad \text{and} \quad \inf_{1 \leq j \leq n} \frac{1 - z_{j+1}}{1 - z_j} \gg \varepsilon_n^2. \tag{12}$$

Let $\omega = i\omega_0$, whose image does not approach -1. Equation (12) shows that $Z = (z_j)$ verifies Newman's condition and is therefore an interpolation sequence in $H^2(\mathbb{D})$, as well as $\omega(Z)$ since $\omega(z_j) = i\, z_{j-1}$ for $j \geq 2$. Therefore, the lower bound of Theorem 7.7 provides, thanks to (11) and (12):

$$a_n(C_\varphi) \gg \inf_{1 \leq j \leq n} \left(\frac{1 - |z_j|^2}{1 - |\omega(z_j)|^2} \right)^{1/2} \gg \inf_{1 \leq j \leq n} \left(\frac{1 - z_j^2}{1 - z_{j-1}^2} \right)^{1/2} \gg \varepsilon_n.$$

Finally, Theorem 7.7 shows that the operator $C_\varphi : \mathscr{H}^2 \to \mathscr{H}^2$ associated with $\varphi = T \circ \omega \circ I$ is compact, since C_{ω_0} and $C_\omega : H^2(\mathbb{D}) \to H^2(\mathbb{D})$ are. □
We may thus prescribe any slow rate of decay (a negative power of $\log n$ or slower) and find a symbol φ such that the approximation numbers $a_n(C_\varphi)$ of the compact operator C_φ descend accordingly. A particular consequence is that there exist compact composition operators on $\mathscr{H}^2$ belonging to no Schatten class S_p, $p < \infty$. The search for composition operators on $H^2(\mathbb{D})$ with this property began with a question of Sarason, followed by Carroll and Cowen's affirmative answer [**9**]. See also [**1**] and [**17**].

8. Proof of one theorem

We finally give the proof of Theorem 7.5. We separate three cases:
(a) $c_0 = 0$. We will apply Bayart's theorem (Theorem 7.3). To this end, we need the following lemma:

LEMMA 8.1. *Let a and b be arbitrary points in $\mathbb{C}_{1/2}$. Then there exists a Dirichlet polynomial χ such that $\chi(\mathbb{C}_0) \subset \mathbb{C}_{1/2}$ and*

$$\chi(a) = b \quad \text{and} \quad \chi'(a) \neq 0.$$

Proof. The Dirichlet polynomial χ will be of the form $\chi(s) = c_1 + c_2 2^{-s}$ for suitable c_1, c_2. Choose $\varepsilon > 0$ such that

$$\operatorname{Re} b - 1/2 \geq \varepsilon(1 + 2^{\operatorname{Re} a});$$

this is possible since by assumption b is in $\mathbb{C}_{1/2}$. Take any c_1 such that $|b - c_1| = \varepsilon$ and choose $c_2 = (b - c_1)2^a$. By construction, we have $\chi(a) = b$ and also

$$\chi'(a) = -(\log 2)c_2 2^{-a} \neq 0.$$

We observe that χ maps $\mathbb{C}_0$ into $\mathbb{C}_{1/2}$ because

$$\operatorname{Re} c_1 - |c_2| - \frac{1}{2} \geq \operatorname{Re} b - \varepsilon - |b - c_1| 2^{\operatorname{Re} a} - \frac{1}{2} = \big(\operatorname{Re} b - \frac{1}{2}\big) - \varepsilon(1 + 2^{\operatorname{Re} a}) \geq 0.$$

□

We now continue to the proof of part (a) of Theorem 7.5. Since φ is assumed to be non-constant, we may fix a point b in $\mathbb{C}_{1/2}$ such that $\varphi'(b) \neq 0$. Set $a = \varphi(b)$ and $\varphi_1 = \chi \circ \varphi$, where χ is as in Lemma 8.1. We have

$$(13) \qquad \varphi_1(b) = \chi(a) = b \quad \text{and} \quad \varphi_1'(b) = \chi'(a)\varphi'(b) \neq 0.$$

Since $C_{\varphi_1} = C_\varphi \circ C_\chi$, the ideal property of approximation numbers gives that

$$(14) \qquad a_n(C_{\varphi_1}) \leq \|C_\chi\| a_n(C_\varphi).$$

This implies that it suffices to establish the lower bound $r^n \ll a_n(C_{\varphi_1})$. We set $\lambda = |\varphi_1'(b)|$, $a_j = a_j(C_{\varphi_1})$, and note that $0 < \lambda < 1$ since b is a fixed point of φ_1, which is not an automorphism of $\mathbb{C}_{1/2}$. We now invoke Bayart's theorem (Theorem 7.3) and Weyl's lemma under the form (2) which together give (with $N = 2$)

$$a_n \geq (a_1)^{-1}(\lambda^{1+\cdots+2n})^{1/n} = (a_1)^{-1}\lambda^{2n+1}.$$

(b) $c_0 = 1$. Set $\gamma_1 = \operatorname{Re} c_1$ and apply Bayart's theorem with $|\lambda_j| = j^{-\gamma_1}$, as well as (2), to get

$$a_n \geq (a_1)^{-1/(N-1)} \Big[(N\,n)^{-Nn\gamma_1}\Big]^{1/(N-1)n} =: \delta_N\, n^{-\frac{N\gamma_1}{N-1}}.$$

This gives the result, by choosing N large enough depending only on ε.

(c) $c_0 \geq 2$. We can no longer apply Bayart's theorem because the spectrum of C_φ becomes very poor, and will instead appeal to Theorem 6.2 in the special case

$$\Omega_1 = \Omega_2 = \mathbb{C}_{1/2},\ H_1 = H_2 = \mathscr{H}^2,\ \Phi = \varphi,\ \|K_z\|^2 = \zeta(2\operatorname{Re} z).$$

We recall that

$$a_n(C_\varphi) \geq M_{\mathcal{H}^2}(S)^{-1} \|\mu_{S',\mathcal{H}^2}\|_{\mathcal{C},\mathcal{H}^2}^{-1/2} \inf_{1\leq j\leq n} \left(\frac{\|K_{s_j}\|}{\|K_{s_j'}\|}\right).$$

It remains to choose S and S', which is done as follows: we first fix $\sigma_0 > 1/2$. Then, we fix $a > 0$ so large that (recall the notation $\varphi(s) = c_0 s + \psi(s)$)

$$a \geq 3 \sup_{\sigma\geq\sigma_0} \Big|\sum_{k=1}^{\infty} c_k k^{-s}\Big| = 3 \sup_{\sigma\geq\sigma_0} |\psi(s)|.$$

Finally, we choose

$$S' = \{s_j' = \sigma_0 + ija,\ 1 \leq j \leq n\} \quad \text{and} \quad S = \{s_j = \varphi(s_j'),\ 1 \leq j \leq n\}.$$

The finite sets S' and $c_0\, S'$ are Carleson sequences with constants independent of n. Indeed, the hypotheses of Carleson's embedding theorem are easily checked for those sequences. Moreover, since $|\varphi(s') - c_0\, s'| \leq a/3$ for $s' \in S'$, the same holds for $S = \varphi(S')$. Next, if s'_j and s'_k are two distinct points of S', we see that

$$|\varphi(s'_j) - \varphi(s'_k)| \geq c_0 |s'_j - s'_k| - |\psi(s'_j)| - |\psi(s'_k)| \geq c_0\, a - 2a/3 \geq a/3.$$

We also have

$$\operatorname{Re} \varphi(s') - 1/2 \leq c_0\, a - 1/2 + a/3 \leq 2c_0\, a$$

for every $s' \in S'$. We therefore get, using $\varphi(s'_j) + \overline{\varphi(s'_k)} - 1 = \varphi(s'_j) - \varphi(s'_k) + 2 \operatorname{Re} \varphi(s'_k) - 1$:

$$\left| \frac{\varphi(s'_j) - \varphi(s'_k)}{\varphi(s'_j) + \overline{\varphi(s'_k)} - 1} \right| \geq \frac{|\varphi(s'_j) - \varphi(s'_k)|}{4c_0\, a + |\varphi(s'_j) - \varphi(s'_k)|} \geq \frac{a/3}{4c_0\, a + a/3} = \frac{1}{12c_0 + 1}.$$

This shows that the sequence $S = \varphi(S')$ is both Carleson and separated. In view of the Shapiro-Shields theorem, it is interpolating in H^2 with constants independent of n. Now, the transference Lemmas 7.2 and 7.3 give, with $\theta = \sigma_0$ and $R = an$:

$$\|\mu_{S', \mathscr{H}^2}\|_{C, \mathscr{H}^2} \leq C n^2 \quad \text{and} \quad M_{\mathscr{H}^2}(S) \leq C n^c$$

where $c = 2\theta + 7/2$. Finally, we note that, for $1 \leq j \leq n$:

$$\frac{\zeta(2 \operatorname{Re} s_j)}{\zeta(2 \operatorname{Re} s'_j)} \geq \frac{1}{\zeta(2 \operatorname{Re} s'_j)} = \frac{1}{\zeta(2\sigma_0)}.$$

So that Theorem 6.2 finally gives $a_n(C_\varphi) \gg n^{-c} \times n^{-1}$, which ends the proof. □

Of course, the constant $7/2$ is fancy and we did not try to find a nearly optimal value for the parameter A of Theorem 7.5.

9. Acknowledgement

We would like to thank E.Fricain, J.Mashreghi and B.Ross, the organizers of the Montréal workshop in August 2013, for the excellent atmosphere they provided during the meeting and for giving us the opportunity to write that survey.

References

[1] John R. Akeroyd, *On Shapiro's compactness criterion for composition operators*, J. Math. Anal. Appl. **379** (2011), no. 1, 1–7, DOI 10.1016/j.jmaa.2010.11.054. MR2776448 (2012c:47068)

[2] Frédéric Bayart, *Hardy spaces of Dirichlet series and their composition operators*, Monatsh. Math. **136** (2002), no. 3, 203–236, DOI 10.1007/s00605-002-0470-7. MR1919645 (2003i:42032)

[3] Frédéric Bayart, *Compact composition operators on a Hilbert space of Dirichlet series*, Illinois J. Math. **47** (2003), no. 3, 725–743. MR2007233 (2004j:47049)

[4] H. Bohr, *Über die gleichmässige Konvergenz Dirichletscher Reihen*, J. Reine Angew. Math. **143** (1912), 203–211.

[5] Bernd Carl and Irmtraud Stephani, *Entropy, compactness and the approximation of operators*, Cambridge Tracts in Mathematics, vol. 98, Cambridge University Press, Cambridge, 1990. MR1098497 (92e:47002)

[6] F. Carlson, *Contributions à la théorie des séries de Dirichlet*, Ark. Mat. (Note I) **16** (1922), 1–19.

[7] Lennart Carleson, *An interpolation problem for bounded analytic functions*, Amer. J. Math. **80** (1958), 921–930. MR0117349 (22 #8129)

[8] Lennart Carleson, *Interpolations by bounded analytic functions and the corona problem*, Ann. of Math. (2) **76** (1962), 547–559. MR0141789 (25 #5186)

[9] Tom Carroll and Carl C. Cowen, *Compact composition operators not in the Schatten classes*, J. Operator Theory **26** (1991), no. 1, 109–120. MR1214923 (94c:47045)

[10] Catherine Finet, Hervé Queffélec, and Alexander Volberg, *Compactness of composition operators on a Hilbert space of Dirichlet series*, J. Funct. Anal. **211** (2004), no. 2, 271–287, DOI 10.1016/j.jfa.2004.01.003. MR2056832 (2005c:47029)

[11] John B. Garnett, *Bounded analytic functions*, 1st ed., Graduate Texts in Mathematics, vol. 236, Springer, New York, 2007. MR2261424 (2007e:30049)

[12] Julia Gordon and Håkan Hedenmalm, *The composition operators on the space of Dirichlet series with square summable coefficients*, Michigan Math. J. **46** (1999), no. 2, 313–329, DOI 10.1307/mmj/1030132413. MR1704209 (2000f:47046)

[13] Håkan Hedenmalm, Peter Lindqvist, and Kristian Seip, *A Hilbert space of Dirichlet series and systems of dilated functions in* $L^2(0,1)$, Duke Math. J. **86** (1997), no. 1, 1–37, DOI 10.1215/S0012-7094-97-08601-4. MR1427844 (99i:42033)

[14] Hermann König, *Eigenvalue distribution of compact operators*, Operator Theory: Advances and Applications, vol. 16, Birkhäuser Verlag, Basel, 1986. MR889455 (88j:47021)

[15] P.Lefèvre, D. Li, H. Queffélec and L. Rodríguez-Piazza, *Approximation numbers of composition operators on the Dirichlet space,* Ark.Mat., to appear.

[16] Pascal Lefèvre, Daniel Li, Hervé Queffélec, and Luis Rodríguez-Piazza, *Some new properties of composition operators associated with lens maps*, Israel J. Math. **195** (2013), no. 2, 801–824, DOI 10.1007/s11856-012-0164-3. MR3096575

[17] Daniel Li, Hervé Queffélec, and Luis Rodríguez-Piazza, *On approximation numbers of composition operators*, J. Approx. Theory **164** (2012), no. 4, 431–459, DOI 10.1016/j.jat.2011.12.003. MR2885418

[18] Daniel Li, Hervé Queffélec, and Luis Rodríguez-Piazza, *Estimates for approximation numbers of some classes of composition operators on the Hardy space*, Ann. Acad. Sci. Fenn. Math. **38** (2013), no. 2, 547–564, DOI 10.5186/aasfm.2013.3823. MR3113094

[19] Hugh L. Montgomery, *Ten lectures on the interface between analytic number theory and harmonic analysis*, CBMS Regional Conference Series in Mathematics, vol. 84, Published for the Conference Board of the Mathematical Sciences, Washington, DC; by the American Mathematical Society, Providence, RI, 1994. MR1297543 (96i:11002)

[20] H. L. Montgomery and R. C. Vaughan, *Hilbert's inequality*, J. London Math. Soc. (2) **8** (1974), 73–82. MR0337775 (49 #2544)

[21] N. K. Nikol′skiĭ, *Treatise on the shift operator*, Grundlehren der Mathematischen Wissenschaften [Fundamental Principles of Mathematical Sciences], vol. 273, Springer-Verlag, Berlin, 1986. Spectral function theory; With an appendix by S. V. Hruščev [S. V. Khrushchëv] and V. V. Peller; Translated from the Russian by Jaak Peetre. MR827223 (87i:47042)

[22] Jan-Fredrik Olsen and Eero Saksman, *On the boundary behaviour of the Hardy spaces of Dirichlet series and a frame bound estimate*, J. Reine Angew. Math. **663** (2012), 33–66, DOI 10.1515/CRELLE.2011.093. MR2889706

[23] Jan-Fredrik Olsen and Kristian Seip, *Local interpolation in Hilbert spaces of Dirichlet series*, Proc. Amer. Math. Soc. **136** (2008), no. 1, 203–212 (electronic), DOI 10.1090/S0002-9939-07-08955-1. MR2350405 (2008j:30004)

[24] Albrecht Pietsch, *s-numbers of operators in Banach spaces*, Studia Math. **51** (1974), 201–223. MR0361883 (50 #14325)

[25] H. Queffélec and K. Seip, *Approximation numbers of composition operators on the* H^2 *space of Dirichlet series*, J. Funct. Anal. **268** (2015), no. 6, 1612–1648. MR3306358

[26] H. Queffélec and K. Seip, *Decay rates for approximation numbers of composition operators*, J.d'Analyse mathématique, to appear.

[27] Alexander P. Schuster and Kristian Seip, *A Carleson-type condition for interpolation in Bergman spaces*, J. Reine Angew. Math. **497** (1998), 223–233, DOI 10.1515/crll.1998.041. MR1617432 (99f:46034)

[28] Kristian Seip, *Zeros of functions in Hilbert spaces of Dirichlet series*, Math. Z. **274** (2013), no. 3-4, 1327–1339, DOI 10.1007/s00209-012-1118-5. MR3078269

[29] H. S. Shapiro and A. L. Shields, *On some interpolation problems for analytic functions*, Amer. J. Math. **83** (1961), 513–532. MR0133446 (24 #A3280)

[30] J. H. Shapiro and P. D. Taylor, *Compact, nuclear, and Hilbert-Schmidt composition operators on H^2*, Indiana Univ. Math. J. **23** (1973/74), 471–496. MR0326472 (48 #4816)
[31] Joel H. Shapiro, *Composition operators and classical function theory*, Universitext: Tracts in Mathematics, Springer-Verlag, New York, 1993. MR1237406 (94k:47049)

Université Lille Nord de France, USTL, Laboratoire Paul Painlevé U. M. R. CNRS 8524, F–59 655 Villeneuve d'Ascq Cedex, France
E-mail address: herve.queffelec@univ-lille1.fr

Contemporary Mathematics
Volume **638**, 2015
http://dx.doi.org/10.1090/conm/638/12803

A short introduction to de Branges–Rovnyak spaces

Dan Timotin

ABSTRACT. The notes provide a short introduction to de Branges–Rovnyak spaces. They cover some basic facts and are intended to give the reader a taste of the theory, providing sufficient motivation to make it interesting.

CONTENTS

1. Introduction

The purpose of these notes is to provide a short introduction to de Branges–Rovnyak spaces, that have been introduced in [**6**, **7**], an area that has seen significant research activity in the last years. They are intended to give to a casual reader a taste of the theory, providing sufficient motivation and connections with other domains to make it, hopefully, interesting. There exist two comprehensive references on the subject: the older book of Sarason [**22**] that contains most of the basic facts, and the more recent monograph of Fricain and Mashreghi [**14**]. The interested reader may study in depth the subject from there.

The prerequisites are basic facts in operator theory on Hilbert space and in Hardy spaces. Many books contain them, but we have preferred to give as a comprehensive reference Nikolski's treaty [**19**], where all can be found (see the beginning of Section 2).

2010 *Mathematics Subject Classification.* Primary 47B32, 47A45.
The author is partially supported by a grant of the Romanian National Authority for Scientific Research, CNCS - UEFISCDI, project number PN-II-ID-PCE-2011-3-0119.

So the plan of the notes is the following. We introduce the de Branges–Rovnyak spaces as natural reproducing kernel spaces, generalizing the model spaces that appear prominently in the theory of contractions. The challenge here is to justify in a sufficient manner a rather exotic object of study, namely contractively included subspaces; we have considered the reproducing kernel approach as especially convenient. We give next some basic properties, following closely [**22**].

The dichotomy b extreme/nonextreme appears soon. The general idea is that the extreme case has many features that are not far from the case of b inner (the classical model spaces), while the nonextreme case is more exotic from this point of view.

Originally, the de Branges–Rovnyak spaces have been developed in view of model theory, that is, giving a universal model for certain class of operators on Hilbert space. They have been in the shade for a few decades, as the much more popular and well developed model theory of Sz–Nagy and Foias [**24**] has gained the upper hand. It is known to experts that the two theories are equivalent, and we thought that a justification of the study of de Branges–Rovnyak spaces should include some presentation of their role as model spaces. It turns out that this is easier done for the extreme case, and we have chosen to present this case at the end of the notes.

As noted above, the basic reference for de Branges–Rovnyak spaces, that has been frequently used in these notes, is the book of Sarason [**22**]. Some simple unproved results appear in the text as exercises; for the others, references are indicated in the text at the relevant places, occasionally with a hint of the proof.

2. Preliminaries

A comprehensive reference for all facts in this section is [**19**], which has the advantage to contain both the necessary prerequisites from function theory (Part A, chapters 1–3) and from operator theory (part C, chapter 1).

If H is a Hilbert space and $H' \subset H$ a closed subspace, we will write $P_{H'}$ for the orthogonal projection onto H'. The space of bounded linear operators from H_1 to H_2 is denoted $\mathcal{B}(H_1, H_2)$; in case $H_1 = H_2 = H$ we write just $\mathcal{B}(H)$. If $T \in \mathcal{B}(H_1, H_2)$ is a contraction, we will denote by D_T the selfadjoint operator $(I - T^*T)^{1/2}$ and by $\mathcal{D}_T$ the closed subspace $\overline{D_T H_1} = \ker T^\perp$. Thus $\mathcal{D}_T \subset H_1$ and $\mathcal{D}_{T^*} \subset H_2$. Obviously $D_T|\mathcal{D}_T$ is one-to-one.

EXERCISE 2.1. We have $TD_T = D_{T^*}T$. In particular, T maps $\mathcal{D}_T$ into $\mathcal{D}_{T^*}$.

We may also consider the domain and/or the range of D_T to be $\mathcal{D}_T$; by an abuse of notation all these operators will still be denoted by D_T. Note that the adjoint of $D_T : \mathcal{D}_T \to H_1$ is $D_T : H_1 \to \mathcal{D}_T$.

EXERCISE 2.2. If $T \in \mathcal{B}(H)$ is a contraction, $H_1 \subset H$ is a closed subset invariant by T, and we denote $T_1 = T|H_1$, then $\mathcal{D}_{T_1} = \overline{P_{H_1}\mathcal{D}_T}$.

We denote by L^2, L^∞ the Lebesgue spaces on the unit circle $\mathbb{T}$; we will also meet their closed subspaces $H^2 \subset L^2$ and $H^\infty \subset L^\infty$ (the *Hardy spaces*). The corresponding norms will be denoted by $\|\cdot\|_2$ and $\|\cdot\|_\infty$ respectively. As usual, H^2 and H^∞ can be identified with their analytic extension inside the unit disc $\mathbb{D}$. We assume known basic facts about inner and outer functions. We will write

$P_+ := P_{H^2}$ (the orthogonal projection in L^2). The symbols $\mathbf{0}$ and $\mathbf{1}$ will denote the constant functions that take this value.

Each function ϕ in L^∞ acts as multiplication on L^2; the corresponding operator will be denoted by M_ϕ, and we have $\|M_\phi\| = \|\phi\|_\infty$. In particular, if $\phi(z) = z$, we will write $Z = M_\phi$. Actually, the commutant of Z (the class of all operators T on L^2 such that $ZT = TZ$) coincides precisely with the class of all M_ϕ for $\phi \in L^\infty$. (Obviously we have to define $\phi = T\mathbf{1} \in L^2$; a little work is needed to show that it is in L^∞.)

The compression $P_+ M_\phi P_+$ restricted to the space H^2 is called the Toeplitz operator with symbol ϕ and is denoted by T_ϕ. Again we have $\|T_\phi\| = \|\phi\|_\infty$; moreover, if $\phi \in H^\infty$, then T_ϕ is one-to-one (this is a consequence of the brothers Riesz Theorem: a function in H^2 is $\neq 0$ a.e.). In particular, if $\phi(z) = z$, we will write $S = T_\phi$; its adjoint S^* acts as

$$(S^* f)(z) = \frac{f(z) - f(0)}{z}. \tag{2.1}$$

We have $T_\phi^* = T_{\bar{\phi}}$.

As noted above, the multiplication operators commute; this is in general not true for the Toeplitz operators.

EXERCISE 2.3. If $\phi \in H^\infty$, or $\psi \in \overline{H^\infty}$, then $T_\psi T_\phi = T_{\psi\phi}$.

If $k_\lambda(z) = \frac{1}{1-\bar{\lambda} z}$ (a *reproducing vector* in H^2—see more on reproducing kernels in Subsection 3.2 below), then for any $\phi \in H^\infty$ we have

$$T_{\bar{\phi}} k_\lambda = \overline{\phi(\lambda)} k_\lambda. \tag{2.2}$$

3. Introducing de Branges–Rovnyak spaces

3.1. Model spaces. Beurling's theorem says that any subspace of H^2 invariant by S is of the form uH^2, with u an inner function.

EXERCISE 3.1. $S|uH^2$ is unitarily equivalent to S.

From some points of view, the orthogonal $\mathbf{K}_u = H^2 \ominus uH^2$ is more interesting: it is a *model space*. It is invariant by S^*, but $S^*|\mathbf{K}_u$ may behave very differently. Actually, we know exactly *how differently*:

THEOREM 3.2. *If T is a contraction on a Hilbert space H, then the following are equivalent:*

(1) *$I - T^*T$ and $I - TT^*$ have rank one and T^n tends strongly to 0.*
(2) *T is unitarily equivalent to $S_u := S^*|\mathbf{K}_u \in \mathcal{B}(\mathbf{K}_u)$ for some inner function u.*

Theorem 3.2 is a particular case of the general Sz.-Nagy–Foias theory of contractions (see, for instance, the revised edition [**24**] of the original monography); one can find it also in [**19**].

So S_u is a *model operator* for a certain class of contractions. We will meet in this course model operators for a more general class.

3.2. Reproducing kernels. We introduce a larger class of spaces that include $\mathbf{K}_u$ for inner u; this will be done by means of reproducing kernels. A Hilbert space $\mathcal{R}$ of functions on a set X is called a *reproducing kernel space* (RKS) if the evaluations at points of X are continuous; we will always have $X = \mathbb{D}$. By Riesz's representation theorem it follows then that for each $\lambda \in \mathbb{D}$ there exists a function $\mathfrak{l}_\lambda^{\mathcal{R}} \in \mathcal{R}$, called the *reproducing vector* for λ, such that $f(\lambda) = \langle f, \mathfrak{l}_\lambda^{\mathcal{R}} \rangle$. The function of two variables $\mathfrak{L}^{\mathcal{R}}(z,\lambda) := \mathfrak{l}_\lambda^{\mathcal{R}}(z) = \langle \mathfrak{l}_\lambda^{\mathcal{R}}, \mathfrak{l}_z^{\mathcal{R}} \rangle$ is called the *reproducing kernel* of the space $\mathcal{R}$. There is a one-to-one correspondence between RKS's and positive definite kernels (see for instance [**1**]).

EXERCISE 3.3. (1) If $\mathcal{R}$ is a RKS, and $\mathcal{R}_1 \subset \mathcal{R}$ is a closed subspace, then $\mathcal{R}_1$ is also a RKS, and $\mathfrak{l}_\lambda^{\mathcal{R}_1} = P_{\mathcal{R}_1} \mathfrak{l}_\lambda^{\mathcal{R}}$.
(2) If $\mathcal{R} = \mathcal{R}_1 \oplus \mathcal{R}_2$, then

$$\mathfrak{L}^{\mathcal{R}} = \mathfrak{L}^{\mathcal{R}_1} \oplus \mathfrak{L}^{\mathcal{R}_2}. \tag{3.1}$$

All three spaces discussed above have reproducing kernels, namely:

$$\begin{aligned} H^2 &\longrightarrow \frac{1}{1-\bar\lambda z}, \\ uH^2 &\longrightarrow \frac{\overline{u(\lambda)}u(z)}{1-\bar\lambda z}, \\ \mathbf{K}_u &\longrightarrow \frac{1-\overline{u(\lambda)}u(z)}{1-\bar\lambda z}, \end{aligned}$$

and one can check that equality (3.1) is satisfied.

Our plan is to obtain RKSs with similar formulas, but replacing the *inner* function u with an *arbitrary* function b in the unit ball of H^∞. That is, we want spaces with kernels $\frac{\overline{b(\lambda)}b(z)}{1-\bar\lambda z}$ and $\frac{1-\overline{b(\lambda)}b(z)}{1-\bar\lambda z}$.

Of course it is not obvious that such RKSs exist. Then, if they exist, we want to identify them concretely, hoping to relate them to the familiar space H^2.

Things are simpler for the first kernel. Note first the next (general) exercise.

EXERCISE 3.4. If $\mathfrak{L}(z,\lambda)$ is a positive kernel on $X \times X$ and $\phi : X \to \mathbb{C}$, then $\phi(z)\overline{\phi(\lambda)}\mathfrak{L}(z,\lambda)$ is a positive kernel.

So $\frac{\overline{b(\lambda)}b(z)}{1-\bar\lambda z}$ is the kernel of *some* space $\mathcal{R}$; but we would like to know it more concretely. The case $b = u$ inner suggests that a good candidate might be bH^2. Now, we already have a problem: if b is a general function, bH^2 might not be closed in H^2, so it is not a genuine Hilbert space. But let us be brave and go on: we want

$$\mathfrak{l}_\lambda^{\mathcal{R}}(z) = \overline{b(\lambda)}b(z)k_\lambda.$$

Since the reproducing kernel property should be valid in $\mathcal{R}$, we must have, for any $f \in H^2$,

$$b(\lambda)f(\lambda) = \langle bf, \mathfrak{l}_\lambda^{\mathcal{R}}(z)\rangle_{\mathcal{R}} = b(\lambda)\langle bf, bk_\lambda\rangle_{\mathcal{R}}$$

and therefore

$$f(\lambda) = \langle bf, bk_\lambda\rangle_{\mathcal{R}}.$$

On the other hand, since $f \in H^2$, we have $f(\lambda) = \langle f, k_z\rangle_{H^2}$.

We have now arrived at the crucial point. If b is inner, then $\langle bf, bk_z\rangle_{H^2} = \langle f, k_z\rangle_{H^2}$ and everything is fine: the scalar product in bH^2 is the usual scalar product in H^2. But, in the general case, we have to define a *different* scalar product on $\mathcal{R} = bH^2$, by the formula

$$\langle bf, bg\rangle_{\mathcal{R}} := \langle f, g\rangle_{H^2}. \tag{3.2}$$

This appears to solve the problem. Since $bf_1 = bf_2$ implies $f_1 = f_2$, formula (3.2) is easily shown to define a scalar product on bH^2. We will denote the corresponding Hilbert space by $\mathcal{M}(b)$. Let us summarize the results obtained.

THEOREM 3.5. *$\mathcal{M}(b)$, defined as bH^2 endowed with the scalar product (3.2), is a Hilbert space, which as a set is a linear subspace (in general not closed) of H^2. Its reproducing kernel is $\frac{\overline{b(\lambda)}b(z)}{1-\bar\lambda z}$, and the inclusion $\iota : \mathcal{M}(b) \to H^2$ is a contraction. $\mathcal{M}(b)$ is invariant by S, and S acts as an isometry on $\mathcal{M}(b)$.*

PROOF. From (3.2) it follows that the map $f \mapsto bf$ is isometric from H^2 onto $\mathcal{M}(b)$, whence $\mathcal{M}(b)$ is complete. The formula for the reproducing kernel has been proved (in fact, it lead to the definition of the space $\mathcal{M}(b)$). We have

$$\|\iota(bf)\|_{H^2} = \|bf\|_{H^2} \le \|f\|_{H^2} = \|bf\|_{\mathcal{M}(b)},$$

and thus ι is a contraction.

Finally $\mathcal{M}(b)$ is invariant by S since $z(bf) = b(zf)$, and

$$\|zbf\|_{\mathcal{M}(b)} = \|b(zf)\|_{\mathcal{M}(b)} = \|zf\|_{H^2} = \|f\|_{H^2} = \|bf\|_{\mathcal{M}(b)},$$

and thus the restriction of S is an isometry. □

This settles the case of the kernel $\frac{\overline{b(\lambda)}b(z)}{1-\bar\lambda z}$. To discuss $\frac{1-\overline{b(\lambda)}b(z)}{1-\bar\lambda z}$ is slightly more complicated, and a preliminary discussion is needed.

3.3. Contractively included subspaces. Let $T : E \to H$ be a bounded one-to-one operator. Define on the image $T(E)$ a scalar product $\langle\cdot,\cdot\rangle'$ by the formula

$$\langle T\xi, T\eta\rangle' := \langle \xi, \eta\rangle_E. \tag{3.3}$$

Then T is a unitary operator from E to $T(E)$ endowed with $\langle\cdot,\cdot\rangle'$. The space obtained is complete; we will denote it by $M(T)$. The linear space $M(T)$ is contained as a set in H, and the inclusion is a contraction if and only if T is a contraction. In this case the space $M(T)$ will be called a *contractively included subspace* of H, and the scalar product will be denoted $\langle\cdot,\cdot\rangle_{M(T)}$.

A slight modification is needed in case T is not one-to-one; then (3.3) cannot be used directly since there $T\xi$ does not determine ξ uniquely. We may however recapture this uniqueness and use (3.3) if we require that $\xi, \eta \in \ker T^\perp$; then T becomes a unitary from $\ker T^\perp$ to $M(T)$.

In almost all cases in this course the corresponding contraction T will be one-to-one, and thus we will apply directly (3.3). The only exception appears in Theorem 5.3, when we will use Lemma 3.6 below, with no direct reference to the scalar product.

We have already met a particular case of this notion: with the above notations, we have $\mathcal{M}(b) = M(T_b)$. Remember that, since $b \in H^\infty$, the operator T_b is one-to-one.

The following is a basic result that is used when we have to deal with two contractively embedded subspaces. It is essentially contained in [**10**].

LEMMA 3.6. *Suppose $T_1 : E_1 \to H$, $T_2 : E_2 \to H$ are two contractions. Then:*

(1) *The space* $M(T_1)$ *is contained contractively in* $M(T_2)$ *(meaning that* $M(T_1) \subset M(T_2)$ *and the inclusion is a contraction) if and only if* $T_1T_1^* \leq T_2T_2^*$.
(2) *The spaces* $M(T_1)$ *and* $M(T_2)$ *coincide as Hilbert spaces (that is, they are equal as sets, and the scalar product is the same) if and only if* $T_1T_1^* = T_2T_2^*$.
(3) $T : H \to H$ *acts contractively on* $M(T_1)$ *(meaning that* $T(M(T_1)) \subset M(T_1)$ *and* $T|M(T_1)$ *is a contraction) if and only if* $TT_1T_1^*T^* \leq T_1T_1^*$.

It is worth at this point to note the following theorem of de Branges and Rovnyak [**6**], which is an analogue of Beurling's theorem.

THEOREM 3.7. *Suppose* $X \subset H^2$ *is a contractively included subspace of* H^2. *The following are equivalent*

(1) X *is invariant by* S *and the restriction* $S|X$ *is an isometry (in the norm of* X*).*
(2) *There exist a function* b *in the unit ball of* H^∞, *such that* $X = \mathcal{M}(b)$.

The function b *is determined up to a multiplicative constant of modulus 1.*

Suppose now that H is a reproducing kernel Hilbert space, with kernel $\mathfrak{L}(z, \lambda)$. We may obtain for the reproducing vectors of $M(T)$ a formula similar to the particular case from the previous subsection.

LEMMA 3.8. *Suppose* $T : E \to H$ *is one-to-one. With the above notations, we have*

$$\mathfrak{l}_\lambda^{M(T)} = TT^*\mathfrak{l}_\lambda^H.$$

PROOF. We have, using (3.3),

$$\langle Tf, TT^*\mathfrak{l}_\lambda^H\rangle_{M(T)} = \langle f, T^*\mathfrak{l}_\lambda^H\rangle_H = \langle Tf, \mathfrak{l}_\lambda^H\rangle_H = (Tf)(\lambda) = \langle Tf, \mathfrak{l}_\lambda^{M(T)}\rangle_{M(T)},$$

which proves the theorem. □

In case $H = H^2$, $T = T_b$, we recapture the previous result: $M(T_b) = \mathcal{M}(b)$ and, using (2.2),

$$\mathfrak{l}_\lambda^{\mathcal{M}(b)} = T_bT_b^*k_\lambda = b\overline{b(\lambda)}k_\lambda.$$

3.4. The complementary space. Remember that our current purpose is to find, if possible, an RKS with kernel $\frac{1-\overline{b(\lambda)}b(z)}{1-\bar\lambda z}$. Let us note that

$$\frac{1-\overline{b(\lambda)}b(z)}{1-\bar\lambda z} = \frac{1}{1-\bar\lambda z} - \frac{\overline{b(\lambda)}b(z)}{1-\bar\lambda z} = \mathfrak{L}^{H^2}(z,\lambda) - \mathfrak{L}^{\mathcal{M}(b)}(z,\lambda).$$

Can we obtain in the general case a formula for such a difference? The answer is positive.

LEMMA 3.9. *With the above notations,*

$$\mathfrak{L}^H(z,\lambda) - \mathfrak{L}^{M(T)}(z,\lambda) = \mathfrak{L}^{M(D_{T^*})}(z,\lambda),$$

where $D_{T^*} : \mathcal{D}_{T^*} \to H$.

PROOF. Using Lemma 3.8, we have

$$\begin{aligned}\mathfrak{L}^H(z,\lambda) - \mathfrak{L}^{M(T)}(z,\lambda) &= \langle \mathfrak{l}_\lambda^H, \mathfrak{l}_z^H \rangle_H - \langle \mathfrak{l}_\lambda^{M(T)}, \mathfrak{l}_z^{M(T)} \rangle_{M(T)} \\ &= \langle \mathfrak{l}_\lambda^H, \mathfrak{l}_z^H \rangle_H - \langle TT^* \mathfrak{l}_\lambda^{H)}, TT^* \mathfrak{l}_z^H \rangle_{M(T)} \\ &= \langle \mathfrak{l}_\lambda^H, \mathfrak{l}_z^H \rangle_H - \langle T^* \mathfrak{l}_\lambda^{H)}, T^* \mathfrak{l}_z^H \rangle_E \\ &= \langle \mathfrak{l}_\lambda^H, (I - TT^*) \mathfrak{l}_z^H \rangle_H = \langle D_{T^*} \mathfrak{l}_\lambda^H, D_{T^*} \mathfrak{l}_z^H \rangle_H \\ &= \langle D_{T^*}^2 \mathfrak{l}_\lambda^H, D_{T^*}^2 \mathfrak{l}_z^H \rangle_{M(D_{T^*})},\end{aligned}$$

(the last equality being a consequence of the fact that D_{T^*} is one-to-one as an operator from $\mathcal{D}_{T^*}$ into H). Then Lemma 3.8, applied to D_{T^*} instead of T, says that the last quantity is precisely $\mathfrak{L}^{M(D_{T^*})}(z,\lambda)$. □

Since D_{T^*} is a contraction, the RKS corresponding to $\mathfrak{L}^H(z,\lambda) - \mathfrak{L}^{M(T)}(z,\lambda)$ is also a space contractively included in H; it is called the space *complementary* to $M(T)$ and will be denoted by $C(T)$.

EXERCISE 3.10. If T is an isometry, then $M(T)$ is a usual subspace of H (with the norm restricted), and $C(T)$ is its orthogonal complement.

If $x \in H$, then one can write

$$x = TT^* x + D_{T^*}^2 x. \tag{3.4}$$

The first term in the right hand side is in $M(T)$, while the second is in $C(T)$. Moreover,

$$\|TT^* x\|_{M(T)}^2 = \|T^* x\|_H^2, \qquad \|D_{T^*}^2 x\|_{C(T)}^2 = \|D_{T^*} x\|_H^2,$$

whence

$$\|x\|^2 = \|TT^* x\|_{M(T)}^2 + \|D_{T^*}^2 x\|_{C(T)}^2.$$

In case T is an isometry, $M(T)$ and $C(T)$ form an orthogonal decomposition of H, and (3.4) is the corresponding orthogonal decomposition of H. In the general case $M(T)$ and $C(T)$ may have a nonzero intersection, and so a decomposition $x = x_1 + x_2$ with $x_1 \in M(T)$, $x_2 \in C(T)$ is no more unique.

EXERCISE 3.11. If T is a contraction, and $x = x_1 + x_2$ with $x_1 \in M(T)$, $x_2 \in C(T)$, then

$$\|x\|^2 \le \|x_1\|_{M(T)}^2 + \|x_2\|_{C(T)}^2,$$

and equality implies $x_1 = TT^* x$, $x_2 = D_{T^*}^2 x$.

At this point we have achieved our first purpose. Lemma 3.9 applied to the case $H = K = H^2$ and $T = T_b$ yields the identification of the reproducing kernel corresponding to $\frac{1-\overline{b(\lambda)}b(z)}{1-\bar\lambda z}$.

THEOREM 3.12. *The RKS with kernel* $\frac{1-\overline{b(\lambda)}b(z)}{1-\bar\lambda z}$ *is* $M(D_{T_b^*})$. *It is a contractively included subspace of* H^2 *that will be denoted* $\mathcal{H}(b)$ *and called the* de Branges–Rovnyak *space associated to the function* b *in the inner ball of* H^∞.

As noted above, if $b = u$ is inner, then $\mathcal{H}(b) = \mathbf{K}_u$.

EXERCISE 3.13. (1) $H^2 = \mathcal{H}(\mathbf{0}) = \mathcal{M}(\mathbf{1})$.
(2) If $\|b\|_\infty < 1$, then $\mathcal{H}(b)$ is a renormed version of H^2.
(3) If $\inf_{z\in\mathbb{D}} |b(z)| > 0$, then $\mathcal{M}(b)$ is a renormed version of H^2.

We will denote from now on $k_\lambda^b = \mathfrak{l}_\lambda^{\mathcal{H}(b)} = \frac{1-\overline{b(\lambda)}b(z)}{1-\bar\lambda z}$.

4. More about contractively included subspaces

LEMMA 4.1. *If $T : E \to H$ is a contraction, then:*

(1) *$\xi \in H$ belongs to $C(T)$ if and only if $T^*\xi \in C(T^*)$.*

(2) *If $\xi_1, \xi_2 \in C(T)$, then*

$$\langle \xi_1, \xi_2 \rangle_{C(T)} = \langle \xi_1, \xi_2 \rangle_H + \langle T^*\xi_1, T^*\xi_2 \rangle_{C(T^*)}.$$

PROOF. The inclusion $T^*(C(T)) \subset C(T^*)$ follows from the intertwining relation in Exercise 2.1. On the other hand, if $T^*\xi \in C(T^*)$, we have $T^*\xi = D_T\eta$ for some $\eta \in H$, and, using again Exercise 2.1,

$$\xi = TT^*\xi + D_{T^*}^2\xi = TD_T\eta + D_{T^*}^2\xi = D_{T^*}(D_{T^*}\xi + T\eta),$$

which shows that $\xi \in C(T)$.

To prove (2), write $\xi_1 = D_{T^*}\eta_1$, $\xi_2 = D_{T^*}\eta_2$, with $\eta_1, \eta_2 \in \mathcal{D}_{T^*}$; then $T^*\eta_1, T^*\eta_2 \in \mathcal{D}_T$. Since both $D_T : \mathcal{D}_T \to H$ and $D_{T^*} : \mathcal{D}_{T^*} \to H$ are one-to-one, we have, using (3.3) and Exercise 2.1,

$$\begin{aligned}
\langle \xi_1, \xi_2 \rangle_{C(T)} &= \langle \eta_1, \eta_2 \rangle = \langle D_{T^*}\eta_1, D_{T^*}\eta_2 \rangle + \langle T^*\eta_1, T^*\eta_2 \rangle \\
&= \langle \xi_1, \xi_2 \rangle + \langle D_T T^*\eta_1, D_T T^*\eta_2 \rangle_{C(T^*)} \\
&= \langle \xi_1, \xi_2 \rangle + \langle T^* D_{T^*}\eta_1, T^* D_{T^*}\eta_2 \rangle_{C(T^*)} \\
&= \langle \xi_1, \xi_2 \rangle + \langle T^*\xi_1, T^*\xi_2 \rangle_{C(T^*)}.
\end{aligned}$$ □

There is a more direct way in which complementarity is related to orthogonality. If $T \in \mathcal{B}(E, H)$ is a contraction, we define the *Julia operator* $J(T) : E \oplus \mathcal{D}_{T^*} \to H \oplus \mathcal{D}_T$ by

$$J(T) = \begin{pmatrix} T & D_{T^*} \\ D_T & -T^* \end{pmatrix}.$$

EXERCISE 4.2. The Julia operator is unitary.

LEMMA 4.3. *Suppose $T : E \to H$ is one-to-one. Denote*

$$\mathcal{X}_1 = J(T)(E \oplus \{0\}), \quad \mathcal{X}_2 = (H \oplus \mathcal{D}_T) \ominus \mathcal{X}_1 = J(T)(\{0\} \oplus \mathcal{D}_{T^*}),$$

and by P_1 the projection of $H \oplus \mathcal{D}_T$ onto its first coordinate H. Then $P_1|\mathcal{X}_1$ is unitary from $\mathcal{X}_1$ onto $M(T)$, and $P_1|\mathcal{X}_2$ is unitary from $\mathcal{X}_2$ onto $C(T)$.

PROOF. We have

$$P_1\mathcal{X}_1 = P_1(\{Tx \oplus D_Tx : x \in E\}) = \{Tx : x \in E\} = M(T).$$

Moreover, if $x_1 \in \mathcal{X}_1$, then $x_1 = Tx \oplus D_Tx$ for some $x \in E$, and

$$\|P_1x_1\|_{M(T)} = \|Tx\|_{M(T)} = \|x\|_E = \|J(T)(x \oplus 0)\| = \|x_1\|,$$

which proves the first part of the lemma.

Now $\mathcal{X}_2 = J(T)(\{0\} \oplus \mathcal{D}_{T^*})$, so

$$P_1\mathcal{X}_2 = P_1(\{D_{T^*}y \oplus -T^*y : y \in \mathcal{D}_{T^*}\}) = \{D_{T^*}y : y \in \mathcal{D}_{T^*}\} = C(T).$$

If $x_2 \in \mathcal{X}_2$, then $x_2 = D_{T^*}y \oplus -T^*y$ for some $y \in \mathcal{D}_{T^*}$, and

$$\|P_1x_2\|_{C(T)} = \|D_{T^*}y\|_{C(T)} = \|y\|_{\mathcal{D}_{T^*}} = \|J(T)(0 \oplus y)\| = \|x_2\|,$$

as required. □

We can view this result as saying that the orthogonal decomposition of $H\oplus\mathcal{D}_T$ as $\mathcal{X}_1\oplus\mathcal{X}_2$ is mapped by projecting onto the first coordinate into the complementary decomposition $H=M(T)+C(T)$ (which is not, in general, a direct sum). So the rather exotic definition of complementary spaces is in fact the projection of a more familiar geometric structure.

5. Back to $\mathcal{H}(b)$

5.1. Some properties of $\mathcal{H}(b)$; definition of X_b. We denote $\mathcal{H}(\bar b):=\mathcal{H}(T_{\bar b})$. Although our focus is on $\mathcal{H}(b)$, the space $\mathcal{H}(\bar b)$ is a useful tool for its study.

LEMMA 5.1. *$\mathcal{H}(\bar b)$ is contained contractively in $\mathcal{H}(b)$.*

PROOF. We have

$$\begin{aligned}T_bT_{\bar b}&=P_+M_bP_+M_{\bar b}P_+|H^2\le P_+M_bM_{\bar b}P_+|H^2\\&=P_+M_{\bar b}M_bP_+|H^2=P_+M_{\bar b}P_+M_bP_+|H^2=T_{\bar b}T_b.\end{aligned}$$

Therefore

$$D^2_{T_b}\le D^2_{T_b^*},$$

whence Lemma 3.6(1) implies that $\mathcal{H}(\bar b)$ is contained contractively in $\mathcal{H}(b)$. □

Lemma 4.1 applied to the case $T=T_b$ yields the following result.

LEMMA 5.2. *If $h\in H^2$, then $h\in\mathcal{H}(b)$ if and only if $T_{\bar b}h\in\mathcal{H}(\bar b)$. If $h_1,h_2\in\mathcal{H}(b)$, then*

$$\langle h_1,h_2\rangle_{\mathcal{H}(b)}=\langle h_1,h_2\rangle_{H^2}+\langle T_{\bar b}h_1,T_{\bar b}h_2\rangle_{\mathcal{H}(\bar b)}. \tag{5.1}$$

We now show that, similarly to model spaces, de Branges–Rovnyak spaces are invariant by adjoints of Toeplitz operators.

THEOREM 5.3. *If $\phi\in H^\infty$, then $\mathcal{H}(b)$ and $\mathcal{H}(\bar b)$ are both invariant under $T_\phi^*=T_{\bar\phi}$, and the norm of this operator in each of these spaces is at most $\|\phi\|_\infty$.*

PROOF. We may assume that $\|\phi\|_\infty\le 1$. By Lemma 3.6(3), in order to show that $T_{\bar\phi}$ acts as a contraction in $\mathcal{H}(\bar b)$ we have to prove the inequality

$$T_{\bar\phi}(I-T_{\bar b}T_b)T_\phi\le I-T_{\bar b}T_b,$$

or

$$\begin{aligned}0\le I-T_{\bar b}T_b-T_{\bar\phi}(I-T_{\bar b}T_b)T_\phi&=I-T_{\bar b}T_b-T_{\bar\phi}T_\phi+T_{\bar\phi}T_{\bar b}T_bT_\phi\\&=I-T_{|b|^2}-T_{|\phi|^2}+T_{|b|^2|\phi|^2}=T_{(1-|b|^2)(1-|\phi|^2)}.\end{aligned}$$

But the last operator is the compression to H^2 of $M_{(1-|b|^2)(1-|\phi|^2)}$, which is positive, since $(1-|b|^2)(1-|\phi|^2)\ge 0$.

This proves the statement for $\mathcal{H}(\bar b)$. Take now $h\in\mathcal{H}(b)$. Lemma 5.2 implies that $T_{\bar b}h\in\mathcal{H}(\bar b)$. By what has been just proved, $T_{\bar b}T_{\bar\phi}h=T_{\bar\phi}T_{\bar b}h\in\mathcal{H}(\bar b)$, and then applying again Lemma 5.2 we obtain $T_{\bar b}h\in\mathcal{H}(b)$.

Finally, using (5.1) and the contractivity of $T_{\bar\phi}$ on H^2 as well as on $\mathcal{H}(\bar b)$, we have

$$\|T_{\bar\phi}h\|^2_{\mathcal{H}(b)}=\|T_{\bar\phi}h\|^2_{H^2}+\|T_{\bar\phi}T_{\bar b}h\|^2_{\mathcal{H}(\bar b)}\le\|h\|^2_{H^2}+\|T_{\bar b}h\|^2_{\mathcal{H}(\bar b)}=\|h\|^2_{\mathcal{H}(b)},$$

so $T_{\bar\phi}$ acts as a contraction in $\mathcal{H}(b)$. □

The most important case is obtained when $\phi(z) = z$. Theorem 5.3 says then that $\mathcal{H}(b)$ is invariant under S^* and the restriction of S^* is a contraction. We will denote by X_b this restriction $S^*|\mathcal{H}(b)$.

COROLLARY 5.4. *The function S^*b is in $\mathcal{H}(b)$.*

PROOF. We have

$$T_{\bar{b}}S^*b = S^*T_{\bar{b}}b = S^*T_{\bar{b}}T_b 1 = -S^*(I - T_{\bar{b}}T_b)1$$

(for the last equality we have used the fact that $S^*1 = 0$). Obviously $(I - T_{\bar{b}}T_b)1 \in \mathcal{H}(\bar{b})$, so Theorem 5.3 implies that $T_{\bar{b}}S^*b = -S^*(I - T_{\bar{b}}T_b)1 \in \mathcal{H}(\bar{b})$. By Lemma 5.2 it follows that $S^*b \in \mathcal{H}(b)$. □

Note that if $b \neq \mathbf{1}$, then $S^*b \neq \mathbf{0}$. Besides S^*b, we know as inhabitants of $\mathcal{H}(b)$ the reproducing vectors k_λ^b. Other elements may be obtained, for instance, by applying to these elements powers or functions of X_b.

EXERCISE 5.5. Show that, if $\lambda \in \mathbb{D}$, then

$$((I - \lambda X_b)^{-1}(S^*b))(z) = \frac{b(z) - b(\lambda)}{z - \lambda}.$$

Therefore the functions in the right hand side belong to $\mathcal{H}(b)$.

In general b itself may not be in $\mathcal{H}(b)$; we will see later exactly when this happens. Let us also compute the adjoint of X_b.

LEMMA 5.6. *If $h \in \mathcal{H}(b)$, then*

$$X_b^* h = Sh - \langle h, S^*b\rangle_{\mathcal{H}(b)} b.$$

PROOF. A computation shows that $X_b k_\lambda^b = S^* k_\lambda^b = \bar{\lambda} k_\lambda^b - \overline{b(\lambda)} S^*b$. Then, if $h \in \mathcal{H}(b)$ and $\lambda \in \mathbb{D}$, then

$$\begin{aligned}(X_b^* h)(\lambda) &= \langle X_b^* h, k_\lambda^b\rangle_{\mathcal{H}(b)} = \langle h, X_b k_\lambda^b\rangle_{\mathcal{H}(b)} = \lambda\langle h, k_\lambda^b\rangle_{\mathcal{H}(b)} - b(\lambda)\langle h, S^*b\rangle_{\mathcal{H}(b)} \\ &= \lambda h(\lambda) - \langle h, S^*b\rangle_{\mathcal{H}(b)} b(\lambda),\end{aligned}$$

which proves the lemma. □

5.2. Another representation of $\mathcal{H}(b)$ and X_b. In the sequel of the course we will use the notation $\Delta = (1 - |b|^2)^{1/2}$. The spaces $\overline{\Delta H^2}$ and $\overline{\Delta L^2}$ are closed subspaces of L^2 invariant with respect to Z. We will denote by V_Δ and Z_Δ the corresponding restrictions of Z.

EXERCISE 5.7. V_Δ is isometric, while Z_Δ is unitary.

The next result, a slight modification of Lemma 4.3, provides another representation of $\mathcal{H}(b)$.

THEOREM 5.8. *Suppose that b is a function in the unit ball of H^∞. Then:*

1. *$S \oplus V_\Delta$ is an isometry on $H^2 \oplus \overline{\Delta H^2}$.*
2. *The space $\mathcal{K}_b := (H^2 \oplus \overline{\Delta H^2}) \ominus \{bh \oplus \Delta h : h \in H^2\}$ is a subspace of $H^2 \oplus \overline{\Delta H^2}$ invariant with respect to $S^* \oplus V_\Delta^*$.*
3. *The projection $P_1 : H^2 \oplus \overline{\Delta H^2} \to H^2$ on the first coordinate maps $\mathcal{K}_b$ unitarily onto $\mathcal{H}(b)$, and $P_1(S^* \oplus V_\Delta^*) = X_b P_1$.*

PROOF. The proof of (1) is immediate. Also, the map $h \mapsto bh \oplus \Delta h$ is an isometry of H^2 onto $\{bh \oplus \Delta h : h \in H^2\}$, which is therefore a closed subspace. Since $(S \oplus V_\Delta)(bh \oplus \Delta h) = b(zh) \oplus \Delta(zh)$, it is immediate that $\{bh \oplus \Delta h : h \in H^2\}$ is invariant by $S \oplus V_\Delta$, whence its orthogonal $\mathcal{K}$ is invariant by $S^* \oplus V_\Delta^*$; thus (2) is proved.

To prove (3), let us apply Lemma 4.3 to the case $T = T_b$, when $C(T) = \mathcal{H}(b)$. It says that, if $\mathcal{X}_2 = (H^2 \oplus \mathcal{D}_{T_b}) \ominus \{T_b h \oplus D_{T_b} h\}$, then the projection onto the first coordinate maps $\mathcal{X}_2$ unitarily onto $\mathcal{H}(b)$. Since, for any $h \in H^2$,

$$\|D_{T_b} h\|^2 = \|h\|^2 - \|bh\|^2 = \frac{1}{2\pi}\int_{-\pi}^{\pi} |h(e^{it})|^2\,dt - \frac{1}{2\pi}\int_{-\pi}^{\pi} |b(e^{it})h(e^{it})|^2\,dt = \|\Delta h\|^2,$$

the map $D_{T_b} h \mapsto \Delta h$ extends to a unitary U from $\mathcal{D}_{T_b}$ onto the closure of $\overline{\Delta H^2}$. Then $I_{H^2} \oplus U$ maps unitarily $H^2 \oplus \mathcal{D}_{T_b}$ onto $H^2 \oplus \overline{\Delta H^2}$, $\mathcal{X}_1$ onto $\{bh \oplus \Delta h : h \in H^2\}$, $\mathcal{X}_2$ onto $\mathcal{K}_b$, and it commutes with the projection on the first coordinate. Therefore P_1 maps $\mathcal{K}_b$ unitarily onto $\mathcal{H}(b)$, and

$$P_1(S^* \oplus V_\Delta^*)|\mathcal{K}_b = P_1(S^* \oplus V_\Delta^*)P_1|\mathcal{K}_b = (S^* \oplus 0)P_1|\mathcal{K}_b = (S^*|\mathcal{H}(b))P_1 = X_b P_1. \ \square$$

5.3. The dichotomy extreme/nonextreme. The study of the spaces $\mathcal{H}(b)$ splits further into two mutually exclusive cases: when b is an extreme point of the unit ball of H^∞ and when it is not. The first case includes $b = u$ inner, and thus will be more closely related to model spaces, while the second includes the case $\|b\| < 1$, and thus there will be properties similar to the whole of H^2. Actually, we will not use extremality directly, but rather through one of the equivalent characterizations given by the next lemma (for which again [**19**] can be used as a reference).

LEMMA 5.9. *If h is a function in the unit ball of H^∞, then the following are equivalent:*

(1) *b is extreme.*
(2) $\frac{1}{2\pi}\int_{-\pi}^{\pi} \log \Delta(e^{it})\,dt = -\infty$.
(3) $\overline{\Delta H^2} = \overline{\Delta L^2}$.

6. The nonextreme case

When $\frac{1}{2\pi}\int_{-\pi}^{\pi} \log \Delta(e^{it})\,dt > -\infty$, there exists a uniquely defined outer function a with $|a| = \Delta$ and $a(0) > 0$; thus $|a|^2 + |b|^2 = 1$. This function is a basic tool in the theory of $\mathcal{H}(b)$ in the nonextreme case. Since (see Exercise 2.3)

$$T_{\bar a} T_a = T_{\bar a a} = I - T_{\bar b b} = I - T_{\bar b} T_b$$

Lemma 3.6(2) implies that $\mathcal{H}(\bar b) = \mathcal{M}(\bar a)$.

EXERCISE 6.1. If a is an outer function, then $\ker T_{\bar a} = \{0\}$.

We can apply Lemma 5.2 to the current situation.

LEMMA 6.2. (1) *We have $h \in \mathcal{H}(b)$ if and only if $T_{\bar b} h \in \mathcal{M}(\bar a)$; when this happens there is a unique (by Exercise 6.1) function $h^+ \in H^2$ such that $T_{\bar b} h = T_{\bar a} h^+$.*
(2) *If $h_1, h_2 \in \mathcal{H}(b)$, then*

$$\langle h_1, h_2\rangle_{\mathcal{H}(b)} = \langle h_1, h_2\rangle_{H^2} + \langle h_1^+, h_2^+\rangle_{H^2}.$$

(3) *If $h \in \mathcal{H}(b)$ and $\phi \in H^\infty$, then $(T_{\bar\phi} h)^+ = T_{\bar\phi} h^+$.*

PROOF. (1) is a consequence of Lemma 5.2 and the equality $\mathcal{H}(\bar{b}) = \mathcal{M}(\bar{a})$; the uniqueness of h^+ follows from Exercise 6.1.

The formula for the scalar product in Lemma 5.2 becomes

$$\langle h_1, h_2\rangle_{\mathcal{H}(b)} = \langle h_1, h_2\rangle_{H^2} + \langle T_{\bar{a}}h_1^+, T_{\bar{a}}h_2^+\rangle_{\mathcal{H}(\bar{a})} = \langle h_1^+, h_2^+\rangle_{H^2},$$

the last equality being a consequence of the fact that $T_{\bar{a}}$ is one-to-one. This proves (2).

Finally,

$$T_{\bar{b}}T_{\bar{\phi}}h = T_{\bar{\phi}}T_{\bar{b}}h = T_{\bar{\phi}}T_{\bar{a}}h^+ = T_{\bar{a}}T_{\bar{\phi}}h^+,$$

proving (3). □

We gather in a theorem some properties of $\mathcal{H}(b)$ for b nonextreme.

THEOREM 6.3. *Suppose b is nonextreme.*

(1) *The polynomials belong to $\mathcal{M}(\bar{a})$ and are dense in $\mathcal{M}(\bar{a})$.*
(2) *$\mathcal{M}(\bar{a})$ is dense in $\mathcal{H}(b)$.*
(3) *The polynomials are dense in $\mathcal{H}(b)$.*
(4) *The function b is in $\mathcal{H}(b)$, and $\|b\|^2_{\mathcal{H}(b)} = |a(0)|^{-2} - 1$.*
(5) *The space $\mathcal{H}(b)$ is invariant by the unilateral shift S.*

PROOF. By checking the action on monomials, it is immediate that the space $\mathcal{P}_n$ of polynomials of degree less or equal to n is invariant by $T_{\bar{a}}$. But $T_{\bar{a}}$ is one-to-one, and so $T_{\bar{a}}|\mathcal{P}_n$ is also onto. So all polynomials belong to the image of $T_{\bar{a}}$, which is $\mathcal{M}(\bar{a})$. Moreover, since $T_{\bar{a}}$ is one-to-one, it is unitary as an operator from H^2 to $\mathcal{M}(\bar{a})$. Then the image of all polynomials, which form a dense set in H^2, is a dense set in $\mathcal{M}(\bar{a})$ which proves (1).

Suppose that $h \in \mathcal{H}(b)$ is orthogonal in $\mathcal{H}(b)$ to all $\mathcal{M}(\bar{a})$. In particular, h is orthogonal to $T_{\bar{a}}S^{*n}h$ for every $n \geq 0$. By Lemma 6.2(3), $(T_{\bar{a}}S^{*n}h)^+ = T_{\bar{a}}S^{*n}h^+$; applying then 6.2(2), we have, for any $n \geq 0$,

$$\begin{aligned} 0 &= \langle h, T_{\bar{a}}S^*nh\rangle_{\mathcal{H}(b)} \\ &= \langle h, T_{\bar{a}}S^*nh\rangle_{H^2} + \langle h^+, T_{\bar{a}}S^*nh^+\rangle_{H^2} \\ &= \frac{1}{2\pi}\int_{-\pi}^{\pi} a(e^{it})(|h(e^{it})|^2 + |h^+(e^{it})|^2)e^{int}\,dt. \end{aligned}$$

Therefore, the function $a(|h|^2 + |h^+|^2)$ belongs to H^1_0. A classical fact about outer functions (see, for instance, [**19**]) implies that we also have $|h|^2 + |h^+|^2 \in H^1_0$. But the only real-valued function in H^1_0 is the zero function, so $h = 0$, which proves (2). Obviously, (1) and (2) imply (3).

We have

$$T_{\bar{b}}b = P_+(\bar{b}b) = P_+(1 - \bar{a}a) = T_{\bar{a}}(1/\overline{a(0)} - a).$$

By Lemma 6.2 it follows that $b \in \mathcal{H}(b)$ and $b^+ = 1/\overline{a(0)} - a$; moreover,

$$\begin{aligned} \|b\|^2_{\mathcal{H}(b)} &= \|b\|^2_{H^2} + \|1/\overline{a(0)} - a\|^2_{H^2} \\ &= \|b\|^2_{H^2} + \|a - a(0)\|^2_{H^2} + \|1/\overline{a(0)} - a(0)\|^2_{H^2} \\ &= \|b\|^2_{H^2} + \|a\|^2_{H^2} - |a(0)|^2 + |a(0)|^{-2} + |a(0)|^2 - 2 \\ &= |a(0)|^{-2} - 1, \end{aligned}$$

which proves (4).

Finally, Lemma 5.6 together with (4) prove the invariance of $\mathcal{H}(b)$ to S. □

7. The extreme case

We point out first some differences with respect to the nonextreme case.

THEOREM 7.1. *Suppose b is extreme. Then:*

(1) *The function b does not belong to $\mathcal{H}(b)$.*
(2) *If $b \neq \mathbf{1}$, then $\mathcal{H}(b)$ is not invariant by S.*

PROOF. Suppose $b \in \mathcal{H}(b)$. By Theorem 5.8, it follows that there exists $\psi \in \overline{\Delta H^2} \subset L^2$, such that $b \oplus \psi \perp \{bh \oplus \Delta h : h \in H^2\}$. So

$$\langle b \oplus \psi, bh \oplus \Delta h \rangle = 0, \text{ for all } h \in H^2,$$

which is equivalent to $|b|^2 + \Delta\psi \in \overline{H_0^2}$. This is equivalent to $1 - \Delta^2 + \Delta\psi \in \overline{H_0^2}$, whence $f := \Delta^2 - \Delta\bar{\psi}$ is a nonzero (note that its zeroth Fourier coefficient is 1) function in H^2. Thus $\Delta^{-1} f = \Delta - \bar{\psi} \in L^2$, or $\Delta^{-2}|f|^2 \in L^1$.

We assert that this is not possible. Indeed,

$$\Delta^{-2}|f|^2 \geq \log(\Delta^{-2}|f|^2) = 2\log|f| - 2\log\Delta.$$

Integrating, we obtain

$$\frac{1}{2\pi}\int \Delta^{-2}(e^{it})|f(e^{it})|^2\,dt \geq 2\frac{1}{2\pi}\int \log|f(e^{it})|\,dt + 2\frac{1}{2\pi}\int(-\log\Delta(e^{it}))\,dt,$$

which cannot be true, since the first two integrals are finite, while the third is infinite by Lemma 5.9.

We have thus proved (1). Then (2) follows from Lemma 5.6, which can be restated as

$$\langle h, S^* b_{\mathcal{H}(b)}\rangle b = Sh - X_b^* h.$$

So, if we choose h not orthogonal (in $\mathcal{H}(b)$) to S^*b (in particular, $h = S^*b$), we must have $Sh \notin \mathcal{H}(b)$. □

In the sequel we will use the geometrical representation of $\mathcal{H}(b)$ given by Theorem 5.8. Using Lemma 5.9, we may replace in its statement $\overline{\Delta H^2}$ by $\overline{\Delta L^2}$.

Denote $\tilde{b}(z) = \overline{b(\bar{z})}$, and $\tilde{\Delta} = (1 - |\tilde{b}|^2)^{1/2}$. The map $f \mapsto \tilde{f}$ is a unitary involution that maps L^2 onto L^2, H^2 onto H^2, and $\overline{\tilde{\Delta} L^2}$ onto $\overline{\Delta L^2}$.

EXERCISE 7.2. b is extreme if and only if $\tilde{b}$ is extreme.

THEOREM 7.3. *Suppose b is extreme. Define*

$$\Omega : L^2 \oplus \overline{\tilde{\Delta} L^2} \to L^2 \oplus \overline{\Delta L^2}$$

by the formula

$$\Omega(f \oplus g) = \Big(b(z)\bar{z}f(\bar{z}) + \Delta\bar{z}g(\bar{z})\Big) \oplus \Big(\Delta\bar{z}f(\bar{z}) - \bar{b}\bar{z}g(\bar{z})\Big).$$

Then Ω is unitary, it maps $\mathcal{K}_{\tilde{b}}$ onto $\mathcal{K}_b$, and

$$\Omega X_{\tilde{b}} = X_b^* \Omega. \tag{7.1}$$

PROOF. Ω acts as the unitary $f \oplus g \mapsto \bar{z}f(\bar{z}) \oplus \bar{z}g(\bar{z})$ followed by the unitary $J(M_b)$, so it is unitary. We have

$$\mathcal{K}_b = (L^2 \oplus \overline{\Delta L^2}) \ominus \big[(H^2_- \oplus \{0\}) \oplus (\{bf \oplus \Delta f : f \in H^2\})\big],$$
$$\mathcal{K}_{\tilde{b}} = (L^2 \oplus \overline{\tilde{\Delta} L^2}) \ominus \big[(H^2_- \oplus \{0\}) \oplus (\{\tilde{b}f \oplus \tilde{\Delta} f : f \in H^2\})\big].$$

If $f \in H^2_-$, then

$$\Omega(f \oplus 0) = b\bar{z}f(\bar{z}) \oplus \Delta\bar{z}f(\bar{z}).$$

But the map $f \mapsto \bar{z}f(\bar{z})$ is a unitary from H^2_- onto H^2, whence it follows that $\Omega(H^2_- \oplus \{0\}) = \{bf \oplus \Delta f : f \in H^2\}$. Similarly we obtain $\Omega(\{bf \oplus \tilde{\Delta} f : f \in H^2\}) = H^2_- \oplus \{0\}$. Therefore

$$\Omega((H^2_- \oplus \{0\}) \oplus (\{\tilde{b}f \oplus \tilde{\Delta} f : f \in H^2\})) = (H^2_- \oplus \{0\}) \oplus (\{bf \oplus \Delta f : f \in H^2\}),$$

whence $\Omega(\mathcal{K}_{\tilde{b}}) = \Omega(\mathcal{K}_b)$.

Finally, we have $X_b = P_{\mathcal{K}_b}(Z^* \oplus Z^*_\Delta)P_{\mathcal{K}_b}|\mathcal{K}_b$ and $X_{\tilde{b}} = P_{\mathcal{K}_{\tilde{b}}}(Z^* \oplus Z^*_{\tilde{\Delta}})P_{\mathcal{K}_{\tilde{b}}}|\mathcal{K}_{\tilde{b}}$. But $\Omega(\mathcal{K}_{\tilde{b}}) = \Omega(\mathcal{K}_b)$ implies $P_{\mathcal{K}_{\tilde{b}}} = \Omega^* P_{\mathcal{K}_b}\Omega$. Therefore

$$\begin{aligned}\Omega X_{\tilde{b}} &= \Omega P_{\mathcal{K}_{\tilde{b}}}(Z^* \oplus Z^*_{\tilde{\Delta}})P_{\mathcal{K}_{\tilde{b}}}|\mathcal{K}_{\tilde{b}} = \Omega\Omega^* P_{\mathcal{K}_b}\Omega(Z^* \oplus Z^*_{\tilde{\Delta}})\Omega^* P_{\mathcal{K}_b}\Omega|\mathcal{K}_{\tilde{b}}\\ &= P_{\mathcal{K}_b}\Omega(Z^* \oplus Z^*_{\tilde{\Delta}})\Omega^* P_{\mathcal{K}_b}\Omega|\mathcal{K}_{\tilde{b}}.\end{aligned}$$

But one checks easily that $\Omega(Z^* \oplus Z^*_{\tilde{\Delta}})\Omega^* = Z \oplus Z_\Delta$, and therefore

$$\begin{aligned}P_{\mathcal{K}_b}\Omega(Z^* \oplus Z^*_{\tilde{\Delta}})\Omega^* P_{\mathcal{K}_b}\Omega|\mathcal{K}_{\tilde{b}} &= P_{\mathcal{K}_b}(Z \oplus Z_\Delta)P_{\mathcal{K}_b}\Omega|\mathcal{K}_{\tilde{b}}\\ &= (P_{\mathcal{K}_b}(Z^* \oplus Z^*_\Delta)P_{\mathcal{K}_b})^*\Omega|\mathcal{K}_{\tilde{b}} = X^*_b\Omega,\end{aligned}$$

which ends the proof of the theorem. □

We note that in case b is nonextreme X^*_b is *never* unitarily equivalent to $X_{\tilde{b}}$.

THEOREM 7.4. *Suppose b is extreme. Then* $\dim \mathcal{D}_{X_b} = \dim \mathcal{D}_{X^*_b} = 1$, *and there exists no subspace of $\mathcal{H}_b$ invariant by X and such that its restriction therein is an isometry.*

PROOF. From Theorem 5.8(3) it follows that we may prove the properties for the restriction $S^* \oplus Z^*_\Delta|\mathcal{K}_b$. Since S^* acts isometrically on H^2_0 and Z^*_Δ is unitary, we have $\mathcal{D}_{S^*\oplus Z^*_\Delta} = \mathbb{C}(1\oplus 0)$. By Exercise 2.2, $\mathcal{D}_{X_b} = \mathbb{C}P_{\mathcal{K}_b}(1\oplus 0)$. But $P_{\mathcal{K}_b}(1\oplus 0) \neq 0$; indeed, otherwise we would have $1\oplus 0 = bh \oplus \Delta h$ for some $h \in H^2$; since $h \neq 0$ a.e., this would imply $\Delta = 0$ a.e., or b inner, which is impossible if $1 = bh$. Therefore $\dim \mathcal{D}_X = 1$.

Applying the same argument to $\tilde{b}$ and using (7.1), it follows that $\dim \mathcal{D}_{X^*} = 1$.

Finally, suppose $\mathcal{Y} \subset \mathcal{K}$ is a closed subspace on which X_b acts isometrically, and $h \oplus g \in \mathcal{Y}$, we have, for any $n \geq 0$,

$$\|h\|^2 + \|g\|^2 = \|h\oplus g\|^2 = \|S^{*n}h \oplus Z^{*\,n}_\Delta g\|^2 = \|S^* h^n\|^2 + \|Z^{*\,n}_\Delta g\|^2 \to \|Z^{*\,n}_\Delta g\|^2 = \|g\|^2,$$

whence $h = 0$. But then we must have $0 \oplus g \perp bf \oplus \Delta f$ for all $f \in H^2$, or $g \perp \overline{\Delta H^2} = \overline{\Delta L^2}$. Since, on the other hand, $g \in \overline{\Delta L^2}$, it follows that $g = 0$, which ends the proof of the theorem. □

8. $\mathcal{H}(b)$ as a model space

The purpose of this section is to prove the converse of Theorem 7.4. This will show that in the extreme case the de Branges–Rovnyak spaces are model spaces for a large class of operators. The theorem below (as well as its proof) is in fact a particular case of the much more general analysis of contractions done in the Sz.Nagy–Foias theory (see [**24**]). Here we have adapted the argument to a "minimal" self-contained form.

THEOREM 8.1. *Suppose $T \in \mathcal{B}(H)$ is a contraction such that* $\dim \mathcal{D}_T = \dim \mathcal{D}_{T^*} = 1$*, and there exists no subspace of H invariant by T and such that its restriction therein is an isometry. Then there exists an extreme b in the unit ball of H^∞ such that T is unitarily equivalent to X_b.*

PROOF. Since the proof is rather long, we divide it in several steps.

Step 1. Dilation of T. To find the required function b, we will develop a certain geometrical construction. Changing the order of the components in the range of the Julia operator yields a unitary operator mapping $H \oplus \mathcal{D}_{T^*}$ into $\mathcal{D}_T \oplus H$ according to the matrix $\left(\begin{smallmatrix} D_T & -T^* \\ T & D_{T^*} \end{smallmatrix}\right)$. We can extend this unitary to a unitary W acting on the single enlarged space

$$\mathfrak{H} = \cdots \oplus \mathcal{D}_{T^*} \oplus \mathcal{D}_{T^*} \oplus H \oplus \mathcal{D}_T \oplus \mathcal{D}_T \oplus \dots,$$

that can be written as an bi-infinite operator matrix:

$$W = \begin{pmatrix} \ddots & & & & & & \\ & 1 & & & & & \\ & & 1 & & & & \\ & & & D_T & -T^* & & \\ & & & \boxed{\mathrm{T}} & D_{T^*} & & \\ & & & & & 1 & \\ & & & & & & 1 \\ & & & & & & & \ddots \end{pmatrix} \tag{8.1}$$

where the boxed entry corresponds to the central entry $T : H \to H$. If we write $\mathfrak{H} = \mathfrak{H}_- \oplus H \oplus \mathfrak{H}_+$, with

$$\mathfrak{H}_- = \cdots \oplus \mathcal{D}_{T^*} \oplus \mathcal{D}_{T^*}, \quad \mathfrak{H}_+ = \mathcal{D}_T \oplus \mathcal{D}_T \oplus \dots,$$

then $\mathfrak{H}_-$ is invariant by W, which acts therein as translation to the left, while $\mathfrak{H}_+$ is invariant by W^*, whose restriction is translation to the right. (This is a consequence of the fact that the 1 entries in the definition of W are all located *immediately above* the main diagonal.)

Step 2. Two embeddings of L^2 into $\mathfrak{H}$. Take a unit vector ϵ_{-1}^- in the $\mathcal{D}_{T^*}$ component of $\mathfrak{H}_-$ which is mostly to the right, and define, for $n \in \mathbb{Z}$, $\epsilon_n^- = W^{-n-1}\epsilon_{-1}^-$. Since $\dim \mathcal{D}_{T^*} = 1$, the family $(\epsilon_n^-)_{n\leq -1}$ forms an orthonormal basis of $\mathfrak{H}_-$. Moreover, the whole family $(\epsilon_n^-)_{n\in\mathbb{Z}}$ is an orthonormal set in $\mathfrak{H}$ (exercise!).

As $(e^{int})_{n\in\mathbb{Z}}$ is an orthonormal basis in L^2, we may define $\omega_- : L^2 \to \mathfrak{H}$ to be the unique isometry that satisfies $\omega_-(e^{int}) = \epsilon_n^-$ for all $n \in \mathbb{Z}$. One checks easily that its image $\omega_-(L^2)$ is a reducing space for W, and $\omega_-^* W \omega_- = M_{e^{-it}}$. The orthogonal projection onto $\omega_-(L^2)$ is $\omega_-\omega_-^*$, and it commutes with W.

An analogous construction can be made for $\mathfrak{H}_+$. We obtain an orthonormal set $(\epsilon_n^+)_{n\in\mathbb{Z}}$ in $\mathfrak{H}$, such that $(\epsilon_n^+)_{n\geq 0}$ is a basis for $\mathfrak{H}_+$. Then $\omega_+ : L^2 \to \mathfrak{H}$ is the isometry that satisfies $\omega_+(e^{int}) = \epsilon_n^+$ for all $n \in \mathbb{Z}$; $\omega_+(L^2)$ is also a reducing space for W, $\omega_+^* W \omega_+ = M_{e^{-it}}$, and $\omega_+\omega_+^* W = W\omega_+\omega_+^*$.

Step 3. Finding b**.** Consider then the map $\omega_-^*\omega_+ : L^2 \to L^2$. We have, using the above remarks as well as the equalities $\omega_+^*\omega_+ = \omega_-^*\omega_- = I_{L^2}$,

$$
\begin{aligned}
\omega_-^*\omega_+ M_{e^{-it}} &= \omega_-^*\omega_+\omega_+^* W\omega_+ = \omega_-^* W\omega_+\omega_+^*\omega_+ = \omega_-^* W\omega_+ \\
&= (\omega_-^*\omega_-)\omega_-^* W\omega_+ = \omega_-^* W\omega_-\omega_-^*\omega_+ = M_{e^{-it}}\omega_-^*\omega_+.
\end{aligned} \tag{8.2}
$$

So $\omega_-^*\omega_+$ commutes with $M_{e^{-it}}$; it follows that it commutes also with its inverse $M_{e^{it}}$ (exercise!). We have noticed in the introduction that in this case we must have $\omega_-^*\omega_+ = M_b$ for some function $b \in L^\infty$, and $\|\omega_-^*\omega_+\| \leq 1$ implies $\|b\|_\infty \leq 1$.

Now, $b = M_b\mathbf{1} = \omega_-^*\omega_+ 1 = \omega_-^*\epsilon_0^+$. Since $\epsilon_0^+ \in \mathfrak{H}_+ \perp \mathfrak{H}_- = \omega_-(\overline{H_0^2})$, it follows that $\omega_-^*\epsilon_0^+ \in H^2$. Thus $b \in L^\infty \cap H^2 = H^\infty$, and we have found our candidate for the function in the unit ball of H^∞. It remains now to check that it satisfies the required properties. As above, we will denote $\Delta = (1 - |b|^2)^{1/2} \in L^\infty$.

Step 4. Constructing the unitary equivalence. Let us now note that the closed linear span $\omega_+ L^2 \vee \omega_- L^2$ equals $\mathfrak{H}$. Indeed, it reduces W and contains $\mathfrak{H}_+$ and $\mathfrak{H}_-$; thus its orthogonal Y has to be a reducing subspace of W contained in H (more precisely, in its embedding in $\mathfrak{H}$). From (8.1) it follows then that $W|Y = T|Y$, so Y should be a subspace of H invariant by T and such that the restriction is isometric (even unitary!), which contradicts the hypothesis. Thus $Y = \{0\}$.

We define then a mapping $U : \omega_+ L^2 \vee \omega_- L^2 \to L^2 \oplus \overline{\Delta L^2}$ by

$$
U(\omega_+ f_+ + \omega_- f_-) = (f_- + bf_+) \oplus \Delta f_+.
$$

We have

$$
\begin{aligned}
\|\omega_+ f_+ + \omega_- f_-\|^2 &= \|\omega_+ f_+\|^2 + \|\omega_- f_-\|^2 + 2\Re\langle \omega_+ f_+, \omega_- f_-\rangle \\
&= \|f_+\|_2^2 + \|f_-\|_2^2 + 2\Re\langle \omega_-^*\omega_+ f_+, f_-\rangle_2 = \|f_+\|_2^2 + \|f_-\|_2^2 \\
&\quad + 2\Re\langle bf_+, f_-\rangle_2,
\end{aligned}
$$

and

$$
\begin{aligned}
\|(f_- + bf_+) \oplus \Delta f_+\|^2 &= \int |f_- + bf_+|^2 + \int \Delta^2 |f_+|^2 \\
&= \int |f_-|^2 + |bf_+|^2 + 2\Re bf_+\bar{f}_- + \Delta^2|f_+|^2 \\
&= \int |f_-|^2 + |f_+|^2 + 2\Re bf_+\bar{f}_-,
\end{aligned}
$$

whence U is an isometry, and it is easy to see that the image is dense. It can be extended to a unitary operator, that we will denote by the same letter,

$$
U : \mathfrak{H} \to L^2 \oplus \overline{\Delta L^2}.
$$

The commutation relations satisfied by $\omega_\pm$ imply that $UW = (Z^* \oplus Z_\Delta^*)U$.

Step 5. Final checks. Now $U(\mathfrak{H}) = U(\mathfrak{H}_-) \oplus U(H) \oplus U(\mathfrak{H}_+)$, and

$$
\begin{aligned}
U(\mathfrak{H}_-) &= U(\omega_-(\overline{H_0^2})) = \{f_- \oplus 0 : f_- \in \overline{H_0^2} = \overline{H_0^2} \oplus \{0\}, \\
U(\mathfrak{H}_+) &= U(\omega_+(H^2)) = \{bf_+ \oplus \Delta f_+ : f_+ \in H^2\},
\end{aligned}
$$

so

$$\begin{aligned}U(H) &= (L^2 \oplus \overline{\Delta L^2}) \ominus \Big((\overline{H_0^2} \oplus \{0\}) \oplus (\{bf_+ \oplus \Delta f_+ : f_+ \in H^2\})\Big)\\ &= (H^2 \oplus \overline{\Delta L^2}) \ominus (\{bf_+ \oplus \Delta f_+ : f_+ \in H^2\}).\end{aligned}$$

As shown by (8.1), T can be viewed as the compression of W to H, so it is unitarily equivalent through U to the compression of $Z^* \oplus Z_\Delta^*$ to $U(H)$. This last is easily seen to be the restriction of $S^* \oplus Z_\Delta^*$ to $(H^2 \oplus \overline{\Delta L^2}) \ominus (\{bf_+ \oplus \Delta f_+ : f_+ \in H^2\})$.

We are very close to the end: Theorem 5.8 would end the proof, provided we could replace in the formula for $U(H)$ the space $H^2 \oplus \overline{\Delta L^2}$ with $H^2 \oplus \overline{\Delta H^2}$ and thus Z_Δ with V_Δ. But the space

$$Y := (H^2 \oplus \overline{\Delta L^2}) \ominus (H^2 \oplus \overline{\Delta H^2}) = \{0\} \oplus (\overline{\Delta L^2} \ominus \overline{\Delta H^2})$$

is invariant with respect to $S^* \oplus Z_2^*$, which acts on it isometrically. By assumption, we must have $Y = \{0\}$, which means that $H^2 \oplus \overline{\Delta L^2} = H^2 \oplus \overline{\Delta H^2}$; this finishes the proof. □

EXERCISE 8.2. Show that Theorem 8.1 implies Theorem 3.2.

9. Further reading

We discuss in this section some directions in which the study of de Branges–Rovnyak spaces has developed.

The model spaces $\mathbf{K}_u$ have no nonconstant multipliers. However, the theory of multipliers is interesting for the case of de Branges–Rovnyak spaces corresponding to nonextreme b; see references [**16**–**18**].

Integral representations of de Branges–Rovnyak spaces appear in [**4**], and have further been developed in [**12, 13, 21**]. The last paper is used in [**5**] to obtain weighted norm inequalities for functions in de Branges–Rovnyak spaces.

The connection between de Branges–Rovnyak and Dirichlet spaces is exploited in [**8, 9**]; that between de Branges–Rovnyak spaces and composition operators in [**15**].

Finally, it is natural from many points of view (including that of model spaces) to consider also matrix or operator valued de Branges–Rovnyak spaces. These have already been introduced in [**4**]; see [**2, 3**] for some recent developments.

References

[1] N. Aronszajn, *Theory of reproducing kernels*, Trans. Amer. Math. Soc. **68** (1950), 337–404. MR0051437 (14,479c)

[2] Joseph A. Ball, Vladimir Bolotnikov, and Sanne ter Horst, *Interpolation in de Branges-Rovnyak spaces*, Proc. Amer. Math. Soc. **139** (2011), no. 2, 609–618, DOI 10.1090/S0002-9939-2010-10505-1. MR2736342 (2012c:47043)

[3] Joseph A. Ball, Vladimir Bolotnikov, and Sanne ter Horst, *Abstract interpolation in vector-valued de Branges-Rovnyak spaces*, Integral Equations Operator Theory **70** (2011), no. 2, 227–263, DOI 10.1007/s00020-010-1844-1. MR2794390 (2012h:46048)

[4] Joseph A. Ball and Thomas L. Kriete III, *Operator-valued Nevanlinna-Pick kernels and the functional models for contraction operators*, Integral Equations Operator Theory **10** (1987), no. 1, 17–61, DOI 10.1007/BF01199793. MR868573 (88a:47013)

[5] Anton Baranov, Emmanuel Fricain, and Javad Mashreghi, *Weighted norm inequalities for de Branges-Rovnyak spaces and their applications*, Amer. J. Math. **132** (2010), no. 1, 125–155, DOI 10.1353/ajm.0.0094. MR2597508 (2011a:46047)

[6] Louis de Branges and James Rovnyak, *Square summable power series*, Holt, Rinehart and Winston, New York-Toronto, Ont.-London, 1966. MR0215065 (35 #5909)

[7] Louis de Branges and James Rovnyak, *Canonical models in quantum scattering theory*, Perturbation Theory and its Applications in Quantum Mechanics (Proc. Adv. Sem. Math. Res. Center, U.S. Army, Theoret. Chem. Inst., Univ. of Wisconsin, Madison, Wis., 1965), Wiley, New York, 1966, pp. 295–392. MR0244795 (39 #6109)

[8] Nicolas Chevrot, Dominique Guillot, and Thomas Ransford, *De Branges-Rovnyak spaces and Dirichlet spaces*, J. Funct. Anal. **259** (2010), no. 9, 2366–2383, DOI 10.1016/j.jfa.2010.07.004. MR2674117 (2012m:46029)

[9] Constantin Costara and Thomas Ransford, *Which de Branges-Rovnyak spaces are Dirichlet spaces (and vice versa)?*, J. Funct. Anal. **265** (2013), no. 12, 3204–3218, DOI 10.1016/j.jfa.2013.08.015. MR3110499

[10] R. G. Douglas, *On majorization, factorization, and range inclusion of operators on Hilbert space*, Proc. Amer. Math. Soc. **17** (1966), 413–415. MR0203464 (34 #3315)

[11] E. Fricain, *Bases of reproducing kernels in de Branges spaces*, J. Funct. Anal. **226** (2005), no. 2, 373–405, DOI 10.1016/j.jfa.2004.11.014. MR2159461 (2006j:30093)

[12] Emmanuel Fricain and Javad Mashreghi, *Boundary behavior of functions in the de Branges-Rovnyak spaces*, Complex Anal. Oper. Theory **2** (2008), no. 1, 87–97, DOI 10.1007/s11785-007-0028-8. MR2390675 (2009a:46054)

[13] Emmanuel Fricain and Javad Mashreghi, *Integral representation of the n-th derivative in de Branges-Rovnyak spaces and the norm convergence of its reproducing kernel* (English, with English and French summaries), Ann. Inst. Fourier (Grenoble) **58** (2008), no. 6, 2113–2135. MR2473631 (2009k:46050)

[14] Javad Mashreghi and Emmanuel Fricain, *Exceptional sets for the derivatives of Blaschke products*, Proceedings of the St. Petersburg Mathematical Society. Vol. XIII, Amer. Math. Soc. Transl. Ser. 2, vol. 222, Amer. Math. Soc., Providence, RI, 2008, pp. 163–170. MR2433525 (2009m:30062)

[15] Michael T. Jury, *Reproducing kernels, de Branges-Rovnyak spaces, and norms of weighted composition operators*, Proc. Amer. Math. Soc. **135** (2007), no. 11, 3669–3675 (electronic), DOI 10.1090/S0002-9939-07-08931-9. MR2336583 (2008h:47050)

[16] B. A. Lotto and D. Sarason, *Multiplicative structure of de Branges's spaces*, Rev. Mat. Iberoamericana **7** (1991), no. 2, 183–220, DOI 10.4171/RMI/110. MR1133377 (92k:46035)

[17] B. A. Lotto and Donald Sarason, *Multipliers of de Branges-Rovnyak spaces*, Indiana Univ. Math. J. **42** (1993), no. 3, 907–920, DOI 10.1512/iumj.1993.42.42042. MR1254125 (95a:46039)

[18] Benjamin A. Lotto and Donald Sarason, *Multipliers of de Branges-Rovnyak spaces. II*, Harmonic analysis and hypergroups (Delhi, 1995), Trends Math., Birkhäuser Boston, Boston, MA, 1998, pp. 51–58. MR1614726 (99c:46022)

[19] N.K. Nikolski: *Operators, Functions, and Systems: An Easy Reading*, AMS, 2002.

[20] N. K. Nikolʹskiĭ and V. I. Vasyunin, *Notes on two function models*, The Bieberbach conjecture (West Lafayette, Ind., 1985), Math. Surveys Monogr., vol. 21, Amer. Math. Soc., Providence, RI, 1986, pp. 113–141, DOI 10.1090/surv/021/11. MR875237 (88f:47008)

[21] A. G. Poltoratskiĭ, *Boundary behavior of pseudocontinuable functions* (Russian, with Russian summary), Algebra i Analiz **5** (1993), no. 2, 189–210; English transl., St. Petersburg Math. J. **5** (1994), no. 2, 389–406. MR1223178 (94k:30090)

[22] Donald Sarason, *Sub-Hardy Hilbert spaces in the unit disk*, University of Arkansas Lecture Notes in the Mathematical Sciences, 10, John Wiley & Sons, Inc., New York, 1994. A Wiley-Interscience Publication. MR1289670 (96k:46039)

[23] Donald Sarason, *Local Dirichlet spaces as de Branges-Rovnyak spaces*, Proc. Amer. Math. Soc. **125** (1997), no. 7, 2133–2139, DOI 10.1090/S0002-9939-97-03896-3. MR1396993 (98h:46023)

[24] Béla Sz.-Nagy, Ciprian Foias, Hari Bercovici, and László Kérchy, *Harmonic analysis of operators on Hilbert space*, Revised and enlarged edition, Universitext, Springer, New York, 2010. MR2760647 (2012b:47001)

Institute of Mathematics of the Romanian Academy, P.O. Box 1-764, Bucharest 014700, Romania

E-mail address: `Dan.Timotin@imar.ro`

Contemporary Mathematics
Volume **638**, 2015
http://dx.doi.org/10.1090/conm/638/12804

Asymptotic Bohr radius for the polynomials in one complex variable

Cheng Chu

ABSTRACT. We consider the Bohr radius R_n for the class of complex polynomials in one variable of degree at most n. It was conjectured by R. Fournier in 2008 that $R_n = \frac{1}{3} + \frac{\pi^2}{3n^2} + o(\frac{1}{n^2})$. We prove this conjecture is true.

1. Introduction

Let $\mathbb{D}$ be the open unit disk in the complex plane $\mathbb{C}$ and H^∞ be the Banach space of bounded analytic functions on $\mathbb{D}$ with the norm

$$||f||_\infty = \sup_{z\in\mathbb{D}} |f(z)|.$$

Also let $\mathcal{P}_n$ denote the subspace of H^∞ consisting of all the complex polynomials of degree at most n. The Bohr radius R for H^∞ is defined as

$$R = \sup\{r \in (0,1) : \sum_{k=0}^{\infty} |a_k| r^k \leqslant ||f||_\infty, \text{ for all } f(z) = \sum_{k=0}^{\infty} a_k z^k \in H^\infty\}.$$

Bohr's famous power series theorem [**1**] shows that $R = \frac{1}{3}$.

In 2005, Guadarrama [**4**] considered the Bohr type radius for the class $\mathcal{P}_n$ defined by

$$R_n = \sup\{r \in (0,1) : \sum_{k=0}^{n} |a_k| r^k \leqslant ||p||_\infty, \text{ for all } p(z) = \sum_{k=0}^{n} a_k z^k \in \mathcal{P}_n\}, \tag{1.1}$$

and gave the estimate

$$\frac{C_1}{3^{n/2}} < R_n - \frac{1}{3} < C_2 \frac{\log n}{n},$$

for some positive constants C_1 and C_2. Later in 2008, Fournier obtained an explicit formula for R_n by using the notion of *bounded preserving* functions. He proved the following theorem [**2**]

2010 *Mathematics Subject Classification.* Primary 41-XX.

Partially supported by National Science Foundation Grant DMS 1300280.

Theorem 1.1. *For each $n \geqslant 1$, let $T_n(r)$ be the following $(n+1)\times(n+1)$ symmetric Toeplitz matrix*

$$(1.2)\qquad \begin{pmatrix} 1 & r & -r^2 & r^3 & \cdots & (-1)^{n-1}r^n \\ r & 1 & r & -r^2 & \cdots & (-1)^{n-2}r^{n-1} \\ -r^2 & r & 1 & r & & \\ r^3 & -r^2 & r & 1 & \ddots & \vdots \\ \vdots & & \ddots & \ddots & \ddots & \\ (-1)^{n-1}r^n & & \cdots & & r & 1 \end{pmatrix}.$$

Then R_n is equal to the smallest root in $(0,1)$ of the equation

$$\det T_n(r) = 0.$$

Based on numerical evidence, he conjectured that

$$R_n = \frac{1}{3} + \frac{\pi^2}{3n^2} + \ldots$$

The purpose of this note is to provide a positive answer. We shall prove

Theorem 1.2. *Let R_n be as in (1.1), then*

$$\lim_{n\to\infty} n^2\left(R_n - \frac{1}{3}\right) = \frac{\pi^2}{3}.$$

2. Main Theorem

In this section, we prove Theorem 1.2. The methods we use are similar to that in [**3**, Chapter 5].

Proof of Theorem 1.2. Fix $r \in (0,1)$; we consider the eigenvalues of $T_n(r)$, the symmetric Toeplitz matrix (1.2). Let $\Delta_n(\lambda) = \det(T_n(r) - \lambda I)$, the characteristic polynomial of $T_n(r)$.

For $n \geqslant 2$, multiplying the second row of $\Delta_n(\lambda)$ by r, adding it to the first row and performing a similar operation with the columns, we have

$$\Delta_n(\lambda) = \det \begin{pmatrix} 1-\lambda+(3-\lambda)r^2 & (2-\lambda)r & 0 & & \cdots & 0 \\ (2-\lambda)r & 1-\lambda & r & -r^2 & \cdots & (-1)^{n-2}r^{n-1} \\ 0 & r & 1-\lambda & r & & \\ & -r^2 & r & 1-\lambda & \ddots & \vdots \\ \vdots & & \ddots & \ddots & \ddots & \\ 0 & & \cdots & & r & 1-\lambda \end{pmatrix}$$

$$(2.1)\qquad = [(3-\lambda)r^2+1-\lambda]\Delta_{n-1}(\lambda) - (2-\lambda)^2\Delta_{n-2}(\lambda).$$

If we set $\Delta_{-1}(\lambda) = 1$, then the recurrence relation (2.1) holds for all $n \geqslant 1$.

Consider the function associated with these Toeplitz matrices $T_n(r)$

$$(2.2)\qquad f(x) = 1 + \sum_{|n|>0} (-1)^{n-1} r^n e^{inx} = \frac{3r^2 + 4r\cos x + 1}{r^2 + 2r\cos x + 1}.$$

Suppose

$$\lambda = f(x),\ x \in [0, \pi].$$

Then the second order recurrence relation (2.1) becomes

$$\Delta_n(\lambda) = [-2(2-\lambda)r\cos x]\Delta_{n-1}(\lambda) - (2-\lambda)^2 r^2 \Delta_{n-2}(\lambda).$$

Its characteristic equation has the roots $(\lambda - 2)re^{\pm ix}$. Adding the initial conditions

$$\Delta_{-1}(\lambda) = 1,\ \Delta_0(\lambda) = 1 - \lambda,$$

we have

$$\Delta_n(\lambda) = \frac{[(\lambda-2)r]^{n+1}}{1-r^2}\left(\frac{\sin(n+2)x}{\sin x} + 2r\frac{\sin(n+1)x}{\sin x} + r^2\frac{\sin nx}{\sin x}\right).$$

Denote

$$p_n(x) = \frac{\sin(n+2)x}{\sin x} + 2r\frac{\sin(n+1)x}{\sin x} + r^2\frac{\sin nx}{\sin x}.$$

Then it is easy to verify that p_n is a polynomial of degree $n+1$ in $\cos x$. Let

$$t_\nu^{(n)} = \frac{\nu\pi}{n+2}, \quad \nu = 1, 2, \cdots, n+1.$$

Direct computation shows that

$$p_n(t_\nu^{(n)}) = (-1)^{\nu+1} 2r(1 + r\cos\nu),$$

thus

$$\operatorname{sgn} p_n(t_\nu^{(n)}) = (-1)^{\nu+1}.$$

Also

$$\lim_{x\to\pi^-} p_n(x) = 2(-1)^{n+1}(1-r)^2.$$

So p_n has exactly $n+1$ distinct zeros $\{x_\nu^{(n)} | \nu = 1, 2, \cdots, n+1\}$ on $[0, \pi]$ such that

$$0 < t_1^{(n)} < x_1^{(n)} < t_2^{(n)} < x_2^{(n)} < \cdots < t_{n+1}^{(n)} < x_{n+1}^{(n)} < \pi. \tag{2.3}$$

That means for each n,

$$\lambda_\nu^{(n)} = f(x_\nu^{(n)}),\ \nu = 1, 2, \cdots, n+1$$

are all the eigenvalues of $T_n(r)$. Since f is decreasing on $[0, \pi]$, $\lambda_{n+1}^{(n)} = f(x_{n+1}^{(n)})$ is the smallest eigenvalue of $T_n(r)$.

Next, we will find an asymptotic expression for $x_{n+1}^{(n)}$. Notice that

$$\lim_{n\to\infty} (-1)^{n+1} \frac{p_n(\pi - \frac{z}{n+2})}{n+2} = (1-r)^2 \frac{\sin z}{z}, \tag{2.4}$$

where (2.4) holds uniformly for $|z| < 2\pi$. Let

$$x_{n+1}^{(n)} = \frac{(n+1)\pi + \epsilon_n}{n+2},$$

then $\epsilon_n \in (0,\pi)$ by relation (2.3). Thus

$$\begin{aligned}
0 &= \lim_{n\to\infty}(-1)^{n+1}\frac{p_n(x_{n+1}^{(n)})}{n+2} \\
&= \lim_{n\to\infty}(-1)^{n+1}\frac{p_n(\pi - \frac{\pi-\epsilon_n}{n+2})}{n+2} \\
&= \lim_{n\to\infty}(1-r)^2\frac{\sin(\pi-\epsilon_n)}{\pi-\epsilon_n}.
\end{aligned}$$

Hence the accumulation points of $\{\epsilon_n\}$ are either 0 or π. Let

$$y^{(n)} = \frac{t_{n+1}^{(n)}+\pi}{2} = \pi - \frac{\frac{\pi}{2}}{n+2}.$$

Using (2.4) again, we have

$$\begin{aligned}
&\lim_{n\to\infty}(-1)^{n+1}\frac{p_n(y^{(n)})}{n+2} \\
&= \lim_{n\to\infty}(-1)^{n+1}\frac{p_n(\pi - \frac{\frac{\pi}{2}}{n+2})}{n+2} \\
&= \lim_{n\to\infty}(1-r)^2\frac{\sin(\frac{\pi}{2})}{\frac{\pi}{2}} > 0.
\end{aligned}$$

When n is sufficiently large,

$$\operatorname{sgn} p_n(y^{(n)}) = (-1)^{n+1} = \operatorname{sgn} p_n(\pi^-),$$

which implies $x_{n+1}^{(n)} \in (t_{n+1}^{(n)}, y^{(n)})$. Consequently, $\epsilon_n \to 0$ as $n\to\infty$, and then

$$x_{n+1}^{(n)} = \pi - \frac{\pi}{n} + o(\frac{1}{n}), \quad \text{as } n\to\infty. \tag{2.5}$$

Now we are ready to find the asymptotic expression for R_n. Notice that $r = R_n$ is the root in $(0,1)$ of the equation

$$\lambda_{n+1}^{(n)} = f(x_{n+1}^{(n)}) = 0.$$

By (2.2), that means

$$3R_n^2 + 4R_n\cos x_{n+1}^{(n)} + 1 = 0.$$

Thus

$$R_n = \frac{1}{3}\Big(-2\cos x_{n+1}^{(n)} - \sqrt{4\cos^2 x_{n+1}^{(n)} - 3}\Big).$$

Using (2.5), we have

$$R_n = \frac{1}{3} + \frac{\pi^2}{3n^2} + o(\frac{1}{n^2}).$$

□

References

[1] H. Bohr, *A theorem concerning power series*, Proc. Lond. Math. Soc. (2) **13** (1914), 1–5.

[2] R. Fournier, *Asymptotics of the Bohr radius for polynomials of fixed degree*, J. Math. Anal. Appl. **338** (2008), 1100–1107.

[3] Ulf Grenander and Gabor Szegö, *Toeplitz forms and their applications*, California Monographs in Mathematical Sciences, University of California Press, Berkeley-Los Angeles, 1958. MR0094840 (20 #1349)

[4] Z. Guadarrama, *Bohr's radius for polynomials in one complex variable*, Comput. Methods Funct. Theory **5** (2007), 143–151.

Department of Mathematics, Washington University in Saint Louis, Saint Louis, Missouri

E-mail address: chengchu@math.wustl.edu

Contemporary Mathematics
Volume **638**, 2015
http://dx.doi.org/10.1090/conm/638/12810

A survey on preservers of spectra and local spectra

Abdellatif Bourhim and Javad Mashreghi

Abstract. This article presents a survey of preservers of spectra and local spectra of operators. Our main goal is to explain Kaplansky's conjecture, attempts that have been made to solve this open conjecture, and the new venues of research that have been opened as the result of such attempts. We will mainly focus on the preserver problems of spectra and local spectra.

Contents

2010 *Mathematics Subject Classification.* Primary 47B49; Secondary 47B48, 47A10, 47A11, 47A53, 47A55, 47A65, 46L10.

Key words and phrases. Linear and nonlinear preservers; Banach algebras; Kaplansky's conjecture; Spectral isometries; Matrices; Operators; Spectra; Spectral radius; Local spectra; Local spectral radius; Inner local spectral radius; Fredholm theory; Generalized invertibility; Minimum and reduced minimum moduli.

This work was supported by NSERC (Canada), FRQNT (Québec) and CNRS (France).

1. Introduction

Linear preserver problems, in the most general setting, demands the characterization of maps between algebras that leave a certain property, a particular relation, or even a subset *invariant*. In all cases that have been studied by now, the maps are either supposed to be linear, or proved to be so.

The earliest result on linear preserver problems was established by Frobenius in 1896 [**99**] who characterized linear maps that preserve the determinant of matrices. His result opened the gate for many researchers who described linear/additive maps that preserve some spectral functions and quantities of matrices and even bounded linear operators.

These attempts led to the one of the most intractable unsolved problems in the theory of Banach algebras and linear preserver problems, i.e., the famous *Kaplansky's conjecture*. This conjecture asserts that every surjective unital invertibility preserving linear map between two semisimple Banach algebras is a Jordan homomorphism. This problem has not yet fully solved and remains open even for general C^*-algebras, but it has been confirmed for von Neumann algebras [**6**] and for the algebra of all bounded linear operators on a Banach spaces [**135, 194, 223**].

The proofs offered therein [**6, 135, 194, 223**] are based on the reduction of Kaplansky's question to the problem of characterizing linear maps preserving idempotents or rank one operators. This kind of techniques can be extended only to some special classes of semisimple Banach algebras [**9, 49, 142, 174**], and for all algebras not belonging to one of these classes the problem is quite challenging. For more details, we refer the reader to [**46**] where Brešar and Šemrl confirmed Kaplansky's conjecture for invertibility preserving maps that, additionally, locally preserves commutativity and then provided a weaker equivalent formulation of Kaplansky's conjecture.

Beside the above cited papers, there are several partial positive answers of Kaplansky's conjecture [**65, 133, 159, 196**]. Moreover, many related preserver problems were introduced and studied afterwards. Some of these approaches are listed below.

(i) The theory of *functional identities* was developed in the book of Brešar, Chebotar and Martindale [**47**]. Some authors used this technic and considered linear preserver problems about maps between algebras or rings preserving certain properties like commutativity, zero-product, zero Jordan product or normality in rings with involution; see for instance [**20**, **21**, **47**, **55**, **56**, **64**] and the references therein.

(ii) Another approach was to characterize preservers of different kinds of generalized numerical ranges and numerical radii; see for instance [**62**, **68**, **100**, **112**, **151**, **154**–**156**, **158**, **160**, **161**, **195**] and the reference therein.

(iii) Generalized invertibility linear preserver problems were initiated by Mbekhta, Šemrl and Rodman in [**186**] and continued by several authors [**26**, **27**, **31**, **32**, **41**] and the references therein.

(iv) Local spectra preservers were initiated by Bourhim and Ransford in [**43**] and have attracted numerous authors who investigated these problems; see for instance [**40**, **42**, **44**, **72**, **76**, **128**] and the references therein.

There are many more interesting results and unsolved preserver problems in the rich area of research. But, unfortunately, we can not talk about all these and we shall only discuss some preserver problems of spectra and local spectra and related spectral quantities.

2. Our approach

In this article, we discuss some old and new results about preservers of spectrum, local spectrum and related spectral quantities of matrices and operators.

In Section 3, we set the stage by defining the notations and concepts that we need in the rest of paper. In particular, we present some background from the theory of Banach algebras and discuss some definitions and concepts of Jordan homomorphisms.

Then in Section 4, we discuss the classical Kaplansky's conjecture. We start by introducing Aupetit's approach. Then we discuss the original Kaplansky's question on invertibility preservers along with some known counterexamples. We present Kaplansky's conjecture on algebras of matrices over an arbitrary field and state. We also discuss such a conjecture on the algebra of all bounded linear operators on a complex Banach space and present some results of nonlinear spectrum preservers of operators.

In Section 5, we mainly consider the problem of characterizing surjective linear maps on the algebra $\mathcal{L}(\mathcal{H})$ of all bounded linear operators on an infinite-dimensional complex Hilbert space $\mathcal{H}$ preserving in both directions generalized invertibility, Fredholm, semi-Fredholm operators. We also present some results on complete description of linear maps on a C^*-algebra of real rank zero that preserve different essential spectral sets and quantities. Furthermore, we discuss the complete characterization of surjective essential spectral isometries and essentially spectrally bounded maps on $\mathcal{L}(\mathcal{H})$.

In Section 6, we discuss additive maps preserving the minimum, surjectivity, maximum and reduced minimum moduli of Banach space operators. We then continue with the general framework in which these concept are translated in Banach algebras and corresponding result of linear preservers of these quantities are presented.

In Section 7, we study the local spectra preservers. More explicitly, we discuss linear maps preserving local spectrum of Banach space operators at a fixed nonzero vector. For example, we consider linear map preserving, increasing or decreasing the inner local spectral radius operators at a fixed nonzero vector. In particular, we give a complete description of surjective maps preserving the local spectrum of the product or triple product of operators at a nonzero fixed point, and present some results on nonlinear local spectra preservers of matrices.

3. Basic notations

In the sequel, the term Banach algebra means a unital complex associative Banach algebra, with unit $\mathbf{1}$, and a C^*-algebra means a unital Banach algebra with an innovation $*$. Let $\mathcal{A}$ be a Banach algebra, and $\text{Inv}(\mathcal{A})$ be the group of all invertible elements of $\mathcal{A}$. For an element a in $\mathcal{A}$, let

$$\sigma(a) := \{\lambda \in \mathbb{C} : a - \lambda\mathbf{1} \notin \text{Inv}(\mathcal{A})\}$$

denote the *spectrum* of a. It is well known that $\sigma(a)$ is a nonempty compact set and, according to a result of Beurling, its maximum modulus coincides with the spectral radius of a defined by

$$\text{r}(a) = \lim_{n\to\infty} \|a^n\|^{\frac{1}{n}}.$$

The *Jacobson radical* of $\mathcal{A}$, denoted by $\text{Rad}(\mathcal{A})$, is the intersection of all maximal left (or right) ideals of $\mathcal{A}$ and coincides with the collection of all elements $x \in \mathcal{A}$ satisfying $\text{r}(xa) = 0$ for all $a \in \mathcal{A}$. The Banach algebra $\mathcal{A}$ is said to be *semisimple* if its Jacobson radical is zero, or equivalently 0 is the only element $x \in \mathcal{A}$ satisfying $\text{r}(xa) = 0$ for all $a \in \mathcal{A}$, and it is said to be *prime* if for any two elements a and b of $\mathcal{A}$, the identity $a\mathcal{A}b = \{0\}$ implies that either $a = 0$ or $b = 0$. The Banach algebra $\mathcal{L}(X)$ of all bounded linear operators on a complex Banach space X is a semisimple prime algebra; see for instance [**11**, Theorem 3.1.4]. Any C^*-algebra $\mathcal{A}$ is semisimple since for any $x \in \mathcal{A}$ we have $0 = \text{r}(xx^*) = \|x\|^2$ only if $x = 0$.

If $\mathcal{A}$ has a minimal left (or right) ideal, then its *socle*, denoted by $\text{Soc}(\mathcal{A})$, is the sum of all minimal left (or right) ideals of $\mathcal{A}$. If $\mathcal{A}$ has no minimal one-sided ideal, the socle is then defined to be trivial, i.e., $\text{Soc}(\mathcal{A}) = \{0\}$. Note that $\text{Soc}(\mathcal{A})$ is an ideal of $\mathcal{A}$ consisting of all elements $a \in \mathcal{A}$ for which $\sigma(xa)$ is finite for all $x \in \mathcal{A}$, and that all its elements are algebraic.

An ideal P of $\mathcal{A}$ is said to be *inessential* if the spectrum of every element $x \in P$ has at most 0 as limit point. An ideal P is *essential* if it has a nonzero intersection with every nonzero ideal of $\mathcal{A}$. If $\mathcal{A}$ is semisimple, then P is essential if and only if 0 is the only element a of $\mathcal{A}$ for which $aP = 0$.

Let $\mathcal{A}$ and $\mathcal{B}$ be complex Banach algebras. Let φ be an additive map from $\mathcal{A}$ into $\mathcal{B}$. We define three categories of such maps. We say that φ is

(i) *homomorphism* if

$$\varphi(ab) = \varphi(a)\varphi(b), \qquad a,\ b \in \mathcal{A};$$

(ii) *anti-homomorphism* if

$$\varphi(ab) = \varphi(b)\varphi(a), \qquad a,\ b \in \mathcal{A};$$

(iii) *Jordan homomorphism* if it preserves squares, i.e.,

$$\varphi(a^2) = \varphi(a)^2, \qquad a \in \mathcal{A}.$$

Equivalently, the map φ is a Jordan homomorphism if and only if it preserves the Jordan structure. More explicitly, this means that

$$\varphi(ab + ba) = \varphi(a)\varphi(b) + \varphi(b)\varphi(a), \qquad a, b \in \mathcal{A}.$$

An injective Jordan homomorphism is called a *Jordan monomorphism*, a surjective Jordan homomorphism is called a *Jordan epimorphism*, and a bijective Jordan homomorphism is called a *Jordan isomorphism*. If $\mathcal{A}$ is a C^*-algebra, then an additive map φ from $\mathcal{A}$ into $\mathcal{B}$ is a Jordan homomorphism if and only if φ preserves squares of self-adjoint elements, i.e., $\varphi(h^2) = \varphi(h)^2$ for all self-adjoint elements $h \in \mathcal{A}$. Clearly, every homomorphism and every anti-homomorphism is a Jordan homomorphism. Conversely, in [**132**], Jacobson and Rickart proved that a Jordan homomorphism from an arbitrary ring into a domain is either a homomorphism or an antihomomorphism. The same conclusion holds for Jordan homomorphisms onto prime rings, as shown by Herstein [**114**] and Smiley [**221**]. In fact, this problem of whether Jordan homomorphisms can be expressed through homomorphisms and antihomomorphisms was considered by several authors; see for instance [**19, 45**] and the reference therein.

For basic concepts and notion from the classical Banach algebra theory, we refer the reader to [**11, 201**].

4. Various forms of Kaplansky's problem

In this section, we discuss Kaplansky's question [**140**]. This essential problem asks for conditions under which a linear map between two Banach algebras preserving invertibility is a Jordan homomorphism. We then study further motivations and some exceptional results regarding this question and also some other related problems.

4.1. Kaplansky's conjecture. Let $\mathcal{A}$ and $\mathcal{B}$ be complex Banach algebras. If φ is a Jordan epimomorphism from $\mathcal{A}$ into $\mathcal{B}$, then φ is necessarily a unital map preserving the invertibility, i.e., $\varphi(\mathbf{1}) = \mathbf{1}$ and $\varphi(x)$ is invertible whenever x is; see for instance [**222**]. In 1970, Kaplansky [**140**] asked when a unital invertibility preserving additive map on rings is a Jordan homomorphism. His question was motivated by two classical results, the result of Marcus and Moyls [**170**] on linear maps preserving eigenvalues of matrices and the Gleason-Kahane-Żelazko theorem [**129, 139, 227**] stating that that every unital invertibility preserving linear map from a Banach algebra $\mathcal{A}$ to a semisimple commutative Banach algebra $\mathcal{B}$ is multiplicative. This result was obtained independently by Gleason in [**129**] and Kahane-Żelazko in [**139**], and was refined by Żelazko in [**227**]. Their proofs are analytic in nature but a different proof has been given by Roitman and Sternfeld in [**203**]. We refer the reader to [**14, 67, 143**] for related results.

In its original form, Kaplansky's question has a negative answer and several counterexamples were constructed for linear unital maps on Banach algebras; see for instance [**9, 13, 135, 140, 204, 222**]. The first example, quoted from [**9, 13**], shows that if $\mathcal{A}$ or $\mathcal{B}$ is not semisimple, then the answer to the Kaplansky's question is negative in general.

EXAMPLE 4.1. Let

$$\mathcal{A} := \left\{ W := \begin{bmatrix} a & c \\ 0 & b \end{bmatrix} : a,\ b,\ c \in M_2(\mathbb{C}) \right\}$$

and define the linear map $\varphi : \mathcal{A} \to \mathcal{A}$ by

$$\varphi(W) := \begin{bmatrix} a & c \\ 0 & b^{tr} \end{bmatrix}, \qquad W = \begin{bmatrix} a & c \\ 0 & b \end{bmatrix} \in \mathcal{A},$$

where b^{tr} denotes the transpose of b. Hence, we have $\sigma(\varphi(W)) = \sigma(W)$. At the same time,

$$\varphi(W^2) - \varphi(W)^2 = \begin{bmatrix} 0 & c(b - b^{tr}) \\ 0 & 0 \end{bmatrix}$$

for all $W \in \mathcal{A}$, and thus φ is not a Jordan homomorphism.

The next example taken from [**204**] shows that without the assumption of surjectivity of the map φ, the answer to Kaplansky's question is negative even when the algebras $\mathcal{A}$ and $\mathcal{B}$ are semisimple.

EXAMPLE 4.2. For the unital linear map $\varphi : M_2(\mathbb{C}) \to M_4(\mathbb{C})$, defined by

$$\varphi(a) := \begin{bmatrix} a & a - a^{tr} \\ 0 & a \end{bmatrix}, \qquad a \in M_2(\mathbb{C}),$$

we have $\sigma(\varphi(a)) = \sigma(a)$. But,

$$\varphi(a^2) - \varphi(a)^2 = \begin{bmatrix} 0 & (a - a^{tr})^2 \\ 0 & 0 \end{bmatrix}$$

for all $a \in M_2(\mathbb{C})$, and thusφ is not a Jordan homomorphism.

Based on several partial positive results and counterexamples, the following conjecture known as Kaplansky's conjecture has been formulated by Aupetit [**7**]. For a more detailed explanation, we refer the reader to [**7**, **50**, **53**, **102**, **104**, **126**, **133**, **196**].

Conjecture 4.3. Every surjective unital invertibility preserving linear map between two semisimple Banach algebras is a Jordan homomorphism.

Obviously, if φ is a unital invertibility preserving linear map from $\mathcal{A}$ into $\mathcal{B}$, then φ compresses the spectrum, i.e.,

$$\sigma(\varphi(x)) \subset \sigma(x), \qquad x \in \mathcal{A}. \tag{4.1}$$

Thus, this conjecture can be formulated as spectrum compressing problem. In some cases, one considers the problem of describing unital surjective linear maps φ between two semisimple Banach algebras $\mathcal{A}$ and $\mathcal{B}$ preserving invertibility in both directions, i.e., $\varphi(x)$ is invertible in $\mathcal{B}$ if and only if x is invertible in $\mathcal{A}$. Obviously, such a unital linear map φ preserves invertibility in both directions if and only it preserves the spectrum, i.e.,

$$\sigma(\varphi(x)) = \sigma(x), \qquad x \in \mathcal{A}. \tag{4.2}$$

Therefore, we arrive at the following variant of Kaplansky's conjecture.

Conjecture 4.4. Every surjective linear map between two semisimple Banach algebras that preserves the spectrum is a Jordan isomorphism.

Kaplansky's conjecture and its variant remain open even for C^*-algebras [**105**]. However, there are a number of partial results in the literature; see for instance [**3**, **4**, **6**, **9**, **28**, **51**, **54**, **63**, **92**, **97**, **98**, **105**, **135**, **153**, **170**, **171**, **191**, **194**, **211**, **216**, **217**, **222**] and the references therein. Some results of these papers will be addressed in what follows.

We can still consider a strong question than Kaplansk's conjecture by asking to describe linear/additive maps φ between Banach algebras $\mathcal{A}$ and $\mathcal{B}$ that strongly preserve invertibility. More explicitly, we look for surjective linear/additive maps $\varphi : \mathcal{A} \to \mathcal{B}$ such that

$$\varphi(x^{-1}) = \varphi(x)^{-1}, \qquad x \in \mathrm{Inv}(\mathcal{A}).$$

Hua's theorem on fields states that if a unital additive map φ from a field $\mathbb{K}$ into another one $\mathbb{F}$ strongly preserves invertibility, then φ is an isomorphism or an anti-isomorphism; see [**127**]. This result has been extended to additive map between Banach algebras.

THEOREM 4.5 (Boudi–Mbekhta [**31**]). *Let $\mathcal{A}$ and $\mathcal{B}$ be Banach algebras and φ be an additive map from $\mathcal{A}$ into $\mathcal{B}$. Then φ strongly preserves invertibility if and only if $\varphi(\mathbf{1})\varphi$ is a unital Jordan homomorphism and $\varphi(\mathbf{1})$ commutes with the range of φ.*

The proof is purely algebraic and uses Hua's identity [**127**] that asserts that if a, b and $a - b^{-1}$ are invertible in a ring, then $a^{-1} - (a - b^{-1})^{-1}$ is also invertible and its inverse is in fact $a - aba$. However, as mentioned by Aupetit and Mouton in the introduction of their paper [**9**], exponentials can be used to prove this result if φ is in addition supposed to be continuous. Indeed, for every $x \in \mathcal{A}$ and real scalar t, we have e^{tx} is invertible and its inverse is e^{-tx}, and thus

$$\begin{aligned}
\mathbf{1} &= \varphi(e^{tx})\,\varphi(e^{-tx}) \\
&= \varphi\left(\mathbf{1} + \frac{x}{1!}t + \frac{x^2}{2!}t^2 + \cdots\right)\varphi\left(\mathbf{1} - \frac{x}{1!}t + \frac{x^2}{2!}t^2 - \cdots\right) \\
&= \left(\varphi(\mathbf{1}) + \frac{\varphi(x)}{1!}t + \frac{\varphi(x^2)}{2!}t^2 + \cdots\right)\left(\varphi(\mathbf{1}) - \frac{\varphi(x)}{1!}t + \frac{\varphi(x^2)}{2!}t^2 - \cdots\right) \\
&= \varphi(\mathbf{1})^2 + (\varphi(x)\varphi(\mathbf{1}) - \varphi(\mathbf{1})\varphi(x))\,t + \left(\frac{\varphi(\mathbf{1})\varphi(x^2)}{2} + \frac{\varphi(x^2)\varphi(\mathbf{1})}{2} - \varphi(x)^2\right)t^2 \\
&+ \cdots .
\end{aligned}$$

It follows, in particular, that

$$\varphi(\mathbf{1})^2 = \mathbf{1}, \tag{4.3}$$

$$\varphi(x)\varphi(\mathbf{1}) = \varphi(\mathbf{1})\varphi(x), \tag{4.4}$$

and

$$\frac{\varphi(\mathbf{1})\varphi(x^2)}{2} + \frac{\varphi(x^2)\varphi(\mathbf{1})}{2} = \varphi(x)^2. \tag{4.5}$$

Replacing x by x^2 in (4.4) and taking into account (4.3), the identity (4.5) becomes

$$\varphi(\mathbf{1})\varphi(x^2) = \varphi(x)^2 = \{\varphi(\mathbf{1})\varphi(x)\}^2 .$$

Since x is an arbitrary element of $\mathcal{A}$, we see that $\varphi(\mathbf{1})\varphi$ is a Jordan homomorphism.

Beside Theorem 4.5, Boudi and Mbekhta characterized in [**31**] additive maps between unital complex Banach algebras which preserve strongly generalized inverses. For related results on strongly preserver problems in Banach algebras, we also refer the reader to [**59, 60**].

4.2. Preservers on matrices. Let $M_n(\mathbb{C})$ be the algebra of all $n\times n$-complex matrices, and $\mathrm{ev}(A)$ be the set of the n eigenvalues of a matrix $A \in M_n(\mathbb{C})$ counting multiplicities. In [**170**], Marcus and Moyls proved that a linear map φ on $M_n(\mathbb{C})$ preserves the set of eigenvalues of matrices counting multiplicities, i.e.,

$$\mathrm{ev}\big(\varphi(A)\big) = \mathrm{ev}(A), \qquad A \in M_n(\mathbb{C}),$$

is either an automorphism or an antiautomorphism of $M_n(\mathbb{C})$. This result had been proved already in 1949 by Dieudonné [**91**], and that even much earlier, in 1897, Frobenius [**99**] had obtained a similar result for determinant preserving linear maps. These results have been extended in several ways and settings, and many linear and nonlinear spectrum-preserver problems on matrices have been studied; see for instance [**3, 17, 28, 30, 63, 66, 71, 73–75, 79, 92, 96, 97, 171, 191, 210**] and the references therein. Let us just mention the results of [**3, 28, 74, 90, 92**] and refer the reader to [**30, 133, 159, 162, 169, 196**] for more results and details.

In [**171**], using a density argument, Marcus and Purves showed that every linear unital invertibility preserving map φ on $M_n(\mathbb{C})$ necessarily fulfills $\mathrm{ev}(\varphi(A)) = \mathrm{ev}(A)$ for all A in $M_n(\mathbb{C})$, and thus Marcus and Moyls' result [**170**] shows that such a map φ must be either an inner automorphism or an inner antiautomorphism. This result remains true when φ is linear bijective unital invertibility preserving map on a matrix algebra $M_n(\mathbb{F})$ over an arbitrary field $\mathbb{F}$. However, without the bijectivity condition, the complete description of linear invertibility preserving maps on $M_n(\mathbb{F})$ was unknown for a long period and was obtained just recently by de Segunis Pazzis in [**90**] who established the following result.

THEOREM 4.6 (de Seguins Pazzis [**90**]). *For any field $\mathbb{F}$, a linear unital map φ on $M_n(\mathbb{F})$, $n \geq 2$, preserves the invertibility if and only if one of the following statements holds.*

(i) There is an invertible matrix $A \in M_n(\mathbb{F})$ such that either

$$\varphi(T) = ATA^{-1}, \qquad T \in M_n(\mathbb{F}),$$

or

$$\varphi(T) = AT^{tr}A^{-1}, \qquad T \in M_n(\mathbb{F}).$$

(ii) There exist an n-dimensional subspace $\mathcal{V} \subset M_n(\mathbb{F})$, called nonsingular, in which each nonzero element is invertible, a linear bijective map ϕ from $\mathbb{F}^n$ into $\mathcal{V}$ and a nonzero $n \times 1$-matrix x such that either

$$\varphi(T) = \phi(Tx), \qquad T \in M_n(\mathbb{F}),$$

or

$$\varphi(T) = \phi(T^{tr}x), \qquad T \in M_n(\mathbb{F}).$$

In his recent paper [**208**], Šemrl extended this result and determined the structure of linear preservers of invertibility maps on central simple algebras. His proof is completely diferent from the proof given in [**90**] and uses the localization technique for linear preservers that was introduced in [**202**], but only for matrices over fields.

Among other results in [**3**], Akbari and Aryapoor proved that the conclusion of Marcus and Moyls' result [**170**] remains valid for linear maps on the algebra $M_n(\mathbb{F})$ of all $n \times n-$matrices over an algebraically closed field $\mathbb{F}$ that preserve at least one eigenvalue of each matrix.

THEOREM 4.7 (Akbari–Aryapoor [**3**]). *If $\mathbb{F}$ is an algebraically closed field, then a linear map φ on $M_n(\mathbb{F})$ preserves at least one eigenvalue of any matrix in $M_n(\mathbb{F})$, i.e.,*

$$\sigma\big(\varphi(T)\big) \cap \sigma(T) \neq \emptyset, \qquad T \in M_n(\mathbb{F}),$$

if and only if there is an invertible matrix $A \in M_n(\mathbb{F})$ such that

$$\text{either} \quad \varphi(T) = ATA^{-1}, \qquad\qquad \text{or} \quad \varphi(T) = AT^{tr}A^{-1},$$

for all $T \in M_n(\mathbb{F})$.

This result has been extended in [**75**] where Costara and Repovš showed, in particular, that for a Lipschitz mapping φ on $M_n(\mathbb{C})$ for which

$$\sigma\big(\varphi(T) - \varphi(S)\big) \cap \sigma(T - S) \neq \emptyset, \qquad T,\ S \in M_n(\mathbb{C}),$$

there exists an invertible matrix $A \in M_n(\mathbb{C})$ such that

$$\text{either} \quad \varphi(T) = ATA^{-1}, \qquad\qquad \text{or} \quad \varphi(T) = AT^{tr}A^{-1},$$

for all $T \in M_n(\mathbb{C})$. Their proof uses complex-analytic techniques in spectral theory and spectral characterizations of the radical and the center.

The classical Mazur-Ulam theorem [**180**] asserts that every surjective isometry between two real normed spaces is real linear up to a translation. In [**28**], Bhatia, Šemrl and Sourour exploited this result and considered a similar problem of whether a surjective map φ between two semisimple Banach algebras $\mathcal{A}$ and $\mathcal{B}$ for which

$$\mathrm{r}(x - y) = \mathrm{r}\big(\varphi(x) - \varphi(y)\big), \qquad x,\ y \in \mathcal{A}, \tag{4.6}$$

must be real linear up to a translation. They answered affirmatively this question when $\mathcal{A}$ and $\mathcal{B}$ coincide with $M_n(\mathbb{C})$, and thus they provided an extension of Marcus and Moyls' result [**170**] in the absence of the linearity of φ but with the additional assumption that φ is surjective.

THEOREM 4.8 (Bhatia–Šemrl–Sourour [**28**]). *A surjective map φ on $M_n(\mathbb{C})$ satisfies*

$$\mathrm{r}\big(\varphi(T) - \varphi(S)\big) = \mathrm{r}(T - S), \qquad T,\ S \in M_n(\mathbb{C}),$$

if and only if there exist a unimodular scalar $\alpha \in \mathbb{C}$, and matrices $A,\ S \in M_n(\mathbb{C})$ such that A is invertible and

$$\varphi(T) = \alpha AT^{\#}A^{-1} + S$$

for all $T \in M_n(\mathbb{C})$, where $T^{\#}$ stands for one of the mappings T, or T^{tr}, or T^*, or $\overline{T}$, the complex conjugation of T.

In [**75**], Costara and Repovš established a similar result for maps of class C^1 on a domain in $M_n(\mathbb{C})$ containing the null matrix. But the description of which maps between semisimple Banach algebras satisfying (4.6) is unknown even when $\mathcal{A}$ and $\mathcal{B}$ coincide with the algebra of all bounded linear operators on an infinite-dimensional Banach space.

Next result describes maps on $M_n(\mathbb{C})$ that preserve spectrum of the difference and, unlike the previous theorems, it imposes no linearity or surjectivity condition on such maps.

THEOREM 4.9 (Costara [**74**], Dolinar–Hou–Kuzma–Qi [**92**]). *For a map φ on $M_n(\mathbb{C})$, the following statements are equivalent.*

(i) $\sigma\big(\varphi(T) - \varphi(S)\big) = \sigma(T - S)$ *for all $T,\ S \in M_n(\mathbb{C})$.*

(ii) $\sigma\big(\varphi(T)-\varphi(S)\big)\subset\sigma(T-S)$ *for all* $T,\ S\in M_n(\mathbb{C})$.
(iii) $\sigma(T-S)\subset\sigma\big(\varphi(T)-\varphi(S)\big)$ *for all* $T,\ S\in M_n(\mathbb{C})$.
(iv) There exists an invertible matrix A *in* $M_n(\mathbb{C})$ *such that either*

$$\varphi(T)=ATA^{-1}+\varphi(0),\qquad T\in M_n(\mathbb{C}),$$

or

$$\varphi(T)=AT^{tr}A^{-1}+\varphi(0),\qquad T\in M_n(\mathbb{C}).$$

This result has been shown before under additional assumptions such as continuity or surjectivity of the map φ. In this form and as stated above, the equivalence between last two statements is due to Costara [**74**] and the equivalence between the second and fourth statements was recently proved by Dolinar, Hou, Kuzma and Qi in [**92**]. We also mention that in [**111**], Havlicek and Šemrl obtained, in particular, a complete description of bijective maps φ on the algebra $M_n(\mathbb{F})$ of all $n\times n-$matrices over a field $\mathbb{F}$ with at least three elements that satisfy $\varphi(a)-\varphi(b)$ is invertible if and only if $a-b$ is.

4.3. Preservers of operators. Throughout this and the following sections, X and Y denote infinite dimensional complex Banach spaces, and $\mathcal{L}(X,Y)$ denotes the space of all bounded linear maps from X into Y. As usual, when $X=Y$, we simply write $\mathcal{L}(X)$ instead of $\mathcal{L}(X,X)$. The dual space of X will be denoted by X^* and the Banach space adjoint of an operator $T\in\mathcal{L}(X)$ will be denoted by T^*. It is well known that every rank one operator has the form given by

$$(x\otimes f)(z):=f(z)x,\qquad z\in X,$$

for some $x\in X$ and $f\in X^*$, and every every finite rank operator T can be written as a finite sum $T=\sum_{k=1}^{n}x_k\otimes f_k$ for some $x_k\in X$ and $f_k\in X^*$, $k=1,2,\ldots,n$. Note that this representation is not unique. Denote the ideal of all finite rank operators on X by $\mathcal{F}(X)$.

Jafarian and Sourour confirmed Kaplansky's conjecture for spectrum-preserving linear maps from $\mathcal{L}(X)$ onto $\mathcal{L}(Y)$ and extended Marcus and Moyls' result [**170**] to the infinite-dimensional case.

THEOREM 4.10 (Jafarian–Sourour [**135**]). *A linear map* φ *from* $\mathcal{L}(X)$ *onto* $\mathcal{L}(Y)$ *preserves the spectrum if and only if one of the following happens.*

(i) There is a bijective mapping $A\in\mathcal{L}(X,Y)$ *such that*

$$\varphi(T)=ATA^{-1},\qquad T\in\mathcal{L}(X).$$

(ii) There is a bijective mapping $B\in\mathcal{L}(X^*,Y)$ *such that*

$$\varphi(T)=BT^*B^{-1},\qquad T\in\mathcal{L}(X).$$

This result has been extended in several directions, and several results about linear/additive maps preserving, compressing, expanding spectrum and its parts or preserving different spectral quantities of operators were obtained; see for instance [**15, 16, 50, 51, 53, 54, 61, 63, 85, 86, 92, 98, 102, 120, 121, 125, 125, 133, 189, 193, 194, 209, 222**]. In [**194**], Omladič and Šemrl showed that the conclusion of this result is valid under the less restrictive condition that φ is only additive. The proofs of the main results of [**135**] and [**194**] are based on the following spectral characterization of operators of rank one.

LEMMA 4.11 (Jafarian–Sourour, Omladič–Šemrl). *For a nonzero operator $R \in \mathcal{L}(X)$, the following statements are equivalent.*

(i) R has rank one.
(ii) $\sigma(T+R) \cap \sigma(T+\lambda R) \subset \sigma(T)$ for every $T \in \mathcal{L}(X)$ and $\lambda \in \mathbb{C}$ with $\lambda \neq 1$.
(iii) $\sigma(T+R) \cap \sigma(T+2R) \subset \sigma(T)$ for every $T \in \mathcal{L}(X)$.

This lemma tells you that every linear/additive spectrum-preserving map from $\mathcal{L}(X)$ onto $\mathcal{L}(Y)$ maps rank one operators into rank one operators. The equivalence between the first two statements is due to Jafarian and Sourour [**135**] and equivalence between the first and last statements was established by Omladič and Šemrl in [**194**] which was needed to extend Jafarian and Sourour's result to additive setting. Omladič and Šemrl used this lemma to show that additive spectrum-preserving maps are, in fact, linear, and then they got their main result from Theorem 4.10.

Many linear/additive preserver problems on $\mathcal{L}(X)$ can be reduced to the problem of determining all linear/additive maps that carry the set of rank one operators onto themselves. This approach has been developed in the finite dimensional situation, and has later proved to be useful also in infinite dimensions as shown in several papers including [**135**]. Beside [**135, 194**], a number of authors described linear/additive maps on different spaces or algebras of matrices or operators that preserve or nonincrease rank one, rank one nilpotent, or rank one idempotent matrices or operators; see for instance [**81, 84, 103, 117, 123, 124, 145, 163**] and the references therein. The following result describes linear/additive maps preserving operators of rank one.

THEOREM 4.12 (Jafarian–Sourour [**135**], Omladič–Šemrl [**194**]). *If φ is a bijective linear (resp. additive) map from $\mathcal{F}(X)$ into $\mathcal{F}(Y)$ preserving rank one operators, then either there are bijective linear (resp. both linear or both conjugate linear) mappings $A : X \to Y$ and $B : X^* \to Y^*$ such that*

$$\varphi(x \otimes f) = Ax \otimes Bf, \qquad x \in X,\ f \in X^*, \tag{4.7}$$

or there are bijective linear (resp. both linear or both conjugate linear) mappings $C : X^* \to Y$ and $D : X \to Y^*$ such that

$$\varphi(x \otimes f) = Cf \otimes Dx, \qquad x \in X,\ f \in X^*. \tag{4.8}$$

Later in [**222**], Sourour showed that Theorem 4.10 remains valid for bijective linear maps between $\mathcal{L}(X)$ and $\mathcal{L}(Y)$ that preserve invertibility in only one direction.

THEOREM 4.13 (Sourour [**222**]). *If φ is a unital surjective linear map from $\mathcal{L}(X)$ into $\mathcal{L}(Y)$ preserving invertibility (equivalently, φ compresses the spectrum), then either φ vanishes on $\mathcal{F}(X)$ or φ is injective. In the latter case, either there is a bijective mapping $A \in \mathcal{L}(X,Y)$ such that $\varphi(T) = ATA^{-1}$ for all $T \in \mathcal{L}(X)$, or there is a bijective mapping $B \in \mathcal{L}(X^*,Y)$ such that $\varphi(T) = BT^*B^{-1}$ for all $T \in \mathcal{L}(X)$.*

One may wonders if every unital surjective linear map from $\mathcal{L}(X)$ into $\mathcal{L}(Y)$ preserving invertibility is automatically injective. The following explanation shows that there are unital linear functional on $\mathcal{L}(X)$ that preserves the invertibility and vanishes on $\mathcal{K}(X)$. In [**165**], Lindenstrauss asked if there exists a complex separable and infinite-dimensional Banach space X such that $\mathcal{L}(X) = \mathbb{C}\mathbf{1} + \mathcal{K}(X)$. In [**101**], Gowers and Maurey provided a remarkable example of a Banach space X not containing an unconditional basic sequence such that the space all operators has the form $\lambda + S$, where S is strictly singular. But the "scalar plus compact" problem of Lindenstrauss was still open and has been solved only recently by Argyros and Haydon in [**5**], where they constructed a Banach space for which $\mathcal{L}(X) = \mathbb{C}\mathbf{1} + \mathcal{K}(X)$. For a such space X, the unital linear functional on $\mathcal{L}(X)$ defined by

$$f(\lambda\mathbf{1} + K) = \lambda, \qquad \lambda \in \mathbb{C},\ K \in \mathcal{K}(X)$$

preserves invertibility and vanishes on $\mathcal{K}(X)$. This example shows that, unlike spectrum-preserving linear maps, a surjective linear map between $\mathcal{L}(X)$ and $\mathcal{L}(Y)$ that preserves invertibility need not be injective. However, using a spectral characterization of compact operators, it is shown in [**222**] that if $X = \mathcal{H}$ is a separable infinite-dimensional Hilbert space, then every surjective linear map from $\mathcal{L}(\mathcal{H})$ into $\mathcal{L}(Y)$ preserving invertibility is injective.

The following result shows that a bijective linear maps between $\mathcal{L}(X)$ and $\mathcal{L}(Y)$ that preserves invertibility in one direction maps rank one operators of $\mathcal{L}(X)$ into rank one operators of $\mathcal{L}(Y)$.

THEOREM 4.14 (Sourour [**222**]). *For an operator $R \in \mathcal{L}(X)$, the following conditions are equivalent.*

(i) R has at most rank one.

(ii) For every $T \in \mathcal{L}(X)$, there exists a compact subset K_T of $\mathbb{C}$ such that

$$\sigma(T + \alpha R) \cap \sigma(T + \beta R) \subset K_T \tag{4.9}$$

for all $\alpha,\ \beta \in \mathbb{C}$ with $\alpha \neq \beta$.

(iii) For every operator $T \in \mathcal{L}(X)$ of rank at most two, (4.9) holds.

Both Theorem 4.13 and Theorem 4.14 were established in [**222**] for complex locally convex topological vector spaces and several variants of these theorems were given therein. The proof of Theorem 4.14 uses some analytic function results such as Picard's Big Theorem and works when the spectrum is replaced by the point spectrum. In fact, results and proofs from [**135, 194, 222**] and their techniques were employed by several authors to extend and describe linear maps preserving or compressing different spectra such as right and left spectra; see for instance [**15, 16, 61, 85, 86, 121, 125, 144, 146, 213–216**]. We also mention that several results on linear preservers have been extended to the setting of additive preservers, and, in many cases, their extensions demonstrated to be nontrivial as the forms of additive preservers are some time not as nice as the ones of the corresponding linear preservers. For results on additive preserver problems, we refer the interested reader, for example, to [**16, 43, 86, 120, 121, 125, 193, 194**] and the references therein.

4.4. Preservers of spectrum. There has also been considerable interest in studying nonlinear maps on operators or matrices preserving the spectrum and its parts. Instead of studying linear maps preserving a certain property one can study maps completely preserving this property without assuming any algebraic condition like linearity or additivity or multiplicativity and prove that such maps are of the standard form, that is, linear or conjugate-linear isomorphisms or anti-isomorphisms, possibly multiplied by a nonzero scalar. In [**190**], Molnár studied maps preserving the spectrum of operator or matrix products. He established, in particular, the following result.

THEOREM 4.15 (Molnár [**190**]). *Let $\mathcal{H}$ be an infinite-dimensional complex Hilbert space. Then a surjective map φ on $\mathcal{L}(\mathcal{H})$ satisfies*

$$\sigma(\varphi(T)\varphi(S)) = \sigma(TS), \qquad T,\ S \in \mathcal{L}(\mathcal{H}),$$

if and only if there exists an invertible operator $A \in \mathcal{L}(\mathcal{H})$ such that

$$\text{either } \varphi(T) = ATA^{-1}, \qquad \text{or } \varphi(T) = -ATA^{-1},$$

for all $T \in \mathcal{L}(\mathcal{H})$.

This result has been extended in several direction for uniform algebras and semisimple commutative Banach algebras; see for instance [**63, 100, 108–110, 115, 119, 122, 136, 147, 166, 167, 188, 198, 199, 224**]. Let us mention one of these results. For the following result, let $\sigma_\pi(T) := \{\lambda \in \sigma(T) : |\lambda| = \mathrm{r}(T)\}$ be the peripheral spectrum of operator T in $\mathcal{L}(X)$.

THEOREM 4.16 (Miura–Honma [**188**]). *Let φ and ϕ be two surjective maps from $\mathcal{L}(X)$ into $\mathcal{L}(Y)$. Then they satisfy*

$$\sigma_\pi\big(\phi(S)\varphi(T)\big) = \sigma_\pi(ST), \qquad S, T \in \mathcal{L}(X),$$

if and only if either there are bijective mappings $A_1,\ A_2 \in \mathcal{L}(X, Y)$ such that

$$\phi(T) = A_1TA_2^{-1} \quad \text{and} \quad \varphi(T) = A_2TA_1^{-1}, \qquad T \in \mathcal{L}(X),$$

or there are bijective mappings $B_1,\ B_2 \in \mathcal{L}(X^*, Y)$ such that

$$\phi(T) = B_1T^*B_2^{-1} \quad \text{and} \quad \varphi(T) = B_2T^*B_1^{-1}, \qquad T \in \mathcal{L}(X).$$

Instead of the usual product, certain authors investigated maps preserving spectra of triple or Jordan products of matrices or operators; see for instance [**80, 228**]. In [**228**], Zhang and Hou investigated maps preserving peripheral spectrum of triple product of operators.

THEOREM 4.17 (Zhang–Hou [**228**]). *A surjective map φ from $\mathcal{L}(X)$ into $\mathcal{L}(Y)$ satisfies*

$$\sigma_\pi(STS) = \sigma_\pi(\varphi(S)\varphi(T)\varphi(S)), \qquad S, T \in \mathcal{L}(X),$$

if and only if there exists a scalar $\lambda \in \mathbb{C}$ with $\lambda^3 = 1$ and either there exists an invertible operator $A \in \mathcal{L}(X, Y)$ such that

$$\varphi(T) = \lambda ATA^{-1}, \qquad T \in \mathcal{L}(X),$$

or there exists a bijective mapping $B \in \mathcal{L}(X^*, Y)$ such that

$$\varphi(T) = \lambda BT^*B^{-1}, \qquad T \in \mathcal{L}(X).$$

Cui and Li characterized maps preserving peripheral spectrum of Jordan products of operators on standard operator algebras.

THEOREM 4.18 (Cui–Li [**80**]). *A surjective map φ from $\mathcal{L}(X)$ into $\mathcal{L}(Y)$ satisfies*

$$\sigma_\pi(\varphi(S)\varphi(T) + \varphi(T)\varphi(S)) = \sigma_\pi(ST + TS), \qquad S, T \in \mathcal{L}(X),$$

if and only if either there exists a bijective mapping $A \in \mathcal{L}(X, Y)$ such that

$$\varphi(T) = \pm ATA^{-1}, \qquad T \in \mathcal{L}(X),$$

or there exists a bijective mapping $B \in \mathcal{L}(X^*, Y)$ such that

$$\varphi(T) = \pm BT^*B^{-1}, \qquad T \in \mathcal{L}(X).$$

A number of authors studied maps between certain operator algebras preserving the spectrum of a general product of operators that includes the usual product, the triple product and Jordan product; see for instance [**116**, **119**]. It is also worth mentioning that it is unknown the full description of surjective maps φ from $\mathcal{L}(X)$ into $\mathcal{L}(Y)$ satisfying either

$$\mathrm{r}(\varphi(S)\varphi(T)) = \mathrm{r}(ST), \quad (S, T \in \mathcal{L}(X)), \tag{4.10}$$

or

$$\mathrm{r}(\varphi(S)\varphi(T)\varphi(S)) = \mathrm{r}(STS), \quad (S, T \in \mathcal{L}(X)), \tag{4.11}$$

or

$$\mathrm{r}(\varphi(S)\varphi(T) + \varphi(T)\varphi(S)) = \mathrm{r}(ST + TS), \quad (S, T \in \mathcal{L}(X)), \tag{4.12}$$

or

$$\mathrm{r}(\varphi(S) \pm \varphi(T)) = \mathrm{r}(S \pm T), \quad (S, T \in \mathcal{L}(X)). \tag{4.13}$$

One may ask similar questions for different spectra or spectral quantities. We mention that Havlicek and Šemrl [**111**] gave a complete characterization of bijective maps φ on the algebra $\mathcal{L}(\mathcal{H})$ of all bounded linear bounded operators on an infinite-dimensional Hilbert space $\mathcal{H}$ satisfying that $\varphi(S) - \varphi(T)$ is invertible if and only if $S - T$ is. This result has been extended by Hou and Li in [**118**] to the case of Banach spaces setting.

THEOREM 4.19 (Havlicek–Šemrl [**111**], Hou–Li in [**118**]). *For a bijective map φ from $\mathcal{L}(X)$ into $\mathcal{L}(Y)$, the following statements are equivalent.*

(i) For every pair of operators S, $T \in \mathcal{L}(X)$, we have $\varphi(S) - \varphi(T)$ is invertible if and only if $S - T$ is.

(ii) There exists an operator $R \in \mathcal{L}(Y)$ and either there are bijective continuous mappings $A : X \to Y$ and $B : Y \to X$ both linear or both conjugate linear such that

$$\varphi(T) = ATB + R, \qquad T \in \mathcal{L}(X),$$

or there are bijective continuous mappings $A : X^* \to Y$ and $B : Y \to X^*$ both linear or both conjugate linear such that

$$\varphi(T) = AT^*B + R, \qquad T \in \mathcal{L}(X).$$

The last case may occur only if X is reflexive.

4.5. Preservers on Banach algebras. Several authors investigated finite rank and compact elements in Banach algebras; see for instance [**8**, **9**, **48**, **52**, **57**, **95**, **152**, **200**, **226**]. Recall that an element a of a semisimple Banach algebra $\mathcal{A}$ is said to have rank one if $\sigma(xa)$ contains at most one nonzero point for all x in $\mathcal{A}$. In [**9**], Aupetit and Mouton established several spectral characterizations of the rank one elements, the socle and the kernel of the hull of the socle in Banach algebras. They used such characterizations to extend the above result of Jafarian and Sourour from [**135**] to semisimple Banach algebras with an essential socle and primitive Banach algebras with minimal ideals. Among other results, they established the following theorem.

THEOREM 4.20 (Aupetit–Mouton [**9**]). *Assume that $\mathcal{A}$ and $\mathcal{B}$ are semisimple Banach algebras such that one of them has an essential socle. Then a unital linear* *map φ from $\mathcal{A}$ into $\mathcal{B}$ preserves the polynomially convex hull of the spectrum $\widehat{\sigma}(\cdot)$, i.e.,*

$$\widehat{\sigma}(\varphi(a)) = \widehat{\sigma}(a), \qquad a \in \mathcal{A},$$

if and only if φ is a Jordan isomorphism.

In [**6**], Aupetit confirmed Kaplansky's conjecture for von Neumann algebras and established, in particular, the following result.

THEOREM 4.21 (Aupetit [**6**]). *A linear map φ from a von Neumann algebra $\mathcal{A}$ onto another von Neumann algebra $\mathcal{B}$ preserves the spectrum if and only if φ is a Jordan isomorphism.*

The methods used in [**6**] are purely spectral and analytic, and similar results in case of Jordan-Banach algebras and Jordan-von Neumann algebras were established therein. As matter of fact, one of the basic tools employed in the proof of this theorem is the following purely spectral characterization of the idempotent elements of of semisimple Banach algebras, which is interesting in its own right.

THEOREM 4.22 (Aupetit [**6**]). *An element a of a semisimple Banach algebra $\mathcal{A}$ is idempotent if and only if $\sigma(a) \subset \{0, 1\}$ and there are scalars $r > 0$ and $C > 0$ such that*

$$\sigma(x) \subset \sigma(a) + C\|x - a\|$$

for all $x \in \mathcal{A}$ with $\|x - a\| < r$.

This characterization is used by Aupetit to prove that a spectrum-preserving linear mapping from a semisimple Banach algebra $\mathcal{A}$ onto another semisimple Banach algebra $\mathcal{B}$ transforms a set of orthogonal idempotents in $\mathcal{A}$ into a set of orthogonal idempotents in $\mathcal{B}$. We also mention that the only property of von Neumann algebras which was used in the arguments given in [**6**] is the fact that every self-adjoint element of a von Neumann algebra can be approximated in norm by a self-adjoint element with finite spectrum and thus Aupetit's arguments are straightforwardly extendable to the setting of C^*-algebras of real rank zero; see for example [**142, 174**]. Recall that a C^*-algebra is of real rank zero if every self-adjoint element can be approximated by self-adjoint elements with finite spectrum [**58**]. The spectral resolution theorem [**137**, Theorem 5.5.2] ensures that all von Neumann algebras, in particular the algebra $\mathcal{L}(\mathcal{H})$, have real rank zero. Other examples of these kind of algebras include the C^*-algebra generated by the identity on $\mathcal{L}(\mathcal{H})$ and the closed ideal $\mathcal{K}(\mathcal{H})$ of all compact operators on $\mathcal{H}$, the Calkin algebra $\mathcal{C}(\mathcal{H})$, the Bunce-Deddens algebras, the Cuntz algebras, AF-algebras, and irrational rotation algebras; see for instance [**89**].

Aupetit's argument from [**6**] and Theorem 4.22 also show that any surjective linear mapping between semisimple Banach algebras $\mathcal{A}$ and $\mathcal{B}$ preserving different spectra transforms idempotents in $\mathcal{A}$ onto idempotents in $\mathcal{B}$. According to [**182**], a map Λ from $\mathcal{A}$ to the closed subsets of $\mathbb{C}$ is called a *∂-spectrum* in $\mathcal{A}$ if

$$\partial\sigma(a) \subseteq \Lambda(a) \subset \sigma(a)$$

for all $a \in \mathcal{A}$. Purely topological arguments show that if Λ is a ∂-spectrum in $\mathcal{A}$, then the polynomial convex hull of $\Lambda(a)$ coincides with the polynomial convex hull of $\sigma(a)$ for all $a \in \mathcal{A}$. The following result is quoted from [**85**, Theorem 2.2] and its proof relies on Theorem 4.22.

THEOREM 4.23. *Let φ be a linear map from a semisimple Banach algebra $\mathcal{A}$ onto another one $\mathcal{B}$, and Λ be a ∂-spectrum in $\mathcal{B}$ such that*

$$\Lambda(\varphi(a)) \subset \sigma(a), \qquad a \in \mathcal{A}.$$

Then φ is unital and maps every idempotent in $\mathcal{A}$ to an idempotent in $\mathcal{B}$.

Brešar, Fošner and Šemrl extended Sourour's result "Theorem 4.13" from [**222**] to semisimple Banach algebras with an essential socle.

THEOREM 4.24 (Brešar–Fošner–Šemrl [**49**]). *Let $\mathcal{A}$ and $\mathcal{B}$ be two semisimple Banach algebras, and let φ be a unital bijective linear from $\mathcal{A}$ into $\mathcal{B}$. If φ preserves invertibility, then*

$$\varphi^{-1}\left(\varphi(a^2) - \varphi(a)^2\right) \cdot \mathrm{Soc}(\mathcal{A}) = \{0\}, \qquad a \in \mathcal{A}.$$

If, furthermore, $\mathrm{Soc}(\mathcal{A})$ *is essential, then φ is a Jordan isomorphism.*

As a consequence of their main result, Brešar, Fošner and Šemrl obtained the following corollary. This is based on two facts. First, in primitive algebras every nonzero ideal is essential. Second, by a well-known theorem of Herstein [**114**], Jordan homomorphisms from a primitive algebra onto another algebra is either an homomorphism or an anti-homomorphism.

COROLLARY 4.25 (Brešar–Fošner–Šemrl [**49**]). *Assume that $\mathcal{A}$ and $\mathcal{B}$ are semisimple Banach algebras and one of them is primitive with nonzero socle. If a unital bijective linear map φ from $\mathcal{A}$ into $\mathcal{B}$ preserves invertibility, then φ is either an isomorphism or anti-isomorphism.*

From what have been discussed above, there are two approaches that are based on the reduction of Kaplansky's problem to the problem of characterizing linear maps preserving idempotents or operators/elements of rank one. Unfortunately, these kind of approaches can not be used in the general case since there are semisimple Banach algebras without nontrivial idempotent or without elements of rank one. For example, any Banach algebra of analytic functions on the open unit disc $\mathbb{D}$, in particular the infamous Banach algebra $H^\infty(\mathbb{D})$, is without nontrivial idempotent. In fact, by the identity principal for the zeros of analytic functions, the relation $f^2 = f$ implies that either $f \equiv 1$ or $f \equiv 0$. Also, any semisimple Banach algebra with zero socle such as the Calkin algebra is without elements of rank one.

In [**46**], Brešar and Šemrl confirmed positively Kaplansky's conjecture for invertibility preserving maps that, additionally, locally preserves commutativity. Recall that a map φ from $\mathcal{A}$ into $\mathcal{B}$ is said to *locally preserve commutativity* if $\varphi(x^2)$ and $\varphi(x)$ commute for all $x \in \mathcal{A}$. If $\mathcal{A}$ is a C^*-algebra, then φ locally preserves commutativity if and only if $\varphi(h^2)$ and $\varphi(h)$ commute for all self adjoint elements $h \in \mathcal{A}$.

THEOREM 4.26 (Brešar–Šemrl [**46**]). *Let $\mathcal{A}$ and $\mathcal{B}$ be Banach algebras with $\mathcal{B}$ semisimple. A unital linear map φ from $\mathcal{A}$ onto $\mathcal{B}$ preserves invertibility and locally preserves commutativity if and only if φ is a Jordan homomorphism.*

Note that, since the surjectivity and locally commutativity preserving condition of φ are automatically fulfilled when $\mathcal{B} = \mathbb{C}$, this theorem is a noncommutative generalization of the Gleason-Kahane-Żelazko theorem. Recall that in order to confirm Conjecture 4.3 one has to show, in principle, that $\varphi(x^2)$ is equal to the square of $\varphi(x)$ for every $x \in \mathcal{A}$. Theorem 4.26 now tells us that it is enough to establish a much weaker assertion that $\varphi(x^2)$ and $\varphi(x)$ always commute for any $x \in \mathcal{A}$, and thus Conjecture 4.3 is now equivalent to the following one. For a neat exposition and details, we refer the reader to [**46**].

Conjecture 4.27. Let $\mathcal{A}$ and $\mathcal{B}$ be semisimple unital Banach algebras and let φ be a unital invertibility preserving linear map from $\mathcal{A}$ onto $\mathcal{B}$. Then φ locally preserves commutativity.

Another interesting result where Conjecture 4.4 is answered positively is obtained by Alaminos, Brešar, Šemrl and Villena in [**4**]. They established the following result.

THEOREM 4.28 (Alaminos–Brešar–Šemrl–Villena [**4**]). *Let $\mathcal{A}$ and $\mathcal{B}$ be semisimple Banach algebras. Every bijective linear map from $M_2(\mathcal{A})$ into $\mathcal{B}$ preserving the spectrum is a Jordan isomorphism.*

Note that if X is a square Banach space, i.e., it is isomorphic to $Y \oplus Y$ for some Banach space Y, then $\mathcal{L}(X)$ is obviously isomorphic to $M_2(\mathcal{L}(Y))$. Thus, in case X is a square Banach space (as most classical Banach spaces are), this theorem implies that every bijective linear spectrum-preserving map from $\mathcal{L}(X)$ into a semisimple Banach algebra $\mathcal{B}$ is a Jordan isomorphism. This extends Jafarian and Sourour's result Theorem 4.10 whose proof, as explained previously, is based on operators of finite rank. The approach offered in [**4**] is different and uses techniques of the general Banach algebra theory. Finally, let us mention that the problem and the above result considered in [**4**] is more general than the one on describing linear maps φ between two semisimple Banach algebra $\mathcal{A}$ and $\mathcal{B}$ such that the linear map φ_2 defined from $M_2(\mathcal{A})$ into $M_2(\mathcal{B})$ by $\varphi([a_{ij}]) = [\varphi(a_{ij})]$ is spectrum-preserving; see [**83, 118**].

4.6. Spectral isometries and spectrally bounded maps. In the sequel, assume that $\mathcal{A}$ and $\mathcal{B}$ are two semisimple Banach algebras. A linear map φ from $\mathcal{A}$ into $\mathcal{B}$ is said to be

(i) a *spectral isometry* if
$$\mathrm{r}\big(\varphi(a)\big) = \mathrm{r}(a), \qquad a \in \mathcal{A};$$
(ii) *spectrally bounded* if there is a constant M such that
$$\mathrm{r}\big(\varphi(a)\big) \le M\mathrm{r}(a), \qquad a \in \mathcal{A};$$
(iii) *spectrally bounded from below* if there is a constant m such that
$$\mathrm{r}\big(\varphi(a)\big) \ge m\mathrm{r}(a), \qquad a \in \mathcal{A}.$$

Note that if φ preserves (resp. compresses, expands) the spectrum, then φ is a spectral isometry (resp. spectrally bounded, spectrally bounded from below).

Spectral isometries and spectrally bounded maps have been studied by several authors when treating different problems; see for instance [**10, 28, 54, 88, 212**]. In [**178**], Mathieu and Schick initiated a systematic study of spectral isometries and spectrally bounded maps, and the question when such maps have to be a Jordan homomorphism has been studied in great detail by a number of authors including Mathieu and his collaborators; see for instance [**10, 77, 87, 88, 98, 164, 174, 175, 177, 178**]. Note that every bounded linear map on a commutative C^*-algebra is automatically spectrally bounded and that there are spectrally bounded linear maps on $M_n(\mathbb{C})$ and on $\mathcal{L}(X)$ that are not Jordan homomorphisms; see [**98, 212**].

In [**54**], Brešar and Šemrl characterized surjective spectral isometries between $\mathcal{L}(X)$ and $\mathcal{L}(Y)$, and thus generalized Jafarian and Sourour's result (Theorem 4.10). Spectral isometries on $M_n(\mathbb{C})$ were previously described by Aupetit in [**10**, Proposition2], and in [**176**] Mathieu and Sourour showed that every unital surjective spectral isometry between finite-dimensional semisimple Banach algebras is a Jordan isomorphism. The description of spectral isometries on $M_n(\mathbb{C})$ can be also deduce from Akbari–Aryapoor's result "Theorem 4.7" [**3**] since spectral isometries cavernn be assumed to be unital and every unital spectral isometry preserves the peripheral spectrum; see [**54**, The proof of Theorem 1].

THEOREM 4.29 (Brešar–Šemrl [**54**]). *A linear map φ from $\mathcal{L}(X)$ onto $\mathcal{L}(Y)$ is a spectral isometry if and only if there is a scalar α of modulus one and either there is a bijective mapping $A \in \mathcal{L}(X,Y)$ such that*

$$\varphi(T) = \alpha ATA^{-1}, \qquad T \in \mathcal{L}(X),$$

or there is a bijective mapping $B \in \mathcal{L}(X^*,Y)$ such that

$$\varphi(T) = \alpha BT^*B^{-1}, \qquad T \in \mathcal{L}(X).$$

For the proof of this theorem, Brešar and Šemrl used Vesentini's theorem on subharmonity of the spectral radius when composed with a $\mathcal{L}(X)$-analytic function and showed that if φ is a spectral isometry from $\mathcal{L}(X)$ onto $\mathcal{L}(Y)$ then φ bijectively maps nilpotent operators of $\mathcal{L}(X)$ into nilpotent operators of $\mathcal{L}(Y)$. This allowed them to apply the main result of Šemrl [**214**] that describes surjective linear maps between $\mathcal{L}(X)$ and $\mathcal{L}(Y)$ preserving nilpotent operators when restricted on the linear span of all nilpotent operators of $\mathcal{L}(X)$. We also point out that this theorem itself has been extended in [**15**, **98**]. In [**98**], Fošner and Šemrl obtained a characterization of the surjective linear maps on $\mathcal{L}(X)$ that are both spectrally bounded and spectrally bounded from below. The obtained forms are somehow different from the ones appeared in Theorem 4.29. In [**15**], Bai and Hou characterized, particular, surjective additive maps preserving the spectral radius of Banach space operators, extending the above result to the additive setting.

When restricted to $\mathcal{L}(X)$, the surjectivity spectrum, the approximate point spectrum, the generalized spectrum, the left/right spectrum and the intersection or the union of any two these spectra are ∂-spectra. Thus, the following corollary is an immediate consequence of Theorem 4.29.

COROLLARY 4.30. *If a linear map φ from $\mathcal{L}(X)$ onto $\mathcal{L}(Y)$ preserves a ∂-spectrum of operators, then either there is a bijective mapping $A \in \mathcal{L}(X,Y)$ such that*

$$\varphi(T) = ATA^{-1}, \qquad T \in \mathcal{L}(X),$$

or there is a bijective mapping $B \in \mathcal{L}(X^*,Y)$ such that

$$\varphi(T) = BT^*B^{-1}, \qquad T \in \mathcal{L}(X).$$

An additive version of this corollary can be obtained using Bai and Hou's result from [**15**] instead of Theorem 4.29. In fact, a stronger result than this corollary was obtained in [**85**] where Cui and Hou described linear maps compressing different spectral sets. An additive version of this theorem can be found in [**146**].

THEOREM 4.31. *If Λ is a ∂-spectrum of operators and φ is a linear map from $\mathcal{L}(X)$ onto $\mathcal{L}(Y)$ such that*

$$\Lambda(\varphi(T)) \subset \sigma(T), \qquad T \in \mathcal{L}(X),$$

then either φ *vanishes on* $\mathcal{F}(X)$ *or* φ *is injective. In the latter case, either there is a bijective mapping* $A \in \mathcal{L}(X, Y)$ *such that*

$$\varphi(T) = ATA^{-1}, \qquad T \in \mathcal{L}(X),$$

or there is a bijective mapping $B \in \mathcal{L}(X^*, Y)$ *such that*

$$\varphi(T) = BT^*B^{-1}, \qquad T \in \mathcal{L}(X).$$

In [**212**], Šemrl described spectrally bounded maps on $\mathcal{L}(\mathcal{H})$ when $\mathcal{H}$ is an infinite-dimensional complex Hilbert space and provided an example showing that there are infinite-dimensional Banach spaces X and spectrally bounded maps on $\mathcal{L}(X)$ that are not Jordan homomorphisms. In general, the full description of spectrally bounded maps on $\mathcal{L}(X)$ is still unknown and even the question of which linear maps φ on $\mathcal{L}(X)$ preserve the quasinilpotency in both directions (i.e., for every $T \in \mathcal{L}(X)$, we have $\mathrm{r}(T) = 0$ if and only if $\mathrm{r}(\varphi(T)) = 0$) is still open.

THEOREM 4.32 (Šemrl [**212**]). *A unital bijective linear map* φ *on* $\mathcal{L}(\mathcal{H})$ *is spectrally bounded if and only if there is an invertible operator* $A \in \mathcal{L}(\mathcal{H})$ *such that either*

$$\varphi(T) = ATA^{-1}, \qquad T \in \mathcal{L}(\mathcal{H}),$$

or

$$\varphi(T) = AT^{tr}A^{-1}, \qquad T \in \mathcal{L}(\mathcal{H}).$$

Here, T^{tr} *denotes the transpose of* T *relative to a fixed but arbitrary orthonormal basis of* $\mathcal{H}$.

Back to spectral isometries and spectrally bounded maps between the semisimple Banach algebras $\mathcal{A}$ and $\mathcal{B}$. We note that if φ is a spectrally bounded map from $\mathcal{A}$ onto $\mathcal{B}$, then φ is continuos but need not be injective as shown by the example followed Theorem 4.13. However, linear surjective maps between $\mathcal{A}$ and $\mathcal{B}$ that are spectrally bounded from below are always injective as shown by the following result quoted from [**42**, Lemma 2.6].

LEMMA 4.33 (Bourhim–Miller [**42**]). *If* $\varphi : \mathcal{A} \to \mathcal{B}$ *is surjective and spectrally bounded from below, then* φ *is a bijective continuous map. In particular,* φ^{-1} *is spectrally bounded.*

We also mention that Mathieu and Schick showed in [**177**, Lemma 3.1] that any spectrally bounded map from $\mathcal{A}$ onto $\mathcal{B}$ preserves nilpotent elements. As shown therein this is a consequence of the open mapping theorem and the following spectral characterization of nilpotent elements of a semisimple Banach algebra.

THEOREM 4.34 (Aupetit–Zemánek, Ransford–White). *Let* a *be an element in a Banach algebra* $\mathcal{A}$ *and* $n \geq 1$. *The following assertions are equivalent.*

(i) $a^n \in \mathrm{Rad}(\mathcal{A})$.

(ii) *For every bounded neighborhood* $U \subset \mathcal{A}$ *of zero, there exists a constant* C_U *such that*

$$\mathrm{r}(a + x) \leq C_U \|x\|^{\frac{1}{n}}, \qquad x \in U.$$

(iii) *There is a bounded neighborhood* $U \subset \mathcal{A}$ *of zero and a constant* C *such that*

$$\mathrm{r}(a + x) \leq C \|x\|^{\frac{1}{n}}, \qquad x \in U.$$

The implication $(i) \Rightarrow (ii)$ is due to Aupetit and Zemánek [**12**] and implication $(iii) \Rightarrow (i)$ is established by Ransford and White in [**197**].

A C^*-algebra $\mathcal{A}$ is said to be *purely infinite* if it has no characters and if for every pair of positive elements a and b in $\mathcal{A}$ with $a \in \mathcal{A}b\mathcal{A}$ there is a sequence $(x_n)_n$ in $\mathcal{A}$ such that $a = \lim_{n\to\infty} x_n^* b x_n$. Recall that any purely infinite simple C^*-algebra has real rank zero; see for instance [**141**]. Mathieu and his various co-authors established several results about spectral isometries and spectrally bounded maps; see for instance [**172**–**178**] and the references therein. In [**177**, Theorem 3.6], Mathieu and Schick showed that a unital surjective spectrally bounded map from a properly infinite von Neumann algebra $\mathcal{A}$ onto a semisimple Banach algebra $\mathcal{B}$ is a Jordan epimorphism. In [**175**], Mathieu showed that this result remains valid if $\mathcal{A}$ is purely infinite simple C^*-algebra and in its turn a such result from [**175**] has been generalized by Lin and Mathieu [**164**, Corollary 2.5] to case of unital spectrally bounded linear maps from a C^*-algebra purely infinite with rank real zero onto a semisimple Banach algebra. The nonunital case was independently observed latter in [**25**, **173**].

THEOREM 4.35 (Bendaoud–Bourhim [**25**], Mathieu [**173**]). *Let $\mathcal{A}$ be a purely infinite C^*-algebra with rank real zero and let $\mathcal{B}$ be a semisimple Banach algebra. If $\varphi : \mathcal{A} \to \mathcal{B}$ is a surjective spectrally bounded linear map, then $\varphi(\mathbf{1})$ is a central invertible element of $\mathcal{B}$ and $\varphi(\mathbf{1})^{-1}\varphi$ is a Jordan epimorphism.*

Evidently any surjective linear map between $\mathcal{A}$ and $\mathcal{B}$ preserving the spectrum is a unital surjective spectral isometry. In [**73**], Costara showed, in particular, that if φ is unital surjective spectral isometry from $\mathcal{A}$ into $\mathcal{B}$ then $\sigma(\varphi(a)) = \sigma(a)$ for all $a \in \mathcal{A}$ for which $\sigma(a)$ has exactly two elements. But it is still unknown whether every unital surjective spectral isometry between C^*-algebras is a Jordan isomorphism and it is still unknown if any unital surjective spectral isometry preserves the spectrum.

THEOREM 4.36 (Costara [**78**]). *Any unital spectral isometry from a semisimple Banach algebra $\mathcal{A}$ onto a primitive Banach algebra $\mathcal{B}$ which locally preserves commutativity is a Jordan homomorphism.*

This result partially extends theorem 4.26 in which $\mathcal{B}$ was assumed to be only semisimple. He also showed that if both $\mathcal{A}$ and $\mathcal{B}$ are von Neumann algebras, then the assumption that $\mathcal{B}$ is primitive can be omitted.

4.7. Left and right invertibility preservers. For any element $a \in \mathcal{A}$, let us denote respectively by $\sigma_l(a)$ and $\sigma_r(a)$ the left spectrum and the right spectrum of a. The element a is said to be *semi-invertible* if it is right or left invertible. Its semi-spectrum is defined by

$$\sigma_{l,r}(a) := \{\lambda \in \mathbb{C} : a - \lambda\mathbf{1} \text{ is not semi-invertible}\},$$

and coincides with $\sigma_l(a) \cap \sigma_r(a)$. A linear map φ between two $\mathcal{A}$ and $\mathcal{B}$ *preserves left invertibility* if $\varphi(a)$ is left invertible in $\mathcal{B}$ whenever a is left invertible in $\mathcal{A}$. The definition of linear maps preserving right invertibility, and semi-invertibility is analogous.

In [**144**], Kuzma was motivated by Sourour's question [**222**] which asks whether a linear unital map from $\mathcal{L}(X)$ onto $\mathcal{L}(Y)$ which preserves invertibility is necessarily injective, and showed that the answer is affirmative when 'invertibility' is replaced by 'noninvertibility' by using of scarcity lemma; see [**11**, Theorem 3.4.25, and Corollary 3.4.18]. As observed in [**42**], this holds at once by applying the above lemma. Indeed, let φ be a linear map from $\mathcal{A}$ onto $\mathcal{B}$ preserving the noninvertibility, i.e., $\varphi(a)$ is not invertible in $\mathcal{B}$ whenever a is not invertible in $\mathcal{A}$. Note that, since φ is surjective, there is $a \in \mathcal{A}$ such that $\varphi(a) = \mathbf{1}$ and thus the map ϕ, defined by $\phi(x) = \varphi(ax)$, $(x \in \mathcal{A})$, is a linear unital surjective mapping preserving the noninvertibility. In particular, $\mathrm{r}(x) \leq \mathrm{r}(\phi(x))$, $(x \in \mathcal{A})$, and ϕ is injective and so is φ, by Lemma 4.33.

This combined with Theorem 4.13 yields the following result that describes surjective linear maps between $\mathcal{L}(X)$ and $\mathcal{L}(Y)$ that preserve noninvertibility.

THEOREM 4.37. *For a surjective linear map φ from $\mathcal{L}(X)$ into $\mathcal{L}(Y)$, the following are equivalent.*

(i) *φ is unital and preserves the noninvertibility.*
(ii) *φ expands the spectrum, i.e., $\sigma(T) \subset \sigma(\varphi(T))$ for all $T \in \mathcal{L}(X)$.*
(iii) *Either $\varphi(T) = ATA^{-1}$ for some $A \in \mathcal{L}(X, Y)$ or $\varphi(T) = BT^*B^{-1}$ for some $B \in \mathcal{L}(X^*, Y)$.*

The following result, quoted from [**26**], shows that a surjective linear map between two semisimple complex Banach algebras preserves invertibility in both directions whenever it preserves any kind of invertibility in both directions. Its proof uses some arguments taken from [**38**].

THEOREM 4.38 (Bendaoud–Bourhim–Burgos–Sarih [**26**]). *Let $\mathcal{A}$ and $\mathcal{B}$ be two semisimple Banach algebras, and let φ be a linear map from $\mathcal{A}$ onto $\mathcal{B}$. If φ either preserves left invertibility, right invertibility or semi-invertibility in both directions, then φ is a bijective map preserving invertibility in both directions.*

One may wonders if this result remains valid if these type of invertibility is preserved in one direction. This is not true in general as shown by the following elementary example. Let $S \in \mathcal{L}(\mathcal{H})$ be a noninvertible isometry on an infinite-dimensional complex Hilbert space $\mathcal{H}$, and consider the surjective linear maps Φ and Ψ defined on $\mathcal{L}(\mathcal{H})$ by

$$\Phi(T) = TS \qquad \text{and} \qquad \Psi(T) = S^*T.$$

Then Φ preserves the left invertibility and Ψ preserves the right invertibility, but none of them preserves invertibility.

However, it is true that a surjective linear map between two semisimple complex Banach algebras preserves noninvertibility whenever it preserves any kind of noninvertibility in one direction.

THEOREM 4.39. *Let $\mathcal{A}$ and $\mathcal{B}$ be two semisimple Banach algebras, and let φ be a linear map from $\mathcal{A}$ onto $\mathcal{B}$. If φ either preserves nonleft invertibility, nonright invertibility or nonsemi-invertibility, then φ is a bijective map preserving noninvertibility.*

PROOF. We shall only deal with the case when φ preserves nonleft invertibility as the other cases follow analogously. By the surjectivity of φ, there is a left invertible element $u \in \mathcal{A}$ such that $\varphi(u) = \mathbf{1}$. Let ψ be the unital linear map from $\mathcal{A}$ into $\mathcal{B}$ defined by $\psi(a) := \varphi(ua)$, $(a \in \mathcal{A})$, and note that a such map is a unital linear map preserving nonleft invertibility. Clearly, we have $\mathrm{r}(a) \leq \mathrm{r}(\psi(a))$ for all $a \in A$ and thus ψ is injective, by Lemma 4.33. Since u is left invertible, there is

$v \in \mathcal{A}$ such that $vu = \mathbf{1}$. Note that, since $u(\mathbf{1} - vu) = u - uvu = 0$, we have $\psi(\mathbf{1} - vu) = \varphi(u(\mathbf{1} - vu)) = \varphi(0) = 0$, and thus the injectivity of ψ shows that $\mathbf{1} - vu = 0$. Hence, u is invertible and thus φ is injective as well.

Now, take an arbitrary element $a \in \mathcal{A}$ such that $b := \varphi(a)$ is invertible in $\mathcal{B}$ and let us show that a is invertible in $\mathcal{A}$. Note that the map ϕ from $\mathcal{A}$ onto $\mathcal{B}$ given by $\phi(x) := b^{-1}\varphi(x)$, $(x \in \mathcal{A})$, preserves nonleft invertibility and $\phi(a) = \mathbf{1}$. By what has been shown above, $a = \varphi^{-1}(b)$ is invertible in $\mathcal{A}$; as desired. □

Combining Theorem 4.13, Theorem 4.37, Theorem 4.38 and Theorem 4.39, one gets immediately the following result.

Corollary 4.40. *For a surjective linear map φ from $\mathcal{L}(X)$ into $\mathcal{L}(Y)$, the following are equivalent.*

(i) *φ is unital and preserves the nonleft-invertibility.*
(ii) *φ expands the left spectrum, i.e., $\sigma_l(T) \subset \sigma_l(\varphi(T))$ for all $T \in \mathcal{L}(X)$.*
(iii) *φ is a unital injective map preserving the left-invertibility.*
(iv) *φ is a injective map compressing the left spectrum, i.e., $\sigma_l(\varphi(T)) \subset \sigma_l(T)$ for all $T \in \mathcal{L}(X)$.*
(v) *φ is unital and preserves the nonright-invertibility.*
(vi) *φ expands the right spectrum, i.e., $\sigma_r(T) \subset \sigma_r(\varphi(T))$ for all $T \in \mathcal{L}(X)$.*
(vii) *φ is a unital injective map preserving the right-invertibility.*
(viii) *φ is a injective map compressing the right spectrum, i.e., $\sigma_r(\varphi(T)) \subset \sigma_r(T)$ for all $T \in \mathcal{L}(X)$.*
(ix) *φ is unital and preserves the nonsemi-invertibility.*
(x) *φ expands the semi-spectrum, i.e., $\sigma_{l,r}(T) \subset \sigma_{l,r}(\varphi(T))$ for all $T \in \mathcal{L}(X)$.*
(xi) *φ is a unital injective map preserving the semi-invertibility.*
(xii) *φ is a injective map compressing the semi spectrum, i.e., $\sigma_{l,r}(\varphi(T)) \subset \sigma_{l,r}(T)$ for all $T \in \mathcal{L}(X)$.*
(xiii) *Either $\varphi(T) = ATA^{-1}$ for some $A \in \mathcal{L}(X, Y)$ or $\varphi(T) = BT^*B^{-1}$ for some $B \in \mathcal{L}(X^*, Y)$. The second case occurs only if both X and Y are reflexive and every left-invertible operator in $\mathcal{L}(X)$ is invertible.*

Variants of this result were established by a number of authors; see for instance [**85**, **216**].

5. Linear maps preserving semi-Fredholm operators and generalized invertibility

Let $\mathcal{L}(\mathcal{H})$ stand for the algebra of all linear bounded operators on an infinite-dimensional complex Hilbert space $\mathcal{H}$ and $\mathcal{K}(\mathcal{H})$ denotes the closed ideal of all *compact operators* on $\mathcal{H}$. The collection of *finite rank operators* is denoted by $\mathcal{F}(\mathcal{H})$. For an operator $T \in \mathcal{L}(\mathcal{H})$, let T^* denote the adjoint, $\mathrm{Ker}(T)$ the kernel or the null space, and $\mathcal{R}(T)$ the range. The operator T is left (resp. right) semi-Fredholm if $\mathcal{R}(T)$ is closed and $\mathrm{Ker}(T)$ is finite-dimensional (resp. $\mathrm{Ker}(T^*)$ is finite-dimensional). The operator T is Fredholm if it is both left and right semi-Fredholm. The *essential norm* of T is given by

$$\|T\|_e := \mathrm{dist}(\mathrm{T}, \mathcal{K}(\mathcal{H}))$$

and the *essential spectral radius*, is

$$\mathrm{r}_e(T) = \lim_{n\to\infty} \|T^n\|_e^{1/n}.$$

This quantity coincides with $\mathrm{r}(\pi(T))$ the classical spectral radius of $\pi(T)$, where π denotes the canonical quotient map from $\mathcal{L}(\mathcal{H})$ onto the *Calkin algebra*

$$\mathcal{C}(\mathcal{H}) := \mathcal{L}(\mathcal{H})/\mathcal{K}(\mathcal{H}).$$

An operator $T \in \mathcal{L}(\mathcal{H})$ admits a generalized inverse if there exists an operator $S \in \mathcal{L}(\mathcal{H})$, called a generalized inverse of T, such that $TST = T$ and $STS = S$; see for instance [**149**]. Equivalently, T has a generalized inverse if and only if T has a closed range. In general, the generalized inverse need not exist and is not unique even if it exists. However, if T is invertible then T^{-1} is the unique generalized inverse of T. For the classical background of Fredholm theory of bounded linear operators, a good reference is Aiena's book [**1**].

Let $\varphi : \mathcal{L}(\mathcal{H}) \to \mathcal{L}(\mathcal{H})$ be a mapping. Then we say that φ is

(i) *surjective up to finite rank operators* if

$$\mathcal{L}(\mathcal{H}) = \mathcal{R}(\varphi) + \mathcal{F}(\mathcal{H}).$$

(ii) *surjective up to compact operators* if

$$\mathcal{L}(\mathcal{H}) = \mathcal{R}(\varphi) + \mathcal{K}(\mathcal{H}).$$

(iii) *preserves semi-Fredholm operators* of $\mathcal{L}(\mathcal{H})$ in both directions provided that T is Fredholm if and only if so is $\varphi(T)$. In a similar way, maps preserving Fredholm operators or generalized invertibility are defined.

(iv) *preserves the essential spectral radius* if

$$\mathrm{r}_e\big(\varphi(T)\big) = \mathrm{r}_e(T), \qquad T \in \mathcal{L}(\mathcal{H}).$$

(v) *essentially spectrally bounded* if there exists a positive constant M such that

$$\mathrm{r}_e(\varphi(T)) \leq M\mathrm{r}_e(T), \qquad T \in \mathcal{L}(\mathcal{H}).$$

Finally, if φ is such that $\varphi\big(\mathcal{K}(\mathcal{H})\big) \subset \mathcal{K}(\mathcal{H})$ then we define the *induced map* $\widehat{\varphi} : \mathcal{C}(\mathcal{H}) \to \mathcal{C}(\mathcal{H})$ by

$$\widehat{\varphi}\big(\pi(T)\big) := \pi\big(\varphi(T)\big), \qquad T \in \mathcal{L}(\mathcal{H}).$$

5.1. Linear maps preserving Generalized invertibility, and semi-Fredholm operators. The problem of characterizing surjective linear maps φ on $\mathcal{L}(\mathcal{H})$ preserving generalized invertibility has was initiated by Mbekhta, Rodman, and Šemrl in [**186**]. They proved that if $\mathcal{H}$ is separable and φ is a bijective continuous unital linear map on $\mathcal{L}(\mathcal{H})$ preserving generalized invertibility in both directions then $\mathcal{K}(\mathcal{H})$ is invariant under φ and the induced linear map by φ on the Calkin algebra $\mathcal{C}(\mathcal{H})$ is either an automorphism or an anti-automorphism. But they left the conjecture whether the same conclusion holds true without the continuity assumption and if an analogous result can be obtained without assuming that φ is unital. This conjecture was completely settled and affirmatively answered by Mbekhta and Šemrl. They established the following result.

THEOREM 5.1 (Mbekhta–Šemrl [**183**]). *Assume that $\mathcal{H}$ is separable and φ is a linear surjective up to finite rank operators mapping on $\mathcal{L}(\mathcal{H})$. If φ preserves generalized invertibility in both directions, then $\varphi(\mathcal{K}(\mathcal{H})) \subset \mathcal{K}(\mathcal{H})$ and the induced map $\widehat{\varphi} : \mathcal{C}(\mathcal{H}) \to \mathcal{C}(\mathcal{H})$ is either an automorphism or an anti-automorphism multiplied by an invertible element of $\mathcal{C}(\mathcal{H})$.*

This theorem holds only in the infinite-dimensional case since every $n \times n$-complex matrix has a generalized inverse, and therefore, every linear map on $M_n(\mathbb{C})$ preserves generalized invertibility. Its proof is shorter than the one appeared in [**186**], and uses the following characterization of semi-Fredholm operators of $\mathcal{L}(\mathcal{H})$ in term of generalized invertibility of operators.

LEMMA 5.2 (Mbekhta–Šemrl [**183**]). *For an operator $T \in \mathcal{L}(\mathcal{H})$, the following statements are equivalent.*

(i) T is semi-Fredholm.
(ii) For every $S \in \mathcal{L}(\mathcal{H})$, there exists $\delta > 0$ such that $T + \lambda S$ has a generalized inverse for all $\lambda \in \mathbb{C}$ with $|\lambda| < \delta$.

This lemma shows that every linear surjective up to finite rank operators map φ on $\mathcal{L}(\mathcal{H})$ preserving generalized invertibility in both directions preserves semi-Fredholm operators of $\mathcal{L}(\mathcal{H})$ in both direction. Clearly, Theorem 5.1 is now a consequence of the following result established by Mbekhta and Šemrl in [**183**].

THEOREM 5.3 (Mbekhta–Šemrl [**183**]). *Assume that $\mathcal{H}$ is separable and φ is a linear map on $\mathcal{L}(\mathcal{H})$ surjective up to compact operators. If φ preserves semi-Fredholm operators of $\mathcal{L}(\mathcal{H})$ in both directions, then $\varphi(\mathcal{K}(\mathcal{H})) \subset \mathcal{K}(\mathcal{H})$ and the induced map $\widehat{\varphi} : \mathcal{C}(\mathcal{H}) \to \mathcal{C}(\mathcal{H})$ is either an automorphism or an anti-automorphism multiplied by an invertible element of $\mathcal{C}(\mathcal{H})$.*

In [**184**], Mbekhta characterized surjective linear maps on $\mathcal{L}(\mathcal{H})$ preserving the set of Fredholm operators in both directions, and then, as a consequence, he obtained a complete characterization of surjective linear maps on $\mathcal{L}(\mathcal{H})$ that preserve the essential spectrum.

THEOREM 5.4 (Mbekhta [**184**]). *Assume that φ is a linear map on $\mathcal{L}(\mathcal{H})$ surjective up to compact operators. Then φ preserves Fredholm operators of $\mathcal{L}(\mathcal{H})$ in both directions if and only if $\varphi(\mathcal{K}(\mathcal{H})) \subset \mathcal{K}(\mathcal{H})$ and the induced map $\widehat{\varphi} : \mathcal{C}(\mathcal{H}) \to \mathcal{C}(\mathcal{H})$ is either an automorphism or an anti-automorphism multiplied by an invertible element of $\mathcal{C}(\mathcal{H})$.*

The papers [**183, 186**] contain many good ideas and elegant results which opened the way for several other authors to further study maps preserving different essential spectra and quantities and generalize the results from [**183, 186**]. In [**82**], Cui and Hou established independently similar results and characterized linear maps on $\mathcal{L}(\mathcal{H})$ preserving certain essential spectral sets such as the set of (left, right) Fredholm operators and the set of semi-Fredholm operators.

5.2. Linear maps preserving Fredholm and Atkinson elements of Banach algebras. The algebras appearing throughout this section are assumed to be infinite-dimensional. Let $\mathcal{A}$ be a semisimple Banach algebra, and let

$$\mathrm{I}(\mathcal{A}) := \left\{a \in \mathcal{A} \colon a + \overline{\mathrm{Soc}(\mathcal{A})} \in \mathrm{Rad}\left(\mathcal{A}/\overline{\mathrm{Soc}(\mathcal{A})}\right)\right\}$$

be the *ideal of inessential elements* of $\mathcal{A}$. It is a closed ideal of $\mathcal{A}$ and contains $\mathrm{Soc}(\mathcal{A})$. It is well known that $\mathrm{Soc}(\mathcal{L}(X)) = \mathcal{F}(X)$ and that $\mathrm{I}(\mathcal{L}(X))$ is the inverse canonical image of the radical of the Calkin algebra and also

$$\mathrm{I}(\mathcal{L}(X)) = \{T \in \mathcal{L}(X) : \dim(\mathrm{Ker}(1 - ST)) < \infty \text{ for every } S \in \mathcal{L}(X)\},$$

and satisfies $\mathcal{K}(X) \subset \mathrm{I}(\mathcal{L}(X))$ with strict inclusion in general; see [**2**, Theorem 1]. However, in the case of a complex Hilbert space $\mathcal{H}$, the ideal of inessential operators $\mathrm{I}(\mathcal{L}(\mathcal{H}))$ coincides with $\mathcal{K}(\mathcal{H})$.

The *generalized Calkin algebra* of $\mathcal{A}$ is defined by

$$\mathcal{C}(\mathcal{A}) := \mathcal{A}/\mathrm{I}(\mathcal{A}),$$

and the quotient map from $\mathcal{A}$ onto $\mathcal{C}(\mathcal{A})$ is denote by π. The generalized Calkin algebra of a C^*-algebra was introduced by Mathieu in [**179**]. We also note that a semisimple Banach algebra is finite dimensional if and only if it coincides with its socle; see for instance [**11**, Theorem 5.4.2]. Since in the above definition our algebra $\mathcal{A}$ is supposed to be infinite-dimensional, the generalized Calkin algebra introduced above is not trivial. An element $a \in \mathcal{A}$ is said to be *left semi-Fredholm* (respectively *right semi-Fredholm*) if $\pi(a)$ is left (respectively right) invertible in $\mathcal{C}(\mathcal{A})$, and is called *Fredholm* if $\pi(a)$ is invertible in $\mathcal{C}(\mathcal{A})$. The element a is said to be *Atkinson* if it is left or right semi-Fredholm [**206, 207**]. A Hilbert space operator $T \in \mathcal{L}(\mathcal{H})$ is Atkinson if and only if it is a semi-Fredholm operator.

The following lemma shows that the ideal of inessential elements in a semisimple Banach algebra coincides with the perturbation class associated with any of the sets of Fredholm, Atkinson, left semi-Fredholm and right semi-Fredholm elements.

LEMMA 5.5 (Bendaoud–Bourhim–Burgos–Sarih [**26**]). *Let $\mathcal{A}$ be a semisimple Banach algebra, and let $a \in \mathcal{A}$. Then the following statements are equivalent.*

(i) $a \in \mathrm{I}(\mathcal{A})$
(ii) $a + b$ *is Fredholm for all Fredholm elements* $b \in \mathcal{A}$.
(iii) $a + b$ *is Atkinson for all Atkinson elements* $b \in \mathcal{A}$.
(iv) $a + b$ *is left semi-Fredholm for all left semi-Fredholm elements* $b \in \mathcal{A}$.
(v) $a + b$ *is right semi-Fredholm for all right semi-Fredholm elements* $b \in \mathcal{A}$.

Let $\mathcal{A}$ and $\mathcal{B}$ be semisimple Banach algebras. A linear map $\varphi : \mathcal{A} \to \mathcal{B}$ is said to be *surjective up to inessential elements* if

$$\mathcal{B} = \mathcal{R}(\varphi) + \mathrm{I}(\mathcal{B}),$$

and is said to *preserve the set of Fredholm elements in both directions* provided that $a \in \mathcal{A}$ is Fredholm if and only if so is $\varphi(a)$. Observe that if $\phi : \mathcal{A} \to \mathrm{I}(\mathcal{B})$ is an arbitrary linear map, then a linear map $\varphi : \mathcal{A} \to \mathcal{B}$ preserves the set of Fredholm elements if and only if $\varphi + \phi$ does. Thus, linear maps $\varphi : \mathcal{A} \to \mathcal{B}$ preserving the set of Fredholm elements only can be determined up to linear perturbation maps from A to $\mathrm{I}(\mathcal{B})$. In a similar way, we define linear maps preserving the set of left semi-Fredholm elements, right semi-Fredholm elements and Atkinson elements in both directions; see for instance [**26, 82, 183, 184**].

THEOREM 5.6 (Bendaoud–Bourhim–Burgos–Sarih [**26**]). *Let $\mathcal{A}$ and $\mathcal{B}$ be two semisimple Banach algebras. Let $\varphi : \mathcal{A} \to \mathcal{B}$ be a surjective up to inessential elements linear map, and consider the following statements.*

(i) φ *preserves set of left semi-Fredholm elements in both directions.*
(ii) φ *preserves set of right semi-Fredholm elements in both directions.*
(iii) φ *preserves set of Atkinson elements in both directions.*

If any of these statements holds, then $\varphi\big(\mathrm{I}(\mathcal{A})\big) \subseteq \mathrm{I}(\mathcal{B})$, *and* φ *preserves the set of Fredholm elements.*

Lemma 5.5 tells us that if any of the above statements holds, then it is immediate that $\varphi(\mathrm{I}(A)) \subseteq \mathrm{I}(B)$ and thus the natural induced map $\hat{\varphi} : \mathcal{C}(A) \to \mathcal{C}(B)$ defined by $\pi(a) \mapsto \pi(\varphi(a))$, is well defined and is a surjective linear map preserving left (resp. right, semi-) invertibility in both directions if (a) (resp. (b), (c)) holds. Thus, Theorem 4.38 applied to $\hat{\varphi}$ shows that $\hat{\varphi}$ preserves invertibility in both directions and consequently φ preserves the set of Fredholm elements in both directions.

THEOREM 5.7 (Bendaoud–Bourhim–Burgos–Sarih [**26**]). *Let $\mathcal{A}$ be a C^*-algebra with real rank zero and let $\mathcal{B}$ be a semisimple Banach algebra. A surjective up to inessential elements linear map φ from $\mathcal{A}$ into $\mathcal{B}$ preserves Fredholm elements in both directions if and only if $\varphi(\mathrm{I}(\mathcal{A})) \subseteq \mathrm{I}(\mathcal{B})$ and the induced mapping $\hat{\varphi} : \mathcal{C}(\mathcal{A}) \to \mathcal{C}(\mathcal{B})$ is a Jordan isomorphism multiplied by an invertible element in $\mathcal{C}(\mathcal{B})$.*

An immediate consequence of Theorem 5.6 and Theorem 5.7 is the following result which extends the main results of [**82, 183, 184**] and shows that the separability of the Hilbert space $\mathcal{H}$ in Theorem 5.1 and Theorem 5.3 can be dropped.

COROLLARY 5.8 (Bendaoud–Bourhim–Burgos–Sarih [**26**]). *Let $\mathcal{A}$ be a C^*-algebra with real rank zero and let $\mathcal{B}$ be a semisimple Banach algebra. Let $\varphi : \mathcal{A} \to \mathcal{B}$ be a surjective up to inessential elements linear map. If φ preserves any of the set of left semi-Fredholm, right semi-Fredholm and Atkinson elements in both directions, then $\varphi(\mathrm{I}(\mathcal{A})) \subseteq \mathrm{I}(\mathcal{B})$, and the induced mapping $\hat{\varphi} : \mathcal{C}(\mathcal{A}) \to \mathcal{C}(\mathcal{B})$ is a Jordan isomorphism multiplied by an invertible element in $\mathcal{C}(\mathcal{B})$.*

5.3. Linear maps preserving the essential spectral radius. The following result, quoted from [**27**], characterizes linear maps from $\mathcal{L}(\mathcal{H})$ onto itself which preserve the essential spectral radius. In view of Theorem 5.6, it extends the main results of [**82, 183, 184, 186**] and also shows that the separability of the Hilbert space $\mathcal{H}$ in Theorem 5.1 and Theorem 5.3 can be dropped.

THEOREM 5.9 (Bendaoud-Bourhim-Sarih [**27**]). *A linear surjective up to compact operators map φ from $\mathcal{L}(\mathcal{H})$ into itself preserves the essential spectral radius if and only if $\varphi(\mathcal{K}(\mathcal{H})) \subseteq \mathcal{K}(\mathcal{H})$ and the induced map $\widehat{\varphi} : \mathcal{C}(\mathcal{H}) \to \mathcal{C}(\mathcal{H})$ is either an automorphism or an anti-automorphism multiplied by a nonzero scalar of modulus one.*

This result has been extended in [**25**] where it is shown that a linear surjective up to compact operators map φ on $\mathcal{L}(\mathcal{H})$ is essentially spectrally bounded if and only if $\varphi(\mathcal{K}(\mathcal{H})) \subseteq \mathcal{K}(\mathcal{H})$ and the induced map $\widehat{\varphi}$ on $\mathcal{C}(\mathcal{H})$ is either an epimorphism or an anti-epimorphism multiplied by a nonzero scalar. Of course, the "if part" is obvious but for the "only if part", the authors first used Vesentini's theorem on subharmonity of the spectral radius when composed with an holomorphic function together with Liouville's theorem to show that if φ is a surjective essentially spectrally bounded linear maps on $\mathcal{L}(\mathcal{H})$, then φ leaves invariant the ideal of all compact operators on $\mathcal{H}$ and therefore the induced mapping on the Calkin algebra $\mathcal{C}(\mathcal{H})$ is a well defined spectrally bounded linear map. To conclude the proof, they applied Theorem 4.35 since the Calkin algebra $\mathcal{C}(\mathcal{H})$ is a purely infinite C^*-algebra with rank real zero.

6. Minimum, surjectivity and reduced minimum moduli preservers

In [**182, 185**], Mbekhta described unital surjective linear maps on $\mathcal{L}(\mathcal{H})$, the algebra of all bounded linear operators on an infinite dimensional complex Hilbert space $\mathcal{H}$, preserving several spectral quantities such as the minimum, the surjectivity and the reduced minimum moduli. In [**182**], he showed that a surjective unital map φ on $\mathcal{L}(\mathcal{H})$ preserves either the minimum modulus or the surjectivity modulus if and only if there exists a unitary operator $U \in \mathcal{L}(\mathcal{H})$ such that $\varphi(T) = UTU^*$ for all $T \in \mathcal{L}(\mathcal{H})$. While, in [**185**], he proved that a unital surjective linear map φ on $\mathcal{L}(\mathcal{H})$ preserves the reduced minimum modulus if and only if there exists a unitary operator $U \in \mathcal{L}(\mathcal{H})$ such that either $\varphi(T) = UTU^*$ for all $T \in \mathcal{L}(\mathcal{H})$ or $\varphi(T) = UT^{tr}U^*$ for all $T \in \mathcal{L}(\mathcal{H})$, where T^{tr} denotes the transpose of T with respect to an arbitrary but a fixed orthonormal basis in $\mathcal{H}$. For the nonunital case, Mbekhta closed his papers [**182, 185**] with a natural conjecture that a surjective linear map on $\mathcal{L}(\mathcal{H})$ preserves either the minimum or the surjectivity or the reduced minimum moduli is an isometry.

Mbekhta's articles [**182**] and [**185**] were followed quickly by several papers treating related problems and his results have been independently extended by several authors to different settings; see for instance [**37–39, 41, 181, 218–220**] and the references therein. Even more, they were extended to a more general setting by characterizing (not necessarily unital) surjective linear maps between C^*-algebras preserving the minimum, surjectivity, maximum, and reduced minimum moduli and his conjecture was positively settled; see [**39**].

6.1. Minimum and surjectivity moduli preservers.

The *minimum modulus* of an operator $T \in \mathcal{L}(X)$ is defined by

$$\mathrm{m}(T) := \inf\{\, \|Tx\| : x \in X,\ \|x\| = 1\,\},$$

and is positive precisely when T is injective and has a closed range. The *surjectivity modulus* of T is given by

$$\mathrm{q}(T) := \sup\{\, \varepsilon \geq 0 : \varepsilon B_X \subseteq T(B_X)\,\},$$

and is positive if and only if T is surjective. While, the *maximum modulus* of T is defined by

$$\mathrm{M}(T) := \max\{\mathrm{m}(T), \mathrm{q}(T)\},$$

and is positive when either T is injective with a closed range or T is surjective. Note that $\mathrm{M}(T) = \mathrm{m}(T) = \mathrm{q}(T) = \|T^{-1}\|^{-1}$ whenever T is invertible and that

$$\mathrm{m}(T) = \inf\{\, \|TS\| : s \in \mathcal{L}(X),\ \|S\| = 1\,\} \tag{6.14}$$

and

$$\mathrm{q}(T) = \inf\{\, \|ST\| : S \in \mathcal{L}(X),\ \|S\| = 1\,\} \tag{6.15}$$

for all operators $T \in \mathcal{L}(X)$; see [**192**]. We also note that

$$|\mathrm{m}(T) - \mathrm{m}(S)| \leq \|T - S\|,$$

and

$$|\mathrm{q}(T) - \mathrm{q}(S)| \leq \|T - S\|$$

for all operators $T,\ S \in \mathcal{L}(X)$. This tells us that the spectral function m, q and M are continuous and implies, in particular, that the sets

$$\{T \in \mathcal{L}(X) : T \text{ is not bounded below}\} = \mathrm{m}^{-1}(\{0\})$$

and

$$\{T \in \mathcal{L}(X) : T \text{ is not onto}\} = \mathrm{q}^{-1}(\{0\})$$

are closed in $\mathcal{L}(X)$.

The following result, which was proved independently by Bourhim and Mbekhta–Oudghiri, describes surjective additive maps from $\mathcal{L}(X)$ onto $\mathcal{L}(Y)$ preserving the minimum and the surjectivity moduli of Banach space operators. Recall that a map φ from $\mathcal{L}(X)$ into $\mathcal{L}(Y)$ preserves a spectral quantity c if

$$\mathrm{c}(\varphi(T)) = \mathrm{c}(T), \qquad T \in \mathcal{L}(X).$$

THEOREM 6.1 (Bourhim [**37**], Mbekhta–Oudghiri [**181**]). *If $\varphi : \mathcal{L}(X) \to \mathcal{L}(Y)$ is an additive surjective map preserving either the minimum modulus or the surjectivity modulus, then either there are bijective isometries $U : X \to Y$ and $V : Y \to X$ both linear or both conjugate linear such that*

$$\varphi(T) = UTV, \qquad T \in \mathcal{L}(X),$$

or there are bijective isometries $U : X^* \to Y$ and $V : Y \to X^*$ both linear or both conjugate linear such that

$$\varphi(T) = UT^*V, \qquad T \in \mathcal{L}(X).$$

From the definitions of the minimum and surjectivity moduli, these quantities are always preserved by maps of the form $\varphi(T) = UTV$, $(T \in \mathcal{L}(X))$, where $U : X \to Y$ and $V : Y \to X$ are both linear or both conjugate linear bijective isometries. While if φ preserves the minimum modulus (resp. the surjectivity modulus), then the second conclusion of the previous theorem can not occur if any one of X and Y is not reflexive or if there is a non invertible surjective (resp. non invertible bounded below) operator in $\mathcal{L}(X)$. We also mention that for the infinite-dimensional separable complex Banach spaces X constructed in [**101**] and [**165**], the spectrum $\sigma(T)$ of any operator $T \in \mathcal{L}(X)$ is countable. Therefore, $\sigma(T) = \sigma_{ap}(T) = \sigma_{su}(T)$ for all $T \in \mathcal{L}(X)$, and every surjective or bounded below linear operator in $\mathcal{L}(X)$ is invertible. Here, $\sigma_{su}(T) := \{\lambda \in \mathbb{C} : \mathrm{q}(T - \lambda) = 0\}$ and $\sigma_{ap}(T) := \{\lambda \in \mathbb{C} : \mathrm{m}(T - \lambda) = 0\}$ are the surjectivity spectrum and the approximate point spectrum, respectively, of T.

The next result characterizes surjective additive maps from $\mathcal{L}(X)$ onto $\mathcal{L}(Y)$ preserving the maximum modulus.

THEOREM 6.2 (Bourhim [**37**]). *An additive surjective map $\varphi : \mathcal{L}(X) \to \mathcal{L}(Y)$ preserves the maximum modulus if and only if either there are bijective isometries $U : X \to Y$ and $V : Y \to X$ both linear or both conjugate linear such that*

$$\varphi(T) = UTV, \qquad T \in \mathcal{L}(X),$$

or there are bijective isometries $U : X^* \to Y$ and $V : Y \to X^*$ both linear or both conjugate linear such that

$$\varphi(T) = UT^*V, \qquad T \in \mathcal{L}(X).$$

The second case can not occur if any one of X and Y is not reflexive.

6.2. Reduced minimum modulus preservers. The *reduced minimum modulus* of an operator $T \in \mathcal{L}(X)$ is defined by

$$\gamma(T) := \begin{cases} \inf\{\|Tx\| : \operatorname{dist}(x, \operatorname{Ker}(T)) \geq 1\} & \text{if } T \neq 0 \\ \infty & \text{if } T = 0. \end{cases}$$

The reduced minimum modulus measures the closedness of the range of operators in the sense that $\gamma(T) > 0$ precisely when T has a closed range; see for instance [**192**, II.10]. Note that $\mathrm{M}(T) \leq \gamma(T)$ and that $\mathrm{M}(T) = \mathrm{M}(T^*) = \gamma(T) = \gamma(T^*)$ provided that $\mathrm{M}(T) > 0$. Moreover, if T is invertible, then

$$\gamma(T) = \mathrm{m}(T) = \mathrm{q}(T) = \mathrm{M}(T) = \|T^{-1}\|^{-1}. \tag{6.16}$$

In [**219**, Theorem 4.2], Skhiri generalized Mbekhta's result by characterizing surjective linear maps on $\mathcal{L}(X)$ preserving the reduced minimum modulus but when $\varphi(\mathbf{1})$ is invertible. His result has been also proved in [**39**, **220**] for the Hilbert space operators case but without the extra condition that $\varphi(\mathbf{1})$ is invertible. In fact, as shall be discussed below, much more has been established in [**39**] where it is shown that a surjective linear map φ between C^*-algebras preserves the reduced minimum modulus if and only if it is a selfadjoint Jordan isomorphism multiplied by a unitary element. This result clearly shows that the condition that $\varphi(\mathbf{1})$ is invertible in the above theorem is superfluous even for the more general setting of the reduced minimum modulus preservers between C^*-algebras.

The following result describes additive surjective maps preserving the reduced minimum modulus of Banach space operators and shows that the condition that $\varphi(\mathbf{1})$ is invertible in [**219**, Theorem 4.2] is superfluous.

THEOREM 6.3 (Bourhim [**37**]). *An additive surjective map $\varphi : \mathcal{L}(X) \to \mathcal{L}(Y)$ preserves the reduced minimum modulus if and only if either there are bijective isometries $U : X \to Y$ and $V : Y \to X$ both linear or both conjugate linear such that*

$$\varphi(T) = UTV, \qquad T \in \mathcal{L}(X),$$

or there are bijective isometries $U : X^* \to Y$ and $V : Y \to X^*$ both linear or both conjugate linear such that

$$\varphi(T) = UT^*V, \qquad T \in \mathcal{L}(X).$$

The second case can not occur if any one of X and Y is not reflexive.

The proof of this theorem depends on some arguments quoted from [**39**, Proof of Theorem 7.2] and the following two results.

THEOREM 6.4 (Bourhim [**37**]). *For an additive surjective map φ on $\mathcal{L}(X)$, there are positive constants m and M such that*

$$m\gamma(T) \leq \gamma(\varphi(T)) \leq M\gamma(T), \qquad T \in \mathcal{L}(X) \tag{6.17}$$

if and only if either there are bijective continuous mappings $A : X \to X$ and $B : X \to X$ both linear or both conjugate linear such that

$$\varphi(T) = ATB, \qquad T \in \mathcal{L}(X),$$

or there are bijective continuous mappings $A : X^* \to X$ and $B : X \to X^*$ both linear or both conjugate linear such that

$$\varphi(T) = AT^*B, \qquad T \in \mathcal{L}(X).$$

The last case may occur only if X is reflexive.

The "if part" of this theorem is obvious but for the "only if part", just as in [**39**], it is shown that if φ satisfies (6.17) then for every operator $T \in \mathcal{L}(X)$, one has $\mathrm{M}(T) = 0$ if and only if $\mathrm{M}(\varphi(T)) = 0$. Thus Theorem 6.7 applies and the desired conclusion follows.

The proof of Theorem 6.3 uses also the following lemma quoted from [**219**, Theorem 3.1 and Corollary 3.2]. The proofs presented therein are long and require several computations and applications of Hahn-Banach Theorem. However, a simple and shorter proof was given in [**37**].

LEMMA 6.5 (Skhiri [**219**]). *For a bijective mapping $A \in \mathcal{L}(X, Y)$, the following statements are equivalent.*

(i) $\|ATA^{-1}\| = \|T\|$ for all invertible operators $T \in \mathcal{L}(X)$.
(ii) $\|ATA^{-1}\| \leq \|T\|$ for all invertible operators $T \in \mathcal{L}(X)$.
(iii) $\|ATA^{-1}\| \geq \|T\|$ for all invertible operators $T \in \mathcal{L}(X)$.
(iv) A is an isometry multiplied by a scalar.

6.3. The surjectivity and inner spectral radii preservers. For an operator $T \in \mathcal{L}(X)$, we have

$$\mathrm{m}(T^m)\mathrm{m}(T^n) \leq \mathrm{m}(T^{m+n})$$

and

$$\mathrm{q}(T^m)\mathrm{q}(T^n) \leq \mathrm{q}(T^{m+n})$$

for all nonnegative integers m, n. Thus, both sequences $(\mathrm{m}(T^n)^{16/n})_{n\geq 1}$ and $(\mathrm{q}(T^n)^{1/n})_{n\geq 1}$ converge and their limits satisfy

$$\mathrm{r}_1(T) := \lim_{n\to\infty} \mathrm{m}(T^n)^{1/n} = \sup\{\mathrm{m}(T^n)^{1/n} : n \geq 1\} = \min\{|\lambda| : \lambda \in \sigma_{ap}(T)\},$$

and

$$\delta(T) := \lim_{n\to\infty} \mathrm{q}(T^n)^{1/n} = \sup\{\mathrm{q}(T^n)^{1/n} : n \geq 1\} = \min\{|\lambda| : \lambda \in \sigma_{su}(T)\},$$

where $\sigma_{ap}(T) := \{\lambda \in \mathbb{C} : \mathrm{m}(T - \lambda) = 0\}$ is the *approximate point spectrum* of T and $\sigma_{su}(T) := \{\lambda \in \mathbb{C} : \mathrm{q}(T - \lambda) = 0\}$ is the *surjectivity spectrum* of T; see [**168**]. These radii $\mathrm{r}_1(T)$ and $\delta(T)$ are called respectively the *inner spectral radius* and the *surjectivity radius* of T. Note that $\mathrm{r}_1(T) = \delta(T) = \mathrm{r}(T^{-1})^{-1}$ whenever T is invertible.

The following result describes surjective additive maps preserving the surjectivity and inner spectral radii.

LEMMA 6.6 (Bourhim [**40**]). *Assume that* c *stands for any one of the spectral functions δ and r_1, and let $\varphi : \mathcal{L}(X) \to \mathcal{L}(Y)$ be an additive surjective map. Then there are m, $M > 0$ such that*

$$m\mathrm{c}(T) \leq \mathrm{c}(\varphi(T)) \leq M\mathrm{c}(T), \qquad T \in \mathcal{L}(X), \tag{6.18}$$

if and only if there is a nonzero scalar α and either there is a continuous linear or conjugate linear bijection $A : X \to Y$ such that

$$\varphi(T) = \alpha ATA^{-1}, \qquad T \in \mathcal{L}(X),$$

or there is a continuous linear or conjugate linear bijection $A : X^* \to Y$ such that

$$\varphi(T) = \alpha AT^*A^{-1}, \qquad T \in \mathcal{L}(X).$$

The last case can not occur if X or Y is not reflexive, or if there is a non invertible surjective operator in $\mathcal{L}(X)$.

From the definition of the surjectivity and inner spectral radii, we immediately note that if $T \in \mathcal{L}(X)$, then

$$\mathrm{r}_1(T) = 0 \Longleftrightarrow \mathrm{m}(T) = 0 \qquad \text{and} \qquad \delta(T) = 0 \Longleftrightarrow \mathrm{q}(T) = 0.$$

So, to prove the above lemma, we used the following result which describes surjective additive maps preserving the zeros of the minimum, surjectivity and and maximum moduli. We say that the mapping $\varphi : \mathcal{L}(X) \to \mathcal{L}(Y)$ *preserves the zeros* of the spectral quantity c if, for each $T \in \mathcal{L}(X)$, we have

$$\mathrm{c}(T) = 0 \iff \mathrm{c}\big(\varphi(T)\big) = 0.$$

LEMMA 6.7 (Bourhim [**37**]). *Assume that* c *stands for any one of the spectral quantities* m, q *or* M. *If* $\varphi : \mathcal{L}(X) \to \mathcal{L}(Y)$ *is an additive surjective map preserving the zeros of* c, *then either there are bijective continuous mappings* $A : X \to Y$ *and* $B : Y \to X$ *both linear or both conjugate linear such that*

$$\varphi(T) = ATB, \qquad T \in \mathcal{L}(X),$$

or there are bijective continuous mappings $A : X^* \to Y$ *and* $B : Y \to X^*$ *both linear or both conjugate linear such that*

$$\varphi(T) = AT^*B, \qquad T \in \mathcal{L}(X).$$

This case may occur only if X *and* Y *are reflexive.*

This lemma and its proof were sitting in [**125**, Theorem 3.1 and its proof] and needed only a simple step to be discovered therein. In [**125**], Hou and Cui characterized unital additive surjective maps on standard operator algebras preserving, in both directions, different concepts including the boundedness from below and surjectivity of operators. In the proof of their result, they showed that if, however, a such map φ is not necessarily unital, then φ is injective and preserves rank-one operators in both directions. This together with Theorem 4.12 gives the possible forms of φ when restricted on $\mathcal{F}(X)$. As observed by several authors such possible forms of the restriction allow one to show easily that $\varphi(\mathbf{1})$ is invertible, and the global forms of φ can be deduced from the above mentioned paper [**125**].

6.4. Minimum, surjectivity and reduced minimum moduli preservers in C^*-algebras. The identities (6.14) and (6.15) opened the way for several authors to generalize and translate the definition and properties of the minimum, surgectivity and reduced minimum moduli to the setting of Banach algebras. For an element a of a Banach algebra $\mathcal{A}$, the *minimum modulus* and the *surjectivity modulus* are defined respectively by

$$\mathrm{m}(a) := \mathrm{m}(L_a) = \inf\{\|ax\| : x \in \mathcal{A},\ \|x\| = 1\},$$

and

$$\mathrm{q}(a) := \mathrm{m}(R_a) = \inf\{\|xa\| : x \in \mathcal{A},\ \|x\| = 1\},$$

where L_a and R_a are the left and right multiplication operators by a. The *maximum modulus* of a is defined by

$$\mathrm{M}(a) := \max\{\mathrm{m}(a), \mathrm{q}(a)\}.$$

Obviously, $\mathrm{m}(a) = 0$ if and only if a is a left topological divisor of zero. At the same token, $\mathrm{q}(a) = 0$ if and only if a is a right topological divisor of zero. Therefore, $\mathrm{M}(a) = 0$ if and only if a is a topological divisor of zero.

The *reduced minimum modulus* (or the *conorm*) of an element a in $\mathcal{A}$ is

$$\gamma(a) := \gamma(L_a) = \begin{cases} \inf\{\|ax\| : \operatorname{dist}(x, \operatorname{Ker}(L_a)) \geq 1\} & \text{if } a \neq 0 \\ \infty & \text{if } a = 0. \end{cases}$$

Clearly, $\gamma(a) > 0$ if and only the left ideal $a\mathcal{A}$ generated by a is closed in $\mathcal{A}$. For further discussion on the reduced minimum modulus of elements in a Banach algebra, we refer the reader to [**106**, **107**].

Assume that $\mathcal{H}$ is an infinite dimensional complex Hilbert space, and let T^{tr} denote the transpose of all operators $T \in \mathcal{L}(\mathcal{H})$ with respect to an arbitrary but a fixed orthonormal basis in $\mathcal{H}$. Mbekhta's results, mentioned previously, assert that every unital surjective linear map $\varphi : \mathcal{L}(\mathcal{H}) \to \mathcal{L}(\mathcal{H})$ which preserves any of above the spectral quantities is an isometry. It is worth recalling that a surjective linear isometry φ on $\mathcal{L}(\mathcal{H})$ is either

$$\varphi(T) = UTV, \qquad T \in \mathcal{L}(\mathcal{H}),$$

or

$$\varphi(T) = UT^{tr}V, \qquad T \in \mathcal{L}(\mathcal{H}),$$

for some unitary operators U, $V \in \mathcal{L}(\mathcal{H})$. More generally, Kadison, in his celebrated paper [**138**], proved that a surjective linear map between two C^*-algebras $\mathcal{A}$ and $\mathcal{B}$ is an isometry if and only it is a Jordan $*$-isomorphism multiplied by a unitary element in $\mathcal{B}$. Recall that a linear map φ between two C^*-algebras $\mathcal{A}$ and $\mathcal{B}$ is said to be *Jordan $*$-isomorphism* if it is a Jordan isomorphism and

$$\varphi(a^*) = \varphi(a)^*, \qquad a \in \mathcal{A}.$$

Based on the Mbekhta's results and conjecture, the following arises in a natural way.

Conjecture 6.8. Let $\mathcal{A}$ and $\mathcal{B}$ be two C^*-algebras, and let φ be a linear map from $\mathcal{A}$ onto $\mathcal{B}$. If φ preserves any of the above defined spectral quantities, then φ is an isometry.

The following results give a positive solution to this conjecture. The first one describes unital linear maps between C^*-algebras increasing any of the spectral quantities defined above.

THEOREM 6.9 (Bourhim–Burgos–Shulman [**39**]). *Assume that* c *denotes any of the spectral quantities* m, q, M *or* γ. *Let* $\mathcal{A}$ *and* $\mathcal{B}$ *be two* C^**-algebras and let* $\varphi : \mathcal{A} \to \mathcal{B}$ *be a unital linear map (not assumed to be surjective) such that*

$$\mathrm{c}(x) \leq \mathrm{c}(\varphi(x)), \qquad x \in \mathcal{A}.$$

Then φ *is a Jordan* $*$*-homomorphism. Moreover, if* φ *is injective, then* φ *is isometric.*

The following result describes surjective linear maps between C^*-algebras preserving any of the spectral quantities defined above.

THEOREM 6.10 (Bourhim–Burgos–Shulman [**39**]). *Assume that* d *denotes any of the spectral quantities* m, q, M *or* γ. *Let* $\mathcal{A}$ *be a semisimple Banach algebra and let* $\mathcal{B}$ *be a* C^**-algebra. If* $\varphi : \mathcal{A} \to \mathcal{B}$ *is a surjective linear map for which*

$$\mathrm{c}(\varphi(x)) = \mathrm{c}(x), \qquad x \in \mathcal{A},$$

then $\mathcal{A}$ *(for its norm and some involution) is a* C^**-algebra, and* φ *is an isometric Jordan* $*$*-isomorphism multiplied by a unitary element of* $\mathcal{B}$.

The key tools for proof of the above results is the following lemma which characterizes hermitian elements in a Banach algebra in terms of the minimum, surjectivity, maximum and reduced minimum moduli. Recall that an element h of $\mathcal{A}$ is called *hermitian* if it has real numerical range, that is, if $f(h) \in \mathbb{R}$, for all f in the Banach dual space of $\mathcal{A}$ with $f(\mathbf{1}) = 1 = \|f\|$. We shall denote by $H(\mathcal{A})$ the closed real subspace of $\mathcal{A}$ of all hermitian elements of $\mathcal{A}$. It is well known that an element in a C^*-algebra is hermitian if and only if it is selfadjoint. By the Vidav-Palmer theorem, a unital Banach algebra $\mathcal{A}$ is a C^*-algebra if and only if it is generated by its hermitian elements, that is, if $\mathcal{A} = H(\mathcal{A}) + iH(\mathcal{A})$. In this case, the involution is given by

$$(h + ik)^* := h - ik \qquad h, k \in H(\mathcal{A}). \tag{6.19}$$

See [**29**, § 5, § 6].

LEMMA 6.11 (Bourhim–Burgos–Shulman [**39**]). *Assume that* c *denotes any of the spectral quantities* m, q, M *or* γ, *and let* a *be an element of the Banach algebra* $\mathcal{A}$. *Then the following assertions are equivalent.*

(i) a *is hermitian.*
(ii) $\|\mathbf{1} + ita\| = 1 + o(t)$ *as* $t \to 0$.
(iii) $\mathrm{c}(\mathbf{1} + ita) = 1 + o(t)$ *as* $t \to 0$.
(iv) $\mathrm{c}(\mathbf{1} + ita) \geq 1 + o(t)$ *as* $t \to 0$.

We conclude this section by the description of inner automorphisms on C^*-algebras preserving the minimum, surjectivity, maximum or reduced minimum modulus.

THEOREM 6.12 (Bourhim–Burgos [**38, 41**]). *Assume that* c *denotes any of the spectral quantities* m, q, M *or* γ *and let* a *and* b *be invertible elements in a* C^*-*algebra* $\mathcal{A}$. *Then the following statements are equivalent.*

(i) ab *is a unitary element, and* $|a|$ *is central in* $\mathcal{A}$.
(ii) $\mathrm{c}(x) = \mathrm{c}(axb)$ *for all* $x \in \mathcal{A}$.

Some variant results of the above theorems can be found in [**39**] where the authors described surjective linear maps between C^*-algebras increasing or decreasing the above defined spectral quantities.

7. Local spectra preservers

In this Section, we start with an introduction in Section 7.1 and collect what seems to be known about local spectra of Banach space bounded linear operators. In Section 7.2, we discuss linear maps preserving local spectrum of Banach space operators at a fixed nonzero vector. Such a section also contains the precise statements and proofs about linear maps compressing or expanding the local spectrum. In Section 7.3 results about linear map preserving, increasing or decreasing the inner local spectral radius operators at a fixed nonzero vector. While, in Section 7.4, we focus on linear maps preserving, increasing or decreasing the outer local spectral radius operators at a fixed nonzero vector. We, in particular, describe linear surjective maps on $\mathcal{L}(\mathcal{H})$ that are locally spectrally bounded. In Section 7.5, we give a complete description of surjective maps φ (not assumed to be linear or continuous) on $\mathcal{L}(X)$ preserving the local spectrum of the product or triple product of operators at a nonzero fixed element of X, and present some results on nonlinear local spectra preservers of matrices. In particular, a local version of Theorem 4.8 that describes surjective maps on matrices preserving the local spectral radius distance. In Section 7.6, we present some results on linear and nonlinear local spectra preservers at nonfixed vectors.

7.1. Background from local spectral theory. For every open subset U of $\mathbb{C}$, we let $\mathcal{O}(U,X)$ denote the space of analytic X-valued functions defined on U. It is a Fréchet space when endowed with the topology of uniform convergence on compact subsets of U. If $T \in \mathcal{L}(X)$ is an operator and U, an open subset of $\mathbb{C}$, we define on $\mathcal{O}(U,X)$ the mapping T_U by

$$(T_U f)(\lambda) := (T-\lambda)f(\lambda), \qquad f \in \mathcal{O}(U,X),\ \lambda \in U.$$

The local resolvent set, $\rho_T(x)$, of an operator $T \in \mathcal{L}(X)$ at a point $x \in X$ is the union of all open subsets U of $\mathbb{C}$ for which there is an analytic function $\varphi : U \to X$ such that $(T-\lambda)\varphi(\lambda) = x$, $(\lambda \in U)$. When regarding x as a constant function, $\rho_T(x)$ is nothing but the union of all open subsets U of $\mathbb{C}$ for which x in $\mathcal{R}(T_U)$, the range of T_U. The local spectrum of T at x is defined by $\sigma_T(x) := \mathbb{C}\backslash\rho_T(x)$, and is obviously a closed subset of $\sigma(T)$, the spectrum of T.

The operator $T \in \mathcal{L}(X)$ is said to have the *single-valued extension property* (SVEP) provided that for every open subset U of $\mathbb{C}$, the equation

$$(T-\lambda)\varphi(\lambda) = 0, \qquad \lambda \in U,$$

has no nontrivial analytic solution φ. Equivalently, T has SVEP precisely when T_U is injective for all open subsets U of $\mathbb{C}$. Every operator $T \in \mathcal{L}(X)$ for which the interior of its point spectrum, $\sigma_p(T)$, is empty enjoys this property. In particular, every compact operator enjoys this property. We also know that if x is a nonzero vector, then $\sigma_T(x)$ is not empty provided that T has the SVEP.

For a positive scalar r, let $\overline{\mathrm{D}}(0,r)$ (resp. $\mathrm{D}(0,r)$) denotes the closed (resp. the open) disc centered at the origin with radius r, and for a closed subset F of $\mathbb{C}$ and an operator $T \in \mathcal{L}(X)$, the subspace

$$\mathcal{X}_T(F) := \{x \in X : (T-\lambda)f(\lambda) = x \text{ has an analytic solution } f \text{ on } \mathbb{C}\backslash F\}$$

is the so-called *glocal spectral subspace* of T, and is, in fact, nothing but $X \cap \mathcal{R}(T_{\mathbb{C}\backslash F})$ where $\mathcal{R}(T_{\mathbb{C}\backslash F})$ is the range of the mapping $T_{\mathbb{C}\backslash F}$. It is a hyperinvariant subspace of T but not necessarily closed. Recall also that The *local spectral radius* of T at $x \in X$ is defined by

$$\mathrm{r}_T(x) := \limsup_{n\to+\infty} \|T^n x\|^{\frac{1}{n}},$$

and coincides with

$$\mathrm{r}_T(x) = \inf\{r \geq 0 : x \in \mathcal{X}_T(\overline{\mathrm{D}}(0,r))\}.$$

See [**148**, Proposition 3.3.13]. Analogously, the *inner local spectral radius* of T at x is defined by

$$\iota_T(x) := \sup\{r \geq 0 : x \in \mathcal{X}_T(\mathbb{C}\backslash\mathrm{D}(0,r))\}.$$

See [**187**]. These concepts are somehow not entirely optimal because the maximum and minimum moduli of the local spectrum of T at x could justifiably also be called the *local spectral radius* and the *inner spectral radius* of T, respectively. These are denoted by

$$\Gamma_T(x) := \max\{|\lambda| : \lambda \in \sigma_T(x)\},$$

and

$$\gamma_T(x) := \min\{|\lambda| : \lambda \in \sigma_T(x)\},$$

with the convention that $\Gamma_T(x) = -\infty$ and $\gamma_T(x) = \infty$ precisely when $\sigma_T(x) = \emptyset$. Note that

$$\iota_T(x) \leq \gamma_T(x) \leq \Gamma_T(x) \leq \mathrm{r}_T(x)$$

when $\sigma_T(x) \neq \emptyset$. These inequalities can be strict but if T has SVEP, then $\gamma_T(x) = \iota_T(x)$ and $\Gamma_T(x) = \mathrm{r}_T(x)$.

The local spectra of matrices is well understood; see for instance [**42**, **225**]. Let $\lambda_1, \ldots, \lambda_r$ be the distinct eigenvalues of a matrix $T \in M_n(\mathbb{C})$ and denote by $E_1, \ldots, E_r$ the corresponding root spaces and by T_i the restriction of T over E_i. It follows that $\mathbb{C}^n = E_1 \oplus E_2 \oplus \cdots \oplus E_r$, and $T = T_1 \oplus T_2 \oplus \cdots \oplus T_r$, and thus for every $x \in \mathbb{C}^n$, we have

$$\sigma_T(x) = \bigcup_{1 \leq i \leq r} \sigma_{T_i}(P_i x) = \bigcup_{1 \leq i \leq r} \{\lambda_i : P_i(x) \neq 0\},$$

where $P_i : \mathbb{C}^n \to E_i$ is the canonical projection. Therefore,

$$\Gamma_T(x) = \mathrm{r}_T(x) = \max\{|\lambda_i| : 1 \leq i \leq r \text{ with } P_i(x) \neq 0\},$$

and

$$\gamma_T(x) = \iota_T(x) = \min\{|\lambda_i| : 1 \leq i \leq r \text{ with } P_i(x) \neq 0\}.$$

The remarkable books by Aiena [**1**], and by Laursen and Neumann [**148**] provide an excellent exposition as well as a rich bibliography of the local spectral theory.

We now turn our attention to some elementary lemmas that summarize some properties of the above local spectral quantities. The first lemma summarizes some basic properties of the local spectrum.

LEMMA 7.1. *For an operator $T \in \mathcal{L}(X)$, vectors $x, y \in X$ and a nonzero scalar $\alpha \in \mathbb{C}$, the following statements hold.*

(i) If T has SVEP, then $\sigma_T(x) \neq \emptyset$ provided that $x \neq 0$.
(ii) $\sigma_T(\alpha x) = \sigma_T(x)$ if $\alpha \neq 0$, and $\sigma_{\alpha T}(x) = \alpha \sigma_T(x)$.
(iii) $\sigma_T(x+y) \subset \sigma_T(x) \cup \sigma_T(y)$. The equality holds if $\sigma_T(x) \cap \sigma_T(y) = \emptyset$.
(iv) If T has SVEP, $x \neq 0$ and $Tx = \lambda x$ for some $\lambda \in \mathbb{C}$, then $\sigma_T(x) = \{\lambda\}$.
(v) If T has SVEP and $Tx = \alpha y$, then $\sigma_T(y) \subset \sigma_T(x) \subset \sigma_T(y) \cup \{0\}$.
(vi) If $R \in \mathcal{L}(X)$ commutes with T, then $\sigma_T(Rx) \subset \sigma_T(x)$.
(vii) $\sigma_{T^n}(x) = \{\sigma_T(x)\}^n$ for all $x \in X$ and $n \geq 1$.

PROOF. See for instance [**1**, **148**]. □

The second lemma gives the connection between the local spectra and the surjectivity spectrum. Racall that for an operator $T \in \mathcal{L}(X)$, the *surjectivity spectrum* is defined by

$$\sigma_{su}(T) := \{\lambda \in \mathbb{C} : T - \lambda \text{ is not onto}\}.$$

It is a nonempty compact subset of $\sigma(T)$ and contains $\partial\sigma(T)$, the boundary of the spectrum of T.

LEMMA 7.2 (Laursen–Neumann [**148**]). *For any operator $T \in \mathcal{L}(X)$, the following statements hold.*

(i) $\sigma_{su}(T) = \cup_{x \in X} \sigma_T(x)$.
(ii) $\mathrm{r}(T) = \sup\{\mathrm{r}_T(x) : x \in X\}$.

(iii) The set $\{x \in X : \sigma_{su}(T) = \sigma_T(x)\}$ is of second category in X.

The last lemma shows, in particular, that no anti-isomorphism preserves local spectra at a fixed vector x_0 in X and that an isomorphism (up to a multiple factor of modulus one) on $\mathcal{L}(X)$ preserves a local spectra at a fixed vector x_0 in X if and only if x_0 is an eigenvector of the intertwining operator.

LEMMA 7.3. *Let $x_0 \in X$ and $y_0 \in Y$ be fixed nonzero elements, let $T \in \mathcal{L}(X)$, and let $\mathrm{c}_T(x_0)$ stands for any of the quantities $\sigma_T(x_0)$, $\mathrm{r}_T(x_0)$, $\iota_T(x_0)$, $\Gamma_T(x_0)$ or $\gamma_T(x_0)$. For bijective maps $A \in \mathcal{L}(X,Y)$ and $B \in \mathcal{B}(X^*,Y)$, the following statements hold.*

(i) $\mathrm{c}_{ATA^{-1}}(Ax_0) = \mathrm{c}_T(x_0)$ for all $T \in \mathcal{L}(X)$.
(ii) If Ax_0 and y_0 are linearly independent, then there are T_1, $T_2 \in \mathcal{L}(X)$ such that $\mathrm{c}_{AT_1A^{-1}}(y_0) = \mathrm{c}_{T_2}(x_0) = 0$ and $\mathrm{c}_{AT_2A^{-1}}(y_0) = \mathrm{c}_{T_1}(x_0) = 1$.
(iii) There is $T \in \mathcal{L}(X)$ such that $\mathrm{c}_{BT^*B^{-1}}(y_0) = 0$ and $\mathrm{c}_T(x_0) = 1$, and vice versa.

PROOF. See for instance the proofs of [**42**, Lemmas 2.1 and 2.2] and [**128**, Lemmas 4 and 5]. □

7.2. Linear preservers of local spectrum. The problem of characterizing linear or additive maps on $\mathcal{L}(X)$ preserving local spectra was initiated by Bourhim and Ransford in [**43**] and continued by a number of authors; see for instance [**40**, **42**, **44**, **72**, **76**, **128**] and the references therein. In [**128**], González and Mbekhta showed that a linear map φ on $M_n(\mathbb{C})$ preserves the local spectrum at a nonzero fixed vector $x_0 \in \mathbb{C}^n$, i.e.,

$$\sigma_{\varphi(T)}(x_0) = \sigma_T(x_0), \qquad T \in M_n(\mathbb{C}),$$

if and only if there is an invertible matrix $A \in M_n(\mathbb{C})$ such that $Ax_0 = x_0$ and $\varphi(T) = ATA^{-1}$ for all $T \in M_n(\mathbb{C})$. In fact, a little bit more could be obtained as an immediate consequence of Theorem 4.7.

COROLLARY 7.4. *Let x_0 be a fixed nonzero vector of $\mathbb{C}^n$. A linear map φ on $M_n(\mathbb{C})$ satisfies*

$$\sigma_{\varphi(T)}(x_0) \cap \sigma_T(x_0) \neq \emptyset, \qquad T \in M_n(\mathbb{C}),$$

if and only if there is an invertible matrix $A \in M_n(\mathbb{C})$ such that $Ax_0 = x_0$ and

$$\varphi(T) = ATA^{-1}, \qquad T \in M_n(\mathbb{C}).$$

In [**44**], Bračič and Müller extended the main result of the aforementioned paper to infinite dimensional Banach spaces; see also [**40**]. Their proof uses a powerful lemma given below, and a density argument to show that if a surjective linear and continuous mapping φ on $\mathcal{L}(X)$ preserves the local spectrum at a fixed nonzero vector x_0 of X, then φ is a spectral isometry and thus it has one of the standard forms, by Theorem 4.29. As no antiautomorphism preserves local spectra at x_0, they deduced that φ is an automorphism and x_0 is an eigenvector of the intertwining operator.

LEMMA 7.5 (Bračič–Müller [**44**]). *Let $T \in \mathcal{L}(X)$, and let $x_0 \in X$ be a nonzero vector. Then, for each $\lambda \in \sigma_{su}(T)$ and every $\epsilon > 0$, there is $S \in \mathcal{L}(X)$ such that $\|T - S\| < \epsilon$ and $\lambda \in \sigma_S(x_0)$.*

This lemma is a powerful and demonstrated to be the backbone of the proofs of several results about continuous linear local spectra preservers. It shows that in the presence of the continuity, a linear local spectra preserving problem can be transferred to a global spectra preserving problem so that standard techniques and known results can be used. As an application of this lemma, we obtain a local version of Theorem 4.34 that gives a local spectral characterization of nilpotent Banach space operators.

LEMMA 7.6. *Let x_0 be a nonzero fixed vector of X, and let $T \in \mathcal{L}(X)$ be a bounded linear operator. For an integer $n \geq 1$, the following statements are equivalent.*

(i) $T^n = 0$.

(ii) For every bounded neighbourhood $U \subset \mathcal{L}(X)$ of zero, there is a constant $C_U > 0$ such that

$$\Gamma_{T+S}(x_0) \leq C_U \|S\|^{\frac{1}{n}}, \qquad S \in U.$$

(iii) For some bounded neighbourhood of zero $U \subset \mathcal{L}(X)$ of zero, there is a constant $C > 0$ such that

$$\Gamma_{T+S}(x_0) \leq C\|S\|^{\frac{1}{n}}, \qquad S \in U.$$

PROOF. By Theorem 4.34 and the fact that $\Gamma_T(x_0) \leq \mathrm{r}(T)$ for all $T \in \mathcal{L}(X)$, we note that the implication $(i) \Rightarrow (ii)$ always holds. Since the implication $(ii) \Rightarrow (iii)$ obviously holds, we only need to show that $(iii) \Rightarrow (i)$.

Assume that the statement (iii) holds, and let $S \in U$ and $\lambda \in \sigma_{su}(T+S)$ be such that $\mathrm{r}(T+S) = |\lambda|$. By Lemma 7.5, there is a sequence of operators $(S_k)_k \subset U$ converging to S such that $\lambda \in \sigma_{T+S_k}(x_0)$ for all k. For every k, we have

$$\mathrm{r}(T+S) = |\lambda| \leq \Gamma_{T+S_k}(x_0)\} \leq C\|S_k\|^{\frac{1}{n}}.$$

Passing to the limit as $k \to \infty$, we get $\mathrm{r}(T+S) \leq C\|S\|^{\frac{1}{n}}$ for all $S \in U$. Applying Theorem 4.34, we deduce that $T^n = 0$ and thus the implication $(iii) \Rightarrow (i)$ is established; as desired. □

Similar argument are given in [**40**] where Bourhim used as well the above lemma and a density argument to show that if a surjective linear and continuous mapping φ on $\mathcal{L}(X)$ preserves the local spectrum at a fixed nonzero vector x_0 of X, then φ preserves the surjectivity modulus so that Lemma 6.6 applies. In fact, both proofs presented in [**40, 44**] show a little bit more than what have been stated therein. The proof given in [**44**] (resp. in [**40**]) gives a complete description of surjective linear continuous maps on $\mathcal{L}(X)$ preserving the maximum modulus (resp. the minimum modulus) of the local spectrum. For details, we refer the reader to [**40, 44**].

THEOREM 7.7 (Bourhim [**40**], Bračič–Müller [**44**]). *Let $x_0 \in X$ and $y_0 \in Y$ be fixed nonzero elements, and φ be a linear continuous map φ from $\mathcal{L}(X)$ onto $\mathcal{L}(Y)$. Then the following statements are equivalent.*

(i) $\Gamma_{\varphi(T)}(y_0) = \Gamma_T(x_0)$ *for all* $T \in \mathcal{L}(X)$.

(ii) $\gamma_{\varphi(T)}(y_0) = \gamma_T(x_0)$ *for all* $T \in \mathcal{L}(X)$.

(iii) There are a bijection $A \in \mathcal{L}(X, Y)$ and a scalar α of modulus one such that $Ax_0 = y_0$ and

$$\varphi(T) = \alpha ATA^{-1}, \qquad T \in \mathcal{L}(X).$$

In [**40**], some results on linear continuous maps compressing or expanding the local spectrum of Hilbert space operators were given. A part of the following result is quoted from [**24**] but the proof given therein contains some errors which may confuse the readers. We therefore present a proof free of errors here.

THEOREM 7.8 (Bendaoud–Sarih [**24**], Bourhim [**40**], Bračič–Müller [**44**]). *Let $x_0 \in X$ and $y_0 \in Y$ be fixed nonzero elements, and φ be a linear continuous map φ from $\mathcal{L}(X)$ onto $\mathcal{L}(Y)$. Then the following statements are equivalent.*

(i) $\sigma_{\varphi(T)}(y_0) = \sigma_T(x_0)$ *for all* $T \in \mathcal{L}(X)$.
(ii) $\sigma_{\varphi(T)}(y_0) \subset \sigma_T(x_0)$ *for all* $T \in \mathcal{L}(X)$.
(iii) $\sigma_T(x_0) \subset \sigma_{\varphi(T)}(y_0)$ *for all* $T \in \mathcal{L}(X)$.
(iv) There exists a bijection $A \in \mathcal{L}(X,Y)$ *such that* $Ax_0 = y_0$ *and* $\varphi(T) = ATA^{-1}$ *for all* $T \in \mathcal{L}(X)$.

PROOF. The equivalence $(i) \iff (iv)$ was established in [**44**]; see also [**40**]. Since the implications $(i) \Rightarrow (ii)$ and $(i) \Rightarrow (iii)$ always hold, we only need to establish the implications $(ii) \Rightarrow (iv)$ and $(iii) \Rightarrow (iv)$.

Assume that $\sigma_{\varphi(T)}(y_0) \subset \sigma_T(x_0)$ for all $T \in \mathcal{L}(X)$, and fix an operator $T \in \mathcal{L}(X)$ and $\lambda \in \sigma_{su}(\varphi(T))$. By Lemma 7.5, there is a sequence $(S_n)_n \subset \mathcal{L}(Y)$ such that $\|S_n - \varphi(T)\| < n^{-1}$ and $\lambda \in \sigma_{S_n}(y_0)$. Applying the open mapping theorem, one can find a sequence $(T_n)_n \subset \mathcal{L}(X)$ converging to T such that $\varphi(T_n) = S_n$ for all n, and thus $\lambda \in \sigma_{T_n}(x_0)$ for all n. Hence, $T_n - \lambda$ is not onto for all n and so is $\lim_n(T_n - \lambda) = T - \lambda$ since the collection of all nonsurjective operators in $\mathcal{L}(X)$ is closed. It follows that $\sigma_{su}(\varphi(T)) \subset \sigma_{su}(T)$ for all $T \in \mathcal{L}(X)$, and thus, by Theorem 4.23, $\varphi(\mathbf{1}) = \mathbf{1}$. Now, to show that φ is injective it suffices, in view of Theorem 4.31, to show that φ does not vanish at an operator of rank one. In fact, we shall show that $\varphi(x_0 \otimes f) \neq 0$ for all linear functionals $f \in X^*$ for which $f(x_0) = 1$. Assume for the sake of contradiction that there is a linear functional $f \in X^*$ such that $f(x_0) = 1$ and $\varphi(x_0 \otimes f) = 0$. We have $(\mathbf{1} + x_0 \otimes f)(x_0) = 2x_0$ and $\varphi(\mathbf{1} + x_0 \otimes f) = \mathbf{1}$, and thus

$$\{1\} = \sigma_{\mathbf{1}}(y_0) = \sigma_{\varphi(\mathbf{1}+x_0\otimes f)}(y_0) \subset \sigma_{\mathbf{1}+x_0\otimes f}(x_0) = \{2\}.$$

This contradiction shows that φ is injective and thus Theorem 4.31 implies that either $\varphi(T) = ATA^{-1}$, $(T \in \mathcal{L}(X))$, for some isomorphism $A \in \mathcal{L}(X,Y)$, or $\varphi(T) = BT^*B^{-1}$, $(T \in \mathcal{L}(X))$, for some isomorphism $B \in \mathcal{B}(X^*,Y)$. Now, by Lemma 7.3, one can show that φ takes only the first form with A can be supposed to satisfy $Ax_0 = y_0$. This establishes the implication $(ii) \Rightarrow (iv)$.

Now, assume that $\sigma_T(x_0) \subset \sigma_{\varphi(T)}(y_0)$ for all $T \in \mathcal{L}(X)$, and let us show that φ is injective. Assume that $\varphi(T_0) = 0$ for some $T_0 \in \mathcal{L}(X)$, and note that $\sigma_{T_0+T}(x_0) \subset \sigma_{\varphi(T_0+T)}(y_0) = \sigma_{\varphi(T)}(y_0)$ for all $T \in \mathcal{L}(X)$. It follows that $\Gamma_{T_0+T}(x_0) \leq \|\varphi\|\|T\|$ for all $T \in \mathcal{L}(X)$ and thus $T_0 = 0$, by Lemma 7.6. Hence, φ is injective and the implication $(ii) \Rightarrow (iv)$ applied to φ^{-1} shows that φ takes the desired form in this case too. This shows that the implication $(iii) \Rightarrow (iv)$ holds and completes the proof. □

Results on surjective additive continuous maps preserving or compressing or expanding the local spectrum can be obtained as trivial generalizations with the same arguments and proofs. We therefore omit the details and wonder if similar results to Theorem 7.7 and Theorem 7.8 remain true when the map φ is linear or additive and, of course, without the continuity condition.

7.3. Inner local spectral radius and preservers. The following result gives a complete description, in particular, of surjective linear continuos maps on $\mathcal{L}(X)$ preserving the inner spectral radius a nonzero fixed point in X.

THEOREM 7.9 (Bourhim [**40**]). *Let $x_0 \in X$ and $y_0 \in Y$ be nonzero elements. A linear continuous map φ from $\mathcal{L}(X)$ onto $\mathcal{L}(Y)$ satisfies*

$$m\iota_T(x_0) \leq \iota_{\varphi(T)}(y_0) \leq M\iota_T(x_0), \ (T \in \mathcal{L}(X)) \tag{7.20}$$

for some positive constants m and M if and only if there is a bijection $A \in \mathcal{L}(X,Y)$ and a nonzero scalar α such that $Ax_0 = y_0$ and $\varphi(T) = \alpha ATA^{-1}$ for all $T \in \mathcal{L}(X)$.

The proof of this theorem uses the fact that

$$\delta(T) \leq \iota_T(x), \qquad T \in \mathcal{L}(X), \qquad x \in X$$

illustrated and clarified in Lemma 7.10, and the lower semi-continuity of the spectral function δ; see [**168**, Proposition 3]. With these facts and Lemma 7.5 in hands, one can show that if a linear continuous map φ from $\mathcal{L}(X)$ onto $\mathcal{L}(Y)$ satisfies (7.20), then φ satisfies (6.18). Thus lemma 6.6 applies and shows that there is a nonzero scalar α and either there is a continuous linear or conjugate linear bijection $A : X \to Y$ such that $\varphi(T) = \alpha ATA^{-1}$, $(T \in \mathcal{L}(X))$, or there is a continuous linear or conjugate linear bijection $A : X^* \to Y$ such that $\varphi(T) = \alpha AT^*A^{-1}$, $(T \in \mathcal{L}(X))$. Finally, Lemma 7.3 tells you that φ takes only the first form with A can be supposed to satisfy $Ax_0 = y_0$. The converse trivially holds.

The following lemma illustrates the connection between the inner local spectral radius and the surjectivity radius.

LEMMA 7.10. *For every operator $T \in \mathcal{L}(X)$, we have*

$$\delta(T) = \min\{\iota_T(x) : x \in X\} = \min\{\gamma_T(x) : x \in X\}.$$

In particular, the set $\{x \in X : \delta(T) = \iota_T(x)\}$ is of second category in X and $\mathcal{L}(X)$.

PROOF. If $\delta(T) = 0$, then there is nothing to prove. If $\delta(T) > 0$, let $D := \{\lambda : |\lambda| < \delta(T)\}$ and note that, by Theorem 5.1 of Leiterer [**150**], the map $T_D : \mathcal{O}(D,X) \to \mathcal{O}(D,X)$ defined by $T_D f(\lambda) := (T-\lambda)f(\lambda), (\lambda \in D)$, is surjective. In particular, for every $x \in X$, there is $f_x \in \mathcal{O}(D,X)$ such that $T_D f_x(\lambda) = (T-\lambda)f_x(\lambda) = x, (\lambda \in D)$, and $x \in \mathcal{X}(\mathbb{C}\backslash D)$. Hence, $\delta(T) \leq \iota_T(x) \leq \gamma_T(x)$ for all $x \in X$.

Conversely, by Lemma 7.10, there exists $x_0 \in X$ such that $\sigma_{su}(T) = \sigma_T(x_0)$ and thus $\delta(T) \leq \iota_T(x_0) \leq \gamma_T(x_0) = \min\{|\lambda| : \lambda \in \sigma_{su}(T)\} = \delta(T)$.

The second part of the lemma holds in view of Lemma 7.2-(*ii*). □

One may naturally ask which linear or additive maps φ from $\mathcal{L}(X)$ onto $\mathcal{L}(Y)$ increase or decrease the inner local spectral radius at a nonzero fixed vector and what are those that satisfy only one of the inequalities in (7.20). If φ is a unital continuous linear, then an answer is given in the following theorem which shows that a little bit more than what was stated in Theorem 7.9 can be obtained.

THEOREM 7.11 (Bourhim [**40**]). *Let $x_0 \in X$ and $y_0 \in Y$ be fixed nonzero vectors. For a linear continuous unital map φ from $\mathcal{L}(X)$ onto $\mathcal{L}(Y)$, the following are equivalent.*

(*i*) *There is a constant $M > 0$ such that $\iota_{\varphi(T)}(y_0) \leq M\iota_T(x_0)$ for all $T \in \mathcal{L}(X)$.*
(*ii*) *There is a constant $m > 0$ such that $m\iota_T(x_0) \leq \iota_{\varphi(T)}(y_0)$ for all $T \in \mathcal{L}(X)$.*

(iii) There is a bijection $A \in \mathcal{L}(X,Y)$ such that $Ax_0 = y_0$ and $\varphi(T) = ATA^{-1}$ for all $T \in \mathcal{L}(X)$.

Similar results on surjective additive continuous maps (which are obviously $\mathbb{R}$-linear continuous maps) preserving or increasing or decreasing the inner local spectral radius at a nonzero fixed vector can be obtained as trivial generalizations with the same arguments and proofs, and without any extra efforts. We therefore omit the details and wonder if similar results to Theorem 7.9 and Theorem 7.11 remain true when the map φ is linear or additive and, of course, without the continuity condition.

7.4. Outer local spectral radius and preservers. González and Mbekhta's main result of [**128**] has been extended by Bourhim and Miller in [**42**] where it is shown that a linear map φ on $M_n(\mathbb{C})$ preserves the local spectral radius at a nonzero vector $x_0 \in \mathbb{C}^n$ if and only if there are a nonzero scalar α and an invertible operator $A \in \mathcal{L}(X)$ such that $Ax_0 = x_0$ and $\varphi(T) = \alpha ATA^{-1}$ for all $T \in \mathcal{L}(X)$. In [**44**], Bračič and Müller extended the main result of the aforementioned paper to infinite dimensional Banach spaces by characterizing surjective linear and continuous maps on $\mathcal{L}(X)$ that preserve the local spectral radius at a fixed nonzero vector of X. In [**76**, Theorem 1.2], Costara showed that every linear surjective map on $\mathcal{L}(X)$ decreasing the local spectral radius at a nonzero vector of X is automatically continuous. In fact, his theorem and its proof show a little bit more.

THEOREM 7.12 (Costara [**76**]). *If $y_0 \in Y$ is a nonzero vector and φ is a linear map from $\mathcal{L}(X)$ onto $\mathcal{L}(Y)$ for which there is a constant $M > 0$ such that*

$$\mathrm{r}_{\varphi(T)}(y_0) \leq M\mathrm{r}(T) \tag{7.21}$$

for all $T \in \mathcal{L}(X)$, then φ is a continuous spectrally bounded map.

In [**33**], it is shown that every linear surjective map on $\mathcal{L}(X)$ increasing the local spectral radius at a nonzero vector of X is automatically continuous. The proof of this result uses similar argument of [**76**].

THEOREM 7.13 (Bourhim–Jari–Mashreghi [**33**]). *Let $\varphi : \mathcal{L}(X) \to \mathcal{L}(Y)$ be a surjective linear map, and let $x_0 \in X$ be a nonzero vector. If there is a constant $M > 0$ such that*

$$\mathrm{r}_T(x_0) \leq M\mathrm{r}(\varphi(T)) \tag{7.22}$$

for all $T \in \mathcal{L}(X)$, then φ is a continuous bijective map spectrally bounded from below.

It is worth mentioning that the characterization of linear or additive maps between $\mathcal{L}(X)$ and $\mathcal{L}(Y)$ satisfying either (7.21) or (7.22) is unknown.

The following result gives a complete description of surjective linear maps on $\mathcal{L}(X)$ that are locally spectrally bounded and locally spectrally bounded below at a fixed nonzero vector in X, and extends [**44**, Theorem 3.4] which describes local spectral isometries on $\mathcal{L}(X)$ at a fixed nonzero point. It is a local version of the main result of [**98**] which describes surjective linear maps on $\mathcal{L}(X)$ that are spectrally bounded and spectrally bounded below.

THEOREM 7.14 (Bourhim [**40**]). *Let $x_0 \in X$ and $y_0 \in Y$ be fixed nonzero elements of X. A linear map φ from $\mathcal{L}(X)$ onto $\mathcal{L}(Y)$ satisfies, for some constants*

$m,\ M > 0$,

$$mr_T(x_0) \le r_{\varphi(T)}(y_0) \le Mr_T(x_0) \tag{7.23}$$

for all $T \in \mathcal{L}(X)$ if and only if there is an invertible operator $A \in \mathcal{L}(X, Y)$ and a nonzero scalar α such that $Ax_0 = y_0$ and $\varphi(T) = \alpha ATA^{-1}$ for all $T \in \mathcal{L}(X)$.

This theorem was established in [**40**] under the additional condition that φ is continuous but Theorem 7.12 or Theorem 7.13 show that the continuity condition is redundant. We also mention that Theorem 7.12 and Theorem 7.13 together with [**40**] allow one to establish the following consequence.

COROLLARY 7.15. *Let $\mathcal{H}$ and $\mathcal{K}$ be two infinite-dimensional complex Hilbert spaces. Let $h_0 \in \mathcal{H}$ and $k_0 \in \mathcal{K}$ be fixed nonzero elements. For a linear map φ from $\mathcal{L}(\mathcal{H})$ onto $\mathcal{L}(\mathcal{K})$, the following are equivalent.*

(i) There are two constants $m,\ M > 0$ such that $mr_T(h_0) \le r_{\varphi(T)}(k_0) \le Mr_T(h_0)$ for all $T \in \mathcal{L}(\mathcal{H})$.
(ii) There is a constant $M > 0$ such that $r_{\varphi(T)}(k_0) \le Mr_T(h_0)$ for all $T \in \mathcal{L}(\mathcal{H})$.
(iii) There is a constant $m > 0$ such that $mr_T(h_0) \le r_{\varphi(T)}(k_0)$ for all $T \in \mathcal{L}(\mathcal{H})$.
(iv) There is a bijection $A \in \mathcal{L}(\mathcal{H}, \mathcal{K})$ and a nonzero scalar α such that $Ah_0 = k_0$ and $\varphi(T) = \alpha ATA^{-1}$ for all $T \in \mathcal{L}(\mathcal{H})$.

7.5. Nonlinear preservers of local spectrum. This section is devoted to nonlinear local spectrum preservers and some related results obtained recently in [**34, 35**]. The following result gives a complete description of surjective maps φ (not assumed to be linear or continuous) on $\mathcal{L}(X)$ preserving the local spectrum of the product of operators at a nonzero fixed element of X .

THEOREM 7.16 (Bourhim–Mashreghi [**35**]). *Let $x_0 \in X$ and $y_0 \in Y$ be two nonzero vectors. A map φ from $\mathcal{L}(X)$ onto $\mathcal{L}(Y)$ satisfies*

$$\sigma_{\varphi(T)\varphi(S)}(y_0) = \sigma_{TS}(x_0),\ (T,\ S \in \mathcal{L}(X)), \tag{7.24}$$

if and only if there exists a bijective bounded linear mapping A from X into Y such that $Ax_0 = y_0$ and either $\varphi(T) = ATA^{-1}$ for all $T \in \mathcal{L}(X)$ or $\varphi(T) = -ATA^{-1}$ for all $T \in \mathcal{L}(X)$.

Note that if X and Y are isomorphic Banach spaces, then the statements of this result can be reduced to the case when $X = Y$ and $x_0 = y_0$. But the fact that "X and Y are isomorphic" is a part of the conclusion of this result rather being a part of its hypothesis. We also mention that the proof of this result uses new ingredients such as the following local spectral identity principle that characterizes in term of the local spectrum when two operators are the same.

THEOREM 7.17 (Bourhim–Mashreghi [**35**]). *Let $x_0 \in X$ be a nonzero vector, and let $A,\ B \in \mathcal{L}(X)$. The following statements are equivalent.*

(i) $A = B$.
(ii) $\sigma_{AT}(x_0) = \sigma_{BT}(x_0)$ for all operators $T \in \mathcal{L}(X)$.
(iii) $\sigma_{AT}(x_0) = \sigma_{BT}(x_0)$ for all rank one operators $T \in \mathcal{L}(X)$.

It also uses a spectral characterization of rank one operators in term of the local spectrum which is new.

THEOREM 7.18 (Bourhim–Mashreghi [**35**]). *For a nonzero vector x_0 of X and a nonzero operator $R \in \mathcal{L}(X)$, the following statements are equivalent.*

(i) *R has rank one.*
(ii) *$\sigma_{RT}(x_0)$ contains at most one nonzero element for all $T \in \mathcal{L}(X)$.*
(iii) *$\sigma_{RT}(x_0)$ contains at most one nonzero element for all $T \in \mathcal{L}(X)$ of at most rank 2.*

The last two theorems tell you that if map φ from $\mathcal{L}(X)$ onto $\mathcal{L}(Y)$ satisfies (7.24), then φ is a bijective linear map preserving the finite rank operators in both directions. Once it is proved that a such map is linear, Theorem 4.12 applies and thus φ takes either (4.7) or (4.8). Next lemmas tell you that φ can not take (4.8) and φ can only take (4.7).

LEMMA 7.19 (Bourhim–Mashreghi [**35**]). *Let $x_0 \in X$ and $y_0 \in Y$ be nonzero vectors, and let $A : X \to Y$ and $B : X^* \to Y^*$ be bijective linear transformations. The following statements are equivalent.*

(i) *For every $x \in X$ and $f \in X^*$, we have $\sigma_{x\otimes f}(x_0) = \sigma_{Ax\otimes Bf}(y_0)$.*
(ii) *A is continuous, $B = A^{*-1}$ and $Ax_0 = \alpha y_0$ for some nonzero scalar $\alpha \in \mathbb{C}$.*

LEMMA 7.20 (Bourhim–Mashreghi [**35**]). *Let $x_0 \in X$ and $y_0 \in Y$ be nonzero vectors, and let $A : X^* \to Y$ and $B : X \to Y^*$ be bijective linear transformations. Then there are $x \in X$ and $f \in X^*$ such that $\sigma_{x\otimes f}(x_0) \neq \sigma_{Af\otimes Bx}(y_0)$.*

The following result is a variant of Theorem 7.16 and its proof uses similar ingredients as above but in terms of triple product of operators.

THEOREM 7.21 (Bourhim–Mashreghi [**34**]). *Let $x_0 \in X$ and $y_0 \in Y$ be two nonzero vectors. A map φ from $\mathcal{L}(X)$ onto $\mathcal{L}(Y)$ satisfies*

$$\sigma_{\varphi(T)\varphi(S)\varphi(T)}(y_0) = \sigma_{TST}(x_0), \quad (T,\ S \in \mathcal{L}(X)), \tag{7.25}$$

if and only if there exists a bijective mapping $A \in \mathcal{B}(X, Y)$ such that $Ax_0 = y_0$ and $\varphi(T) = \lambda ATA^{-1}$ for all $T \in \mathcal{L}(X)$, where λ is a third root of unity, i.e., $\lambda^3 = 1$.

The above results remain true for matrices. Some nonlinear preservers of local spectra of matrices were obtained in [**22, 23, 70**]. In [**70**], Costara described surjective maps φ on $M_n(\mathbb{C})$ preserving the local spectral radius distance and established the following result.

THEOREM 7.22 (Costara [**70**]). *Let x_0 be a nonzero vector of $\mathbb{C}^n$. Then a surjective map φ on $M_n(\mathbb{C})$ satisfies $\varphi(0) = 0$ and*

$$\mathrm{r}_{\varphi(T)-\varphi(S)}(x_0) = \mathrm{r}_{T-S}(x_0), \quad (T,\ S \in M_n(\mathbb{C})) \tag{7.26}$$

if and only if there exists an invertible matrix $A \in M_n(\mathbb{C})$ and unimodular scalar $\alpha \in \mathbb{C}$ such that either $Ax_0 = x_0$ and $\varphi(T) = ATA^{-1}$ for all $T \in M_n(\mathbb{C})$ or $A\overline{x_0} = x_0$ and $\varphi(T) = A\overline{T}A^{-1}$ for all $T \in M_n(\mathbb{C})$. Here, $\overline{x_0}$ is the complex conjugation of x_0.

7.6. Preservers of local spectra at non fixed vectors. In [**43**], Bourhim and Ransford proved that the only additive map Φ from $\mathcal{B}(X)$ to itself for which

$$\sigma_{\Phi(T)}(x) = \sigma_T(x) \qquad (T \in \mathcal{L}(X),\ x \in X),$$

is the identity, and investigated several extensions of this result. While, in [**72**], Costara described surjective linear maps on $\mathcal{L}(X)$ which preserve operators of local spectral radius zero at points of X. He showed, in particular, that if φ is a surjective linear map on $\mathcal{L}(X)$ such that for every $x \in X$ and $T \in \mathcal{L}(X)$, we have

$$\mathrm{r}_T(x) = 0 \quad \text{if and only if} \quad \mathrm{r}_{\varphi(T)}(x) = 0, \tag{7.27}$$

then there exists a nonzero scalar c such that $\varphi(T) = c\,T$ for all $T \in \mathcal{L}(X)$. To prove this result, Costara used the spectral characterizations of rank one operators in terms of the spectrum and the point spectrum, which are due to Sourour [**222**].

This result has been extended in [**36**] where Bourhim and Mashreghi obtained similar result without assuming that φ is linear and providing a self-contained proof based on elementary consideration from local spectral theory and that avoided the spectral characterizations of rank one operators in terms of the spectrum and the point spectrum.

THEOREM 7.23 (Bourhim–Mashreghi [**36**]). *If φ is a surjective (not necessarily linear) map on $\mathcal{L}(X)$ that satisfies*

$$\mathrm{r}_{T-S}(x) = 0 \quad \textit{if and only if} \quad \mathrm{r}_{\varphi(T)-\varphi(S)}(x) = 0, \tag{7.28}$$

for every $x \in X$ and $S,\ T \in \mathcal{L}(X)$, then there is a nonzero scalar $c \in \mathbb{C}$ and an operator $A \in \mathcal{L}(X)$ such that $\varphi(T) = cT + A$ for all $T \in \mathcal{L}(X)$.

By Lemma 7.2, we note that if φ is a map on $\mathcal{L}(X)$ satisfying (7.28), then φ verifies

$$\mathrm{r}(T-S) = 0 \quad \text{if and only if} \quad \mathrm{r}(\varphi(T)-\varphi(S)) = 0,\ (S,\ T \in \mathcal{L}(X)). \tag{7.29}$$

The description of maps φ on $\mathcal{L}(X)$ satisfying (7.29) is unknown and remains an open problem even when φ is supposed to be linear. We don't know either which surjective linear maps φ on $\mathcal{L}(X)$ preserve operators of the local spectral radius zero at a nonzero fixed vector $x_0 \in X$; i.e., those surjective linear maps φ on $\mathcal{L}(X)$ that satisfy

$$\mathrm{r}_{T-S}(x_0) = 0 \quad \text{if and only if} \quad \mathrm{r}_{\varphi(T)-\varphi(S)}(x_0) = 0, \tag{7.30}$$

for all $S,\ T \in \mathcal{L}(X)$. One may also wonders if similar questions and results hold when replacing $\mathrm{r}_T(x)$ by $\Gamma_T(x)$. Note that $\Gamma_T(x) = 0$ means that $\sigma_T(x) = \{0\}$.

In [**69**], Costara characterized linear maps on $M_n(\mathbb{C})$ preserving the local spectrum at non-fixed vectors. He established the following result.

THEOREM 7.24 (Costara [**69**]). *Let φ be a linear map on $M_n(\mathbb{C})$. Then for every $T \in M_n(\mathbb{C})$ there exists a nonzero vector $x_T \in \mathbb{C}^n$ such that*

$$\sigma_{\varphi(T)}(x_T) \cap \sigma_T(x_T) \neq \emptyset$$

if and only if there exists an invertible matrix A in $M_n(\mathbb{C})$ such that either

$$\varphi(T) = ATA^{-1}, \qquad T \in M_n(\mathbb{C}),$$

or

$$\varphi(T) = AT^{tr}A^{-1}, \qquad T \in M_n(\mathbb{C}).$$

He described linear maps on $M_n(\mathbb{C})$ preserving the local spectral radius at non-fixed vectors.

THEOREM 7.25 (Costara [**69**]). *Let φ be a linear map on $M_n(\mathbb{C})$. Then for every $T \in M_n(\mathbb{C})$ there exists a nonzero vector $x_T \in \mathbb{C}^n$ such that*

$$\mathrm{r}_{\varphi(T)}(x_T) = \mathrm{r}_T(x_T)$$

if and only if there exists a unimodular scalar $\alpha \in \mathbb{C}$ and an invertible matrix A in $M_n(\mathbb{C})$ such that either

$$\varphi(T) = \alpha ATA^{-1}, \qquad T \in M_n(\mathbb{C}),$$

or

$$\varphi(T) = \alpha A T^{tr} A^{-1}, \qquad T \in M_n(\mathbb{C}).$$

The above result and its proof remains valid if the local spectral radius is replaced by the inner local spectral radius.

References

[1] Pietro Aiena, *Fredholm and local spectral theory, with applications to multipliers*, Kluwer Academic Publishers, Dordrecht, 2004. MR2070395 (2005e:47001)

[2] Pietro Aiena, *An internal characterization of inessential operators*, Proc. Amer. Math. Soc. **102** (1988), no. 3, 625–626, DOI 10.2307/2047235. MR928992 (89e:47023)

[3] S. Akbari and M. Aryapoor, *On linear transformations preserving at least one eigenvalue*, Proc. Amer. Math. Soc. **132** (2004), no. 6, 1621–1625, DOI 10.1090/S0002-9939-03-07262-9. MR2051122 (2005b:15002)

[4] J. Alaminos, M. Brešar, P. Šemrl, and A. R. Villena, *A note on spectrum-preserving maps*, J. Math. Anal. Appl. **387** (2012), no. 2, 595–603, DOI 10.1016/j.jmaa.2011.09.024. MR2853129 (2012j:46057)

[5] Spiros A. Argyros and Richard G. Haydon, *A hereditarily indecomposable $\mathcal{L}_\infty$-space that solves the scalar-plus-compact problem*, Acta Math. **206** (2011), no. 1, 1–54, DOI 10.1007/s11511-011-0058-y. MR2784662 (2012e:46031)

[6] Bernard Aupetit, *Spectrum-preserving linear mappings between Banach algebras or Jordan-Banach algebras*, J. London Math. Soc. (2) **62** (2000), no. 3, 917–924, DOI 10.1112/S0024610700001514. MR1794294 (2001h:46078)

[7] Bernard Aupetit, *Sur les transformations qui conservent le spectre* (French, with English summary), Banach algebras '97 (Blaubeuren), de Gruyter, Berlin, 1998, pp. 55–78. MR1656598 (99m:46114)

[8] Bernard Aupetit and H. du T. Mouton, *Trace and determinant in Banach algebras*, Studia Math. **121** (1996), no. 2, 115–136. MR1418394 (97i:46086)

[9] B. Aupetit and H. du T. Mouton, *Spectrum preserving linear mappings in Banach algebras*, Studia Math. **109** (1994), no. 1, 91–100. MR1267714 (95c:46070)

[10] Bernard Aupetit, *Spectral characterization of the radical in Banach and Jordan-Banach algebras*, Math. Proc. Cambridge Philos. Soc. **114** (1993), no. 1, 31–35, DOI 10.1017/S0305004100071371. MR1219911 (94e:46081)

[11] Bernard Aupetit, *A primer on spectral theory*, Universitext, Springer-Verlag, New York, 1991. MR1083349 (92c:46001)

[12] Bernard Aupetit and Jaroslav Zemánek, *Local behavior of the spectrum near algebraic elements*, Linear Algebra Appl. **52/53** (1983), 39–44, DOI 10.1016/0024-3795(83)80006-8. MR709343 (84h:46063)

[13] Bernard Aupetit, *Une généralisation du théorème de Gleason-Kahane-Żelazko pour les algèbres de Banach* (French, with English summary), Pacific J. Math. **85** (1979), no. 1, 11–17. MR571623 (82a:46044)

[14] Cătălin Badea, *The Gleason-Kahane-Żelazko theorem and Ransford's generalised spectra* (English, with French summary), C. R. Acad. Sci. Paris Sér. I Math. **313** (1991), no. 10, 679–683. MR1135438 (93a:46086)

[15] Zhao Fang Bai and Jin Chuan Hou, *Additive maps preserving nilpotent operators or spectral radius*, Acta Math. Sin. (Engl. Ser.) **21** (2005), no. 5, 1167–1182, DOI 10.1007/s10114-005-0503-y. MR2176329 (2006e:47073)

[16] Zhaofang Bai and Jinchuan Hou, *Linear maps and additive maps that preserve operators annihilated by a polynomial*, J. Math. Anal. Appl. **271** (2002), no. 1, 139–154, DOI 10.1016/S0022-247X(02)00105-1. MR1923752 (2003g:47064)

[17] Line Baribeau and Thomas Ransford, *Non-linear spectrum-preserving maps*, Bull. London Math. Soc. **32** (2000), no. 1, 8–14, DOI 10.1112/S0024609399006426. MR1718765 (2000j:15008)

[18] Bruce A. Barnes, G. J. Murphy, M. R. F. Smyth, and T. T. West, *Riesz and Fredholm theory in Banach algebras*, Research Notes in Mathematics, vol. 67, Pitman (Advanced Publishing Program), Boston, Mass.-London, 1982. MR668516 (84a:46108)

[19] W. E. Baxter and W. S. Martindale III, *Jordan homomorphisms of semiprime rings*, J. Algebra **56** (1979), no. 2, 457–471, DOI 10.1016/0021-8693(79)90349-1. MR528587 (80f:16008)

[20] K. I. Beidar and Ying-Fen Lin, *On surjective linear maps preserving commutativity*, Proc. Roy. Soc. Edinburgh Sect. A **134** (2004), no. 6, 1023–1040, DOI 10.1017/S0308210500003619. MR2107481 (2005i:16043)

[21] K. I. Beidar, M. Brešar, M. A. Chebotar, and Y. Fong, *Applying functional identities to some linear preserver problems*, Pacific J. Math. **204** (2002), no. 2, 257–271, DOI 10.2140/pjm.2002.204.257. MR1907893 (2003e:46088)

[22] Mohamed Bendaoud, *Preservers of local spectra of matrix sums*, Linear Algebra Appl. **438** (2013), no. 5, 2500–2507, DOI 10.1016/j.laa.2012.10.028. MR3005308

[23] M. Bendaoud, M. Douimi, and M. Sarih, *Maps on matrices preserving local spectra*, Linear Multilinear Algebra **61** (2013), no. 7, 871–880, DOI 10.1080/03081087.2012.716429. MR3175332

[24] M. Bendaoud and M. Sarih, *Additive local spectrum compressors*, Linear Algebra Appl. **435** (2011), no. 6, 1473–1478, DOI 10.1016/j.laa.2011.03.050. MR2807164 (2012f:47106)

[25] M. Bendaoud and A. Bourhim, *Essentially spectrally bounded linear maps*, Proc. Amer. Math. Soc. **137** (2009), no. 10, 3329–3334, DOI 10.1090/S0002-9939-09-09815-3. MR2515402 (2010f:47069)

[26] M. Bendaoud, A. Bourhim, M. Burgos, and M. Sarih, *Linear maps preserving Fredholm and Atkinson elements of C^*-algebras*, Linear Multilinear Algebra **57** (2009), no. 8, 823–838, DOI 10.1080/03081080802253698. MR2571856 (2010m:47051)

[27] M. Bendaoud, A. Bourhim, and M. Sarih, *Linear maps preserving the essential spectral radius*, Linear Algebra Appl. **428** (2008), no. 4, 1041–1045, DOI 10.1016/j.laa.2007.09.010. MR2382111 (2008m:47051)

[28] Rajendra Bhatia, Peter Šemrl, and A. R. Sourour, *Maps on matrices that preserve the spectral radius distance*, Studia Math. **134** (1999), no. 2, 99–110. MR1688218 (2000a:47072)

[29] F. F. Bonsall and J. Duncan, *Numerical ranges of operators on normed spaces and of elements of normed algebras*, London Mathematical Society Lecture Note Series, vol. 2, Cambridge University Press, London-New York, 1971. MR0288583 (44 #5779)

[30] Peter Botta, *Linear maps that preserve singular and nonsingular matrices*, Linear Algebra and Appl. **20** (1978), no. 1, 45–49. MR0485930 (58 #5724)

[31] Nadia Boudi and Mostafa Mbekhta, *Additive maps preserving strongly generalized inverses*, J. Operator Theory **64** (2010), no. 1, 117–130. MR2669430 (2011j:47115)

[32] Nadia Boudi and Youness Hadder, *On linear maps preserving generalized invertibility and related properties*, J. Math. Anal. Appl. **345** (2008), no. 1, 20–25, DOI 10.1016/j.jmaa.2008.03.066. MR2422630 (2009e:47058)

[33] A. Bourhim, T. Jari and J. Mashreghi, *Peripheral local spectrum preservers and maps increasing the local spectral radius*, preprint.

[34] Bourhim, Abdellatif and Mashreghi, Javad, *Maps preserving the local spectrum of triple product of operators*, Linear Multilinear Algebra **63** (2015), no. 4, 765–773.

[35] Abdellatif Bourhim and Javad Mashreghi, *Maps preserving the local spectrum of product of operators*, Glasgow Math. J., available on CJO2014. DOI 10.1017/S0017089514000585.

[36] Abdellatif Bourhim and Javad Mashreghi, *Local spectral radius preservers*, Integral Equations Operator Theory **76** (2013), no. 1, 95–104, DOI 10.1007/s00020-013-2041-9. MR3041723

[37] Abdellatif Bourhim, *Additive maps preserving the reduced minimum modulus of Banach space operators*, J. Operator Theory **67** (2012), no. 1, 279–288. MR2881543

[38] Abdellatif Bourhim and María Burgos, *Linear maps preserving the minimum modulus*, Oper. Matrices **4** (2010), no. 2, 245–256, DOI 10.7153/oam-04-12. MR2667336 (2011g:47081)

[39] A. Bourhim, M. Burgos, and V. S. Shulman, *Linear maps preserving the minimum and reduced minimum moduli*, J. Funct. Anal. **258** (2010), no. 1, 50–66, DOI 10.1016/j.jfa.2009.10.003. MR2557954 (2010j:47052)

[40] Abdellatif Bourhim, *Surjective linear maps preserving local spectra*, Linear Algebra Appl. **432** (2010), no. 1, 383–393, DOI 10.1016/j.laa.2009.08.020. MR2566487 (2010j:47053)

[41] Abdellatif Bourhim and María Burgos, *Linear maps preserving regularity in C^*-algebras*, Illinois J. Math. **53** (2009), no. 3, 899–914 (2010). MR2727361 (2011k:46075)

[42] Abdellatif Bourhim and Vivien G. Miller, *Linear maps on $M_n(\mathbb{C})$ preserving the local spectral radius*, Studia Math. **188** (2008), no. 1, 67–75, DOI 10.4064/sm188-1-4. MR2430550 (2009m:47094)

[43] Abdellatif Bourhim and Thomas Ransford, *Additive maps preserving local spectrum*, Integral Equations Operator Theory **55** (2006), no. 3, 377–385, DOI 10.1007/s00020-005-1392-2. MR2244195 (2007c:47004)

[44] Janko Bračič and Vladimír Müller, *Local spectrum and local spectral radius of an operator at a fixed vector*, Studia Math. **194** (2009), no. 2, 155–162, DOI 10.4064/sm194-2-3. MR2534182 (2011m:47011)

[45] Matej Brešar, *Jordan homomorphisms revisited*, Math. Proc. Cambridge Philos. Soc. **144** (2008), no. 2, 317–328, DOI 10.1017/S0305004107000825. MR2405892 (2009b:47059)

[46] Matej Brešar and Peter Šemrl, *An extension of the Gleason-Kahane-Żelazko theorem: a possible approach to Kaplansky's problem*, Expo. Math. **26** (2008), no. 3, 269–277, DOI 10.1016/j.exmath.2007.11.004. MR2437097 (2009k:47103)

[47] Matej Brešar, Mikhail A. Chebotar, and Wallace S. Martindale III, *Functional identities*, Frontiers in Mathematics, Birkhäuser Verlag, Basel, 2007. MR2332350 (2008j:16068)

[48] Matej Brešar and Yuri V. Turovskii, *Compactness conditions for elementary operators*, Studia Math. **178** (2007), no. 1, 1–18, DOI 10.4064/sm178-1-1. MR2282487 (2007m:47087)

[49] Matej Brešar, Ajda Fošner, and Peter Šemrl, *A note on invertibility preservers on Banach algebras*, Proc. Amer. Math. Soc. **131** (2003), no. 12, 3833–3837 (electronic), DOI 10.1090/S0002-9939-03-07192-2. MR1999931 (2004d:47071)

[50] Matej Brešar and Peter Šemrl, *Spectral characterization of idempotents and invertibility preserving linear maps*, Exposition. Math. **17** (1999), no. 2, 185–192. MR1690403 (2000d:16050)

[51] Matej Brešar and Peter Šemrl, *Invertibility preserving maps preserve idempotents*, Michigan Math. J. **45** (1998), no. 3, 483–488, DOI 10.1307/mmj/1030132294. MR1653251 (2000j:47119)

[52] Matej Brešar and Peter Šemrl, *Finite rank elements in semisimple Banach algebras*, Studia Math. **128** (1998), no. 3, 287–298. MR1611128 (99a:46089)

[53] Matej Brešar and Peter Šemrl, *Linear preservers on $\mathcal{B}(X)$*, Linear operators (Warsaw, 1994), Banach Center Publ., vol. 38, Polish Acad. Sci., Warsaw, 1997, pp. 49–58. MR1457000 (99c:47044)

[54] Matej Brešar and Peter Šemrl, *Linear maps preserving the spectral radius*, J. Funct. Anal. **142** (1996), no. 2, 360–368, DOI 10.1006/jfan.1996.0153. MR1423038 (97i:47070)

[55] Matej Brešar, *Commuting traces of biadditive mappings, commutativity-preserving mappings and Lie mappings*, Trans. Amer. Math. Soc. **335** (1993), no. 2, 525–546, DOI 10.2307/2154392. MR1069746 (93d:16044)

[56] Matej Brešar and C. Robert Miers, *Commutativity preserving mappings of von Neumann algebras*, Canad. J. Math. **45** (1993), no. 4, 695–708, DOI 10.4153/CJM-1993-039-x. MR1227654 (94h:46091)

[57] Rudi M. Brits and Heinrich Raubenheimer, *Finite spectra and quasinilpotent equivalence in Banach algebras*, Czechoslovak Math. J. **62(137)** (2012), no. 4, 1101–1116, DOI 10.1007/s10587-012-0066-x. MR3010259

[58] Lawrence G. Brown and Gert K. Pedersen, *C^*-algebras of real rank zero*, J. Funct. Anal. **99** (1991), no. 1, 131–149, DOI 10.1016/0022-1236(91)90056-B. MR1120918 (92m:46086)

[59] M. Burgos, A. C. Márquez-García, and A. Morales-Campoy, *Strongly preserver problems in Banach algebras and C^*-algebras*, Linear Algebra Appl. **437** (2012), no. 5, 1183–1193, DOI 10.1016/j.laa.2012.04.019. MR2942341

[60] María Burgos, A. C. Márquez-García, and A. Morales-Campoy, *Linear maps strongly preserving Moore-Penrose invertibility*, Oper. Matrices **6** (2012), no. 4, 819–831, DOI 10.7153/oam-06-53. MR3025038

[61] Xiaohong Cao and Shizhao Chen, *Linear maps between operator algebras preserving certain spectral functions*, Banach J. Math. Anal. **8** (2014), no. 1, 39–46. MR3161680

[62] Maria Inez Cardoso Gonccalves and Ahmed Ramzi Sourour, *Isometries of a generalized numerical radius*, Linear Algebra Appl. **429** (2008), no. 7, 1478–1488, DOI 10.1016/j.laa.2008.04.007. MR2444336 (2009m:15028)

[63] Jor-Ting Chan, Chi-Kwong Li, and Nung-Sing Sze, *Mappings preserving spectra of products of matrices*, Proc. Amer. Math. Soc. **135** (2007), no. 4, 977–986 (electronic), DOI 10.1090/S0002-9939-06-08568-6. MR2262897 (2007k:15007)

[64] M. A. Chebotar, W.-F. Ke, P.-H. Lee, and N.-C. Wong, *Mappings preserving zero products*, Studia Math. **155** (2003), no. 1, 77–94, DOI 10.4064/sm155-1-6. MR1961162 (2003m:47066)

[65] M. D. Choi, D. Hadwin, E. Nordgren, H. Radjavi, and P. Rosenthal, *On positive linear maps preserving invertibility*, J. Funct. Anal. **59** (1984), no. 3, 462–469, DOI 10.1016/0022-1236(84)90060-0. MR769376 (86a:46071)

[66] Erik Christensen, *On invertibility preserving linear mappings, simultaneous triangularization and property L*, Linear Algebra Appl. **301** (1999), no. 1-3, 153–170, DOI 10.1016/S0024-3795(99)00195-0. MR1728048 (2000k:15037)

[67] Erik Christensen, *Two generalizations of the Gleason-Kahane-Żelazko theorem*, Pacific J. Math. **177** (1997), no. 1, 27–32, DOI 10.2140/pjm.1997.177.27. MR1444770 (98c:46089)

[68] Sean Clark, Chi-Kwong Li, Jennifer Mahle, and Leiba Rodman, *Linear preservers of higher rank numerical ranges and radii*, Linear Multilinear Algebra **57** (2009), no. 5, 503–521, DOI 10.1080/03081080902823630. MR2532419 (2010g:47016)

[69] Constantin Costara, *Local spectrum linear preservers at non-fixed vectors*, Linear Algebra Appl. **457** (2014), 154–161, DOI 10.1016/j.laa.2014.05.031. MR3230437

[70] Constantin Costara, *Surjective maps on matrices preserving the local spectral radius distance*, Linear Multilinear Algebra **62** (2014), no. 7, 988–994, DOI 10.1080/03081087.2013.801967. MR3232673

[71] Constantin Costara, *Unital invertibility-preserving linear maps into matrix spaces*, J. Math. Anal. Appl. **412** (2014), no. 1, 99–102, DOI 10.1016/j.jmaa.2013.10.045. MR3145783

[72] Constantin Costara, *Linear maps preserving operators of local spectral radius zero*, Integral Equations Operator Theory **73** (2012), no. 1, 7–16, DOI 10.1007/s00020-012-1953-0. MR2913657

[73] Constantin Costara, *On the automorphisms of the spectral unit ball*, Proc. Amer. Math. Soc. **140** (2012), no. 12, 4181–4186, DOI 10.1090/S0002-9939-2012-11266-3. MR2957207

[74] C. Costara, *Maps on matrices that preserve the spectrum*, Linear Algebra Appl. **435** (2011), no. 11, 2674–2680, DOI 10.1016/j.laa.2011.04.026. MR2825274 (2012j:15002)

[75] Constantin Costara and Dušan Repovš, *Nonlinear mappings preserving at least one eigenvalue*, Studia Math. **200** (2010), no. 1, 79–89, DOI 10.4064/sm200-1-5. MR2720208 (2012a:15013)

[76] Constantin Costara, *Automatic continuity for linear surjective mappings decreasing the local spectral radius at some fixed vector*, Arch. Math. (Basel) **95** (2010), no. 6, 567–573, DOI 10.1007/s00013-010-0191-4. MR2745466 (2011j:47021)

[77] Constantin Costara and Dušan Repovš, *Spectral isometries onto algebras having a separating family of finite-dimensional irreducible representations*, J. Math. Anal. Appl. **365** (2010), no. 2, 605–608, DOI 10.1016/j.jmaa.2009.11.040. MR2587062 (2011f:46053)

[78] C. Costara, *Commuting holomorphic maps on the spectral unit ball*, Bull. Lond. Math. Soc. **41** (2009), no. 1, 57–62, DOI 10.1112/blms/bdn104. MR2481989 (2010c:47093)

[79] Constantin Costara and Thomas Ransford, *On local irreducibility of the spectrum*, Proc. Amer. Math. Soc. **135** (2007), no. 9, 2779–2784, DOI 10.1090/S0002-9939-07-08779-5. MR2317952 (2008i:32001)

[80] Jianlian Cui and Chi-Kwong Li, *Maps preserving peripheral spectrum of Jordan products of operators*, Oper. Matrices **6** (2012), no. 1, 129–146, DOI 10.7153/oam-06-09. MR2952440

[81] Jian Lian Cui and Jin Chuan Hou, *Additive maps preserving rank-1 nilpotency on nest algebras* (Chinese, with English and Chinese summaries), Acta Math. Sci. Ser. A Chin. Ed. **27** (2007), no. 2, 193–203. MR2312353 (2008d:47149)

[82] Jinchuan Hou and Jianlian Cui, *Linear maps preserving essential spectral functions and closeness of operator ranges*, Bull. Lond. Math. Soc. **39** (2007), no. 4, 575–582, DOI 10.1112/blms/bdm045. MR2346937 (2008i:47079)

[83] Jian Lian Cui and Jin Chuan Hou, *A characterization of homomorphisms between Banach algebras*, Acta Math. Sin. (Engl. Ser.) **20** (2004), no. 4, 761–768, DOI 10.1007/s10114-004-0312-8. MR2096788 (2005f:47093)

[84] Jianlian Cui, Maozheng Guo, and Jinchuan Hou, *Rank-1 nilpotency linear preservers on nest algebras*, Linear Algebra Appl. **377** (2004), 267–290, DOI 10.1016/j.laa.2003.08.017. MR2022176 (2004k:47068)

[85] Jianlian Cui and Jinchuan Hou, *Linear maps between Banach algebras compressing certain spectral functions*, Rocky Mountain J. Math. **34** (2004), no. 2, 565–584, DOI 10.1216/rmjm/1181069868. MR2072795 (2005b:47072)

[86] Jianlian Cui and Jinchuan Hou, *Additive maps on standard operator algebras preserving parts of the spectrum*, J. Math. Anal. Appl. **282** (2003), no. 1, 266–278, DOI 10.1016/S0022-247X(03)00146-X. MR2000343 (2004f:47050)

[87] Jianlian Cui and Jinchuan Hou, *The spectrally bounded linear maps on operator algebras*, Studia Math. **150** (2002), no. 3, 261–271, DOI 10.4064/sm150-3-4. MR1891847 (2003a:47079)

[88] Raúl Curto and Martin Mathieu, *Spectrally bounded generalized inner derivations*, Proc. Amer. Math. Soc. **123** (1995), no. 8, 2431–2434, DOI 10.2307/2161270. MR1249873 (95j:46055)

[89] Kenneth R. Davidson, *C^*-algebras by example*, Fields Institute Monographs, vol. 6, American Mathematical Society, Providence, RI, 1996. MR1402012 (97i:46095)

[90] Clément de Seguins Pazzis, *The singular linear preservers of non-singular matrices*, Linear Algebra Appl. **433** (2010), no. 2, 483–490, DOI 10.1016/j.laa.2010.03.021. MR2645099 (2011c:15081)

[91] Jean Dieudonné, *Sur une généralisation du groupe orthogonal à quatre variables* (French), Arch. Math. **1** (1949), 282–287. MR0029360 (10,586l)

[92] Gregor Dolinar, Jinchuan Hou, Bojan Kuzma, and Xiaofei Qi, *Spectrum nonincreasing maps on matrices*, Linear Algebra Appl. **438** (2013), no. 8, 3504–3510, DOI 10.1016/j.laa.2012.12.034. MR3023292

[93] Ronald G. Douglas, *Banach algebra techniques in operator theory*, Academic Press, New York-London, 1972. Pure and Applied Mathematics, Vol. 49. MR0361893 (50 #14335)

[94] Driss Drissi, *The single-valued extension property and subharmonicity*, Extracta Math. **22** (2007), no. 3, 235–244. MR2405083 (2008m:47003)

[95] A. Fernández López, E. García Rus, M. Gómez Lozano, and M. Siles Molina, *Jordan canonical form for finite rank elements in Jordan algebras*, Linear Algebra Appl. **260** (1997), 151–167, DOI 10.1016/S0024-3795(97)80008-0. MR1448354 (98e:17040)

[96] Virginia Forstall, Aaron Herman, Chi-Kwong Li, Nung-Sing Sze, and Vincent Yannello, *Preservers of eigenvalue inclusion sets of matrix products*, Linear Algebra Appl. **434** (2011), no. 1, 285–293, DOI 10.1016/j.laa.2010.08.016. MR2737249 (2012d:15059)

[97] Ajda Fošner and Peter Šemrl, *Additive maps on matrix algebras preserving invertibility or singularity*, Acta Math. Sin. (Engl. Ser.) **21** (2005), no. 4, 681–684, DOI 10.1007/s10114-005-0519-3. MR2156942 (2006i:15005)

[98] Ajda Fošner and Peter Šemrl, *Spectrally bounded linear maps on $\mathcal{B}(X)$*, Canad. Math. Bull. **47** (2004), no. 3, 369–372, DOI 10.4153/CMB-2004-036-4. MR2072597 (2005e:47096)

[99] G. Frobenius, *Ueber die Darstellung der endlichen Gruppen durch lineare Substitutionen*, Berl. Ber., (1897) 994-1015.

[100] Hwa-Long Gau and Chi-Kwong Li, *C^*-isomorphisms, Jordan isomorphisms, and numerical range preserving maps*, Proc. Amer. Math. Soc. **135** (2007), no. 9, 2907–2914 (electronic), DOI 10.1090/S0002-9939-07-08807-7. MR2317968 (2008c:47062)

[101] W. T. Gowers and B. Maurey, *The unconditional basic sequence problem*, J. Amer. Math. Soc. **6** (1993), no. 4, 851–874, DOI 10.2307/2152743. MR1201238 (94k:46021)

[102] Alexander Guterman, Chi-Kwong Li, and Peter Šemrl, *Some general techniques on linear preserver problems*, Linear Algebra Appl. **315** (2000), no. 1-3, 61–81, DOI 10.1016/S0024-3795(00)00119-1. MR1774960 (2002a:15003)

[103] Don Hadwin, Jinchuan Hou, and Hassan Yousefi, *Completely rank-nonincreasing linear maps on spaces of operators*, Linear Algebra Appl. **383** (2004), 213–232, DOI 10.1016/j.laa.2004.01.002. MR2073905 (2005e:47097)

[104] Lawrence A. Harris, *Invertibility preserving linear maps of Banach algebras*, Complex analysis and dynamical systems, Contemp. Math., vol. 364, Amer. Math. Soc., Providence, RI, 2004, pp. 59–66, DOI 10.1090/conm/364/06677. MR2097978 (2005g:47067)

[105] Lawrence A. Harris and Richard V. Kadison, *Affine mappings of invertible operators*, Proc. Amer. Math. Soc. **124** (1996), no. 8, 2415–2422, DOI 10.1090/S0002-9939-96-03445-4. MR1340389 (96j:46055)

[106] Robin Harte and Mostafa Mbekhta, *On generalized inverses in C^*-algebras*, Studia Math. **103** (1992), no. 1, 71–77. MR1184103 (93i:46097)

[107] Robin Harte and Mostafa Mbekhta, *Generalized inverses in C^*-algebras. II*, Studia Math. **106** (1993), no. 2, 129–138. MR1240309 (94k:46113)

[108] O. Hatori, T. Miura and H. Takagi, *Multiplicatively spectrum-preserving and norm-preserving maps between invertible groups of commutative Banach algebras*, preprint.

[109] Osamu Hatori, Takeshi Miura, and Hiroyuki Takagi, *Unital and multiplicatively spectrum-preserving surjections between semi-simple commutative Banach algebras are linear and multiplicative*, J. Math. Anal. Appl. **326** (2007), no. 1, 281–296, DOI 10.1016/j.jmaa.2006.02.084. MR2277782 (2007m:47090)

[110] Osamu Hatori, Takeshi Miura, and Hiroyuki Takagi, *Characterizations of isometric isomorphisms between uniform algebras via nonlinear range-preserving properties*, Proc. Amer. Math. Soc. **134** (2006), no. 10, 2923–2930 (electronic), DOI 10.1090/S0002-9939-06-08500-5. MR2231616 (2007h:46063)

[111] Hans Havlicek and Peter Šemrl, *From geometry to invertibility preservers*, Studia Math. **174** (2006), no. 1, 99–109, DOI 10.4064/sm174-1-8. MR2239816 (2007g:47055)

[112] Kan He, Jin Chuan Hou, and Xiu Ling Zhang, *Maps preserving numerical radius or cross norms of products of self-adjoint operators*, Acta Math. Sin. (Engl. Ser.) **26** (2010), no. 6, 1071–1086, DOI 10.1007/s10114-010-9001-y. MR2644047 (2011f:47066)

[113] I. N. Herstein, *Topics in ring theory*, The University of Chicago Press, Chicago, Ill.-London, 1969. MR0271135 (42 #6018)

[114] I. N. Herstein, *Jordan homomorphisms*, Trans. Amer. Math. Soc. **81** (1956), 331–341. MR0076751 (17,938f)

[115] M. Hosseini and F. Sady, *Multiplicatively range-preserving maps between Banach function algebras*, J. Math. Anal. Appl. **357** (2009), no. 1, 314–322, DOI 10.1016/j.jmaa.2009.04.008. MR2526831 (2010m:47047)

[116] Jinchuan Hou, Chi-Kwong Li, and Ngai-Ching Wong, *Maps preserving the spectrum of generalized Jordan product of operators*, Linear Algebra Appl. **432** (2010), no. 4, 1049–1069, DOI 10.1016/j.laa.2009.10.018. MR2577648 (2011a:47077)

[117] Jinchuan Hou and Li Huang, *Maps completely preserving idempotents and maps completely preserving square-zero operators*, Israel J. Math. **176** (2010), 363–380, DOI 10.1007/s11856-010-0032-y. MR2653198 (2011g:47082)

[118] Jinchuan Hou and Li Huang, *Characterizing isomorphisms in terms of completely preserving invertibility or spectrum*, J. Math. Anal. Appl. **359** (2009), no. 1, 81–87, DOI 10.1016/j.jmaa.2009.05.041. MR2542157 (2010g:47073)

[119] Jinchuan Hou, Chi-Kwong Li, and Ngai-Ching Wong, *Jordan isomorphisms and maps preserving spectra of certain operator products*, Studia Math. **184** (2008), no. 1, 31–47, DOI 10.4064/sm184-1-2. MR2365474 (2008i:47080)

[120] Jinchuan Hou and Meiyan Jiao, *Additive maps preserving Jordan zero-products on nest algebras*, Linear Algebra Appl. **429** (2008), no. 1, 190–208, DOI 10.1016/j.laa.2008.02.021. MR2419149 (2009e:47116)

[121] Jin Chuan Hou and Li Huang, *Additive maps between standard operator algebras compressing certain spectral functions*, Acta Math. Sin. (Engl. Ser.) **24** (2008), no. 12, 2041–2048, DOI 10.1007/s10114-008-6428-5. MR2453083 (2009i:47084)

[122] Jinchuan Hou and Qinghui Di, *Maps preserving numerical ranges of operator products*, Proc. Amer. Math. Soc. **134** (2006), no. 5, 1435–1446, DOI 10.1090/S0002-9939-05-08101-3. MR2199190 (2006k:47077)

[123] Jinchuan Hou and Jianlian Cui, *Completely rank nonincreasing linear maps on nest algebras*, Proc. Amer. Math. Soc. **132** (2004), no. 5, 1419–1428, DOI 10.1090/S0002-9939-03-07275-7. MR2053348 (2005b:47073)

[124] Jinchuan Hou and Jianlian Cui, *Rank-1 preserving linear maps on nest algebras*, Linear Algebra Appl. **369** (2003), 263–277, DOI 10.1016/S0024-3795(02)00725-5. MR1988491 (2004c:47078)

[125] Jinchuan Hou and Jianlian Cui, *Additive maps on standard operator algebras preserving invertibilities or zero divisors*, Linear Algebra Appl. **359** (2003), 219–233, DOI 10.1016/S0024-3795(02)00419-6. MR1948445 (2003j:47048)

[126] Jin Chuan Hou and Peter Šemrl, *Linear maps preserving invertibility or related spectral properties*, Acta Math. Sin. (Engl. Ser.) **19** (2003), no. 3, 473–484, DOI 10.1007/s10114-003-0269-z. International Workshop on Operator Algebra and Operator Theory (Linfen, 2001). MR2014028 (2004h:47054)

[127] Loo-Keng Hua, *On the automorphisms of a sfield*, Proc. Nat. Acad. Sci. U. S. A. **35** (1949), 386–389. MR0029886 (10,675d)

[128] Manuel González and Mostafa Mbekhta, *Linear maps on $M_n(\mathbb{C})$ preserving the local spectrum*, Linear Algebra Appl. **427** (2007), no. 2-3, 176–182, DOI 10.1016/j.laa.2007.07.005. MR2351350 (2008i:47078)

[129] Andrew M. Gleason, *A characterization of maximal ideals*, J. Analyse Math. **19** (1967), 171–172. MR0213878 (35 #4732)

[130] T. N. E. Greville, *Some new generalized inverses with spectral properties*, Proc. Sympos. Theory and Application of Generalized Inverses of Matrices (Lubbock, Texas, 1968), Texas Tech. Press, Lubbock, Tex., 1968, pp. 26–46. MR0251048 (40 #4279)

[131] Alexander Guterman, Chi-Kwong Li, and Peter Šemrl, *Some general techniques on linear preserver problems*, Linear Algebra Appl. **315** (2000), no. 1-3, 61–81, DOI 10.1016/S0024-3795(00)00119-1. MR1774960 (2002a:15003)

[132] N. Jacobson and C. E. Rickart, *Jordan homomorphisms of rings*, Trans. Amer. Math. Soc. **69** (1950), 479–502. MR0038335 (12,387h)

[133] A. A. Jafarian, *A survey of invertibility and spectrum preserving linear maps* (English, with English and Persian summaries), Bull. Iranian Math. Soc. **35** (2009), no. 2, 1–10, 267. MR2642921 (2011c:47080)

[134] Ali A. Jafarian, Leiba Rodman, and Peter Šemrl, *Linear maps preserving the isomorphism class of lattices of invariant subspaces*, Proc. Amer. Math. Soc. **126** (1998), no. 12, 3607–3617, DOI 10.1090/S0002-9939-98-04921-1. MR1610913 (99e:47012)

[135] Ali A. Jafarian and A. R. Sourour, *Spectrum-preserving linear maps*, J. Funct. Anal. **66** (1986), no. 2, 255–261, DOI 10.1016/0022-1236(86)90073-X. MR832991 (87m:47011)

[136] A. Jiménez-Vargas, Aaron Luttman, and Moisés Villegas-Vallecillos, *Weakly peripherally multiplicative surjections of pointed Lipschitz algebras*, Rocky Mountain J. Math. **40** (2010), no. 6, 1903–1922, DOI 10.1216/RMJ-2010-40-6-1903. MR2764228 (2012b:46093)

[137] Richard V. Kadison and John R. Ringrose, *Fundamentals of the theory of operator algebras. Vol. I*, Graduate Studies in Mathematics, vol. 15, American Mathematical Society, Providence, RI, 1997. Elementary theory; Reprint of the 1983 original. MR1468229 (98f:46001a)

[138] Richard V. Kadison, *Isometries of operator algebras*, Ann. Of Math. (2) **54** (1951), 325–338. MR0043392 (13,256a)

[139] J.-P. Kahane and W. Żelazko, *A characterization of maximal ideals in commutative Banach algebras*, Studia Math. **29** (1968), 339–343. MR0226408 (37 #1998)

[140] Irving Kaplansky, *Algebraic and analytic aspects of operator algebras*, American Mathematical Society, Providence, R.I., 1970. Conference Board of the Mathematical Sciences Regional Conference Series in Mathematics, No. 1. MR0312283 (47 #845)

[141] Eberhard Kirchberg and Mikael Rørdam, *Non-simple purely infinite C^*-algebras*, Amer. J. Math. **122** (2000), no. 3, 637–666. MR1759891 (2001k:46088)

[142] Istvan Kovacs, *Invertibility-preserving maps of C^*-algebras with real rank zero*, Abstr. Appl. Anal. **6** (2005), 685–689, DOI 10.1155/AAA.2005.685. MR2202956 (2006k:47079)

[143] S. Kowalski and Z. Słodkowski, *A characterization of multiplicative linear functionals in Banach algebras*, Studia Math. **67** (1980), no. 3, 215–223. MR592387 (82d:46070)

[144] Bojan Kuzma, *Noninvertibility preservers on Banach algebras*, Czechoslovak Math. J. **56(131)** (2006), no. 3, 919–921, DOI 10.1007/s10587-006-0066-9. MR2261663 (2007g:47053)

[145] B. Kuzma, *Additive mappings preserving rank-one idempotents*, Acta Math. Sin. (Engl. Ser.) **21** (2005), no. 6, 1399–1406, DOI 10.1007/s10114-005-0598-1. MR2190009 (2006k:47080)

[146] B. Kuzma, *Additive spectrum compressors*, J. Math. Anal. Appl. **304** (2005), no. 1, 13–21, DOI 10.1016/j.jmaa.2004.09.004. MR2124645 (2005k:47081)

[147] Scott Lambert, Aaron Luttman, and Thomas Tonev, *Weakly peripherally-multiplicative mappings between uniform algebras*, Function spaces, Contemp. Math., vol. 435, Amer. Math. Soc., Providence, RI, 2007, pp. 265–281, DOI 10.1090/conm/435/08383. MR2359435 (2009b:46108)

[148] Kjeld B. Laursen and Michael M. Neumann, *An introduction to local spectral theory*, London Mathematical Society Monographs. New Series, vol. 20, The Clarendon Press, Oxford University Press, New York, 2000. MR1747914 (2001k:47002)

[149] David C. Lay, *Spectral properties of generalized inverses of linear operators*, SIAM J. Appl. Math. **29** (1975), 103–109. MR0372645 (51 #8852)

[150] Jürgen Leiterer, *Banach coherent analytic Fréchet sheaves*, Math. Nachr. **85** (1978), 91–109, DOI 10.1002/mana.19780850108. MR517643 (80b:32026)

[151] Gorazd Lešnjak, *Additive preservers of numerical range*, Linear Algebra Appl. **345** (2002), 235–253, DOI 10.1016/S0024-3795(01)00498-0. MR1883275 (2002j:15035)

[152] Gorazd Lešnjak, *More on finite rank elements*, Publ. Math. Debrecen **53** (1998), no. 1-2, 207–215. MR1661042 (99k:46084)

[153] Pengtong Li, Fangyan Lu, and Jipu Ma, *Invertibility preserving linear maps on $\mathcal{J}$-subspace lattice algebras*, Linear Algebra Appl. **372** (2003), 155–166, DOI 10.1016/S0024-3795(03)00499-3. MR1999145 (2004d:47073)

[154] Chi-Kwong Li, Yiu-Tung Poon, and Nung-Sing Sze, *Linear preservers of tensor product of unitary orbits, and product numerical range*, Linear Algebra Appl. **438** (2013), no. 10, 3797–3803, DOI 10.1016/j.laa.2011.07.039. MR3034498

[155] Chi-Kwong Li and Leiba Rodman, *Preservers of spectral radius, numerical radius, or spectral norm of the sum on nonnegative matrices*, Linear Algebra Appl. **430** (2009), no. 7, 1739–1761, DOI 10.1016/j.laa.2008.04.022. MR2494661 (2010a:15032)

[156] Chi-Kwong Li, Leiba Rodman, and Peter Šemrl, *Linear maps on selfadjoint operators preserving invertibility, positive definiteness, numerical range*, Canad. Math. Bull. **46** (2003), no. 2, 216–228, DOI 10.4153/CMB-2003-022-3. MR1981676 (2004b:47063)

[157] Chi-Kwong Li and Peter Šemrl, *Numerical radius isometries*, Linear Multilinear Algebra **50** (2002), no. 4, 307–314, DOI 10.1080/03081080290025480. MR1936954 (2003i:15025)

[158] Chi-Kwong Li, *A survey on linear preservers of numerical ranges and radii*, Taiwanese J. Math. **5** (2001), no. 3, 477–496. MR1849772 (2002g:15055)

[159] Chi-Kwong Li and Stephen Pierce, *Linear preserver problems*, Amer. Math. Monthly **108** (2001), no. 7, 591–605, DOI 10.2307/2695268. MR1862098 (2002g:15005)

[160] Chi-Kwong Li, *C-numerical ranges and C-numerical radii*, Linear and Multilinear Algebra **37** (1994), no. 1-3, 51–82, DOI 10.1080/03081089408818312. Special Issue: The numerical range and numerical radius. MR1313758 (95k:15039)

[161] Chi-Kwong Li and Nam-Kiu Tsing, *Linear preservers on numerical ranges, numerical radii and unitary similarity invariant norms*, Linear and Multilinear Algebra **33** (1992), no. 1-2, 63–73, DOI 10.1080/03081089208818182. A survey of linear preserver problems. MR1346783

[162] Chi-Kwong Li and Nam-Kiu Tsing, *Linear preserver problems: a brief introduction and some special techniques*, Linear Algebra Appl. **162/164** (1992), 217–235, DOI 10.1016/0024-3795(92)90377-M. Directions in matrix theory (Auburn, AL, 1990). MR1148401 (93b:15003)

[163] Ming-Huat Lim, *Linear rank one preservers between spaces of matrices with zero trace*, Linear Algebra Appl. **430** (2009), no. 11-12, 2982–2996, DOI 10.1016/j.laa.2009.01.012. MR2517852 (2010c:15026)

[164] Ying-Fen Lin and Martin Mathieu, *Jordan isomorphism of purely infinite C^*-algebras*, Q. J. Math. **58** (2007), no. 2, 249–253, DOI 10.1093/qmath/hal024. MR2334865 (2008k:46189)

[165] J. Lindenstrauss, *Some open problems in Banach space theory*, Seminaire Choquet. Initiation à l'analyse, **15**, No 18, (1975-1976) 9 pp.

[166] Aaron Luttman and Scott Lambert, *Norm conditions for uniform algebra isomorphisms*, Cent. Eur. J. Math. **6** (2008), no. 2, 272–280, DOI 10.2478/s11533-008-0016-x. MR2393194 (2009c:46072)

[167] Aaron Luttman and Thomas Tonev, *Uniform algebra isomorphisms and peripheral multiplicativity*, Proc. Amer. Math. Soc. **135** (2007), no. 11, 3589–3598 (electronic), DOI 10.1090/S0002-9939-07-08881-8. MR2336574 (2008k:46150)

[168] Endre Makai Jr. and Jaroslav Zemánek, *The surjectivity radius, packing numbers and boundedness below of linear operators*, Integral Equations Operator Theory **6** (1983), no. 3, 372–384, DOI 10.1007/BF01691904. MR701026 (84m:47005)

[169] Marvin Marcus, *Linear operations on matrices*, Amer. Math. Monthly **69** (1962), 837–847. MR0147491 (26 #5007)

[170] Marvin Marcus and B. N. Moyls, *Linear transformations on algebras of matrices*, Canad. J. Math. **11** (1959), 61–66. MR0099996 (20 #6432)

[171] Marvin Marcus and Roger Purves, *Linear transformations on algebras of matrices: the invariance of the elementary symmetric functions*, Canad. J. Math. **11** (1959), 383–396. MR0105425 (21 #4167)

[172] Martin Mathieu and Ahmed R. Sourour, *Spectral isometries on non-simple C^*-algebras*, Proc. Amer. Math. Soc. **142** (2014), no. 1, 129–135, DOI 10.1090/S0002-9939-2013-11704-1. MR3119188

[173] Martin Mathieu, *A collection of problems on spectrally bounded operators*, Asian-Eur. J. Math. **2** (2009), no. 3, 487–501, DOI 10.1142/S1793557109000418. MR2568008 (2011a:47079)

[174] Martin Mathieu, *Spectrally bounded operators on simple C^*-algebras. II*, Irish Math. Soc. Bull. **54** (2004), 33–40. MR2138429 (2005m:47074)

[175] Martin Mathieu, *Spectrally bounded operators on simple C^*-algebras*, Proc. Amer. Math. Soc. **132** (2004), no. 2, 443–446 (electronic), DOI 10.1090/S0002-9939-03-07215-0. MR2022367 (2004i:47065)

[176] Martin Mathieu and Ahmed Ramzi Sourour, *Hereditary properties of spectral isometries*, Arch. Math. (Basel) **82** (2004), no. 3, 222–229, DOI 10.1007/s00013-003-0595-5. MR2053626 (2004m:47073)

[177] Martin Mathieu and Gerhard J. Schick, *Spectrally bounded operators from von Neumann algebras*, J. Operator Theory **49** (2003), no. 2, 285–293. MR1991740 (2004c:47079)

[178] Martin Mathieu and Gerhard J. Schick, *First results on spectrally bounded operators*, Studia Math. **152** (2002), no. 2, 187–199, DOI 10.4064/sm152-2-6. MR1916549 (2003e:47066)

[179] Martin Mathieu, *Elementary operators on prime C^*-algebras. II*, Glasgow Math. J. **30** (1988), no. 3, 275–284, DOI 10.1017/S0017089500007369. MR964574 (90h:46093)

[180] S. Mazur and S. Ulam, *Sur les transformations isométriques d'espaces vectoriels normés*, C. R. Acad. Sci. Paris, **194** (1932) 946-948.

[181] M. Mbekhta and M. Oudghiri, *Additive mappings preserving minimum and surjectivity moduli*, Acta Sci. Math. (Szeged) **76** (2010), no. 1-2, 207–215. MR2668417 (2011f:47067)

[182] Mostafa Mbekhta, *Linear maps preserving the minimum and surjectivity moduli of operators*, Oper. Matrices **4** (2010), no. 4, 511–518, DOI 10.7153/oam-04-27. MR2777295 (2011m:47069)

[183] M. Mbekhta and P. Šemrl, *Linear maps preserving semi-Fredholm operators and generalized invertibility*, Linear Multilinear Algebra **57** (2009), no. 1, 55–64, DOI 10.1080/03081080701559500. MR2536067 (2010f:47029)

[184] Mostafa Mbekhta, *Linear maps preserving the set of Fredholm operators*, Proc. Amer. Math. Soc. **135** (2007), no. 11, 3613–3619 (electronic), DOI 10.1090/S0002-9939-07-08874-0. MR2336577 (2008h:47075)

[185] Mostafa Mbekhta, *Linear maps preserving the generalized spectrum*, Extracta Math. **22** (2007), no. 1, 45–54. MR2368700 (2008i:47077)

[186] Mostafa Mbekhta, Leiba Rodman, and Peter Šemrl, *Linear maps preserving generalized invertibility*, Integral Equations Operator Theory **55** (2006), no. 1, 93–109, DOI 10.1007/s00020-006-1421-9. MR2226639 (2007e:47055)

[187] T. L. Miller, V. G. Miller, and M. M. Neumann, *Local spectral properties of weighted shifts*, J. Operator Theory **51** (2004), no. 1, 71–88. MR2055805 (2005c:47045)

[188] Takeshi Miura and Dai Honma, *A generalization of peripherally-multiplicative surjections between standard operator algebras*, Cent. Eur. J. Math. **7** (2009), no. 3, 479–486, DOI 10.2478/s11533-009-0033-4. MR2534467 (2010g:47074)

[189] L. Molnár, *Selected preserver problems on algebraic structures of linear operators and on function spaces*, Lecture Notes in Mathematics, vol. 1895, Springer-Verlag, Berlin, 2007. MR2267033 (2007g:47056)

[190] Lajos Molnár, *Some characterizations of the automorphisms of $B(H)$ and $C(X)$*, Proc. Amer. Math. Soc. **130** (2002), no. 1, 111–120 (electronic), DOI 10.1090/S0002-9939-01-06172-X. MR1855627 (2002m:47047)

[191] Janez Mrčun, *Lipschitz spectrum preserving mappings on algebras of matrices*, Linear Algebra Appl. **215** (1995), 113–120, DOI 10.1016/0024-3795(93)00078-E. MR1317474 (95k:15014)

[192] Vladimir Müller, *Spectral theory of linear operators and spectral systems in Banach algebras*, Operator Theory: Advances and Applications, vol. 139, Birkhäuser Verlag, Basel, 2003. MR1975356 (2004c:47008)

[193] Matjaž Omladič and Peter Šemrl, *Additive mappings preserving operators of rank one*, Linear Algebra Appl. **182** (1993), 239–256, DOI 10.1016/0024-3795(93)90502-F. MR1207085 (94f:47039)

[194] Matjaž Omladič and Peter Šemrl, *Spectrum-preserving additive maps*, Linear Algebra Appl. **153** (1991), 67–72, DOI 10.1016/0024-3795(91)90210-N. MR1111386 (92e:47061)

[195] V. J. Pellegrini, *Numerical range preserving operators on a Banach algebra*, Studia Math. **54** (1975), no. 2, 143–147. MR0388104 (52 #8941)

[196] S. Pierce, et al., *A survey of linear preserver problems*, Linear and Multilinear Algebra, **33** (1992) 1-129.

[197] Thomas Ransford and Michael White, *Spectral characterization of algebraic elements*, Bull. London Math. Soc. **33** (2001), no. 1, 77–82, DOI 10.1112/blms/33.1.77. MR1798579 (2001j:46066)

[198] N. V. Rao and A. K. Roy, *Multiplicatively spectrum-preserving maps of function algebras. II*, Proc. Edinb. Math. Soc. (2) **48** (2005), no. 1, 219–229, DOI 10.1017/S0013091504000719. MR2117721 (2005k:46125)

[199] N. V. Rao and A. K. Roy, *Multiplicatively spectrum-preserving maps of function algebras*, Proc. Amer. Math. Soc. **133** (2005), no. 4, 1135–1142 (electronic), DOI 10.1090/S0002-9939-04-07615-4. MR2117215 (2005k:46124)

[200] Heinrich Raubenheimer, *On quasinilpotent equivalence of finite rank elements in Banach algebras*, Czechoslovak Math. J. **60(135)** (2010), no. 3, 589–596, DOI 10.1007/s10587-010-0045-z. MR2672403 (2011j:46076)

[201] C. E. Rickart, *General theory of Banach algebras*, Kreiger, New York, 1974.

[202] Leiba Rodman and Peter Šemrl, *A localization technique for linear preserver problems*, Linear Algebra Appl. **433** (2010), no. 11-12, 2257–2268, DOI 10.1016/j.laa.2010.07.032. MR2736150 (2011k:15051)

[203] M. Roitman and Y. Sternfeld, *When is a linear functional multiplicative?*, Trans. Amer. Math. Soc. **267** (1981), no. 1, 111–124, DOI 10.2307/1998572. MR621976 (82j:46061)

[204] B. Russo, *Linear mappings of operator algebras*, Proc. Amer. Math. Soc. **17** (1966), 1019–1022. MR0198269 (33 #6428)

[205] B. Russo and H. A. Dye, *A note on unitary operators in C^*-algebras*, Duke Math. J. **33** (1966), 413–416. MR0193530 (33 #1750)

[206] James W. Rowell, *Unilateral Fredholm theory and unilateral spectra*, Proc. Roy. Irish Acad. Sect. A **84** (1984), no. 1, 69–85. MR771647 (86g:46076)

[207] Christoph Schmoeger, *Atkinson theory and holomorphic functions in Banach algebras*, Proc. Roy. Irish Acad. Sect. A **91** (1991), no. 1, 113–127. MR1173164 (93f:46074)

[208] Peter Šemrl, *Invertibility preservers on central simple algebras*, J. Algebra **408** (2014), 42–60, DOI 10.1016/j.jalgebra.2014.03.012. MR3197170

[209] Peter Šemrl and Ahmed Ramzi Sourour, *Order preserving maps on Hermitian matrices*, J. Aust. Math. Soc. **95** (2013), no. 1, 129–132, DOI 10.1017/S1446788713000104. MR3123746

[210] Peter Šemrl, *Maps on matrix spaces*, Linear Algebra Appl. **413** (2006), no. 2-3, 364–393, DOI 10.1016/j.laa.2005.03.011. MR2198941 (2007a:15002)

[211] Peter Šemrl, *Invertibility preserving linear maps and algebraic reflexivity of elementary operators of length one*, Proc. Amer. Math. Soc. **130** (2002), no. 3, 769–772 (electronic), DOI 10.1090/S0002-9939-01-06177-9. MR1866032 (2002g:47075)

[212] Peter Šemrl, *Spectrally bounded linear maps on $\mathcal{B}(H)$*, Quart. J. Math. Oxford Ser. (2) **49** (1998), no. 193, 87–92, DOI 10.1093/qjmath/49.193.87. MR1617339 (2000b:47090)

[213] Peter Šemrl, *Linear mappings that preserve operators annihilated by a polynomial*, J. Operator Theory **36** (1996), no. 1, 45–58. MR1417185 (97i:47071)

[214] Peter Šemrl, *Linear maps that preserve the nilpotent operators*, Acta Sci. Math. (Szeged) **61** (1995), no. 1-4, 523–534. MR1377381 (97e:47052)

[215] Peter Šemrl, *Linear mappings preserving square-zero matrices*, Bull. Austral. Math. Soc. **48** (1993), no. 3, 365–370, DOI 10.1017/S0004972700015811. MR1248039 (95a:15001)

[216] Peter Šemrl, *Two characterizations of automorphisms on $B(X)$*, Studia Math. **105** (1993), no. 2, 143–149. MR1226624 (94h:47064)

[217] M. H. Shirdarreh Haghighi, *Invertibility preserving linear mappings into $M_2(\mathbb{C})$*, Studia Math. **186** (2008), no. 1, 41–45, DOI 10.4064/sm186-1-5. MR2395303 (2009a:46089)

[218] Haïkel Skhiri, *Linear maps preserving the minimum and surjectivity moduli of Hilbert space operators*, J. Math. Anal. Appl. **358** (2009), no. 2, 320–326, DOI 10.1016/j.jmaa.2009.05.007. MR2532509 (2010g:47076)

[219] Haïkel Skhiri, *Reduced minimum modulus preserving in Banach space*, Integral Equations Operator Theory **62** (2008), no. 1, 137–148, DOI 10.1007/s00020-008-1612-7. MR2442907 (2009g:47094)

[220] Haïkel Skhiri, *A Mbekhta's conjecture for reduced minimum modulus preserving*, Acta Sci. Math. (Szeged) **74** (2008), no. 3-4, 853–862. MR2487950 (2010b:47095)

[221] M. F. Smiley, *Jordan homomorphisms onto prime rings*, Trans. Amer. Math. Soc. **84** (1957), 426–429. MR0083484 (18,715b)

[222] A. R. Sourour, *Invertibility preserving linear maps on $\mathcal{L}(X)$*, Trans. Amer. Math. Soc. **348** (1996), no. 1, 13–30, DOI 10.1090/S0002-9947-96-01428-6. MR1311919 (96f:47069)

[223] A. R. Sourour, *The Gleason-Kahane-Żelazko theorem and its generalizations*, Functional analysis and operator theory (Warsaw, 1992), Banach Center Publ., vol. 30, Polish Acad. Sci., Warsaw, 1994, pp. 327–331. MR1285618 (95i:46066)

[224] Thomas Tonev and Aaron Luttman, *Algebra isomorphisms between standard operator algebras*, Studia Math. **191** (2009), no. 2, 163–170, DOI 10.4064/sm191-2-4. MR2472195 (2010e:46049)

[225] Aleksandar Torgašev, *On operators with the same local spectra*, Czechoslovak Math. J. **48(123)** (1998), no. 1, 77–83, DOI 10.1023/A:1022467611697. MR1614080 (99c:47005)

[226] Klaus Vala, *Sur les éléments compacts d'une algèbre normée* (French), Ann. Acad. Sci. Fenn. Ser. A I No. **407** (1967), 8. MR0222642 (36 #5692)

[227] W. Żelazko, *A characterization of multiplicative linear functionals in complex Banach algebras*, Studia Math. **30** (1968), 83–85. MR0229042 (37 #4620)

[228] Wen Zhang and Jinchuan Hou, *Maps preserving peripheral spectrum of Jordan semi-triple products of operators*, Linear Algebra Appl. **435** (2011), no. 6, 1326–1335, DOI 10.1016/j.laa.2011.03.008. MR2807155 (2012e:47114)

Department of Mathematics, 215 Carnegie Building, Syracuse University, Syracuse, New York 13244

E-mail address: abourhim@syr.edu

Département de mathématiques et de statistique, Université Laval, Québec, QC, G1V 0A6, Canada

E-mail address: javad.mashreghi@mat.ulaval.ca

Contemporary Mathematics
Volume **638**, 2015
http://dx.doi.org/10.1090/conm/638/12805

Commutants of finite Blaschke product multiplication operators

Carl C. Cowen and Rebecca G. Wahl

ABSTRACT. If ψ is a bounded analytic function on the unit disk $\mathbb{D}$, the analytic multiplication operator, or Toeplitz operator, T_ψ, on a Hilbert space of analytic functions on the disk, $\mathcal{H}$, is defined by $(T_\psi f)(z) = \psi(z)f(z)$. The (unilateral) shift on $\mathcal{H}$ is the operator of multiplication by z, that is, T_z. Several authors have studied commutants of such operators on the Hardy Hilbert space, $H^2(\mathbb{D})$, but the topic is interesting in other spaces as well, such as the Bergman space $A^2(\mathbb{D})$. In this paper we characterize operators on $\mathcal{H}$, for many Hilbert spaces of analytic functions, that commute with T_B where B is a finite Blaschke product, including on the Bergman space $A^2(\mathbb{D})$ and standard weight Bergman spaces. In particular, we show that the operators in the commutant map H^∞ into H^∞ and that the commutants on the various spaces are 'the same' in the sense that their description is the same and they have the same restrictions to H^∞. The techniques used are based on a detailed understanding of some of the invariant subspaces of T_B.

1. Introduction

Our goal in this paper is, for a given finite Blaschke product B, to give a complete description of the operators on a Hilbert space of analytic functions $\mathcal{H}$ that commute with T_B, the operator of multiplication by B. In particular, we will show, under quite general conditions, that if S commutes with T_B, then S maps $H^\infty(\mathbb{D})$ into itself (Theorem 15). In addition, we prove that for these spaces, the commutants of T_B are all 'the same' (Theorem 16). Since the commutants of finite Blaschke product Toeplitz operators are 'well known' for the Hardy space H^2, we can say that, in some sense, these are the 'same' operators that commute with T_B on other Hilbert spaces. The primary tool in this study is a detailed understanding of the invariant subspaces of the operators T_B.

In this paper, we consider Hilbert spaces, $\mathcal{H}$, of analytic functions (see [**8**]) on the unit disk $\mathbb{D}$ that satisfy the following conditions:

(I) The constant function $1(z) = 1$ for z in $\mathbb{D}$ is in $\mathcal{H}$ and $\|1\| = 1$

(II) For each α in the unit disk, the linear functional $f \mapsto f(\alpha)$ is continuous on $\mathcal{H}$

2010 *Mathematics Subject Classification.* Primary 47B32; Secondary 47B33, 47B38, 47A15.
Key words and phrases. Weighted composition operator, multiple-valued weighted composition operator, shift-invariant subspace, multiplication operator, analytic Toeplitz operator, commutant.

(III) For each ψ in $H^\infty(\mathbb{D})$, the operator T_ψ given by $(T_\psi f)(z) = \psi(z)f(z)$ for f in $\mathcal{H}$ is bounded on $\mathcal{H}$.

and

(IV) If α is in $\mathbb{D}$ and f is in $\mathcal{H}$ with $f(\alpha) = 0$, then $f/(z-\alpha)$ is also in $\mathcal{H}$.

Since $\mathcal{H}$ is a Hilbert space, condition (II) implies that for each α in the unit disk, there is a vector, K_α, in $\mathcal{H}$ for which we have $f(\alpha) = \langle f, K_\alpha \rangle$ for each f in $\mathcal{H}$. Conditions (I) and (III) say that both $\mathcal{H}$ and the multiplier algebra of $\mathcal{H}$ contain $H^\infty(\mathbb{D})$. It follows from condition (II) that the multiplier algebra of $\mathcal{H}$ is contained in $H^\infty(\mathbb{D})$, so the spaces we consider have multiplier algebra H^∞. If ψ is a bounded analytic function on the unit disk $\mathbb{D}$, the operator T_ψ in condition (III) is called an analytic multiplication operator, or analytic Toeplitz operator on $\mathcal{H}$. The fact that $(\lambda - \psi)^{-1}$ is a bounded analytic function if λ is not in the closure of $\psi(\mathbb{D})$ means that conditions (II) and (III) imply $\|T_\psi\| = \|\psi\|_\infty$ and this means $\|\psi\| \leq \|\psi\|_\infty$ for ψ in $H^\infty(\mathbb{D})$.

The classical Hardy Hilbert space $H^2(\mathbb{D})$, also denoted H^2, is the Hilbert space of functions f analytic on the unit disk $\mathbb{D}$ satisfying

$$\sup_{0<r<1} \int_0^{2\pi} |f(re^{i\theta})|^2 \frac{d\theta}{2\pi} < \infty$$

Similarly, we denote by $A^2(\mathbb{D})$, or A^2, the classical Bergman space, the Hilbert space of analytic functions that satisfy

$$\int_{\mathbb{D}} |f(z)|^2 \frac{dA(z)}{\pi} < \infty$$

When the above inequalities are satisfied, the left-hand side is the square of the norm of f in H^2 or A^2, respectively. The conditions (I), (II), (III), and (IV) hold for $H^2(\mathbb{D})$, $A^2(\mathbb{D})$, and many other spaces, such as many weighted Hardy spaces and the standard weight Bergman spaces. On the other hand, condition (III) does not hold for the Dirichlet space because not all bounded functions are multipliers of the Dirichlet space. Nevertheless, versions of the main theorems are true in that setting as well. Much of the work of this paper is taken directly from the ideas of Cowen's paper "The commutant of an analytic Toeplitz operator" [**3**] written in the Hardy space setting, but it is closely related to ideas of Deddens and Wong [**9**], Shields and Wallen [**12**], Thomson [**13–16**], and more recent authors such as Douglas, Sun, and Zheng [**11**] and Douglas, Putinar, and K. Wang [**10**]. While the statements of Theorem 15 and Theorem 16 date back to the 1970's and the work of Thomson and Cowen just cited, a general framework for this kind of operator was introduced only much more recently in work of Cowen and Gallardo Gutiérrez [**7**] in a different context, that of adjoints of composition operators, and were called "multiple-valued weighted composition operators".

The authors are grateful for the advice of Sheldon Axler in the preparation of this paper, and especially grateful for his willingness to let us include his heretofore unpublished result, Theorem 8, which expands the collection of spaces to which our ideas apply.

It is well known that the kernel functions K_α, arising from condition (II), are $K_\alpha(z) = (1-\overline{\alpha}z)^{-1}$ and $K_\alpha(z) = (1-\overline{\alpha}z)^{-2}$ for the spaces H^2 and A^2, respectively. This motivates the following definition:

Definition For $\kappa \geq 1$, the Hilbert space H^2_κ is the space for which the monomials z^n form an orthogonal basis and the kernel functions are given by

$$K_\alpha(z) = (1-\overline{\alpha}z)^{-\kappa} \qquad \text{for } \alpha \text{ and } z \text{ in } \mathbb{D}$$

It is well known that these spaces are equivalent to the standard weight Bergman spaces and that they satisfy the conditions above (for example, see [**8**, p. 27]).

2. General Theory

In this section, we identify some useful necessary and sufficient conditions for an operator to be in the commutant of an analytic Toeplitz operator.

The kernel functions for evaluating functions in $\mathcal{H}$ play a key role in this work because they are eigenvectors for the adjoints of analytic Toeplitz operators: for each f in $\mathcal{H}$,

$$\langle f, T_\psi^* K_\alpha\rangle = \langle T_\psi f, K_\alpha\rangle = \langle \psi f, K_\alpha\rangle = \psi(\alpha)f(\alpha) = \langle f, \overline{\psi(\alpha)}K_\alpha\rangle$$

Since this is true for all f in $\mathcal{H}$, we see that $T_\psi^* K_\alpha = \overline{\psi(\alpha)}K_\alpha$.

The following, in various forms, appears in papers [**3**, **9**, **13**–**16**], among others, and is the foundation of this paper. For example, in H^2, the set J in (3) of the theorem can be any subset of the disk that contains a non-Blaschke sequence.

THEOREM 1. **(Fundamental Lemma)** *For S a bounded operator on $\mathcal{H}$ and ψ in H^∞, the following are equivalent,*

(1) *S commutes with T_ψ*
(2) *For all α in $\mathbb{D}$, $S^*K_\alpha \perp (\psi - \psi(\alpha))\mathcal{H}$*
(3) *There is a set J in $\mathbb{D}$ for which f in $\mathcal{H}$ and $f(\alpha) = 0$ for α in J implies $f \equiv 0$ and for all α in J, $S^*K_\alpha \perp (\psi - \psi(\alpha))\mathcal{H}$.*

PROOF. (1) $\Longrightarrow$ (2): Let S be in $\{T_\psi\}'$. For all g in $\mathcal{H}$ and α in $\mathbb{D}$, we have

$$\begin{aligned}\langle(\psi - \psi(\alpha))g, S^*K_\alpha\rangle &= \langle ST_\psi g, K_\alpha\rangle - \psi(\alpha)\langle Sg, K_\alpha\rangle \\ &= \langle T_\psi Sg, K_\alpha\rangle - \psi(\alpha)\langle Sg, K_\alpha\rangle \\ &= \psi(\alpha)(Sg)(\alpha) - \psi(\alpha)(Sg)(\alpha) = 0\end{aligned}$$

and S^*K_α is in $((\psi - \psi(\alpha))\mathcal{H})^\perp$.
(2) $\Longrightarrow$ (3): Let $J = \mathbb{D}$.
(3) $\Longrightarrow$ (1): Let S be in $\mathcal{B}(\mathcal{H})$ with $S^*K_\alpha \perp (\psi - \psi(\alpha))\mathcal{H}$ for α in J. For g in $\mathcal{H}$ and α in J, we have

$$\begin{aligned}0 &= \langle(\psi - \psi(\alpha))g, S^*K_\alpha\rangle = (ST_\psi g)(\alpha) - \psi(\alpha)(Sg)(\alpha) \\ &= (ST_\psi g)(\alpha) - (T_\psi Sg)(\alpha) = (ST_\psi g - T_\psi Sg)(\alpha)\end{aligned}$$

Since J is not a zero-set for functions of $\mathcal{H}$, this means $(ST_\psi g - T_\psi Sg)(\alpha)$ for *all* points of the open disk. Since this is true for all g in $\mathcal{H}$, we have $ST_\psi = T_\psi S$ as desired. □

COROLLARY 2. *If ψ is in H^∞ and ω is a bounded analytic function on $\psi(\mathbb{D})$, then $\{T_\psi\}' \subset \{T_{\omega\circ\psi}\}'$.*

PROOF. Let β be in $\psi(\mathbb{D})$. Since $(\omega(z) - \omega(\beta))/(z - \beta)$ is bounded and analytic on $\psi(\mathbb{D})$, the function $(\omega\circ\psi) - \omega(\beta))/(\psi - \beta)$ is in H^∞. For α in $\mathbb{D}$, we have

$$(\omega\circ\psi) - \omega(\psi(\alpha)) = \frac{(\omega\circ\psi) - \omega(\psi(\alpha))}{\psi - \psi(\alpha)}(\psi - \psi(\alpha))$$

so that

$$(\omega \circ \psi) - \omega(\psi(\alpha))\mathcal{H} \subset (\psi - \psi(\alpha))\mathcal{H}$$

If S is in $\{T_\psi\}'$, then for all α in $\mathbb{D}$, $S^*K_\alpha \perp (\psi - \psi(\alpha))\mathcal{H}$ and this means $S^*K_\alpha \perp (\omega \circ \psi) - \omega(\psi(\alpha))\mathcal{H}$ and S is in $\{T_{\omega\circ\psi}\}'$ also. □

COROLLARY 3. *If ψ is in H^∞ and φ is a bounded, univalent, analytic function on $\psi(\mathbb{D})$, then $\{T_\psi\}' = \{T_{\varphi\circ\psi}\}'$.*

Easy examples show the converse of the first corollary is false, so a function for which it is true must be special in some way. This suggestion and the Fundamental Lemma suggest the following definition. If $\mathcal{V}$ is a set of vectors, we let $\bigvee \mathcal{V}$ be the closure of the subspace spanned by $\mathcal{V}$.

Definition For ψ in H^∞, we say ψ *is $\mathcal{H}$-ancestral* if for every α in $\mathbb{D}$, we have

$$\bigvee\{S^*K_\alpha : S \in \{T_\psi\}'\} = [(\psi - \psi(\alpha))\mathcal{H}]^\perp$$

Notice that $\{T_\psi\}' \subset \{T_\varphi\}'$ if and only if T_φ is in $\{T_\psi\}''$, so the following is also a statement about double commutants.

THEOREM 4. *Let ψ and φ be in H^∞. If ψ is ancestral and $\{T_\psi\}' \subset \{T_\varphi\}'$, then there is a function μ analytic on $\psi(\mathbb{D})$ so that $\varphi = \mu \circ \psi$.*

PROOF. For each α in $\mathbb{D}$ we have

$$[(\psi - \psi(\alpha)\mathcal{H}]^\perp = \bigvee\{S^*K_\alpha : S \in \{T_\psi\}'\} \subset \bigvee\{S^*K_\alpha : S \in \{T_\varphi\}'\} \subset [(\varphi - \varphi(\alpha)\mathcal{H}]^\perp$$

For w in $\psi(\mathbb{D})$, choose α in $\mathbb{D}$ with $\psi(\alpha) = w$, let $\mu(w) = \varphi(\alpha)$. The function μ is well-defined, for if $\psi(\alpha) = \psi(\beta)$, then $K_\beta \in [(\psi - \psi(\alpha)\mathcal{H}]^\perp \subset [(\varphi - \varphi(\alpha)\mathcal{H}]^\perp$ so $\varphi(\alpha) = \varphi(\beta)$ also. Moreover, μ is analytic. Indeed, if $w = \psi(\alpha)$ and $\psi'(\alpha) \neq 0$, and if ζ is the branch of ψ^{-1} defined in a neighborhood of w with $\zeta(w) = \alpha$, then $\mu = \varphi \circ \zeta$ in this neighborhood. Since μ is single valued, it follows immediately that μ is analytic on all of $\psi(\mathbb{D})$ and $\varphi = \mu \circ \psi$. □

COROLLARY 5. *If ψ_1 and ψ_2 are ancestral functions in H^∞ and $\{T_{\psi_1}\}' = \{T_{\psi_2}\}'$, then there is a univalent function μ so that $\psi_2 = \mu \circ \psi_1$.*

PROOF. Applying Theorem 4 twice, we see that there are analytic functions μ and $\widetilde{\mu}$ so that $\psi_2 = \mu \circ \psi_1$ and $\psi_1 = \widetilde{\mu} \circ \psi_2$. Therefore $\psi_1 = (\widetilde{\mu} \circ \mu) \circ \psi_1$ which means $\widetilde{\mu} \circ \mu$ is the identity on $\psi_1(\mathbb{D})$, and μ is univalent. □

The corollary shows that if two ancestral functions are not each composites of the other, then the commutants of their multiplication operators are actually different. This suggests a strategy for studying commutants of analytic multiplication operators: for each Hilbert space, $\mathcal{H}$, characterize the $\mathcal{H}$-ancestral functions and find criteria for finding ancestral functions associated with each analytic multiplier. In [**3**] it was shown that every inner function is H^2-ancestral, but that not all H^2-ancestral functions are inner.

Question: Is every inner function A^2-ancestral? H^2_κ-ancestral? $\mathcal{H}$-ancestral?

A consequence of the work below is that every finite Blaschke product *is* $\mathcal{H}$-ancestral.

3. Relations between Finite Blaschke Products and the Structure of $\mathcal{H}$

The first goal of this section is to state and prove the Theorem of Axler, Theorem 8, which is set in a more general setting: $\mathcal{X}$ is a Banach space of analytic functions on some bounded open set $\Omega \subset \mathbb{C}$ that satisfies the obvious generalizations to Ω and $\mathcal{X}$ of the conditions (I), (II), (III), and (IV). The following lemma will be used in the proof of Theorem 8, so it is stated in the Banach space setting as well, even though we will use the results only for Hilbert spaces.

LEMMA 6. *For Ω a bounded open set in $\mathbb{C}$, let $\mathcal{X}$ be a Banach space of analytic functions on Ω satisfying the conditions above. If λ is a point of Ω and m is a positive integer, the bounded operator $T_{(z-\lambda)^m}$ of multiplication of functions in $\mathcal{X}$ by $(z-\lambda)^m$ has closed range and a bounded left inverse that is defined on its range. Moreover, the range of $T_{(z-\lambda)^m}$ has co-dimension m in $\mathcal{X}$.*

PROOF. Because Ω is a bounded subset of the plane, the polynomial $z-\lambda$ is a function in $H^\infty(\Omega)$ which means $T_{z-\lambda}$ is a bounded operator by condition (III). The linear functional Λ defined on $\mathcal{X}$ by $\Lambda(g) = g(\lambda)$ is bounded by condition (II), so K_1=kernel(Λ) is a closed subspace of $\mathcal{X}$. On the other hand, if f belongs to K_1, then $f(\lambda) = 0$ and $g = f/(z-\lambda)$ is in $\mathcal{X}$ by condition (IV). We see that $T_{z-\lambda}g = f$, which means the range of $T_{z-\lambda}$ is the closed subspace K_1, also a Banach space. Define the operator V on K_1 by $Vf = f/(z-\lambda)$. Clearly, V is defined on all of the range of $T_{z-\lambda}$ and V is a left inverse for $T_{z-\lambda}$. Because $T_{z-\lambda}$ is a bounded operator, its graph is a closed subspace of $\mathcal{X} \times \mathcal{X}$. Since the graph of V is isometrically isomorphic to the graph of $T_{z-\lambda}$, the boundedness of V follows from the closed graph theorem. Since range($T_{z-\lambda}$) $= K_1$, the kernel of the linear functional Λ, the co-dimension of range($T_{z-\lambda}$) is 1. If $m = 1$, we are done.

If $m > 1$, then, define $K_2 = T_{z-\lambda}(K_1) = T_{(z-\lambda)^2}\mathcal{X}$. This is a closed subspace of K_1 because it is the kernel of $\Lambda \circ V$ acting on K_1 and V^2 is a left inverse of $T_{(z-\lambda)^2}$. Since K_2 has co-dimension 1 in K_1 and K_1 has co-dimension 1 in $\mathcal{X}$, the closed subspace K_2 which is the range of $T_{(z-\lambda)^2}$ has co-dimension 2 in $\mathcal{X}$. If $m = 2$, we are done. Continuing, we get $K_m = T_{(z-\lambda)^m}\mathcal{X}$ is a closed subspace of $\mathcal{X}$ of co-dimension m and V^m is a left inverse of $T_{(z-\lambda)^m}$ on its range. □

COROLLARY 7. *For Ω a bounded open set in $\mathbb{C}$, let $\mathcal{X}$ be a Banach space of analytic functions on Ω satisfying the conditions above. If p is a polynomial of degree k whose roots are in Ω, the bounded operator T_p defined by $T_p(g) = pg$ for g in $\mathcal{X}$ has closed range and a bounded left inverse that is defined on its range. Moreover, the range of T_p has co-dimension k in $\mathcal{X}$.*

PROOF. Factoring the polynomial p gives

$$p(z) = c(z-\lambda_1)^{m_1}(z-\lambda_2)^{m_2}\cdots(z-\lambda_n)^{m_n}$$

where $c \neq 0$ is a complex number, the λ_j are the roots of p with multiplicity m_j, and $k = \sum_j m_j$. The corollary follows by applying Lemma 6 successively for each of the roots λ_j because the constructions in the lemma work independently for the different roots. □

The following result is an extension of the ideas of and uses the tools from some of the important theorems from Axler's paper [**1**] and the paper of Axler, Conway, and McDonald [**2**]. It follows from their arguments that $A^2(\Omega)$ satisfies condition (IV).

THEOREM 8 (**Theorem of Axler**). *Suppose $\mathcal{X}$ is a Banach space of analytic functions on a bounded open set $\Omega \subset \mathbb{C}$ that satisfies the obvious generalizations to Ω and $\mathcal{X}$ of the conditions (I), (II), (III), and (IV). If f is in $H^\infty(\Omega)$ and f is bounded away from 0 near $\partial\Omega$, then $f\mathcal{X}$, the range of the bounded operator T_f, is a closed subspace of $\mathcal{X}$ with finite codimension and T_f has a bounded left inverse defined on $f\mathcal{X}$.*

PROOF. To say that f is bounded away from 0 near the boundary of Ω means there are $\delta > 0$ and $\epsilon > 0$ so that if z is in Ω such that $|w - z| < \delta$ for some w *not* in Ω, then $|f(z)| \geq \epsilon$. Since Ω is a bounded set, the subset $F = \{z \in \Omega : |z - w| \geq \delta \text{ for all } w \notin \Omega\}$ is compact. This means that there are at most finitely many points $\lambda_1, \lambda_2, \cdots, \lambda_n$ in Ω for which $f(\lambda) = 0$. Let m_j be the multiplicity of the zero of f at λ_j.

This means that f is divisible by the polynomial $p(z) = (z - \lambda_1)^{m_1}(z - \lambda_2)^{m_2} \cdots (z - \lambda_n)^{m_n}$, which has degree $k = \sum_{j=1}^n m_j$, and the function h given by

$$h(z) = \frac{f(z)}{(z - \lambda_1)^{m_1}(z - \lambda_2)^{m_2} \cdots (z - \lambda_n)^{m_n}}$$

is analytic and does not vanish in Ω. Indeed, since F is compact and h is non-zero on F, the minimum of $|h(z)|$ for z in F is positive and since each λ_j is in F and $|f(z)| \geq \epsilon$ for z in $\Omega \setminus F$, we see that there is $\zeta > 0$ so that $|h(z)| \geq \zeta$ for all z in Ω. In addition, since h is analytic on Ω, the largest values of $|h(z)|$ occur near the boundary of Ω. The zeros of p have distance more than δ from the boundary of Ω, so near the boundary $|p(z)| \geq \delta^k$ and near the boundary, $|h(z)| \leq \delta^{-k}\|f\|_\infty < \infty$. This means that h is invertible in $H^\infty(\Omega)$ and T_h is an invertible operator on $\mathcal{X}$.

Thus, we have factored T_f as $T_f = T_p T_h$. It follows that $f\mathcal{X}$, the range of T_f, is range$(T_p T_h)$ which is range(T_p) because T_h is invertible with range$(T_h) = \mathcal{X}$. Now, by Corollary 7, range(T_f) = range(T_p) is a closed subspace of $\mathcal{X}$ with co-dimension k.

Finally, suppose V is a left inverse of T_p defined on range(T_p) = range(T_f), then, because $T_f = T_p T_h = T_h T_p$, we see that

$$(VT_h^{-1})T_f = V(T_h^{-1}T_h)T_p = VT_p = I$$

so VT_h^{-1} is a left inverse of T_f defined on range(T_f). □

COROLLARY 9. *Let $\mathcal{H}$ be a Hilbert space of analytic functions on the disk $\mathbb{D}$ that satisfies the conditions (I), (II), (III), and (IV). If B is a finite Blaschke product of order n, then the bounded operator T_B defined by $T_B(g) = Bg$ for g in $\mathcal{H}$ has closed range and a bounded left inverse that is defined on its range. Moreover, the range of T_B has co-dimension n in $\mathcal{H}$.*

PROOF. A Blaschke product, B, of order n is analytic on the open unit disk and has n zeros in the disk (counting multiplicity), is continuous on the closed disk and has $|B(e^{i\theta})| = 1$ for all real θ. From this, we see that B is in $H^\infty(\mathbb{D})$ and that there is δ with $0 < \delta < .5$ so that if $|z| > 1 - \delta$, then $|B(z)| > .5$, that is, B is bounded away from 0 near the boundary of $\mathbb{D}$.

From Theorem 8, it follows that T_B has closed range with co-dimension n in $\mathcal{H}$ and has a left inverse defined on range(T_B). □

4. Commutant of T_B for B a Finite Blaschke Product

Because Toeplitz operators on H^2 with inner function symbol are pure isometries, their commutants can be easily described as the bounded operators that can be represented as a lower triangular Block Toeplitz matrices with respect to the decomposition of H^2 as $H^2 = \bigoplus_{n=0}^{\infty} \psi^n \mathcal{W}$ where $\mathcal{W}$ is the 'wandering subspace' $\mathcal{W} = H^2 \ominus \psi H^2$. This was done by Deddens and Wong [**9**, Thm. 1], although in more generality. This perspective can be generalized to other spaces, $\mathcal{H}$, but the generalization is less useful because the inner Toeplitz operators are typically not isometries on other spaces.

The perspective we will take, based on the ideas of Section 2, was developed by Thomson [**13**–**16**] and Cowen [**3**–**6**]. The main goal of this section is to prove an extension, Theorem 15, of the Proposition [**3**, p. 6] that asserts boundedness on H^∞ of operators in the commutant of T_B acting on H^2.

Definition If B is finite Blaschke product, E_B, the *exceptional set for* B, is defined by

$$E_B = \{z \in \mathbb{D} : B(z) = B(w) \text{ for some } w \text{ with } B'(w) = 0\}$$

We will write Ω_B for the open set $\Omega_B = \mathbb{D} \setminus E_B$.

If B is a Blaschke product of order n, then B is an n-to-1 map of the disk onto the disk and it has at most $n-1$ critical points in the disk and none on the unit circle. In particular, this implies the exceptional set includes at most $n(n-1)$ points. Thus, B is an n-to-1 map of the set Ω_B onto itself, and it is locally univalent on this set.

For each point α in Ω_B, there are n (distinct) points $\alpha_1 = \alpha, \alpha_2, \cdots, \alpha_n$ such that $B(\alpha_j) = B(\alpha)$ and for ζ in $\mathbb{D}$, with $\zeta \neq \alpha_j$ for any j, $B(\zeta) \neq B(\alpha)$. Moreover, if $U \subset \Omega_B$ is a simply connected open neighborhood of α, there are analytic functions β_j defined on U such that $\beta_j(\alpha) = \alpha_j$ and $B(\beta_j(z)) = B(z)$ for each j and all z in U. Note that for each α and each U and the α_j and β_j as described, the function $\beta_1(z) \equiv z$, but for $j \neq 1$, the numbering would not need to be consistent throughout Ω_B. This is because, although the β_j are locally analytic and arbitrarily continuable in Ω_B, for $j \neq 1$, the β_j are usually not single-valued analytic functions.

In Section 7, we will use a different perspective, that of Riemann surfaces, to give a more detailed description of the operators in the commutant of T_B. We will associate a Riemann surface W_B with the multiple-valued mapping $B^{-1} \circ B$ that is an n-sheeted cover of the open subset Ω_B in the disk. The locally analytic maps β_j are associated with transition maps between the sheets. The non-compact Riemann surface W_B is never connected for $n \geq 2$ because one sheet of W_B is always associated with the single-valued branch $\beta_1(z) = z$ and is isomorphic to Ω_B. The Riemann surface W_B has n components for some Blaschke products, such as $B(z) = z^n$, but usually has a more interesting structure. For example, for most Blaschke products of order 3, there will be only one component besides the trivial component and it will be a 2-sheeted cover over Ω_B. The book of Veech [**17**] has a readable, elementary description of Riemann surfaces.

These ideas lead to the following specialization of the Fundamental Lemma (Thm. 1) to the finite Blaschke product case.

THEOREM 10. **(Fundamental Lemma 2)** *For S a bounded operator on $\mathcal{H}$ and B a finite Blaschke product, the following are equivalent,*

(1) *S commutes with T_B*
(2) *For each α in Ω_B, writing $\alpha_1 = \alpha$, α_2, $\cdots$, α_n for the n points with $B(\alpha_j) = B(\alpha)$, there are constants c_j for $j = 1, \cdots, n$ so that*

$$S^* K_\alpha = \sum_{j=1}^{n} \overline{c_j} K_{\alpha_j} \tag{1}$$

(3) *There is a subset J of Ω_B for which f in $\mathcal{H}$ and $f(\alpha) = 0$ for α in J implies $f \equiv 0$ and for each α in J, writing $\alpha_1 = \alpha$, α_2, $\cdots$, α_n for the n points with $B(\alpha_j) = B(\alpha)$, there are constants c_j for $j = 1, \cdots, n$ so that*

$$S^* K_\alpha = \sum_{j=1}^{n} \overline{c_j} K_{\alpha_j}$$

PROOF. This is an immediate application of Theorem 1 to the case $\psi = B$ because Ω_B or a sufficiently large subset of Ω_B are subsets of $\mathbb{D}$ that fit part (3) of the theorem.

Let α be a point of Ω_B. To see that conditions (2) or (3) of the above statement imply part (3) of Theorem 1, we will apply Corollary 9 to the finite Blaschke product $\widetilde{B}$, also of order n, given by

$$\widetilde{B}(z) = \frac{B(z) - B(\alpha)}{1 - \overline{B(\alpha)}B(z)}$$

For any α in $\mathbb{D}$, the value of B satisfies $|B(\alpha)| < 1$, so the function $h(z) = (1 - \overline{B(\alpha)}B(z))^{-1}$ is an invertible H^∞ function.

Condition (3) of Theorem 1 applied to $\psi = B$ relates $S^* K_\alpha$ to the subspace $(B - B(\alpha))\mathcal{H}$. From the fact that h is an invertible H^∞ function and Corollary 9 and its proof, we see

$$(B - B(\alpha))\mathcal{H} = (B - B(\alpha))T_h\mathcal{H} = \widetilde{B}\mathcal{H} = \text{span}\{K_{\alpha_j} : j = 1, \cdots n\}^\perp$$

Since $\widetilde{B}\mathcal{H}$ has co-dimension n and the $\{K_{\alpha_j} : j = 1, \cdots n\}$ are a basis for the orthogonal complement, the statement $S^* K_\alpha$ is orthogonal to $(B - B(\alpha))\mathcal{H}$ is just the linear combination statement in (2) or (3) of the above Theorem. □

We can easily deduce more information about the coefficients in Equation (1).

THEOREM 11. *For S a bounded operator on $\mathcal{H}$ that commutes with T_B, let be U a simply connected open subset of Ω_B, and let the β_j be the n branches of $B^{-1} \circ B$ that are defined in U. Then the functions $c_j(\alpha)$ for $j = 1, \cdots, n$ that satisfy*

$$S^*(K_\alpha) = \sum_{j=1}^{n} \overline{c_j(\alpha)} K_{\beta_j(\alpha)} \tag{2}$$

are each analytic in U.

PROOF. Because the open set U is simply connected, each of the β_j are single valued analytic functions in U. For each α in U, consider the polynomials $q_k(z, \alpha)$ of degree $n - 1$, with roots $\beta_j(\alpha)$ for $j \neq k$, given by

$$q_k(z, \alpha) = \prod_{j \neq k} (z - \beta_j(\alpha))$$

Because each β_j has values in the disk, these polynomials are in H^∞ and for each h in $\mathcal{H}$, the functions $q_k(\cdot,\alpha)h$ and $S(q_k(\cdot,\alpha)h)$ are also in $\mathcal{H}$ and are analytic functions in the disk. For each α in U

$$\begin{aligned}
(3) \quad S(q_k(\cdot,\alpha)h)(\alpha) &= \langle Sq_k(\cdot,\alpha)h, K_\alpha\rangle \\
&= \langle q_k(\cdot,\alpha)h, S^*K_\alpha\rangle = \left\langle q_k(\cdot,\alpha)h, \sum_{j=1}^{n} \overline{c_j(\alpha)}K_{\beta_j(\alpha)}\right\rangle \\
&= \sum_{j=1}^{n} c_j(\alpha)\langle q_k(\cdot,\alpha)h, K_{\beta_j(\alpha)}\rangle \\
(4) \quad &= c_k(\alpha)q_k(\beta_k(\alpha),\alpha)h(\beta_k(\alpha))
\end{aligned}$$

since $q_k(\beta_j(\alpha),\alpha) = 0$ unless $j = k$.

Now if h is an analytic function in the disk such that h and $1/h$ are in H^∞, then, since h does not vanish in $\mathbb{D}$ and $q_k(\beta_k(\alpha),\alpha) = \prod_{j\neq k}(\beta_k(\alpha) - \beta_j(\alpha)) \neq 0$ in U, we see that

$$c_k(\alpha) = \frac{(S(q_k(\cdot,\alpha)h))(\alpha)}{h(\beta_k(\alpha))\prod_{j\neq k}(\beta_k(\alpha) - \beta_j(\alpha))}$$

Since the right hand side of this equation is analytic in α in the set U, the function c_k is also an analytic function on U and, since k was an arbitrary positive integer not bigger than n, the result is proved. □

COROLLARY 12. *If S is a bounded operator on $\mathcal{H}$ that commutes with T_B, the number α is in Ω_B, and the functions β_j and c_j for $j = 1,\cdots,n$ are as in Theorem* 11, *then for every f in $\mathcal{H}$,*

$$(Sf)(\alpha) = \sum_{j=1}^{n} c_j(\alpha)f(\beta_j(\alpha))$$

PROOF. This follows immediately from Equation (2) and the observation that

$$Sf(\alpha) = \langle Sf, K_\alpha\rangle = \langle f, S^*K_\alpha\rangle = \left\langle f, \sum_{j=1}^{n}\overline{c_j(\alpha)}K_{\beta_j(\alpha)}\right\rangle = \sum_{j=1}^{n} c_j(\alpha)f(\beta_j(\alpha))$$

□

5. A Special Annulus and the Wrapping Function

For B a finite Blaschke product of order n and E_B the exceptional set for B, let $\delta_i = \min\{1 - |z| : z \in E_B\}$. In fact, B is also an n - to - 1 map of the Riemann sphere outside of the closed disk onto itself, with $n - 1$ critical points outside the disk, so we let

$$\widetilde{E}_B = \{1 < |z| \leq \infty : B(z) = B(w) \text{ for some } w \text{ with } B'(w) = 0\}$$

and $\delta_o = \min\{|z| - 1 : z \in \widetilde{E}_B\}$. Since B is analytic in a neighborhood of the closed disk and is an n - to - 1 map of $\mathbb{D}$ onto itself, then B is an n - to - 1 map of the unit circle, $\partial\mathbb{D}$, onto itself and B' is non-zero on $\partial\mathbb{D}$. Choose θ_1 so that $-\pi < \theta_1 \leq \pi$ and $B(e^{i\theta_1}) = 1$ and let 'log' denote the branch of the logarithm function that gives $\log(B(e^{i\theta_1})) = 0$. Define the function ℓ to be the continuous function on the interval $\theta_1 \leq \theta \leq \theta_1 + 2\pi$ for which $\ell(\theta) = -i\log(B(e^{i\theta}))$. We will call ℓ the *wrapping function of* B. It is easily seen that ℓ is a strictly increasing function on the interval $[\theta_1, \theta_1 + 2\pi]$ with $\ell(\theta_1) = 0$ and $\ell(\theta_1 + 2\pi) = 2n\pi$.

Let δ_3 be the minimum of δ_i, δ_o, and $\delta_2 = \min\{|z|-1 : |z| > 1 \text{ and } |B(z)| < 2\}$, and let A_0 be the annulus $A_0 = \{z : |1 - |z|| < \delta_3\}$. Now, let

$$\delta = \min\{|1 - |z|| : z \notin A_0 \cap B(A_0) \cap B^{-1}(A_0)\}$$

and let A be the annulus $A = \{z : 1 - \delta < |z| < 1 + \delta\}$. Note that if z is in A, then z, $B(z)$, and $B^{-1}(z)$ are in A_0, and that $\partial\mathbb{D} \subset A$. In addition, let $\hat{A}$ be the compact annulus $\hat{A} = \{z : 1 - \delta/2 \le |z| \le 1 + \delta/2\}$.

For $j = 1, 2, \cdots, n+1$, let θ_j be the point of $[\theta_1, \theta_1 + 2\pi]$ such that $\ell(\theta_j) = 2\pi(j-1)$. Further, let $\beta_{n+1}(w) = \beta_1(w) = w$, $\beta_2(w)$, $\cdots$, $\beta_n(w)$ be the branches of $B^{-1} \circ B$ defined on A such that $\beta_j(e^{i\theta_1}) = e^{i\theta_j}$.

Since B is an n- to -1 map of the Riemann sphere onto itself and since no point of the exceptional sets E_B or $\widetilde{E}_B$ are in A, for each w in A, the n numbers $\beta_1(w)$, $\beta_2(w)$, $\cdots$, $\beta_n(w)$ are distinct. This means that the β_j are single-valued analytic functions on A and $|\beta_k(w) - \beta_j(w)| \ne 0$ for $1 \le j < k \le n$.

Now, let η be defined by

$$3\eta = \min_{1 \le j < k \le n} \{|\beta_k(w) - \beta_j(w)| : w \in \hat{A}\}$$

As noted above, the β_j's are continuous, none of the numbers is zero, and $\hat{A}$ is compact, so we see that $\eta > 0$. This means that

$$\{w \in \mathbb{C} : B(w) = B(z) \text{ for some } z \text{ with } |z - e^{i\theta}| < \eta \text{ and } z \in \hat{A}\}$$

is an open subset of A consisting of n components, each containing exactly one of the n points ζ on $\partial\mathbb{D}$ for which $B(\zeta) = B(e^{i\theta})$.

We record these observations in the following Proposition.

PROPOSITION 13. *Let B be a finite Blaschke product of order n. There is a number $\delta > 0$ so that on the annulus $A = \{z : 1 - \delta < |z| < 1 + \delta\}$ there are n branches $\beta_1(z) \equiv z$, β_2, $\cdots$, β_n, of $B^{-1} \circ B$, consistently numbered on all of A, that take distinct values at each point of A. Moreover, for the compact annulus $\hat{A} = \{z : 1 - \delta/2 \le |z| \le 1 + \delta/2\}$ there is a number $\eta > 0$ so that for all w in $\hat{A}$ and $j \ne k$, we have $|\beta_j(w) - \beta_k(w)| \ge \eta/3$.*

6. Consequences for Commutants of Operators on H^2_κ

In general, we have been assuming Hilbert spaces of functions that are analytic on the unit disk and have properties (I) to (IV) listed at the beginning. In much of this section, we will need better control over the spaces in order to get the desired results. We begin with just the earlier assumptions on the space $\mathcal{H}$.

In Theorem 11, we showed the coefficients in Equation (1) are conjugates of locally analytic functions. In this section we need to estimate their size for α near the boundary of the disk. Equation (4) includes the equality

$$c_k(\alpha) q_k(\beta_k(\alpha), \alpha) h(\beta_k(\alpha)) = \langle S q_k(\cdot, \alpha) h, K_\alpha \rangle$$

This leads to

$$|c_k(\alpha)||q_k(\beta_k(\alpha), \alpha)||h(\beta_k(\alpha))| = |\langle S q_k(\cdot, \alpha) h, K_\alpha \rangle| \le \|S\| \|q_k(\cdot, \alpha)\|_\infty \|h\| \|K_\alpha\|$$

and this inequality holds for every function h in $\mathcal{H}$ and each k with $1 \le k \le n$. Now $\|q_k(\cdot,\alpha)\|_\infty \le 2^{n-1}$ because q_k has $n-1$ factors and each is a difference between two points of the disk and the term $|q_k(\beta_k(\alpha),\alpha)|$ is $\prod_{j\neq k}|\beta_k(\alpha)-\beta_j(\alpha)|$. We need to explore the consequences of this for each possible k and we will see that an informative choice for h is $h = K_{\beta_k(\alpha)}$. With this choice for h, using α in $\hat{A}\cap\mathbb{D}$, and solving the inequality for $|c_k(\alpha)|$ gives

$$\begin{aligned} |c_k(\alpha)| &\le \frac{\|S\|\|q_k(\cdot,\alpha)\|_\infty\|K_{\beta_k(\alpha)}\|\|K_\alpha\|}{|q_k(\beta_k(\alpha),\alpha)||K_{\beta_k(\alpha)}(\beta_k(\alpha))|} \le \frac{2^{n-1}\|S\|\|K_{\beta_k(\alpha)}\|\|K_\alpha\|}{\prod_{j\neq k}|\beta_k(\alpha)-\beta_j(\alpha)|\|K_{\beta_k(\alpha)}\|^2} \\ &\le \|S\|\left(\frac{6}{\eta}\right)^{n-1}\frac{\|K_\alpha\|}{\|K_{\beta_k(\alpha)}\|} \qquad (5) \end{aligned}$$

Equation (5) holds in all spaces $\mathcal{H}$ considered in this paper, but it is clear that detailed information about the size of c_k near the unit circle will depend on the nature of the kernel functions in $\mathcal{H}$, but for relatively generic spaces this information is not available.

It is for this reason that for the rest of the paper, we restrict attention to the spaces, H^2_κ, which are defined by explicit representations for the kernel functions. In addition to properties (I) to (IV), we will assume the monomials form an orthogonal set whose span is dense in the space and the kernel functions assumed in (II) are the functions $K_\alpha(z) = (1-\overline{\alpha}z)^{-\kappa}$. These spaces include the usual Hardy (H^2, $\kappa = 1$) and Bergman (A^2, $\kappa = 2$) spaces and also the standard weight Bergman spaces ($\kappa > 1$) but do not include smaller spaces than H^2, like the Dirichlet space. The spaces H^2_κ are nested, with H^2 being the smallest and the spaces growing as κ increases. The following lemma will be useful in using Equation (5) to get specific estimates for c_j.

LEMMA 14. *Suppose f is analytic in a neighborhood of z_0, where $|z_0| = 1$, and $|f(z)| = 1$ for $|z| = 1$ and z in the neighborhood of z_0 where f is analytic. Suppose $f'(z_0) \neq 0$ and suppose $|z| < 1$ implies $|f(z)| < 1$. Then*

$$\lim_{r\to 1^-}\frac{1-|f(rz_0)|^2}{1-r^2} = |f'(z_0)|$$

PROOF. Since f is analytic in a neighborhood of z_0, there is $\epsilon > 0$ so that f is analytic on the neighborhood $N(\epsilon) = \{z : |z - z_0| < \epsilon\}$ and f is given, in $N(\epsilon)$, by the power series

$$f(z) = f(z_0) + c(z-z_0) + a_2(z-z_0)^2 + \cdots$$

where $f'(z_0) = c$ is non-zero by assumption.

Since f maps the unit circle near z_0 into the unit circle near $f(z_0)$, conformality at $f(z_0)$ means $f'(z_0)z_0\overline{f(z_0)}$ is real and positive which, since $|z_0| = |f(z_0)| = 1$, means $f'(z_0)z_0\overline{f(z_0)} = |f'(z_0)| = |c|$.

For $1-\epsilon < r < 1$

$$f(rz_0) = f(z_0) + c(rz_0 - z_0) + a_2(rz_0-z_0)^2 + \cdots = f(z_0) + cz_0(r-1) + O((1-r)^2)$$

where we are using the 'big Oh notation' in which $O((1-r)^2)$ means that the missing terms satisfy $\limsup_{r\to 1^-} \frac{|\text{'missing terms'}|}{(1-r)^2} < \infty$. From the series, we see

$$\begin{aligned}|f(rz_0)|^2 &= \left(f(z_0)+cz_0(r-1)+O((1-r)^2)\right)\left(\overline{f(z_0)}+\overline{cz_0}(r-1)+O((1-r)^2)\right)\\ &= |f(z_0)|^2 - 2\mathrm{Re}\left(cz_0\overline{f(z_0)}(1-r)\right)+O((1-r)^2)\\ &= 1-2|c|(1-r)+O((1-r)^2)\end{aligned}$$

Rewriting this, we see

$$\frac{1-|f(rz_0)|^2}{1-r^2} = \frac{2|c|(1-r)}{1-r^2} + \frac{O((1-r)^2)}{1-r^2} = \frac{2|c|}{1+r} + O(1-r)$$

which gives

$$\lim_{r\to 1^-} \frac{1-|f(rz_0)|^2}{1-r^2} = |c| = |f'(z_0)|$$

□

The following is the first main theorem of this paper.

THEOREM 15. *Let B be a finite Blaschke product and let the Toeplitz operator T_B and S be bounded operators acting on H^2_κ for $\kappa \geq 1$. If S is in $\{T_B\}'$, then for all f in H^∞, Sf is in H^∞ and*

$$\|Sf\|_\infty \leq M_{\kappa,B}\|S\|\|f\|_\infty$$

where the constant $M_{\kappa,B}$ depends only on the Blaschke product B and the index κ associated with the Hilbert space.

PROOF. Let n be the order of Blaschke product B and let β_1, β_2, $\cdots$, β_n, δ, η, and $\hat{A}$ be as in Proposition 13 in the previous section and Equation (5) above. Let S be a bounded operator on H^2_κ that commutes with T_B

If f is an H^∞ function, then f is in H^2_κ and Sf is in H^2_κ and in particular, this means Sf is analytic in the open unit disk. Our goal is to show that because S is in $\{T_B\}'$, actually Sf is in H^∞. Since Sf is analytic in the disk, it is enough to show that there is a constant $\widetilde{M}$ so that for all θ with $0 \leq \theta \leq 2\pi$,

$$\limsup_{r\to 1^-} |(Sf)(re^{i\theta})| \leq \widetilde{M}$$

The open set Ω_B is large enough so that $re^{i\theta}$ is in $\hat{A} \subset \Omega_B$ when $1-\delta/2 < r < 1$ and $0 \leq \theta \leq 2\pi$. We saw in Corollary 12 that for each α in Ω_B,

$$(Sf)(\alpha) = \langle f, S^*(K_\alpha)\rangle = \langle f, \sum_{j=1}^n \overline{c_j(\alpha)}K_{\beta_j(\alpha)}\rangle = \sum_{j=1}^n c_j(\alpha)f(\beta_j(\alpha))$$

It follows that to prove that Sf has bounded values near the unit circle, it is sufficient to prove that for each j, we have $\limsup_{r\to 1^-}|c_j(re^{i\theta})|$ is a bounded function of θ. In fact, we see from Equation (5) that

$$\limsup_{r\to 1^-} |c_j(re^{i\theta})| \leq \|S\| \left(\frac{6}{\eta}\right)^{n-1} \lim_{r\to 1^-} \frac{\|K_{re^{i\theta}}\|}{\|K_{\beta_j(re^{i\theta})}\|}$$

The fact that we are working in H^2_κ means $\|K_\alpha\| = (1-|\alpha|^2)^{-\kappa/2}$. For each j, the function β_j is analytic on $\hat{A}$, maps the unit circle into itself, maps the set $\hat{A}\cap\mathbb{D}$ into $\mathbb{D}$, and $\beta_j'(\alpha)\neq 0$ for any α in $\hat{A}$, so we see from Lemma 14 that

$$\lim_{r\to 1^-}\frac{\|K_{re^{i\theta}}\|}{\|K_{\beta_j(re^{i\theta})}\|} = \lim_{r\to 1^-}\left(\frac{1-|\beta_j(re^{i\theta})|^2}{1-r^2}\right)^{\kappa/2} = |\beta_j'(e^{i\theta})|^{\kappa/2}$$

Now, since $|\beta_j'|$ is a continuous function on the unit circle, $|\beta_j'(e^{i\theta})|$ is bounded. It follows that

$$\begin{aligned}\limsup_{r\to 1^-}\left|Sf(re^{i\theta})\right| &= \limsup_{r\to 1^-}\left|\sum_{j=1}^n c_j(re^{i\theta})f(\beta_j(re^{i\theta}))\right| \\ &\leq \|S\|\|f\|_\infty\left(\frac{6}{\eta}\right)^{n-1}\sum_{j=1}^n \max_{0\leq\theta\leq 2\pi}|\beta_j'(e^{i\theta})|^{\kappa/2}\end{aligned}$$

Letting $M_{\kappa,B} = \left(\frac{6}{\eta}\right)^{n-1}\sum_{j=1}^n \max_{0\leq\theta\leq 2\pi}|\beta_j'(e^{i\theta})|^{\kappa/2}$, this proves the theorem. □

7. A More Detailed Description of Operators in the Commutant

For a given finite Blaschke product B, recall the descriptions, in the paragraphs preceding Theorem 10, of the exceptional set E_B, the open subset of the disk $\Omega_B = \mathbb{D}\setminus E_B$, and the non-compact Riemann surface W_B that is an n-sheeted cover over Ω_B in the disk.

THEOREM 16. *Let B be a finite Blaschke product of order n, E_B its exceptional set, and W_B the n-sheeted Riemann surface over the subset Ω_B of the disk, as above.*

If S is a bounded operator on H^2_κ for $\kappa\geq 1$ that commutes with T_B, then there is a bounded analytic function G on the Riemann surface W_B so that for f in H^2_κ,

$$(Sf)(\alpha) = (B'(\alpha))^{-1}\sum G((\beta,\alpha))\beta'(\alpha)f(\beta(\alpha)) \tag{6}$$

where the sum is taken over the n branches of $B^{-1}\circ B$ at α in Ω_B. Moreover, if α_0 is a zero of order m of B', and $\psi_1, \psi_2, \cdots, \psi_n$ is a basis for $((B-B(\alpha_0))\,H^2)^\perp$, then G has the property that the function h_k defined by

$$h_k(\alpha) = \sum G((\beta,\alpha))\beta'(\alpha)\psi_k(\beta(\alpha)) \tag{7}$$

on the punctured disk $\{\alpha : 0<|\alpha-\alpha_0|<\delta(\alpha_0)\}$, where $\delta(\alpha_0)$ is the distance from α_0 to $E_B\setminus\{\alpha_0\}$, has a removable singularity, and after removal, a zero of order m at α_0 for $j=1,2,\cdots,n$.

Conversely, if G is a bounded analytic function on W_B that has properties (7) at each zero of B', then (6) defines a bounded linear operator on H^2_κ with S in $\{T_B\}'$.

PROOF. If f is a function that is analytic in the disk and f' is in H^∞, for any z and w in $\mathbb{D}$, we have

$$\left|\frac{f(z)-f(w)}{z-w}\right| = \left|\frac{1}{z-w}\int_w^z f'(\tau)\,d\tau\right| \leq \sup\{|f'(\tau)| : \tau\in[w,z]\}$$

Of course, for such an f, the function $(f-f(w))/(z-w)$ has a removable singularity at $z=w$ and choosing the value $f'(w)$ for the value of makes the function analytic in $\mathbb{D}$ and we have shown it is in H^∞. We conclude that for each w in $\mathbb{D}$, the function satisfies $\|(f - f(w))/(z-w)\|_\infty \leq \|f'\|_\infty$.

Let S be in $\{T_B\}'$. Since B is a finite Blaschke product, B' is in H^∞, and for each w in $\mathbb{D}$, we have $\|(B - B(w))/(z-w)\|_\infty \leq \|B'\|_\infty$. Now, it follows from Theorem 15 that there is a constant M so that if α is a point of Ω_B and β is a branch of $B^{-1} \circ B$ near α, then

$$\left| S\left(\frac{B - B(\alpha)}{z - \beta(\alpha)}\right)(\alpha)\right| \leq \left\| S\left(\frac{B - B(\alpha)}{z - \beta(\alpha))}\right)\right\|_\infty \leq M$$

We define G on W_B by

$$G(\beta, \alpha) = S\left(\frac{B - B(\alpha)}{z - \beta(\alpha)}\right)(\alpha)$$

and observe that G is in $H^\infty(W_B)$.

If α_0 is in Ω_B and $\beta_1, \beta_2, \cdots, \beta_n$ are the branches of $B^{-1} \circ B$ defined near α_0, then for α near α_0, the subspace $[(B - B(\alpha))H^2_\kappa]^\perp$ is spanned by $K_{\beta_1(\alpha)}$, $K_{\beta_2(\alpha)}$, $\cdots$, $K_{\beta_n(\alpha)}$. Theorem 11 shows that there are analytic functions $c_1, c_2, \cdots, c_n$ defined in a neighborhood of α_0 such that

$$S^* K_\alpha = \sum_{j=1}^{n} \overline{c_j(\alpha)} K_{\beta_j(\alpha)}$$

Now, if α_0 is in Ω_B and β_j and β_k are different branches of $B^{-1} \circ B$ near α_0, then for α near α_0,

$$\left(\frac{B - B(\alpha)}{z - \beta_j(\alpha)}\right)(\beta_k(\alpha)) = 0 \quad \text{and} \quad \left(\frac{B - B(\alpha)}{z - \beta_j(\alpha)}\right)(\beta_j(\alpha)) = B'(\beta_j(\alpha))$$

and

$$S\left(\frac{B - B(\alpha)}{z - \beta_j(\alpha)}\right)(\alpha) = \left\langle \frac{B - B(\alpha)}{z - \beta_j(\alpha)}, S^*(K_\alpha)\right\rangle = c_j(\alpha)B'(\beta_j(\alpha))$$

which means that

$$c_j(\alpha) = \frac{G(\beta_j, \alpha)}{B'(\beta_j(\alpha))}$$

For α a point of Ω_B, if β is a branch of $B^{-1} \circ B$, the equality $B(\beta(\alpha)) = B(\alpha)$ means $B'(\beta(\alpha))\beta'(\alpha) = B'(\alpha)$ and we conclude that for every α in Ω_B and every j, with $1 \leq j \leq n$,

$$c_j(\alpha) = \frac{\beta_j'(\alpha)G(\beta_j, \alpha)}{B'(\alpha)}$$

Corollary 12 implies, for every f in H^2_κ, that

$$(Sf)(\alpha) = (B'(\alpha))^{-1} \sum G((\beta, \alpha))\beta'(\alpha)f(\beta(\alpha))$$

which is Equation (6) as we were to prove.

The properties (7) follow from the representation (6) and the observation that the functions $S\psi_k$ do not have poles in the disk.

Conversely, if G is in $H^\infty(W_B)$, since $B(\beta(\alpha)) = B(\alpha)$ for each branch β the operator defined by Equation (6) commutes formally with T_B. Moreover, the properties described by Equation (7) mean that $(Sf)(\alpha)$ as defined by Equation (6) defines an analytic function on the disk for each f in H^2_κ.

Note that each of the spaces H^2_κ is equivalent to a weighted Bergman space in which the determination of whether an analytic function on the disk is in the space or not is determined by an integrability condition which means that if g_1 is in H^2_κ and $|g_2(\alpha)| \leq M|g_1(\alpha)|$ for each α in $\mathbb{D}$ and some constant M, then g_2 is also in H^2_κ. In particular, the boundedness of G on W_B, the boundedness of $B'(\alpha)^{-1}$ and $\beta'(\alpha)$ near the unit circle, and the fact that the number of terms in the summation defining $(Sf)(\alpha)$ is finite, guarantees that $|(Sf)(\alpha)| \leq M|f(\alpha)|$ for some constant, M, that does not depend on f, so Sf defined by Equation (6) gives a bounded operator on H^2_κ. □

As a consequence of this result, we see that the commutants of T_B on all of the Hardy spaces H^2_κ are the same in the sense that they have the same formulas in their actions on these spaces, and therefore the operators on these spaces are all extensions or restrictions of each other.

COROLLARY 17. *The commutants of T_B as an operator on H^2 and of T_B as an operator on H^2_κ are 'the same'. Specifically, if S is an operator on one of these spaces that commutes with T_B, then there is a bounded analytic function G on the Riemann surface W_B that gives a formula for S acting on other spaces and* Theorem 16 *implies that it is bounded on the other spaces.*

Since algebraic relationships between operators on the various spaces are preserved, we see that bounded projections commuting with T_B on one of these spaces are projections on all of the spaces.

COROLLARY 18. *If P is a bounded operator acting on H^2 such that $P^2 = P$ and $T_BP = PT_B$, then P is a bounded operator acting on H^2_κ such that $P^2 = P$ and $T_BP = PT_B$.*

On the other hand, because the inner products on the spaces are different from each other, the projections that are self-adjoint on one of the spaces are not necessarily self-adjoint on other of the spaces. It is not easy to see which projections on a given space are self-adjoint.

References

[1] Sheldon Axler, *Multiplication operators on Bergman spaces*, J. Reine Angew. Math. **336** (1982), 26–44, DOI 10.1515/crll.1982.336.26. MR671320 (84b:30052)

[2] Sheldon Axler, John B. Conway, and Gerard McDonald, *Toeplitz operators on Bergman spaces*, Canad. J. Math. **34** (1982), no. 2, 466–483, DOI 10.4153/CJM-1982-031-1. MR658979 (83i:47034)

[3] Carl C. Cowen, *The commutant of an analytic Toeplitz operator*, Trans. Amer. Math. Soc. **239** (1978), 1–31. MR0482347 (58 #2420)

[4] Carl C. Cowen, *Commutants of analytic Toeplitz operators with automorphic symbol*, Hilbert space operators (Proc. Conf., Calif. State Univ., Long Beach, Calif., 1977), Lecture Notes in Math., vol. 693, Springer, Berlin, 1978, pp. 71–75. MR526533 (80g:47031)

[5] Carl C. Cowen, *The commutant of an analytic Toeplitz operator. II*, Indiana Univ. Math. J. **29** (1980), no. 1, 1–12, DOI 10.1512/iumj.1980.29.29001. MR554813 (82e:47038)

[6] Carl C. Cowen, *An analytic Toeplitz operator that commutes with a compact operator and a related class of Toeplitz operators*, J. Funct. Anal. **36** (1980), no. 2, 169–184, DOI 10.1016/0022-1236(80)90098-1. MR569252 (81d:47020)

[7] Carl C. Cowen and Eva A. Gallardo-Gutiérrez, *A new class of operators and a description of adjoints of composition operators*, J. Funct. Anal. **238** (2006), no. 2, 447–462, DOI 10.1016/j.jfa.2006.04.031. MR2253727 (2007e:47033)

[8] Carl C. Cowen and Barbara D. MacCluer, *Composition operators on spaces of analytic functions*, Studies in Advanced Mathematics, CRC Press, Boca Raton, FL, 1995. MR1397026 (97i:47056)

[9] James A. Deddens and Tin Kin Wong, *The commutant of analytic Toeplitz operators*, Trans. Amer. Math. Soc. **184** (1973), 261–273. MR0324467 (48 #2819)

[10] Ronald G. Douglas, Mihai Putinar, and Kai Wang, *Reducing subspaces for analytic multipliers of the Bergman space*, J. Funct. Anal. **263** (2012), no. 6, 1744–1765, DOI 10.1016/j.jfa.2012.06.008. MR2948229

[11] Ronald G. Douglas, Shunhua Sun, and Dechao Zheng, *Multiplication operators on the Bergman space via analytic continuation*, Adv. Math. **226** (2011), no. 1, 541–583, DOI 10.1016/j.aim.2010.07.001. MR2735768 (2012b:47084)

[12] A. L. Shields and L. J. Wallen, *The commutants of certain Hilbert space operators*, Indiana Univ. Math. J. **20** (1970/1971), 777–788. MR0287352 (44 #4558)

[13] James E. Thomson, *Intersections of commutants of analytic Toeplitz operators*, Proc. Amer. Math. Soc. **52** (1975), 305–310. MR0399927 (53 #3765)

[14] James E. Thomson, *The commutants of certain analytic Toeplitz operators*, Proc. Amer. Math. Soc. **54** (1976), 165–169. MR0388156 (52 #8993)

[15] James E. Thomson, *The commutant of a class of analytic Toeplitz operators*, Amer. J. Math. **99** (1977), no. 3, 522–529. MR0461196 (57 #1181)

[16] James Thomson, *The commutant of a class of analytic Toeplitz operators. II*, Indiana Univ. Math. J. **25** (1976), no. 8, 793–800. MR0417843 (54 #5891)

[17] William A. Veech, *A second course in complex analysis*, W. A. Benjamin, Inc., New York-Amsterdam, 1967. MR0220903 (36 #3955)

IUPUI (Indiana University – Purdue University, Indianapolis), Indianapolis, Indiana 46202-3216

E-mail address: ccowen@math.iupui.edu

Butler University, Indianapolis, Indiana 46208-3485

E-mail address: rwahl@butler.edu

Contemporary Mathematics
Volume **638**, 2015
http://dx.doi.org/10.1090/conm/638/12806

Complex approximation and extension-interpolation on arbitrary sets in one dimension

P. M. Gauthier

ABSTRACT. While uniform rational approximation is usually on compact sets of the plane, we consider arbitrary plane sets and also discuss traces of holomorphic functions on such sets.

We consider approximation to be the analyst's substitute for extension. Section 1 deals with approximation and Section 2 with extension. In Section 1, our main section, we give a brief and biased survey of rational approximation on compacta and endeavor to say something new concerning approximation on arbitrary sets. An important difference appears in rational approximation on arbitrary bounded sets: the approximating rational functions may have poles on the boundary of the set, whereas in rational approximation on compacta, the approximating rational functions have no poles on the compactum.

Section 2 is motivated by the following two well-known facts, the first of which is essentially the sheaf property of holomorphic functions.

THEOREM 1. *A function $f : U \to \mathbb{C}$ defined on an open set $U \subset \mathbb{C}^n$ is holomorphic if there is an open cover $\{U_\alpha, \alpha \in \mathcal{A}\}$, $\cup_\alpha U_\alpha = U$, such that $f|U_\alpha$ is holomorphic in U_α, for each $\alpha \in \mathcal{A}$.*

The next theorem is the famous theorem of Hartogs on separate holomorphy.

THEOREM 2. *A function $f : U \to \mathbb{C}$ defined on an open set $U \subset \mathbb{C}^n$ is holomorphic if, for each complex line ℓ in a coordinate direction, the restriction of f to $\ell \cap U$ is holomorphic.*

We seek analogous results when the open set is replaced by an arbitrary subset E of $\mathbb{C}$ or, more generally, of a one-dimensional complex analytic subset of $\mathbb{C}^n$. In particular, since every open Riemann surface is biholomorphic to a complex curve in $\mathbb{C}^3$, E may be an arbitrary subset of an open Riemann surface.

1. Rational approximation

Analytic extensions-interpolations are usually unique and 'hence' usually do not exist. Approximation is the analysts substitute for extension-interpolation. For all

2010 *Mathematics Subject Classification.* Primary 30A99; Secondary 32C99.
Key words and phrases. Holomorphic extension, holomorphic interpolation.
Research supported by NSERC (Canada).

applications to the physical world, sufficiently good approximations are as useful as extensions.

The two most natural forms of convergence for functions taking their values in the Riemann sphere $\overline{\mathbb{C}}$ are spherically uniform convergence and uniform convergence.

Let χ be the spherical metric on $\overline{\mathbb{C}}$. Given a set $E \subset \overline{\mathbb{C}}$ and a function $f : E \to \overline{\mathbb{C}}$, we say that a sequence $\{r_j\}$ of rational functions converges *spherically uniformly* to f on E, if

$$\sup_{z\in E} \chi(r_j(z), f(z)) \to 0, \quad j \to \infty.$$

If f is finite-valued, that is $f : E \to \mathbb{C}$, we shall say that the sequence $\{r_j\}$ converges *uniformly* to f on E, if

$$\sup_{z\in E} |r_j(z) - f(z)| \to 0, \quad j \to \infty.$$

This type of convergence implies that, for large j, the rational functions r_j have no poles on E. Nevertheless, the functions r_j may have poles on the closure $\overline{E}$.

Both types of convergence have their advantages and disadvantages. In this note, we shall consider only uniform (not spherically uniform) approximation. Uniform approximations (that is, using uniform convergence) are usually performed on *compact* sets of the Riemann sphere $\overline{\mathbb{C}}$, but we shall also consider approximation on arbitrary subsets of $\overline{\mathbb{C}}$.

By Runge's theorem, every holomorphic function on a compact set K can be uniformly approximated by rational functions (having no poles on K). Since uniform limits on K of such rational functions are in the class $A(K)$ of functions continuous on K and holomorphic on the interior, it is natural to try to approximate functions in $A(K)$ by rational functions. Although rational approximation is an old subject, the *raison d'être* of this subject was put on firm ground, when Alice Roth in 1938 gave the first example of a compact set K for which not every function in $A(K)$ can be approximated by rational functions (see [**6**]). This example was called the Swiss Cheese, because it is the closed disc with open discs removed. It is a happy coincidence that Alice Roth was Swiss. She was a dear friend of mine and I referred to her as Alice in Switzerland (see [**2**]).

As we have said, rational approximation is usually considered on compact sets, but we shall consider approximation on arbitrary subsets of $\overline{\mathbb{C}}$. When E is a compact subset of the Riemann sphere $\overline{\mathbb{C}}$, it is standard to denote by $R(E)$ the family of functions on E which are uniform limits of rational functions (having no poles on E). We shall use the same notation for an arbitrary (not necessarily compact) set $E \subset \overline{\mathbb{C}}$. Let R_E denote the set of rational functions having no poles on E and $A(E)$ the family of functions $f : E \to \mathbb{C}$, which are continuous on E and holomorphic on E^o. Clearly, $R(E) \subset A(E)$. We claim that

$$R(E) = R_E + R(\overline{E}), \tag{1}$$

by which we mean that a function is in $R(E)$ if and only if it is the sum of a rational function having no poles on E plus a function in $R(\overline{E})$. To see this, let $f \in R(E)$. Then, there is a sequence of rational functions $\{r_j\}$, such that $|f - r_j| < 1/j, j = 1, 2, \cdots$. Since f is finite-valued (but maybe unbounded), the sequence $\{r_j\}$ is finite-valued and uniformly Cauchy on E. If one of the rational functions r_j has a pole at a point $z \in \overline{E}$, then, since $r_j - r_k$ is bounded near z on E, it follows that r_k also has a pole at z and with the same principal part. Since the sequence $\{r_j\}$ is uniformly Cauchy on E and has the same poles and principal parts on $\overline{E}$, the sequence $r_1 - r_j$ is a sequence of rational functions, which on $\overline{E}$ has no poles and is uniformly Cauchy and hence converges uniformly to a function $g \in R(\overline{E})$. On E, we have $g = r_1 - f$, so $f = r_1 - g$. We have shown that the left member of (1) is contained in the right member. The reverse inclusion follows from the triangle inequality.

A fundamental result in rational approximation is the so-called Fusion Lemma of Alice Roth (see [**6**]) which states that, under certain conditions on two sets, we can approximate simultaneously two rational functions on the union, with an error measured in terms of how close these functions are on the intersection.

LEMMA 1 (Roth). *Suppose K_1, K_2, k are compacta in the extended complex plane $\overline{\mathbb{C}}$ such that $K_1 \cap K_2 = \emptyset$. Then, there exists a constant A, depending only on K_1 and K_2, such that, if r_1, r_2 are rational functions such that $r_1 - r_2$ has no poles on k and*

$$|r_1(z) - r_2(z)| < \epsilon \quad (z \in k),$$

then there exists a rational function r such that, for $j = 1, 2$,

$$|r(z) - r_j(z)| < A\epsilon \quad (z \in K_j \cup k).$$

Note that the Fusion Lemma is trivial, if one of the compacta is the empty set. In the fusion lemma, we can replace the compact sets by arbitrary sets as follows.

LEMMA 2. *Suppose B_1, B_2, b are sets in the extended complex plane $\overline{\mathbb{C}}$ such that $\overline{B}_1 \cap \overline{B}_2 = \emptyset$. Then, there exists a constant A, depending only on B_1 and B_2, such that, if r_1, r_2 are rational functions such that $r_1 - r_2$ has no poles on b and*

$$|r_1(z) - r_2(z)| < \epsilon \quad (z \in b),$$

then, there exists a rational function r such that, for $j = 1, 2$, the difference $r - r_j$ has no poles on $\overline{B}_j \cup \overline{b}$ and

$$|r(z) - r_j(z)| < A\epsilon \quad (z \in \overline{B}_j \cup \overline{b}).$$

PROOF. Denote by $\overline{A}$ a constant associated to the compacta $K_1 = \overline{B}_1$ and $K_2 = \overline{B}_2$ in Lemma 1. Although on $\overline{b}$ we can assert that $|r_1 - r_2| \le \epsilon$, we cannot be certain that we have strict inequality. However, we do have $|r_1 - r_2| < 2\epsilon$ on $\overline{b}$. By Lemma 1, there is a rational function r such that, for $j = 1, 2$, the difference $r - r_j$ has no poles on $\overline{B}_j \cup \overline{b}$ and we have $|r - r_j| < \overline{A}2\epsilon$, for $z \in \overline{B}_j \cup \overline{b}$. Set $A = 2\overline{A}$. □

In the next lemma and subsequently, the notation $f|E$ signifies the restriction $f|_E$.

LEMMA 3. *Let B_1, B_2, b be as in the previous lemma and $f : (B_1 \cup b \cup B_2) \to \mathbb{C}$. If $f|(B_j \cup b) \in R(B_j \cup b)$, for $j = 1, 2$, then $f \in R(B_1 \cup b \cup B_2)$.*

PROOF. Let A be a constant from Lemma 2 and let $\epsilon > 0$. For $j = 1, 2$, there exist rational functions r_j such that $|r_j - f| < \epsilon$ on $B_j \cup b$. By Lemma 2, there is a rational function r such that $|r - r_j| < A \cdot 2 \cdot \epsilon$ on $\overline{B}_j \cup \overline{b}$. Thus, $|r - f| < (2A + 1)\epsilon$ on $B_j \cup b$, for $j = 1, 2$ and so

$$|r - f| < (2A + 1)\epsilon \quad (z \in B_1 \cup b \cup B_2).$$

Consequently, $f \in R(B_1 \cup b \cup B_2)$. □

Remark. The intersection of sets of rational approximation is again a set of rational approximation. That is, if $K_j, j = 1, 2$, are two compact sets for which $R(K_j) = A(K_j)$, then $R(K_1 \cap K_2) = A(K_1 \cap K_2)$. In [**10**] we gave a short proof, forgetting (*mea culpa*) that we had learned that short proof a few years earlier from Petr Paramonov [**15**].

To better appreciate Lemma 3, we note that, if a function can be approximated on each of two compact sets, it does not automatically follow that it can be approximated on their union. For an example, see [**6**, p. 113, A3].

Let us say that two bounded sets E_1 and E_2 form a *fusible* pair if

$$\overline{(E_1 \setminus E_2)} \cap \overline{(E_2 \setminus E_1)} = \emptyset.$$

LEMMA 4. *Suppose E_1 and E_2 is a pair of fusible bounded sets and $f : (E_1 \cup E_2) \to \mathbb{C}$. If $f|E_j \in R(E_j)$, for $j = 1, 2$, then $f \in R(E_1 \cup E_2)$.*

PROOF. This is just a rephrasing of the previous lemma, with $B_1 = E_1 \setminus E_2$, $B_2 = E_2 \setminus E_1$ and $b = E_1 \cap E_2$. □

EXAMPLE 1. *Let E be a bounded set contained in the union of two overlapping strips $S_j = \{z : a_j \leq \Re z \leq b_j, j = 1, 2, \}$ where $a_1 < a_2 < b_1 < b_2$. Suppose $f : E \to \mathbb{C}$ is such that $f \in R(E \cap S_j), j = 1, 2$. Then, $f \in R(E)$.*

This example easily extends to finitely many strips.

LEMMA 5. *Let E be a bounded set contained in the union of strips $S_j = \{z : x_{j-1} \leq \Re z \leq x_j\}, j = 1, \cdots, n$. Suppose $f : E \to \mathbb{C}$ is such that*

$$f|(E \cap (S_j \cup S_{j+1})) \in R(E \cap (S_j \cup S_{j+1})), \quad j = 1, \cdots, n-1.$$

Then, $f \in R(E)$.

PROOF. If $n = 2$, there is nothing to prove. Suppose $n > 2$. From the previous example, it follows that $f|(E \cap (S_1 \cup S_2 \cup S_3)) \in R(E \cap (S_1 \cup S_2 \cup S_3))$. If $n = 3$, the proof is over. If $n > 3$, we invoke the previous example for the strips $S_1 \cup S_2 \cup S_3$, and $S_3 \cup S_4$. After finitely many steps, we have that $f|(E \cap (S_1 \cup \cdots \cup S_n)) \in R(E \cap (S_1 \cup \cdots \cup S_n))$ That is, $f \in R(E)$. □

It will be convenient to present a different formulation of Lemma 5. Let us say that closed rectangles, $Q_j = \{z : a_j \leq x \leq b_j, c \leq y \leq d\}, j = 1, \cdots n$, form a *horizontal chain of overlapping rectangles* if

$$a_1 < a_2 < b_1 < a_3 < b_2 < a_4 < \cdots < b_{n-1} < b_n.$$

The analogous definition for a vertical chain of overlapping rectangles is obvious. The proof of the following lemma is similar to that of Lemma 5.

LEMMA 6. *Let E be a bounded set contained in the union of a chain of overlapping rectangles $Q_1, \ldots, Q_n$. Suppose $f : E \to \mathbb{C}$ is such that*

$$f|(E \cap Q_j) \in R(E \cap Q_j), \quad j = 1, \cdots, n.$$

Then, $f \in R(E)$.

Roth employed her Fusion Lemma 1 to prove the Bishop Localization Theorem (see [**6**]) for rational approximation on compacta. With the preceding lemmas, we also obtain the following localization result for arbitrary sets.

LEMMA 7. *Let $E \subset \overline{\mathbb{C}}$ and $f : E \to \mathbb{C}$ be such that, for some $\delta > 0$ and every $z \in E$, there is a closed disc $\overline{D}_z \subset \overline{\mathbb{C}}$, centered at z of spherical radius $> \delta$, such that $f|(E \cap \overline{D}_z) \in R(E \cap \overline{D}_z)$. Then, $f \in R(E)$.*

PROOF. Put $E = E_1 \cup E_2$, where $E_1 = E \setminus \{z : |z| < 1\}$ and $E_2 = E \cap \{z : |z| < 2\}$. Let us show that $f|E_1 \in R(E_1)$ and $f|E_2 \in R(E_2)$.

Consider first E_2. Since any smaller δ also works, for each $z \in E_2$, we may replace D_z by a square centered at z with diameter $> \delta$ and, by abuse of notation, we continue to denote this square by D_z. From the compactness of $\overline{E}_2$, and since $\delta > 0$, we may assume that $\overline{E}_2$ has a finite cover by such squares. Since we can do the same with any smaller δ, we may cover $\overline{E}_2$ with a grid of closed squares $Q_{i,j}, 1 \leq i, j \leq n$, such that, $f|(E_2 \cap Q_{i,j}) \in R(E_2 \cap Q_{i,j})$, and for each j, the squares $Q_{i,j}, 1 \leq i \leq n$ form a horizontal chain of overlapping squares and also, the $\cup_i Q_{i,j}, j = 1, \cdots, n$ form a vertical chain of overlapping rectangles.

For each j, we invoke Lemma 6 to conclude that $f|(E_2 \cap \cup_i Q_{i,j}) \in R(E_2 \cap \cup_i Q_{i,j})$. Now, we invoke Lemma 6 again for the chain $\cup_i Q_{i,j}, j = 1, \cdots, n$ to conclude that $f|E_2 \in R(E_2)$.

If we first perform the inversion $z \mapsto 1/z$, the proof that $f|E_1 \in R(E_1)$, is the same.

Set $B_1 = E \cap \{z : |z| \geq 2\}, B_2 = E \cap \{z : |z \leq 1\}, b = E \cap \{z : 1 < |z| < 2\}$. The lemma now follows from Lemma 3. □

We remark that Lemma 7 would not hold, if we dropped the δ hypothesis. For example, consider as E the interval $(0, 1]$ and the function $f(x) = \sin(1/x)$ thereon.

The preceding lemma yields the Bishop Localization Theorem (see [**6**]).

THEOREM 3. *Let K be a compact subset of $\overline{\mathbb{C}}$ and $f : K \to \mathbb{C}$ be such that, for every $z \in K$, there is a closed disc $\overline{D}_z \subset \overline{\mathbb{C}}$, centered at z such that $f|(K \cap \overline{D}_z) \in R(K \cap \overline{D}_z)$. Then, $f \in R(K)$.*

PROOF. For each $z \in K$, there is a closed disc $D(z, 2\delta_z) \subset \overline{\mathbb{C}}$ such that $f|(K \cap D(z, 2\delta_z)) \in R(K \cap D(z, 2\delta_z))$. There is a finite cover of K of the form $D(z_j, \delta_j)$, where $\delta_j = \delta_{z_j}$. For each $z \in K$, there is some j such that $D(z, \delta_j) \subset D(z_j, 2\delta_j)$. Hence, $f|(K \cap D(z, \delta_j)) \in R(K \cap D(z, \delta_j))$. Let $\delta = \min_j \delta_j$. Then, by Lemma 7, $f \in R(K)$. □

For Ω an open subset of $\overline{\mathbb{C}}$, we denote by $\mathcal{M}(\Omega)$ the family of functions meromorphic on Ω. Also, if $E \subset \Omega$, denote by $\mathcal{M}_E(\Omega)$ the functions in $\mathcal{M}(\Omega)$ having no poles on E and denote by $\widetilde{E}$ the closure of E in Ω. Moreover, denote by $\overline{\mathcal{M}}(\Omega, E)$ the set of functions $f : E \to \mathbb{C}$, which are uniform limits of functions in $\mathcal{M}(\Omega)$. Thus, for each $\epsilon > 0$, there is a $g \in \mathcal{M}(\Omega)$, such that $|f - g| < \epsilon$ on E. It follows that g has no poles on E. Thus, $\overline{\mathcal{M}}(\Omega, E)$ is the same as $\overline{\mathcal{M}}_E(\Omega)$, the set of functions $f : E \to \mathbb{C}$, which are uniform limits of functions in $\mathcal{M}_E(\Omega)$.

We claim that

$$\overline{\mathcal{M}}_E(\Omega) \subset \mathcal{M}_E(\Omega) + A(\widetilde{E}), \tag{2}$$

by which is meant that if f is in the left member, then f is the sum of a meromorphic function on Ω having no poles on E and a function which extends to a function in $A(\widetilde{E})$. The proof of this claim is the same as the earlier proof of (1).

In particular, if $\Omega = \overline{\mathbb{C}}$, then $\mathcal{M}(\Omega)$ is just the family R of rational functions, $\mathcal{M}_E(\Omega) = R_E$ and $\overline{\mathcal{M}}_E(\Omega) = R(E)$. Hence, (2) becomes $R(E) \subset R_E + A(\overline{E})$. Since it is not in general true that $A(\overline{E}) \subset R(E)$, the reverse inclusion of (2) does not in general hold.

THEOREM 4. *For an arbitrary set* $E \subset \overline{\mathbb{C}}$ *and a function* $f : E \to \mathbb{C}$, *suppose, for each* $w \in E$, *there is a closed disc* $\overline{D}_w \subset \overline{\mathbb{C}}$ *centered at* w, *such that* $f|(E \cap \overline{D}_w) \in R(E \cap \overline{D}_w)$. *Then,* $f \in \overline{\mathcal{M}}_E(\Omega)$, *for the following open neighborhood of* E :

$$\Omega = \bigcup_{w \in E} D_w.$$

PROOF. We claim that f has a unique spherically continuous extension $\widetilde{f}$ to $\widetilde{E}$. From (1) we have that,

$$f|(E \cap \overline{D}_w) = r_w + g_w,$$

where $r_w \in R_{E \cap \overline{D}_w}$ and $g_w \in R(\overline{E \cap \overline{D}_w})$. Certainly, r_w is spherically continuous everywhere and, since $\widetilde{E} \cap D_w \subset \overline{E \cap \overline{D}_w}$, the function g_w is continuous on $\widetilde{E} \cap D_w$. This extends $f|(E \cap D_w)$ to a function f_w spherically continuous on $\widetilde{E} \cap D_w$ and

$$f_w(z) = r_w(z) + g_w(z), \quad z \in \widetilde{E} \cap D_w. \tag{3}$$

The extension is unique at each point of $\widetilde{E} \cap D_w$, since $E \cap D_w$ is dense in $\widetilde{E} \cap D_w$. This uniqueness shows that the respective extensions f_w on $\widetilde{E} \cap D_w, w \in E$, are compatible. Thus, by setting

$$\widetilde{f}(z) = f_w(z), \quad z \in \widetilde{E} \cap D_w, \quad w \in E,$$

we obtain a unique spherically continuous extension of f to $\widetilde{E}$.

Let $P = \widetilde{f}^{-1}(\infty)$. The local representations f_w of $\widetilde{f}$ show that $P \subset (\Omega \setminus E)$, P is a discrete set in Ω, and these singularities are poles of the functions r_w. There is a meromorphic function s in Ω such that $s - r_w$ has no poles on $E \cap \overline{D}_w$, for each $w \in E$. To see this, consider first the case where $\Omega = \overline{\mathbb{C}}$. Then P is finite and there is a rational function s which does the job. If $\Omega \neq \mathbb{C}$, then, since (each component of) Ω is a Stein manifold, the first Cousin Problem is solvable on Ω and so there indeed exists such an s. Set $g = \widetilde{f} - s$. Then $\widetilde{f} = s + g$ represents $\widetilde{f}$ as the sum of a function $s \in \mathcal{M}_E(\Omega)$ plus a function $g \in A(\widetilde{E})$.

We claim that, for each $z \in \widetilde{E}$, there is a closed disc K_z containing z, such that

$$g|(\widetilde{E} \cap K_z) \in R(\widetilde{E} \cap K_z).$$

To see this, we may choose a $w \in E$, such that $z \in D_w$, since $\widetilde{E} \subset \Omega$. Choose a closed disc K_z containing z and contained in D_w. Then, by (3), $\widetilde{f}|(\widetilde{E} \cap K_z) = r_w + g_w$, with $g_w \in R(\widetilde{E} \cap K_z)$. Thus,

$$g|(\widetilde{E} \cap K_z) = \widetilde{f}|(\widetilde{E} \cap K_z) - s = r_w + g_w - s.$$

Since s removes the poles of $\widetilde{f}$, the function $r_w - s$ is holomorphic on the compact set $\widetilde{E} \cap K_z$ and so, by Runge's theorem, $r_w - s \in R(\widetilde{E} \cap K_z)$. This establishes the claim.

Let K be an arbitrary compact subset of $\widetilde{E}$. For each $z \in K$, choose a $w \in E$ and a closed disc K_z as in the previous paragraph. It follows from the Bishop Localization Theorem 3 that $g|K \in R(K)$. We have shown that, for every compact $K \subset \widetilde{E}$, $g|K \in R(K)$. Now we may invoke the Roth Localization Theorem for closed sets [**6**, page 131] to conclude that g can be uniformly approximated on $\widetilde{E}$ by functions meromorphic on Ω. In particular,

$$g \in \overline{\mathcal{M}}(\Omega, \widetilde{E}) \subset \overline{\mathcal{M}}(\Omega, E).$$

Since, on E, we have $f = s+g$ and trivially $s \in \overline{M}(\Omega, E)$, we have $f \in \overline{M}(\Omega, E)$. □

For E and f satisfying the hypotheses of Theorem 4, we cannot conclude that $f \in R(E)$. An important example is the interval $E = (0, 1]$ and the function $f(x) = e^{1/x}$. Then, f and E satisfy the hypotheses of Theorem 4, but $f \notin R(E)$, because formula (1) is not satisfied.

As we have already mentioned, classical rational approximation on a compact set K is by rational functions having no poles on K, whereas, in approximating on an arbitrary set E, the rational functions might have poles on the boundary of E. An analogous situation has arisen in several complex variables. For a compact set $K \subset \mathbb{C}^n$, it is traditional to approximate continuous functions on K, which are holomorphic on the interior, by rational functions having no poles on K. However, in [**7**, Lemma 3.1], where one uniformly rationally approximates on a bounded open set in $D \subset \mathbb{C}^n$ functions which are holomorphic and uniformly continuous (and hence extend continuously to the compactum $\overline{D}$), the pole set of the rational functions may meet the boundary of D. In the present note, however, the functions to be uniformly approximated on E need not be uniformly continuous.

The theory of complex approximation on compact subsets of the Riemann sphere $\overline{\mathbb{C}}$ has been extended to approximation on closed (not necessarily compact) subsets of $\mathbb{C}$, by Carleman, Keldysh, Lavrentiev, Roth, Arakelian and others (see [**6**]). Complex approximation on compact subsets of $\overline{\mathbb{C}}$ has also been extended to approximation on *compact* subsets of arbitrary Riemann surfaces, however there have been serious obstacles (see [**11**]) to developing an approximation theory for *closed* subsets of Riemann surfaces. Despite what is said in the Mathematical Review for [**11**], necessary and sufficient conditions for approximation on closed subsets of Riemann surfaces are still unknown. It is for this reason that in this section on approximation, we have considered only subsets of the Riemann sphere.

In the next section, however, which concerns extension-interpolation, we shall consider Riemann surfaces and, more generally, one-dimensional (complex) analytic spaces.

2. Extension-interpolation

Let X be an analytic subset of an open set $\Omega \subset \mathbb{C}^n$, one can define several classes of holomorphic and quasi-holomorphic functions on X :

- The holomorphic functions.
- The w-holomorphic (weakly holomorphic) functions.
- The c-holomorphic functions.

The holomorphic functions are sections of the structure sheaf $\mathcal{O}_X$. While holomorphic functions are intrinsically related to the structure of X, w-holomorphic functions appear in a natural way, for example in relation to the Abel Theorem or the Lie-Griffiths Theorem see [**14**]. Weakly holomorphic functions are defined and holomorphic on the regular part of X and bounded in a neighborhood of each point of X. Notice that it may not be defined at singular points of X. On analytic sets, a better generalization of holomorphy was introduced by Reinhold Remmert

[**18**]. A function $f : U \to \mathbb{C}$, defined on an open subset $U \subset X$, is said to be c-holomorphic if it is continuous on U and holomorphic on the regular part of U. The c-holomorphic functions are more friendly in that they are (continuously) defined at all points of X. As an example of the beauty of c-holomorphic functions, there is the result of Denkowski [**3**] that a mapping $f : X \to \mathbb{C}^n$ is c-holomorphic if and only if it is continuous and its graph $\Gamma_j := \{(x, f(x)) : x \in X\}$ is an analytic subset of $\Omega \times \mathbb{C}^n$. This generalizes the well-known fact that a mapping defined an an open subset of $\mathbb{C}^n$ is holomorphic if and only if it is continuous and its graph is an analytic set. The c-holomorphic functions on X are also sections of a sheaf $\tilde{\mathcal{O}}$ located between the structure sheaf $\mathcal{O}_X$ and the sheaf of continuous functions $\mathcal{C}_X$. For a reduced analytic space $(X, \mathcal{O}_X)$, the sheaf $\tilde{\mathcal{O}}$ is $\mathcal{O}_X$-coherent [**13**, Th. 15.3] (see also [**19**, 81]). Holomorphic obviously implies c-holomorphic, and the latter also implies w-holomorphic. We shall work on locally irreducible 1-dimensional spaces, in which case weakly holomorphic and c-holomorphic functions are the same objects.

Consider a pair of complex singular Stein spaces X and Y, such that Y is a subvariety of X. Theorems A and B of Cartan imply that every function holomorphic on Y is the restriction of a function holomorphic on X. However, if Y is an arbitrary subset of X, it seems a difficult problem to identify those functions on Y which are restrictions of holomorphic functions on X. As a preliminary effort in this direction, we shall limit ourselves to the case that X is of dimension 1. Moreover, we shall pose the problem for restrictions of c-holomorphic functions rather than holomorphic or w-holomorphic functions, because c-holomorphic functions are easier to understand. It is our hope that, the investigation of this problem for c-holomorphic functions will lead to some insights as to how to deal with the analogous problem for w-holomorphic and holomorphic functions.

Let E be a subset of on analytic set X. In order to emphasize that a function f has a domain of definition, we shall denote a function $f : E \to \mathbb{C}$ briefly by $f{:}E$. If $F \subset E$, we denote the restriction of $f{:}E$ to F by $f{:}E|F$. We wish to distinguish between two notions of c-holomorphy on F : firstly, c-holomorphy of the *restriction* of $f{:}E$ to F and secondly, c-holomorphy of $f{:}E$ *itself* on F.

We give three definitions.

Definition 1. We say that $f{:}E$ is c-holomorphic on E, if there is an open neighborhood U of E in X and a function $g{:}U$ c-holomorphic on U such that $g = f$ on E. If E is open, this coincides with Remmert's definition of c-holomorphy.

For the next two definitions, we suppose moreover that F is a subset of E.

Definition 2A. We shall say that the restriction $f{:}E|F$ of $f{:}E$ to F is c-holomorphic, if there is an open neighborhood U of F in X and $g{:}U$ c-holomorphic on U such that $g = f$ on F.

Definition 2B. We shall say that $f{:}E$ (itself) is c-holomorphic on F, if there is an open neighborhood U of F in X and $g{:}U$ c-holomorphic on U such that $g = f$ on $E \cap U$.

Definition 1, 2A and 2B all coincide when $F = E$.

EXAMPLE 2. *If $f : E \to \mathbb{C}$ and $x \in E$, then to say that $f{:}E$ is c-holomorphic at x means that there is an open neighborhood U_x of x in X and $g{:}U_x$ c-holomorphic on U_x such that $g = f$ on $E \cap U_x$. We may obviously assume U_x is the intersection of X with an open ball centered at $x \in \mathbb{C}^n$.*

EXAMPLE 3. *For $X = \mathbb{C}$, let F_r be the real axis, F_i be the imaginary axis and $E = F_r \cup F_i$. Define f on E by setting $f(z) = z$ on F_r and $f(z) = 0$ on F_i. Then,*

$f{:}E|F_r$ and $f{:}E|F_i$ are both c-holomorphic, but $f{:}E$ is c-holomorphic on neither F_r nor F_i.

An analytic set X is *reducible* if $X = X_1 \cup X_2$, where X_j are proper analytic subsets of X. Of course, we define X to be irreducible if it is not reducible. X is irreducible if and only if the set of regular points is connected. An analytic set X in $\mathbb{C}^n$ is said to be irreducible at a point $p \in X$, if there is a fundamental system of neighborhood U_j of p in $\mathbb{C}^n$, such that each $U_j \cap X$ is irreducible.

If X is one-dimensional and irreducible at a point p, then there is a neighborhood U of p and a holomorphic mapping of the unit disc $\varphi : \Delta \to U \cap X$ which is a bijection and $\varphi : \Delta \setminus \{0\} \to (U \cap X) \setminus \{p\}$ is biholomorphic. Let us call $U \cap X$ a parametric disc at p. For each $0 < \lambda < 1$, consider the mapping φ_λ on Δ defined as $\varphi_\lambda(z) = \varphi(\lambda z)$. The image $W_\lambda = \varphi_\lambda(\Delta)$ is a neighborhood of p in X. Thus, there is a neighborhood U_λ of p in $\mathbb{C}^n$ such that $W_\lambda = U_\lambda \cap X$. The family $\{W_\lambda : 0 < \lambda < 1\}$ is a fundamental system of neighborhoods of p in X, each of which is a parametric disc at p.

EXAMPLE 4. *In $\mathbb{C}^2$ the set $X = \{(z,w) : z^2 = w^3\}$ has only the origin as singular point. It is (locally) irreducible. It is the image of $\mathbb{C}$ under the mapping $\varphi(t) = (t^3, t^2)$.*

LEMMA 8. *Suppose X is one-dimensional and irreducible at $p \in X$. If f and g are c-holomorphic functions in neighborhoods U_p and V_p respectively of p and f and g agree on a sequence converging to p, then there is a neighborhood W_p of p such that $f = g$ on $X \cap W_p$.*

PROOF. We may choose a neighborhood $W_p \subset U_p \cap V_p$ and a parametrization $\varphi : \Delta \to X \cap W_p$. The functions $f \circ \varphi = g \circ \varphi$ are continuous on Δ and holomorphic on $\Delta \setminus \{0\}$. Thus, 0 is a removable singularity and they are holomorphic on Δ. Since they agree on a sequence converging to 0, they agree on all of Δ. Moreover, since φ is a homeomorphism, $f = g$ on $X \cap W_p$. □

Note that this lemma fails, if X is not irreducible at p. Consider, for example, the analytic set $zw = 0$ in $\mathbb{C}^2$ and the holomorphic functions w and zw.

THEOREM 5. *Let $f{:}E$ be a function defined on a subset E of a locally irreducible 1-dimensional analytic set X. If $f{:}E$ is c-holomorphic at each point of E, then $f{:}E$ is c-holomorphic.*

PROOF. As in Example 1, to say that $f : E$ is c-holomorphic at each point of E means that, for each $z \in E$, there is an open ball B_z centered at z in $\mathbb{C}^n$ and a function F_z c-holomorphic in $B_z \cap X$ such that,

$$F_z(\zeta) = f(\zeta), \quad \text{for all} \quad \zeta \in B_z \cap E.$$

Denote by r_z the radius of B_z. Let E' be the set of non-isolated points of E. According to the local parametrization, for each $z \in E'$, we may choose a parametric disc D_z for X at z, such that $\operatorname{diam} D_z < r_z/2$.

Claim: for every two such discs D_z, D_w, with $z, w \in E'$, we have

$$F_z(\zeta) = F_w(\zeta), \quad \text{for all} \quad \zeta \in D_z \cap D_w.$$

We may suppose that $D_z \cap D_w \neq \emptyset$ and
$\operatorname{diam} D_w \leq \operatorname{diam} D_z$. Then, $D_w \subset B_z$. Indeed, suppose $\zeta \in D_w$ and choose $\xi \in$

$D_z \cap D_w$. Then,

$$|\zeta - z| \leq |\zeta - \xi| + |\xi - z| < \mathrm{diam} D_w + \mathrm{diam} D_z \leq 2\mathrm{diam} D_z < r_z.$$

Thus, $D_w \subset B_z$. Since w is a limit point of E, and both F_z and F_w equal f on $E \cap D_w$, the claim follows from the lemma.

We have

$$E' \subset G' \stackrel{def}{=} \bigcup_{z \in E'} D_z.$$

By the claim, we may define a holomorphic function F on the open set G' of X, by setting $F = F_z$ on each D_z. Moreover, $F = f$ on $G' \cap E$.

Every point of $E \setminus G'$ is isolated, so it looks easy to extend F holomorphically to these points. However, we must be careful about points of E on $\partial G'$. We may arrange the points of $E \setminus G'$ in a sequence (possibly finite) $\{e_j\}$. Let $\overline{B}_j$ be a closed ball in $\mathbb{C}^n$ centered at e_j which contains no other point of E and whose radius is less than $1/j$.

Since the family $\{\overline{B}_j \cap G'\}_j$ is locally finite in G', the set

$$G = G' \setminus \cup_j \overline{B}_j$$

is open and so is the set

$$\mathcal{G} = G \cup \bigcup_j (B_j \cap X).$$

Set $F = f(e_j)$ on $B_j \cap X$. Then, F is holomorphic on the open set $\mathcal{G}$ which contains E. Moreover, $F = f$ on $E \cap \mathcal{G}$. This proves the theorem. □

Definition. For the purposes of this note, a sequence Z convergent in a set E means a convergent sequence of distinct points in E **together with** its limit, provided this limit is also in E:

$$Z = \{z_0\} \cup \{z_j\}_{j=1}^{\infty}, \quad z_j \to z_0.$$

LEMMA 9. *Let $f{:}E$ be a function defined on a subset E of a locally irreducible 1-dimensional analytic set X. Suppose, for some point $z \in E$, $f{:}E|Z$ is c-holomorphic on each sequence Z converging in E to z. Then, $f{:}E$ is c-holomorphic at z.*

PROOF. If z is isolated in E, the lemma is obvious. Suppose Z is a sequence of distinct points of E converging to z. By assumption, there is a parametric disc D_Z at z and a function F_Z c-holomorphic on D_Z, such that $F_Z = f$ on $Z \cap D_Z$. We claim that there is a parametric disc U_Z at z, such that $F_Z = f$ in $E \cap U_Z$. If not, there is a sequence a_j converging to z in $E \cap D_Z$, with $F_Z(a_j) \neq f(a_j)$. Let A be a convergent sequence in E whose terms are the points of Z supplemented by the points a_j and let F_A be the associated function c-holomorphic on a parametric disc $D_A \subset D_Z$ at z, which agrees with f on $A \cap D_A$. Since F_Z and F_A are both c-holomorphic at z and agree on $Z \cap D_A$, we have, by Lemma 8, that $F_Z = F_A$ on D_A. For large j, the a_j are in this neighborhood, so $F_Z(a_j) = F_A(a_j)$. But $F_A(a_j) = f(a_j)$, so $F_Z(a_j) = f(a_j)$. Contradiction. □

COROLLARY 1. . *If $f{:}E|(C \cap E)$ is c-holomorphic at z, for all curves C passing through z, then $f{:}E$ is c-holomorphic at z.*

It is natural to ask to what extent, in the lemma, we may replace *all* sequences converging to z by a smaller family of sequences and to what extent in the corollary, we may replace all curves through z by a smaller family of curves. The following example is relevant.

EXAMPLE 5. *Let $E = X = \mathbb{C}$, let Z_o be a sequence converging in $\mathbb{C}$ to 0 and let $\mathcal{Z}$ be the family of all sequences converging in $\mathbb{C}$ to 0 which are disjoint (except for the common point 0) from Z_o. Set*

$$f(z) = \begin{cases} 1 & \text{if } z \in Z_o \setminus \{0\} \\ 0 & \text{otherwise.} \end{cases}$$

Then $f{:}\mathbb{C}|Z$ is c-holomorphic, for each sequence $Z \in \mathcal{Z}$, but $f{:}\mathbb{C}$ is not c-holomorphic at 0.

If we choose Z_o to be the sequence $\{z_j = 1/j + ie^{-j} : j = 1, 2, \cdots\}$, then for each analytic curve C passing through 0, the point 0 is isolated in $C \cap Z_o$, so $f{:}\mathbb{C}|C$ is c-holomorphic at 0, but clearly $f{:}\mathbb{C}$ is not c-holomorphic at 0. Thus, in the corollary, we cannot replace the family of all curves passing through 0 by the family of all real analytic curves passing through 0.

THEOREM 6. *.Let $f{:}E$ be a function defined on a subset E of a locally irreducible 1-dimensional analytic set X. Then, $f{:}E$ is c-holomorphic if (and only if) it is c-holomorphic on each sequence Z convergent in E.*

PROOF. The theorem easily follows from Lemma 9 and Theorem 5. □

COROLLARY 2. *Let E be a subset of a locally irreducible 1-dimensional analytic set. If for all curves C, $f{:}E$ restricted to $E \cap C$ is c-holomorphic, then $f{:}E$ is c-holomorphic.*

In order to make use of Theorem 6 on the complex plane, it is important to know, for which pairs (Z, W) of convergent sequences in $\mathbb{C}$, there exists f holomorphic on a neighborhood of Z such that

$$f(z_j) = w_j, \quad j = 0, 1, 2, \cdots,$$

where Z and W, as before, are of the form $Z = \{z_0\} \cup \{z_j : j = 1, 2, \cdots ; z_j \to z_0\}$, with z_j distinct, and $W = \{w_0\} \cup \{w_j : j = 1, 2, \cdots ; w_j \to w_0\}$.

This problem is equivalent to the problem of determining for which pairs (Z, W) of convergent sequences in $\mathbb{C}$, the interpolation problem is *eventually* solvable. That is, there exists f holomorphic in a neighborhood of z_0 such that $f(z_j) = w_j$ for all sufficiently large j. Indeed, if there is a holomorphic function $f{:}Z$ such that $f(z_j) = w_j$, for *all* j, then trivially, this holds for sufficiently large j. Conversely, suppose, for some disc $D = \{z : |z - z_0| < r\}$, there is a holomorphic function $f{:}D$ such that $f(z_j) = w_j$ for all $z_j \in D$, with $j > J$. Choose, $0 < r_0 < r$, such that $r_0 < |z_j - z_0|$, for all $j \leq J$ and set $\overline{D}_0 = \{z : |z - z_0| \leq r_0\}$. Now set $f_0 = f$ on $\overline{D}_0$ and for all j such that $|z_j - z_0| > r_0$, let D_j respectively be disjoint discs centered at z_j which are also disjoint from $\overline{D}_0$. Set $f_0 = w_j$ on such D_j. Let E be the union of $\overline{D}_0$ with these D_j. Then $f_0{:}E$ is holomorphic on a neighborhood of Z and $f_0(z_j) = w_j$, for all $j = 0, 1, \cdots$.

This problem was completely solved by Ivar Bendixson [**1**] in the 19th century and independently, almost one hundred years later, by Ronen Peretz [**16**], [**17**] in his Masters' Thesis under Dov Aharonov. Bendixson's paper gives the solution as an extension of the polynomial interpolation formula of Carl Friedrich Gauß.

For data $z_1, \cdots, z_n; w_1, \cdots, w_n$, where the z_j are distinct, the interpolation polynomial p_{n-1} is the unique polynomial of degree $n - 1$ such that $p_{n-1}(z_\nu) =$

w_ν, for $\nu = 1, \cdots, n$. Of course, this polynomial is unique, but it has various representations. Gauß [**9**, p. 274] writes this polynomial as

$$p_{n-1} = a_0 + a_1(z - z_1) + \cdots + a_{n-1}(z - z_1)\cdots(z - z_{n-1}),$$

where the coefficients a_j are given by a very simple formula in terms of $z_1, \cdots, z_{j+1}; w_1, \cdots, w_{j+1}$. It is important to note that this formula is independent of n. That is, if $m < n$, the coefficients a_j of p_{m-1} and p_{n-1} are the same for $j = 0, \cdots, m-1$. It is easy to see that this can be done. Indeed, for $n = 1$, the interpolating polynomial is $p_0(z) = a_0$, where $a_0 = w_1$. Suppose, p_{n-1} have been constructed in this form, for $n = 1, \cdots, k$. Set

$$p_k(z) = p_{k-1}(z) + a_k(z - z_1)\cdots(z - z_k), \quad a_k = \frac{w_{k+1} - p_{k-1}(z_k)}{(z_{k+1} - z_1)\cdots(z_{k+1} - z_k)}.$$

Clearly, p_k is the k-th interpolating polynomial and is of the desired form.

To a convergent sequence of distinct points $z_1, z_2, \cdots; z_j \to z_0$, and a convergent sequence $w_1, w_2, \cdots$, Bendixson associates the formal series

$$\Sigma = \sum_{j=0}^{\infty} a_j(z - z_1)\cdots(z - z_j),$$

where the n-th partial sum is the interpolating polynomial for

$$z_1, \cdots, z_{n+1}; w_1, \cdots, w_{n+1}.$$

in the form given by Gauß. Let us call this the formal series associated to the sequences $z_1, z_2, \cdots; w_1, w_2, \cdots$. Since, for each n, the n-th partial sum is the interpolating polynomial for $z_1, z_2, \cdots, z_{n+1}; w_1, w_2, \cdots, w_{n+1}$, this formal series has the required interpolating properties, namely, it converges at each z_j to the value w_j. However, it is not clear whether this formal series converges at some value not in the sequence $z_1, z_2, \cdots$, much less whether it represents a function holomorphic in a neighborhood of the limit point z_0. Bendixson proves the following result.

THEOREM 7. *(Bendixson) Given a convergent sequence $z_1, z_2, \cdots; z_j \to z_0$ of distinct points and a convergent sequence $w_1, w_2, \cdots$, the following are equivalent.*

1) There exists a function f holomorphic in a neighborhood of z_0, such that $f(z_j) = w_j$, for all sufficiently large j.

2) The formal series Σ, associated to the sequences $z_1, z_2, \cdots; w_1, w_2, \cdots$, converges at some point z different from all the z_j.

Moreover, if these conditions hold, the coefficients of Σ are uniquely determined by the Gauß formula and $\Sigma(z) = f(z)$ in the disc of convergence of the Taylor series of f at z_0.

Bendixson, in condition 2), omitted to stipulate that the point z should be different from all of the z_j. We claim that his theorem is wrong without this condition. Since, for every convergent sequence $z_1, z_1, \cdots; z_j \to z_0$ and every convergent sequence $w_1, w_2, \cdots$, the associated formal series Σ converges at each z_j, the implication 2) $\Longrightarrow$ 1) would give that there exists a function f holomorphic in a neighborhood of z_0, such that $f(z_j) = w_j$, $j = 1, 2, \cdots$. If $z_0 = 0, z_j = 1/j$ and $w_j = \sqrt{1/j}$, an interpolating function would not have a derivative at 0, and so could not be holomorphic. This absurdity verifies our claim. However, Bendixson's proofs are valid with the additional assumption in 2) that z is different from all of the z_j.

In terms of a formal series, Bendixson's result gives a formal solution to the infinite interpolation problem. He also gives a necessary and sufficient condition for the solvability in terms of the convergence of this formal series. It is desireable to characterize solvability more explicitly (more numerically) in terms of the data $z_1, z_2, \cdots; w_1, w_2, \cdots$. Peretz ([**16**], [**17**]) has done this in terms of Vandermonde determinants. In the finite case, the coefficients of the Lagrange interpolating polynomial can be found in terms of $w_1, \cdots, w_{n-1}$ and the Vandermonde matrix for $z_1, \ldots, z_{n-1}$. For the infinite case, Peretz gives the following characterization.

THEOREM 8. *(Peretz) Let $z_1, z_2, \cdots; z_n \to z_0$ be a convergent sequence of distinct points and $w_1, w_2, \cdots$ a convergent sequence of points. Then, there exists f holomorphic in a neighborhood of z_0, such that $f(z_n) = w_n$, for all sufficiently large n, if and only if there exists an N such that*

$$\limsup_{n\to\infty} \left| \frac{\begin{vmatrix} 1 & z_N - z_0 & (z_N - z_0)^{n-1} & \cdots & w_N \\ 1 & z_{N+1} - z_0 & (z_{N+1} - z_0)^{n-1} & \cdots & w_{N+1} \\ \cdot & & & & \\ \cdot & & & & \\ \cdot & & & & \\ 1 & z_{N+n} - z_0 & (z_{N+n} - z_0)^{n-1} & \cdots & w_{N+n} \end{vmatrix}}{\begin{vmatrix} 1 & z_N - z_0 & \cdots & (z_N - z_0)^n \\ \cdot & & & \\ \cdot & & & \\ \cdot & & & \\ 1 & z_{N+n} - z_0 & \cdots & (z_{N+n} - z_0)^n \end{vmatrix}} \right|^{1/n} < \infty.$$

With the exception of the two well known theorems in the introduction, this note has dealt only with functions defined on subsets of the complex plane or of one-dimensional analytic sets. It appears that some of the preceding results have analogues in higher dimensions, replacing points by divisors (analytic sets of codimension 1) and sequences of points by sequences of divisors. However, whereas in one complex dimension we have stated results regarding arbitrary sets E, in higher dimensions our techniques would yield results at most for sets E which are unions of divisors.

The discussion in the present paper can be considered in a more general context of Hartogs type problems. For $\mathcal{L}$ a given family of subsets of a set $E \subset \mathbb{C}^n$, we may ask the following. If $f|L$ is c-holomorphic for all $L \in \mathcal{L}$ according to Definition 2B, does it follow that f is holomorphic? Usually, E is taken to be open. Hartogs' Theorem, mentioned in the introduction, asserts that, if $E \subset \mathbb{C}^n$ is open and $\mathcal{L} = \{L = \ell \cap E\}$, where ℓ runs over the family of all complex lines in the coordinate directions, then indeed f must be holomorphic on E.

If E is not open then the issue is whether or not f extends holomorphically to a neighborhood of E. Our Theorem 6 applies for arbitrary $E \subset \mathbb{C}$, where $\mathcal{L}$ is the family of all convergent sequences in E and, in the corollary, $\mathcal{L}$ is the family of all $C \cap E$, where C runs over the family of all curves.

For Hartogs type results in the above sense, for various families $\mathcal{L}$ on open subsets of $\mathbb{C}$, we refer to the recent paper [**5**] and the references therein.

In the present paper, all functions are complex valued. One can also consider Hartogs type questions for functions taking their values in a complex space Y other than $\mathbb{C}$. The most natural Y to start with is complex projective space $\mathbb{P}$. As usual, we denote by $[a:b]$ the homogeneous coordinates of a point in $\mathbb{P}$. An easy counterexample to the Hartogs theorem on separate holomorphy in this setting is the mapping $g:\mathbb{C}^2\to\mathbb{P}$ given as follows:

$$g(z,w)=[zw:z^2+w^2] \quad \text{if } (z,w)\neq(0,0),$$
$$g(0,0)=[0:1].$$

The mapping g is separately holomorphic, but not even continuous at the origin.

With Eduardo Santillan Zeron [**12**], we obtained the following. Let E be open in $\mathbb{C}^n, n\geq 3$, and $f:E\to\mathbb{P}^m$ such that $f|L$ is holomorphic for each 2-plane in coordinate directions, Then f is holomorphic. We were unaware that Georges Dloussky [**4**] had earlier obtained a similar result with a different proof.

Given a function $f{:}E$ on a set E, this paper has been concerned with what we might call *external* holomorphy, where a function could be said to be externally holomorphic on E if it is the restriction of a function holomorphic in a neighborhood of E. A more delicate subject would be the study of what we might call *internal* holomorphy of $f{:}E$, where we would first say what is meant by saying that E has analytic structure at a point $z\in E$. Then, we would define the regular part of E as the set of points where there is analytic structure. We would then define what it means for a function to be holomorphic on the regular part of E. A function $f{:}E$ could then be said to be c-holomorphic on E if it is continuous and holomorphic on the regular part of E.

In this context, it is interesting to bear in mind (see [**8**]) that a function $f{:}U$, defined on a finely open subset U of $\mathbb{C}$ is finely holomorphic if and only if every point of U has a fine neighborhood $V\subset U$ on which f coincides with a (Euclidean) C^1-function $\widetilde{f}{:}\mathbb{C}$ that satisfies the Cauchy-Riemann equation $\overline{\partial}\widetilde{f}=0$ on V.

Acknowledgement. I thank Charles Fefferman, for useful conversations, Eduardo Santillan Zeron for useful correspondence, Myrto Manolaki for pointing out errors in earlier versions and Ronen Peretz for sending me his masters' thesis as well as its English translation, both unpublished. I also thank the referee for helpful suggestions.

References

[1] I. Bendixson, *Sur une extension à l'infini de la formule d'interpolation de Gauss* (French), Acta Math. **9** (1887), no. 1, 1–34, DOI 10.1007/BF02406727. MR1554706

[2] Ulrich Daepp, Paul Gauthier, Pamela Gorkin, and Gerald Schmieder, *Alice in Switzerland: the life and mathematics of Alice Roth*, Math. Intelligencer **27** (2005), no. 1, 41–54, DOI 10.1007/BF02984813. MR2145826

[3] Maciej P. Denkowski, *Residue calculus for c-holomorphic functions*, Ark. Mat. **47** (2009), no. 1, 73–89, DOI 10.1007/s11512-008-0088-7. MR2480916 (2010c:32004)

[4] G. Dloussky, *Analyticité séparée et prolongement analytique* (French), Math. Ann. **286** (1990), no. 1-3, 153–168, DOI 10.1007/BF01453570. MR1032928 (91i:32005)

[5] Buma L. Fridman and Daowei Ma, *Holomorphic functions on subsets of* $\mathbb{C}$, J. Math. Soc. Japan **65** (2013), no. 1, 1–12, DOI 10.2969/jmsj/06510001. MR3034396

[6] Dieter Gaier, *Lectures on complex approximation*, Birkhäuser Boston, Inc., Boston, MA, 1987. Translated from the German by Renate McLaughlin. MR894920 (88i:30059b)

[7] Theodore W. Gamelin, *Uniform algebras*, Prentice-Hall, Inc., Englewood Cliffs, N. J., 1969. MR0410387 (53 #14137)

[8] Stephen J. Gardiner, *Recent progress on fine differentiability and fine harmonicity*, Complex analysis and potential theory, CRM Proc. Lecture Notes, vol. 55, Amer. Math. Soc., Providence, RI, 2012, pp. 283–291. MR2986908

[9] Gauß, C. F. Theoria interpolationis methodo nova tractata (Latin). Werke **Band 3,** Königlichen Gesellschaft der Wissenschaften, Göttingen, 265-303 (1866).

[10] Gauthier, P. M. Zero-free polynomial approximation on the union of two sets amply intersecting. Transactions of the Institute of Mathematics of the National Academy of Sciences of Ukraine, **10,** 92-114 (2013).
[11] P. M. Gauthier and W. Hengartner, *Uniform approximation on closed sets by functions analytic on a Riemann surface*, Approximation theory (Proc. Conf. Inst. Math., Adam Mickiewicz Univ., Poznań, 1972), Reidel, Dordrecht, 1975, pp. 63–69. MR0486540 (58 #6263)
[12] P. M. Gauthier and E. S. Zeron, *Hartogs' theorem on separate holomorphicity for projective spaces*, Canad. Math. Bull. **52** (2009), no. 1, 84–86, DOI 10.4153/CMB-2009-010-8. MR2494322 (2009k:32006)
[13] *Several complex variables. VII*, Encyclopaedia of Mathematical Sciences, vol. 74, Springer-Verlag, Berlin, 1994. Sheaf-theoretical methods in complex analysis; A reprint of *Current problems in mathematics. Fundamental directions. Vol. 74* (Russian), Vseross. Inst. Nauchn. i Tekhn. Inform. (VINITI), Moscow; Edited by H. Grauert, Th. Peternell and R. Remmert. MR1326617 (96k:32001)
[14] Gennadi Henkin and Mikael Passare, *Abelian differentials on singular varieties and variations on a theorem of Lie-Griffiths*, Invent. Math. **135** (1999), no. 2, 297–328, DOI 10.1007/s002220050287. MR1666771 (99m:32038)
[15] Paramonov, P. Private communication (2003).
[16] Peretz, R. Masters' Thesis (Hebrew), Technion-Israel Institute of Technology (1983).
[17] Peretz, R. Analytic images of a convergent sequence, unpublished manuscript.
[18] Reinhold Remmert, *Projektionen analytischer Mengen* (German), Math. Ann. **130** (1956), 410–441. MR0086353 (19,170c)
[19] Yum-tong Siu, *Analytic sheaf cohomology groups of dimension n of $n-$dimensional complex spaces.*, Trans. Amer. Math. Soc. **143** (1969), 77–94. MR0252684 (40 #5902)

Département de mathématiques et de statistique, Université de Montréal, CP-6128 Centreville, Montréal, H3C3J7, Canada

E-mail address: gauthier@dms.umontreal.ca

Contemporary Mathematics
Volume **638**, 2015
http://dx.doi.org/10.1090/conm/638/12807

Cyclicity in non-extreme de Branges-Rovnyak spaces

Emmanuel Fricain, Javad Mashreghi, and Daniel Seco

Abstract. We characterize cyclicity in non-extreme de Branges-Rovnyak spaces for the backward shift, and in some particular cases for the forward shift.

1. Introduction

To each $\varphi \in L^\infty(\mathbb{T})$, we associate the Toeplitz operator defined on H^2 as

$$T_\varphi(f) = P_+(\varphi f), \qquad f \in H^2,$$

where P_+ denotes the Riesz projection, that is the orthogonal projection of $L^2(\mathbb{T})$ onto H^2. It is well known that T_φ is a bounded operator on H^2 of norm equal to $\|\varphi\|_\infty$. In particular, if $b \in H^\infty$, $\|b\|_\infty \le 1$, then T_b is a contraction and $I - T_bT_{\bar{b}} = I - T_bT_b^*$ is a positive operator. Now, for $b \in H^\infty$, $\|b\|_\infty \le 1$, the *de Branges-Rovnyak space* $\mathscr{H}(b)$ is defined to be

$$\mathscr{H}(b) := (I - T_bT_{\bar{b}})^{1/2}H^2,$$

endowed with the inner product

$$\langle (I - T_bT_{\bar{b}})^{1/2}f, (I - T_bT_{\bar{b}})^{1/2}g\rangle_b := \langle f, g\rangle_2,$$

for $f, g \perp \ker((I - T_bT_{\bar{b}})^{1/2})$. That is to say, $\mathscr{H}(b)$ is normed to make $(I - T_bT_{\bar{b}})^{1/2}$ a partial isometry of H^2 onto $\mathscr{H}(b)$. When $\|b\|_\infty < 1$, the operator $I - T_bT_{\bar{b}}$ is an isomorphism on H^2 and thus $\mathscr{H}(b) = H^2$ with an equivalent norm. On the other extreme, when b is an inner function then $T_bT_{\bar{b}}$ is the orthogonal projection of H^2 onto bH^2 and thus $\mathscr{H}(b)$ turns out to be the model space $(bH^2)^\perp = H^2 \ominus bH^2 = K_b$. In what follows, we will present some well-known general facts concerning $\mathscr{H}(b)$ spaces. See [**6**, **10**] for more details on these spaces.

An important property of $\mathscr{H}(b)$ spaces is that they are invariant under the backward shift operator $S^* = T_{\bar{z}}$. These spaces have been introduced by L. de Branges and J. Rovnyak in the context of model theory [**3**, **4**]. A whole class of Hilbert space contractions is unitarily equivalent to $S^*|\mathscr{H}(b)$, for an appropriate b.

It turns out that the properties of $\mathscr{H}(b)$ spaces depend on whether or not b is an extreme point of the closed unit ball of H^∞, i.e., whether or not $\log(1 - |b|)$ is

---

2010 *Mathematics Subject Classification.* Primary 47A16; Secondary 47B32, 47A15.
Key words and phrases. Cyclicity, de Branges-Rovnyak spaces, shift operator.
For this work, we were supported by grants from Labex CEMPI (ANR-11-LABX-0007-01), NSERC (100756), ERC Grant 2011-ADG-20110209 from EU programme FP2007-2013, MEC/MICINN Project MTM2011-24606, and Generalitat de Catalunya 2009SGR420.

integrable with respect to normalized Lebesgue measure $dm = d\theta/2\pi$ on the unit circle $\mathbb{T}$.

We will now assume that b is a *non-extreme* point of the closed unit ball of H^∞, which, by the Arens–Buck–Carleson–Hoffman–Royden [**8**] theorem, is equivalent to the condition

$$\int_{\mathbb{T}} \log(1 - |b|)dm > -\infty. \tag{1.1}$$

To abbreviate, we will simply say b is non-extreme. In this case there is a unique outer function a with $a(0) > 0$ such that

$$|a(\zeta)|^2 + |b(\zeta)|^2 = 1 \quad m-\text{a.e. } \zeta \in \mathbb{T}. \tag{1.2}$$

We call a pair (a, b) satisfying the above property a *Pythagorean pair.* We say that (a, b) forms a *corona pair* if moreover we have

$$|a(z)| + |b(z)| \geq \delta \qquad (z \in \mathbb{D}),$$

for some constant $\delta > 0$.

For $\phi \in L^\infty(\mathbb{T})$, let

$$\mathscr{M}(\phi) := T_\phi H^2$$

endowed with the norm which makes T_ϕ a partial isometry from H^2 onto $T_\phi H^2$. When b is non-extreme then $\mathscr{M}(a) = aH^2$ is contractively contained in $\mathscr{H}(b)$, that is, for any $f \in H^2$, the function $af \in \mathscr{H}(b)$ and

$$\|af\|_b \leq \|af\|_{\mathscr{M}(a)} = \|f\|_2. \tag{1.3}$$

It is also known, when b is non-extreme, that for every $f \in \mathscr{H}(b)$, we have $T_{\overline{b}} f \in \mathscr{M}(\overline{a})$ and one can obtain the scalar product of $f_1, f_2 \in \mathscr{H}(b)$ via the formula

$$\langle f_1, f_2 \rangle_b = \langle f_1, f_2 \rangle_2 + \langle g_1, g_2 \rangle_2, \tag{1.4}$$

where $g_i \in H^2$ and $T_{\overline{b}} f_i = T_{\overline{a}} g_i$, $i = 1, 2$. Note that g_i is unique since $T_{\overline{a}}$ is one-to-one due to the fact that a is outer .

Still under the hypothesis that b is non extreme, we have $\mathscr{M}(a) = \mathscr{H}(b)$ (with equivalent norms) if and only if (a, b) forms a corona pair and $T_{a/\bar{a}}$ is invertible on H^2. See [**9**, Theorem 3]. Recall that the Toeplitz operator $T_{a/\bar{a}}$ is invertible on H^2 if and only if $|a|^2$ satisfies the so-called Muckenhoupt (A_2) condition.

We have already seen that $\mathscr{H}(b)$ spaces are invariant under the backward shift operator and in the case when b is non-extreme, they are also invariant under the forward shift S. We will denote by X_b (respectively by Y_b) the restriction of S^* (respectively of S) to $\mathscr{H}(b)$, i.e.

$$X_b = S^*_{|\mathscr{H}(b)} \quad \text{and} \quad Y_b = S_{|\mathscr{H}(b)}.$$

In this note, we discuss the problem of characterizing the cyclic vectors for X_b and Y_b. Recall that if T is a bounded operator on a Hilbert space $\mathscr{H}$, then a vector $x \in \mathscr{H}$ is said to be cyclic for T if

$$\text{span}_{\mathscr{H}}(T^n x : n \geq 0) = \mathscr{H},$$

where $\text{span}_{\mathscr{H}}(A)$ denotes the closed linear span of elements of A in $\mathscr{H}$.

2. Cyclicity for X_b

Recall that cyclic vectors for S^* have been characterized by Douglas–Shapiro–Shields. They showed in [**5**] that a function f in H^2 is not cyclic for S^* if and only if it has a bounded type meromorphic pseudo continuation across $\mathbb{T}$ to $\mathbb{D}_e = \{z \in \mathbb{C} : 1 < |z| \leq \infty\}$. This is also equivalent to the existence of two functions $g, h \in \bigcup_{p>0} H^p$ such that

$$f = \frac{\bar{h}}{\bar{g}}, \qquad \text{(a.e. on } \mathbb{T}).$$

It turns out that cyclic vectors of X_b are precisely the cyclic vectors of S^* which live in $\mathscr{H}(b)$.

THEOREM 2.1. *Let b be a non-extreme point of the unit ball of H^∞ and let $f \in \mathscr{H}(b)$. Then the following assertions are equivalent:*

(i) The function f is cyclic for X_b.

(ii) The function f, as an element of H^2, is cyclic for S^*.

PROOF. Assume that f is cyclic for X_b. Then, for any polynomial p, there exists a sequence of polynomials p_n such that

$$\|p_n(X_b)f - p\|_b \to 0, \qquad \text{as } n \to \infty.$$

Since, $\mathscr{H}(b)$ is contained contractively in H^2, then we have

$$\|p_n(S^*)f - p\|_2 \to 0, \qquad \text{as } n \to \infty.$$

Thus, any polynomial p belongs to $\operatorname{span}_{H^2}(S^{*n} f : n \geq 0)$, which is a closed subspace of H^2. Since the polynomials are dense in H^2, we get that

$$\operatorname{span}_{H^2}(S^{*n} f : n \geq 0) = H^2.$$

Conversely assume that f is cyclic for S^* and denote by $\mathcal{J}$ the closed subspace of $\mathscr{H}(b)$ defined by

$$\mathcal{J} = \operatorname{span}_{\mathscr{H}(b)}(X_b^n f : n \geq 0).$$

Since $\mathcal{J}$ is a closed invariant subspace of X_b, then, according to [**9**, Theorem 5], there exists a function u which is either inner or $u \equiv 0$ such that

$$\operatorname{span}_{\mathscr{H}(b)}(X_b^n f : n \geq 0) = \mathscr{H}(b) \cap \mathscr{H}(u).$$

If u is inner since $f \in \mathscr{H}(u)$, then it is not cyclic for S^*, which is contrary to the hypothesis. Thus, $u \equiv 0$ and $\mathscr{H}(u) = H^2$, which implies that

$$\operatorname{span}_{\mathscr{H}(b)}(X_b^n f : n \geq 0) = \mathscr{H}(b).$$

Therefore, the function f is cyclic for X_b. □

3. Cyclicity for Y_b

The situation of cyclic vectors for Y_b appears to be dramatically more difficult, although some of the properties of cyclicity in the Hardy space are preserved. We first begin with a simple observation. Observe that a classical result of Beurling showed that cyclic functions with respect to the forward shift operator in the Hardy space are exactly outer functions.

LEMMA 3.1. *Let b be a non-extreme point of the unit ball of H^∞ and let $f \in \mathscr{H}(b)$. If f is cyclic for Y_b, then f, as an element of H^2, is also cyclic for S.*

PROOF. Since f is cyclic for Y_b, then there exists a sequence of polynomials $(p_n)_n$ such that $\|p_n f - 1\|_b \to 0$ as $n \to \infty$. Using the fact that $\mathscr{H}(b)$ is contained contractively in H^2, we deduce that $\|p_n f - 1\|_2 \to 0$ as $n \to \infty$. That implies that f is a cyclic vector for S. □

In the light of Beurling theorem, Lemma 3.1 implies that cyclic vectors of Y_b are outer functions.

In some cases, we can also provide a sufficient condition. We remind that $\mathscr{M}(a) \subset \mathscr{H}(b)$ and the inclusion map is contractive.

LEMMA 3.2. *Let (a, b) be a Pythagorean pair and assume that the function a^{-1} belongs to $L^2(\mathbb{T})$. Let f be an outer function in $\mathscr{M}(a)$. Then f is cyclic for Y_b.*

PROOF. Since f is an outer function in $\mathscr{M}(a)$, we can write $f = ag$, with $g \in H^2$. It follows from the uniqueness of the Nevanlinna factorization that g is also an outer function. In particular, it should be cyclic for S. Since $a^{-1} \in L^2(\mathbb{T})$ and a is outer, then $a^{-1} \in H^2$ and there exists a sequence of polynomials $(p_n)_n$ such that $\|p_n g - a^{-1}\|_2 \to 0$ as $n \to \infty$. It remains to remember (1.3) to conclude that

$$\|p_n f - 1\|_b = \|p_n a g - 1\|_b \leq \|p_n g - a^{-1}\|_2,$$

and thus f is cyclic for Y_b. □

This allows us to characterize cyclicity under a simple assumption.

THEOREM 3.3. *Let (a, b) be a corona pair, let the operator $T_{a/\bar{a}}$ be invertible, and let $f \in \mathscr{H}(b)$. Then the following assertions are equivalent:*

(i) f is cyclic for Y_b.
(ii) f, as an element of H^2, is cyclic for S.
(iii) f is an outer function.

PROOF. The equivalence between *(ii)* and *(iii)* is well-known and due to Beurling (see [**6**] for instance). *(i)* $\Longrightarrow$ *(ii)*: follows from Lemma 3.1. *(iii)* $\Longrightarrow$ *(i)*: assume that f is an outer function. According to the hypothesis on the Pythagorean pair (a, b) we know that $\mathscr{H}(b) = \mathscr{M}(a)$ (with equivalent norms). In particular, since $1 \in \mathscr{H}(b)$, we must have $a^{-1} \in H^2$. Thus it remains to apply Lemma 3.2. □

4. An important example

We will now give an example showing that the hypothesis that the operator $T_{a/\bar{a}}$ is invertible cannot be dropped in Theorem 3.3. Consider $b(z) = (1+z)/2$ and $a(z) = (1-z)/2$. It is easy to check that (a, b) is a corona pair. Moreover, the operator $T_{a/\bar{a}} = -S$ is not invertible. We show that the outer function $f(z) = 1 - z$ is not cyclic in $\mathscr{H}(b)$. Indeed, let $n \geq 0$. It is easy to check that

$$T_{\bar{b}} 1 = T_{\bar{a}} 1 \quad \text{and} \quad T_{\bar{b}}(z^n(1-z)) = -T_{\bar{a}}(z^n(1+z)).$$

Thus, using (1.4), we get that

$$\begin{aligned} \langle 1, z^n f \rangle_b &= \langle 1, (1-z)z^n \rangle_2 - \langle 1, (1+z)z^n \rangle_2 \\ &= -2\langle 1, z^{n+1} \rangle_2 = 0. \end{aligned}$$

This proves that $1 \perp z^n f$ for any $n \geq 0$ and thus f cannot be cyclic for Y_b.

In fact, in this particular case, we can completely determine the cyclic vectors for Y_b using a description given by Sarason in [**11**]. More precisely, it is proved that in the case when $b(z) = (1+z)/2$, then

$$\mathscr{H}(b) = (z-1)H^2 \dotplus \mathbb{C}$$

where the symbol $\dotplus$ means that the decomposition is direct (not necessarily orthogonal). Moreover, the subspace $(z-1)H^2$ is a closed subspace of $\mathscr{H}(b)$ and we have

$$\|f\|_b \asymp \|g\|_2 + |\lambda|$$

for every $f = (z-1)g + \lambda$ where $g \in H^2$. Since $|g(z)| = O((1-|z|)^{-1/2})$, we easily see that every function $f \in \mathscr{H}(b)$ has a non tangential limit at 1 and $\lambda = f(1)$.

THEOREM 4.1. *Let $b(z) = (1+z)/2$ and $f \in \mathscr{H}(b)$. Then f is cyclic for Y_b if and only if f is an outer function such that $f(1) \neq 0$.*

PROOF. Decompose f as $f = (z-1)g + f(1)$, where $g \in H^2$. Let $(p_n)_n$ be a sequence of polynomials and write $p_n = p_n(1) + (z-1)q_n$ where q_n is a polynomial. We have

$$\begin{aligned} p_n f - 1 &= p_n((z-1)g + f(1)) - 1 = (z-1)p_n g + f(1)p_n - 1 \\ &= (z-1)(p_n g + f(1)q_n) + (f(1)p_n(1) - 1). \end{aligned}$$

Thus, we get

$$\|p_n f - 1\|_b \asymp \|p_n g + f(1)q_n\|_2 + |f(1)p_n(1) - 1|. \tag{4.1}$$

First, assume that $f(1) = 0$. Then, for any sequence of polynomials $(p_n)_n$, we have by (4.1)

$$\|p_n f - 1\|_b \gtrsim 1,$$

and thus, f cannot be cyclic for Y_b.

Conversely, assume now that f is an outer function such that $f(1) \neq 0$ and let us prove that f is cyclic for Y_b. Put $\lambda = 1/f(1)$. Since f is outer, by Beurling theorem, there exists a sequence of polynomials $(q_n)_n$ such that

$$\|q_n f + \lambda g\|_2 \to 0, \quad n \to \infty.$$

If we consider $p_n = \lambda + (z-1)q_n$ then $p_n(1) = \lambda$ and we have $p_n g + f(1)q_n = \lambda g + f q_n$. Hence, $\|p_n g + f(1)q_n\|_2 \to 0$ as $n \to +\infty$. Thus by (4.1), we conclude that $\|p_n f - 1\|_b \to 0$ as $n \to \infty$, which proves that f is cyclic for Y_b. □

In particular, we get the following corollary. Denote by $\mathrm{Hol}(\overline{\mathbb{D}})$, the set of all functions that are holomorphic on an open neighbourhood of the unit disk.

COROLLARY 4.2. *Let $b(z) = (1+z)/2$ and f be a function in $Hol(\overline{\mathbb{D}})$. Then f is cyclic for Y_b if and only if f has no zeros on $\mathbb{D}$ and $f(1) \neq 0$.*

REMARK 4.3. It should be noted that the result of Theorem 4.1 can be rephrased like this: let $b(z) = (1+z)/2$ and let $f \in \mathscr{H}(b)$. Then, f is cyclic for Y_b if and only if f is an outer function such that $f \notin \mathscr{M}(a)$. Note that in the case when $\mathscr{M}(a)$ is a proper closed subspace of $\mathscr{H}(b)$, then the fact that f is an outer function such that $f \notin \mathscr{M}(a)$ still remains a necessary condition for cyclicity.

REMARK 4.4. One should note that if we combine results of [**7**] and [**2**], then we can generalize Theorem 4.1 to the case when (a, b) is a rational pair and the function a has simple zeros on $\mathbb{T}$. However, the result given in [**7**] is based on a difficult description of invariant subspaces of $\mathscr{H}(b)$ (in the case when $b(z) = (1+z)/2$) obtained by Sarason in [**9**] and our approach seems to us more elementary, though a little bit less general.

References

[1] A. Blandignères, E. Fricain, F. Gaunard, A. Hartmann and W.T. Ross, *Direct and reverse Carleson measures for $\mathscr{H}(b)$ spaces*, preprint.

[2] Constantin Costara and Thomas Ransford, *Which de Branges-Rovnyak spaces are Dirichlet spaces (and vice versa)?*, J. Funct. Anal. **265** (2013), no. 12, 3204–3218, DOI 10.1016/j.jfa.2013.08.015. MR3110499

[3] Louis de Branges and James Rovnyak, *Canonical models in quantum scattering theory*, Perturbation Theory and its Applications in Quantum Mechanics (Proc. Adv. Sem. Math. Res. Center, U.S. Army, Theoret. Chem. Inst., Univ. of Wisconsin, Madison, Wis., 1965), Wiley, New York, 1966, pp. 295–392. MR0244795 (39 #6109)

[4] Louis de Branges and James Rovnyak, *Square summable power series*, Holt, Rinehart and Winston, New York-Toronto, Ont.-London, 1966. MR0215065 (35 #5909)

[5] R. G. Douglas, H. S. Shapiro, and A. L. Shields, *Cyclic vectors and invariant subspaces for the backward shift operator.* (English, with French summary), Ann. Inst. Fourier (Grenoble) **20** (1970), no. fasc. 1, 37–76. MR0270196 (42 #5088)

[6] E. Fricain and J. Mashreghi, *An Introduction to $\mathcal{H}(b)$ spaces, Vol.I & II*, New math. Monographs 20 & 21, Cambridge University Press, to appear.

[7] Dominique Guillot, *Fine boundary behavior and invariant subspaces of harmonically weighted Dirichlet spaces*, Complex Anal. Oper. Theory **6** (2012), no. 6, 1211–1230, DOI 10.1007/s11785-010-0124-z. MR3000683

[8] Karel de Leeuw and Walter Rudin, *Extreme points and extremum problems in H_1*, Pacific J. Math. **8** (1958), 467–485. MR0098981 (20 #5426)

[9] Donald Sarason, *Doubly shift-invariant spaces in H^2*, J. Operator Theory **16** (1986), no. 1, 75–97. MR847333 (88d:47014b)

[10] Donald Sarason, *Sub-Hardy Hilbert spaces in the unit disk*, University of Arkansas Lecture Notes in the Mathematical Sciences, 10, John Wiley & Sons, Inc., New York, 1994. A Wiley-Interscience Publication. MR1289670 (96k:46039)

[11] Donald Sarason, *Local Dirichlet spaces as de Branges-Rovnyak spaces*, Proc. Amer. Math. Soc. **125** (1997), no. 7, 2133–2139, DOI 10.1090/S0002-9939-97-03896-3. MR1396993 (98h:46023)

[12] Donald Sarason, *Unbounded Toeplitz operators*, Integral Equations Operator Theory **61** (2008), no. 2, 281–298, DOI 10.1007/s00020-008-1588-3. MR2418122 (2010c:47073)

LABORATOIRE PAUL PAINLEVÉ, UFR DE MATHÉMATIQUES, BÂTIMENT M2, UNIVERSITÉ DES SCIENCES ET TECHNOLOGIES LILLE 1, 59 655 VILLENEUVE D'ASCQ CÉDEX, FRANCE

E-mail address: emmanuel.fricain@math.univ-lille1.fr

DÉPARTAMENT DE MATHEMATIQUES ET DE STATISTIQUE, UNIVERSITÉ LAVAL, QUÉBEC, QC, G1K 7P4, CANADA

E-mail address: javad.mashreghi@mat.ulaval.ca

MATHEMATICS INSTITUTE, ZEEMAN BUILDING, UNIVERSITY OF WARWICK, COVENTRY CV4 7AL, UNITED KINGDOM

E-mail address: d.seco@warwick.ac.uk

Contemporary Mathematics
Volume **638**, 2015
http://dx.doi.org/10.1090/conm/638/12808

Integral representations of the derivatives in $\mathcal{H}(b)$ spaces

Emmanuel Fricain and Javad Mashreghi

ABSTRACT. In this survey, we provide integral representations for the boundary values of derivatives of functions in the de Branges–Rovnyak spaces $\mathcal{H}(b)$, where b is in the unit ball of H^∞. To achieve this goal, we need to study several properties of Blaschke products which are interesting in their own right.

1. Introduction

Let $H^p(\mathbb{D})$, $0 < p \le \infty$, denote the classical Hardy space of analytic functions on the unit disc $\mathbb{D} = \{z \in \mathbb{C} : |z| < 1\}$. As usual, we also treat $H^p(\mathbb{D})$ as a closed subspace of $L^p(\mathbb{T}, m)$, where $\mathbb{T} = \partial\mathbb{D}$ and m is the normalized arc length measure on $\mathbb{T}$. Let b be in the unit ball of $H^\infty(\mathbb{D})$. Then the canonical factorization of b is $b = BF$, where

$$B(z) = \gamma \prod_n \frac{|a_n|}{a_n} \frac{a_n - z}{1 - \overline{a_n} z}, \qquad (z \in \mathbb{D}),$$

is the Blaschke product with zeros $a_n \in \mathbb{D}$ satisfying the Blaschke condition

$$\sum_n (1 - |a_n|) < \infty,$$

γ is a constant of modulus one, and F is of the form

$$F(z) = \exp\left(- \int_{\mathbb{T}} \frac{\zeta + z}{\zeta - z} \, d\sigma(\zeta) \right), \qquad (z \in \mathbb{D}),$$

where $d\sigma = -\log|b|\, dm + d\mu$ and $d\mu$ is a positive singular measure on $\mathbb{T}$. In the definition of B, we assume that $|a_n|/a_n = 1$ whenever $a_n = 0$.

In this paper, we study some aspects of the de Branges–Rovnyak spaces

$$\mathcal{H}(b) = (Id - T_b T_{\overline{b}})^{1/2} H^2(\mathbb{D}).$$

Here T_φ denotes the Toeplitz operator defined on $H^2(\mathbb{D})$ by $T_\varphi(f) = P_+(\varphi f)$, where P_+ is the (Riesz) orthogonal projection of $L^2(\mathbb{T})$ onto $H^2(\mathbb{D})$. In general, $\mathcal{H}(b)$ is not closed with respect to the norm of $H^2(\mathbb{D})$. However, it is a Hilbert space when equipped with the inner product

$$\langle\, (Id - T_b T_{\overline{b}})^{1/2} f,\, (Id - T_b T_{\overline{b}})^{1/2} g \,\rangle_b = \langle f, g \rangle_2,$$

2010 *Mathematics Subject Classification.* Primary 46E22; Secondary 47A15, 33C05.

Key words and phrases. de Branges-Rovnyak spaces, model subspaces of H^2, integral representation.

This work was supported by funds from NSERC (Canada) and the CNRS (France).

where f and g are chosen so that

$$f, g \perp \ker\, (Id - T_b T_{\bar{b}})^{1/2}.$$

As a very special case, if $|b| = 1$ a.e. on $\mathbb{T}$, or equivalently when b is an inner function for the unit disc, then $Id - T_b T_{\bar{b}}$ is an orthogonal projection and the $\mathcal{H}(b)$ norm coincides with the $H^2(\mathbb{D})$ norm. In this case, $\mathcal{H}(b)$ becomes a closed (ordinary) subspace of $H^2(\mathbb{D})$, which coincides with the shift-coinvariant subspace $K_b := H^2(\mathbb{D}) \ominus bH^2(\mathbb{D})$.

These spaces (and more precisely their general vector-valued version) appeared first in L. de Branges and J. Rovnyak [**10, 11**] as universal model spaces for Hilbert space contractions. In particular, this model says that if T is a contraction on a Hilbert space $\mathcal{H}$ such that $I - TT^*$ and $I - T^*T$ are of rank one and that

$$\|T^n f\|_{\mathcal{H}} = \|f\|_{\mathcal{H}}, \quad (n \geq 1) \Longrightarrow f = 0,$$

then there exists a function b which is an extreme point of the closed unit ball of H^∞ such that T is unitarily equivalent to the restriction of the backward shift operator on $\mathcal{H}(b)$. Despite the operator model theory for Hilbert space contractions and in particular the invariant subspace problem, the motivation of de Branges–Rovnyak for the study of these spaces seems to be the quantum scattering theory. The connection with this topic had to do with using the machinery of Hilbert spaces of analytic functions to set up a formalism for the study of the perturbation theory of self–adjoint operators, an important subject in wave–operator approach to scattering theory. Since the work of de Branges–Rovnyak, one discovered that $\mathcal{H}(b)$ spaces have an important role to play in numerous questions of complex analysis and function theory. In particular, the notion of complementary space of a Hilbert space contractively contained into another, which is hidden in the definition of $\mathcal{H}(b)$ space we give here, is one of the key notion involved in the solution of the Bieberbach conjecture given by de Branges [**9**]. Sarason also pointed out a connection with the problem of rigid functions or equivalently exposed points of the unit ball of $H^1(\mathbb{D})$. See [**27, 28**]. They also play an important role in operator theory. Sarason [**25, 29**] used $\mathcal{H}(b)$ spaces in his study of the kernel of Toeplitz operators. Jury in [**19**] found a nice proof of the classical result which says that any composition operator is bounded on H^2 and his proof is only based on the reproducing kernel theory and particularly those of $\mathcal{H}(b)$ spaces. Finally, let us mention the theory of linear systems for which $\mathcal{H}(b)$ spaces represent a useful tool and a natural framework. See for instance the survey of D. Alpay [**3**] and the recent work of Ball–Bolotnikov [**4**] in this direction. For further information on the internal and still mysterious structure of these spaces, see [**16, 26**]. Let us mention for instance the natural question of multipliers of $\mathcal{H}(b)$ spaces which is far from being understood despite numerous work from Davis, Lotto, McCarthy, Sarason, Suarez [**12, 20–23, 32**].

In the case where b is an inner function, H. Helson [**18**] studied the problem of analytic continuation across the boundary for functions in K_b. Then, still when b is an inner function, P. Ahern and D. Clark [**1, 2**] characterized those points of the boundary at which every function $f \in K_b$ and all its derivatives up to order n have a radial limit. Recently, we gave an extension of the preceding results of Helson and of Ahern–Clark to $\mathcal{H}(b)$ spaces [**6, 14, 15**].

Deriving the representation formulas for higher derivatives requires a lot of background on function theory. As a matter of fact, sometimes even tracing the right result to apply is not that easy. Moreover, in certain cases, one needs to combine two results in different papers to get the required conclusion. These facts made us to write this survey. On one hand, the results about boundary behaviour of Blaschke product are interesting in their own right. We gave them with concrete proofs. On the other hand, we use them to obtain our representation formulas for the derivatives of functions in $\mathcal{H}(b)$ spaces.

2. Preliminaries

We first recall some basic well-known facts concerning reproducing kernels in $\mathcal{H}(b)$. For any $w \in \mathbb{D}$, the linear functional $f \longmapsto f(w)$ is bounded on $H^2(\mathbb{D})$ and thus, by Riesz' theorem, it is induced by a unique element k_w of $H^2(\mathbb{D})$. On the other hand, by Cauchy's formula, we have

$$f(w) = \frac{1}{2\pi} \int_0^{2\pi} \frac{f(e^{i\vartheta})}{1 - we^{-i\vartheta}}\, d\vartheta, \qquad (f \in H^2(\mathbb{D}),\ w \in \mathbb{D}),$$

and thus k_w is the so called cauchy kernel

$$k_w(z) = \frac{1}{1 - \overline{w}z}, \qquad (z \in \mathbb{D}).$$

Now, since $\mathcal{H}(b)$ is contained contractively in $H^2(\mathbb{D})$, the restriction to $\mathcal{H}(\mathbb{D})$ of the evaluation functional at $w \in \mathbb{D}$ is a bounded linear functional on $\mathcal{H}(\mathbb{D})$. Hence, relative to the inner product in $\mathcal{H}(b)$, it is induced by a vector k_w^b in $\mathcal{H}(b)$. In other words, for all $f \in \mathcal{H}(b)$, we have

$$f(w) = \langle f, k_w^b \rangle_b.$$

But if $f = (Id - T_b T_{\overline{b}})^{1/2} f_1 \in \mathcal{H}(b)$, we have

$$\langle f, (Id - T_b T_{\overline{b}}) k_w \rangle_b = \langle f_1, (Id - T_b T_{\overline{b}})^{1/2} k_w \rangle_2 = \langle f, k_w \rangle_2 = f(w),$$

which implies that

$$k_w^b = (Id - T_b T_{\overline{b}}) k_w.$$

Finally, using the well known result $T_{\overline{b}} k_w = \overline{b(w)} k_w$, we obtain

$$k_w^b(z) = \frac{1 - \overline{b(w)} b(z)}{1 - \overline{w}z}, \qquad (z \in \mathbb{D}).$$

We know that $\mathcal{H}(b)$ is invariant under the backward shift operator S^* and, in the following, we use extensively the contraction $X_b = X = S^*|\mathcal{H}(b)$. Its adjoint satisfies the important formula

$$X^* h = Sh - \langle h, S^* b \rangle_b\, b, \tag{2.1}$$

for all $h \in \mathcal{H}(b)$. See [**11**, Theorem 13] for the original proof and [**26**, pages 11-12] for another proof.

A point $w \in \overline{\mathbb{D}}$ is said to be regular (for b) if either $w \in \mathbb{D}$ and $b(w) \neq 0$, or $w \in \mathbb{T}$ and b admits an analytic continuation across a neighbourhood $V_w = \{z : |z - w| < \varepsilon\}$ of w with $|b| = 1$ on $V_w \cap \mathbb{T}$. The spectrum of b, denoted by $\sigma(b)$, is then defined as the complement in $\overline{\mathbb{D}}$ of all regular points of b.

For $f \in H^2(\mathbb{D})$, we have

$$f \in \mathcal{H}(b) \Longleftrightarrow T_{\overline{b}} f \in \mathcal{H}(\overline{b}).$$

Moreover, if $f_1, f_2 \in \mathcal{H}(b)$, then

$$\langle f_1, f_2 \rangle_b = \langle f_1, f_2 \rangle_2 + \langle T_{\overline{b}} f_1, T_{\overline{b}} f_2 \rangle_{\overline{b}}. \tag{2.2}$$

See Lotto–Sarason [**22**, Lemma 2.2].

We also mention an integral representation for functions in $\mathcal{H}(\bar{b})$ [**26**, page 16]. Let $\rho(\zeta) = 1 - |b(\zeta)|^2$ a.e. on $\mathbb{T}$, and let $L^2(\rho)$ stand for the usual Hilbert space of measurable functions $f : \mathbb{T} \to \mathbb{C}$ with $\|f\|_\rho < \infty$, where

$$\|f\|_\rho^2 = \int_{\mathbb{T}} |f|^2 \rho \, dm.$$

For each $w \in \mathbb{D}$, the Cauchy kernel k_w belongs to $L^2(\rho)$. Hence, we define $H^2(\rho)$ to be the span in $L^2(\rho)$ of the functions k_w ($w \in \mathbb{D}$). If q is a function in $L^2(\rho)$, then $q\rho$ is in $L^2(\mathbb{T})$, being the product of $q\rho^{1/2} \in L^2(\mathbb{T})$ and the bounded function $\rho^{1/2}$. Finally, we define the operator $C_\rho : L^2(\rho) \longrightarrow H^2(\mathbb{D})$ by

$$C_\rho(q) = P_+(q\rho).$$

Then C_ρ is a partial isometry from $L^2(\rho)$ onto $\mathcal{H}(\bar{b})$ whose initial space equals to $H^2(\rho)$ and it is an isometry if and only if b is an extreme point of the unit ball of $H^\infty(\mathbb{D})$. Recall that by a well–known result of de Leeuw and W. Rudin, b is an extreme point of the closed unit ball of $H^\infty(\mathbb{D})$ if and only if $\log(1 - |b|) \notin L^1(\mathbb{T})$.

3. Derivatives of Blaschke products

Let $(a_n)_{n\geq 1}$ be a Blaschke sequence in $\mathbb{D}$, and let B be the corresponding Blaschke product. Fix a point ζ on the boundary $\mathbb{T}$. If ζ is not an accumulation point of the sequence $(a_n)_{n\geq 1}$, then B is actually analytic at this point and hence, in particular, for any value of $j \geq 0$, both limits

$$\lim_{r\to 1^-} B^{(j)}(r\zeta) \qquad \text{and} \qquad \lim_{R\to 1^+} B^{(j)}(R\zeta)$$

exist and are equal. What is more interesting is that ζ might be an accumulation point of the sequence $(a_n)_{n\geq 1}$ and yet some of the above properties still hold.

The case $N = 0$ and $N = 1$ of the following result is due to Frostman [**17**]. This is in fact the most crucial case. Frostman result was generalized by Cargo [**7**]. Then the most general version was obtained by Ahern and Clark [**1, 2**].

THEOREM 3.1 (Frostman–Cargo–Ahern–Clark). *Let $(a_n)_{n\geq 1}$ be a Blaschke sequence in $\mathbb{D}$, and let B be the corresponding Blaschke product. Assume that for an integer $N \geq 0$ and a point $\zeta \in \mathbb{T}$ we have*

$$\sum_{n=1}^{\infty} \frac{1 - |a_n|}{|\zeta - a_n|^{N+1}} \leq A. \tag{3.1}$$

Then the following hold.

(i) For each $0 \leq j \leq N$, both limits

$$B^{(j)}(\zeta) := \lim_{r\to 1^-} B^{(j)}(r\zeta) \qquad \textit{and} \qquad \lim_{R\to 1^+} B^{(j)}(R\zeta)$$

exist and are equal.

(ii) There is a constant $C = C(N, A)$ such that the estimation

$$|B^{(j)}(r\zeta)| \leq C$$

uniformly holds for $r \in [0, 1]$ and $0 \leq j \leq N$.

PROOF. The essential case is $N = 0$. The rest follows by induction.

Case $N = 0$: Our strategy is to show that, under the proposed condition, $|B(r\zeta)|$ and $\arg B(r\zeta)$ have both finite limits as r tends to 1^-. For the simplicity of notations, without loss of generality, assume that $\zeta = 1$.

In the course of proof, we repeatedly use the inequalities

$$|1 - \overline{a}_n r| > 1 - r \quad \text{and} \quad |1 - \overline{a}_n r| > \tfrac{1}{2}|1 - a_n|,$$

for $r \in (0, 1)$, which are elementary to establish. As the first application, note that

$$\frac{(1-r^2)(1-|a_n|^2)}{|1-\overline{a}_n r|^2} \le 2\frac{(1-r^2)(1-|a_n|^2)}{(1-r)|1-a_n|} \le 8\frac{1-|a_n|}{|1-a_n|}.$$

Therefore, the Weierstrass M-test shows that the series

$$\sum_{n\ge1} \frac{(1-r^2)(1-|a_n|^2)}{|1-\overline{a}_n r|^2}$$

converges uniformly in $r \in [0, 1]$, and thus

$$\lim_{r\to1^-} \sum_{n\ge1} \frac{(1-r^2)(1-|a_n|^2)}{|1-\overline{a}_n r|^2} = 0.$$

But, we have

$$\begin{aligned} |B(r)|^2 &= \prod_{n\ge1} \frac{|a_n - r|^2}{|1-\overline{a}_n r|^2} \\ &= \prod_{n\ge1} \left(1 - \frac{(1-r^2)(1-|a_n|^2)}{|1-\overline{a}_n r|^2}\right) \\ &\ge 1 - \sum_{n\ge1} \frac{(1-r^2)(1-|a_n|^2)}{|1-\overline{a}_n r|^2}, \end{aligned}$$

and this estimation enables us to deduce

$$\liminf_{r\to1^-} |B(r)|^2 \ge 1 - \lim_{r\to1^-} \sum_{n\ge1} \frac{(1-r^2)(1-|a_n|^2)}{|1-\overline{a}_n r|^2} = 1.$$

Since $|B(z)| < 1$, we conclude that

$$\lim_{r\to1^-} |B(r)| = 1.$$

To deal with the argument, write

$$\frac{\overline{a}_n}{|a_n|}\frac{a_n - r}{1-\overline{a}_n r} = \frac{1}{|a_n|}\frac{|a_n|^2 - 1 + 1 - r\overline{a}_n}{1-\overline{a}_n r} = \frac{1}{|a_n|}\left(1 - \frac{1-|a_n|^2}{1-\overline{a}_n r}\right).$$

Thus

$$\arg\left(\frac{\overline{a}_n}{|a_n|}\frac{a_n - r}{1-\overline{a}_n r}\right) = \arg\left(1 - \frac{1-|a_n|^2}{1-\overline{a}_n r}\right).$$

and, for large enough n for which the combination $(1-|a_n|)/(|1-a_n|)$ is small, we have

$$\left|\arg\left(1 - \frac{1-|a_n|^2}{1-\overline{a}_n r}\right)\right| \le M\frac{1-|a_n|^2}{|1-\overline{a}_n r|} \le 4M\frac{1-|a_n|}{|1-a_n|},$$

where M is a positive constant. Thus the series

$$\arg B(r) = \sum_{n\geq 1} \arg \left(1 - \frac{1-|a_n|^2}{1-\overline{a}_n r}\right)$$

converges absolutely and uniformly on $[0,1]$, which proves that $\lim_{r\to 1^-} \arg B(r)$ exists.

The preceding two discussions together show that $L = \lim_{r\to 1^-} B(r)$ exists and has modulus one, i.e $|L| = 1$. The estimation in part (ii) trivially holds with $C = 1$. Finally, the Blaschke product satisfies the functional equation

$$B(z)\,\overline{B(1/\overline{z})} = 1.$$

Therefore,

$$\lim_{R\to 1^+} B(R) = \frac{1}{\lim_{R\to 1^+} \overline{B(1/R)}} = \frac{1}{\lim_{r\to 1^-} \overline{B(r)}} = \frac{1}{\overline{L}} = L.$$

This argument also shows that if $\varepsilon > 0$ is such that $[1-\varepsilon, 1)$ is free from the zeros of B, then B is actually continuous on $[1-\varepsilon, 1+\varepsilon]$.

Case $N \geq 1$: Fix $1 \leq j \leq N$, and suppose that the result holds for $0, 1, \ldots, j-1$. Using the formula of B and taking the logarithmic derivative of both sides gives us

$$\frac{B'(z)}{B(z)} = \sum_{n\geq 1} \frac{(1-|a_n|^2)}{(z-a_n)(1-\overline{a}_n z)}. \tag{3.2}$$

Thus,

$$B'(z) = \sum_{n\geq 1} B_n(z)\,\frac{(1-|a_n|^2)}{(1-\overline{a}_n z)^2}, \tag{3.3}$$

where

$$B_n(z) = \frac{B(z)(1-\overline{a}_n z)}{(z-a_n)}, \qquad (n \geq 1), \tag{3.4}$$

is the subproduct formed with all zeros except a_n. Now, we use the formula for B' and take the derivative of both sides $j-1$ times. Leibnitz's formula tells us

$$B^{(j)}(z) = \sum_{k=0}^{j-1} \binom{j-1}{k} \sum_{n\geq 1} B_n^{(j-1-k)}(z)\,\frac{(k+1)!\overline{a}_n^k(1-|a_n|^2)}{(1-\overline{a}_n z)^{k+2}}.$$

Note that on the right side we have $B_n^{(\ell)}$, where ℓ runs between 0 and $j-1$. Hence, the induction hypothesis applies. To deal with the other term, we consider $r < 1$ and $R > 1$ separately.

If $r < 1$, then

$$\begin{aligned}
\left|\frac{(k+1)!\overline{a}_n^k(1-|a_n|^2)}{(1-\overline{a}_n r)^{k+2}}\right| &\leq \frac{(k+1)!(1-|a_n|^2)}{|(1-a_n)/2|^{k+2}} \\
&\leq \frac{2(k+1)!(1-|a_n|)}{|(1-a_n)/2|^{N+1}} \\
&= 2^{N+2}(k+1)!\,\frac{(1-|a_n|)}{|1-a_n|^{N+1}}.
\end{aligned}$$

But, for $R > 1$, we have

$$\begin{aligned}\left|\frac{(k+1)!\bar{a}_n^k(1-|a_n|^2)}{(1-\bar{a}_n R)^{k+2}}\right| &\leq \frac{(k+1)!(1-|a_n|^2)}{|R^{-1}-a_n|^{k+2}} \\ &\leq M\,\frac{(1-|a_n|)}{|1-a_n|^{N+1}},\end{aligned}$$

where M is a constant. This is because the condition (3.1) ensures that any Stolz domain anchored at ζ can only contain a finite number of the zeros a_n. Take any of these domains anchored at $\zeta = 1$, e.g. the one with opening $\pi/2$ or more explicitly the domain $|\Im z| \leq 1 - \Re z$. Then, for a_n's which are not in this domain but are close to $\zeta = 1$, say at a distance at most 1, we have

$$|R^{-1} - a_n| \leq |1 - a_n|/\sqrt{2}.$$

Thus,

$$\begin{aligned}\frac{(k+1)!(1-|a_n|^2)}{|R^{-1}-a_n|^{k+2}} &\leq \frac{2^{(k+2)/2}(k+1)!(1-|a_n|^2)}{|1-a_n|^{k+2}} \\ &\leq \frac{2^{(N+1)/2}N!(1-|a_n|^2)}{|1-a_n|^{N+1}}.\end{aligned}$$

The other points rest at a uniform positive distance from $\zeta = 1$.

Based on the above discussion and the induction hypothesis, if $\delta > 0$ is such that $[1-\delta, 1)$ is free from the zeros of B, then all the series

$$\sum_{n\geq 1} B_n^{(j-1-k)}(z)\,\frac{(k+1)!\bar{a}_n^k(1-|a_n|^2)}{(1-\bar{a}_n z)^{k+2}}, \qquad (0 \leq k \leq j-1),$$

are uniformly and absolutely convergent for $z \in [1-\delta, 1+\delta]$. Hence, $B^{(j)}(z)$ is also a continuous function on this interval, which can be equally stated as in the theorem based on the right and left limits at $\zeta = 1$.

Appealing to the induction hypothesis, assume that the estimation in part (ii) holds for derivatives up to order $j-1$. Then the above calculation for $r < 1$ shows that

$$\begin{aligned}|B^{(j)}(r)| &\leq \sum_{k=0}^{j-1}\binom{j-1}{k}\sum_{n\geq 1}|B_n^{(j-1-k)}(r)|\,\frac{2^{N+2}(k+1)!(1-|a_n|)}{|1-a_n|^{N+1}} \\ &\leq \left(\sum_{k=0}^{j-1}\binom{j-1}{k}2^{N+2}(k+1)!\right)CA.\end{aligned}$$

Hence, with a bigger constant the result holds for the derivative of order j. We choose the largest constant corresponding to the derivative of order N as the constant C. □

The mere usefulness of the estimation in part (ii) Theorem 3.1 is that the constant C does not depend on the distribution of zeros. It just depend on the upper bound A and the integer N. Hence, it is equally valid for all the subproducts of B.

Theorem 3.1 is also valid if $\zeta \in \mathbb{D}$. As a matter of fact, the proof is simpler in this case, since part (i) is trivial. Hence, we can say that if $\zeta \in \overline{\mathbb{D}}$ and

$$\sum_{n=1}^{\infty} \frac{1-|a_n|}{|1-\bar{a}_n\zeta|^{N+1}} \leq A,$$

then there is a constant $C = C(N, A)$ such that the estimation

$$|B^{(j)}(r\zeta)| \leq C$$

uniformly holds for $r \in [0,1]$ and $0 \leq j \leq N$.

COROLLARY 3.2. *Let $(a_n)_{n\geq 1}$ be a Blaschke sequence in $\mathbb{D}$, and let B be the corresponding Blaschke product. Let $\zeta \in \mathbb{T}$ be such that*

$$\sum_{n=1}^{\infty} \frac{1-|a_n|}{|\zeta - a_n|^2} < \infty.$$

Then B has derivative in the sense of Carathéodory at ζ and

$$|B'(\zeta)| = \sum_{n=1}^{\infty} \frac{1-|a_n|^2}{|\zeta - a_n|^2}.$$

PROOF. That B has derivative in the sense of Carathéodory at ζ is a direct consequence of Theorem 3.1. To obtain the formula for $|B'(\zeta)|$, we use (3.3). Note that our condition implies that the subproducts B_n have radial limits at ζ. Hence, we can let $r \to 1$ in

$$B'(r\zeta) = \sum_{n=1}^{\infty} B_n(r\zeta)\, \frac{(1-|a_n|^2)}{(1-\bar{a}_n r\zeta)^2}$$

to obtain

$$B'(\zeta) = \sum_{n=1}^{\infty} B_n(\zeta)\, \frac{(1-|a_n|^2)}{(1-\bar{a}_n\zeta)^2}.$$

The upper bound

$$\left| \frac{(1-|a_n|^2)}{(1-\bar{a}_n r\zeta)^2} \right| \leq \frac{4(1-|a_n|)}{|\zeta - a|^2}, \qquad (0 < r < 1),$$

allows the passage of limit inside the sum. But, according to (3.4), we have

$$B_n(\zeta) = \frac{B(\zeta)(1-\bar{a}_n\zeta)}{(\zeta - a_n)}, \qquad (n \geq 1).$$

Plugging back this in the formula for $B'(\zeta)$ gives

$$B'(\zeta) = \bar{\zeta} B(\zeta) \sum_{n=1}^{\infty} \frac{1-|a_n|^2}{|a_n - \zeta|^2}.$$

By taking the absolute values of both sides the result follows. □

4. Higher derivatives of b

Let b be an element in the closed unit ball of H^∞. According to the canonical factorization theorem, b can be decomposed as

$$b(z) = B(z)S(z)O(z), \qquad (z \in \mathbb{D}), \tag{4.1}$$

where

$$B(z) = \gamma \prod_n \left(\frac{|a_n|}{a_n} \frac{a_n - z}{1 - \overline{a}_n z} \right)$$

and

$$S(z) = \exp\left(- \int_{\mathbb{T}} \frac{\zeta + z}{\zeta - z} \, d\sigma(\zeta) \right)$$

and

$$O(z) = \exp\left(\int_{\mathbb{T}} \frac{\zeta + z}{\zeta - z} \log|b(\zeta)| \, dm(\zeta) \right).$$

We can also extend the function b outside the unit disk by the identity (4.1) and the formulas provided for B, S and O. The extended function is analytic for $|z| > 1$, $z \neq 1/\overline{a}_n$. At $1/\overline{a}_n$ it has a pole of the same order as a_n, as a zero of B. We denote this function also by b and it is easily verified that it satisfies the functional identity

$$b(z) \, \overline{b(1/\overline{z})} = 1. \tag{4.2}$$

One should be careful in dealing with function b inside and outside the unit disc. For example, if

$$b(z) = \tfrac{1}{2} z^n, \qquad (|z| < 1),$$

it is natural to use the same nice formula for $|z| > 1$. However, the functional equation (4.2) says that

$$b(z) = 2z^n, \qquad (|z| > 1).$$

Hence, b and its derivatives up to order n show a different behaviour if we approach a point $\zeta_0 \in \mathbb{T}$ from with $\mathbb{D}$ or from outside. In Theorem 4.1 below, we show that under certain circumstances, this can be avoided.

For our application in this section, we can merge $S(z)$ and $O(z)$ and write

$$b(z) = B(z) f(z) \tag{4.3}$$

where

$$f(z) = \exp\left(- \int_{\mathbb{T}} \frac{\zeta + z}{\zeta - z} \, d\mu(\zeta) \right), \tag{4.4}$$

and μ is the positive measure $d\mu(\zeta) = -\log|b(\zeta)| \, dm(\zeta) + d\sigma(\zeta)$. Now, Leibnitz's formula says

$$b^{(j)}(z) = \sum_{k=0}^{j} B^{(k)}(z) f^{(j-k)}(z).$$

For the derivatives of B on a ray, we already established Theorem 3.1. However, a similar result holds for function f, and thus similar statements actually hold for b, i.e. for any function in the closed unit ball of H^∞.

A special case of the following result and for $N = 0$ is in [**7**] without proof. The general version was mentioned in [**1, 2**], again without proof.

THEOREM 4.1. *Let b be a in the closed unit ball of H^∞ with the decomposition (4.1). Assume that, for an integer $N \geq 0$ and a point $\zeta_0 \in \mathbb{T}$, we have*

$$\sum_{n=1}^{\infty} \frac{1-|a_n|}{|\zeta_0 - a_n|^{N+1}} + \int_{\mathbb{T}} \frac{d\sigma(\zeta)}{|\zeta_0 - \zeta|^{N+1}} + \int_0^{2\pi} \frac{\big|\log|b(\zeta)|\big|}{|\zeta_0-\zeta|^{N+1}}\, dm(\zeta) \leq A. \tag{4.5}$$

Then the following hold.

(i) For each $0 \leq j \leq N$, both limits

$$b^{(j)}(\zeta_0) := \lim_{r\to 1^-} b^{(j)}(r\zeta_0) \qquad \text{and} \qquad \lim_{R\to 1^+} b^{(j)}(R\zeta)$$

exist and are equal.

(ii) There is a constant $C = C(N, A)$ such that the estimation

$$|b^{(j)}(r\zeta_0)| \leq C$$

uniformly holds for $r \in [0,1]$ and $0 \leq j \leq N$.

PROOF. As discussed before theorem, it is enough to establish the result just for the function $f = SO$ given by (4.4). The proof has the same flavor as the proof of Theorem 3.1. We first do the case is $N = 0$, and then the rest follows by induction.

Case $N = 0$: We show that, under the condition (4.5), which now translates as

$$\int_{\mathbb{T}} \frac{d\mu(\zeta)}{|\zeta_0 - \zeta|} \leq A, \tag{4.6}$$

$|f(r\zeta_0)|$ and $\arg f(r\zeta_0)$ have both finite limits as r tends to 1^-. For the simplicity of notations, without loss of generality, assume that $\zeta_0 = 1$.

A simple computation shows that

$$f(r) = \exp\left(-\int_{\mathbb{T}} \frac{1-r^2}{|\zeta - r|^2} d\mu(\zeta)\right) \exp\left(-i\int_{\mathbb{T}} \frac{r\Im(\zeta)}{|\zeta-r|^2} d\mu(\zeta)\right).$$

Therefore, we have explicit formulas for $|f(r)|$ and $\arg f(r)$.

The assumption (4.6) implies that there is no Dirac mass at $\zeta_0 = 1$, i.e. $\mu(\{1\}) = 0$. Therefore,

$$\lim_{r\to 1^-} \frac{1-r^2}{|\zeta - r|^2} = 0$$

for μ-almost every $\zeta \in \mathbb{T}$. Moreover, we have the upper bound estimation

$$\frac{1-r^2}{|\zeta-r|^2} \leq \frac{2}{|1-\zeta|}, \qquad (\zeta \in \mathbb{T}),$$

which holds uniformly for all values of the parameter $r \in (0,1)$. The condition (4.6) means that the function on the right-hand side belongs to $L^1(\mu)$. Hence, by the dominated convergence theorem, we get

$$\lim_{r\to 1^-} \int_{\mathbb{T}} \frac{1-r^2}{|\zeta-r|^2} d\mu(\zeta) = 0.$$

In return, this observation implies

$$\lim_{r\to 1^-} |f(r)| = 1.$$

In a similar manner,

$$\lim_{r\to 1^-} \frac{r\Im(\zeta)}{|\zeta-r|^2} = \frac{\Im(\zeta)}{|1-\zeta|^2}$$

for μ-almost all $\zeta \in \mathbb{T}$. We also have the upper bound estimation

$$\frac{r|\Im(\zeta)|}{|\zeta - r|^2} \leqslant \frac{2}{|1-\zeta|}, \qquad (\zeta \in \mathbb{T}),$$

which holds uniformly for all values of the parameter $r \in (0,1)$. Finally, again by the dominated convergence theorem, we see that the limit

$$\lim_{r\to 1^-} \int_{\mathbb{T}} \frac{r\Im(\zeta)}{|\zeta - r|^2} d\mu(\zeta) = \int_{\mathbb{T}} \frac{\Im(\zeta)}{|\zeta - 1|^2} d\mu(\zeta)$$

exists and is a finite real number. In return, this implies

$$\lim_{r\to 1^-} \arg f(r)$$

also exists and is a finite real number. Therefore, $L := \lim_{r\to 1^-} f(r)$ exists and, moreover, $|L| = 1$.

Put $L = \lim_{r\to 1^-} f(r)$. By (4.2), the function f satisfies the functional equation

$$f(z)\,\overline{f(1/\bar{z})} = 1.$$

Therefore,

$$\lim_{R\to 1^+} f(R) = \frac{1}{\lim_{R\to 1^+} \overline{f(1/R)}} = \frac{1}{\lim_{r\to 1^-} \overline{f(r)}} = \frac{1}{\bar{L}} = L.$$

This argument also shows that f is actually bounded on $[0, +\infty)$. The estimation in part (ii) trivially holds with $C = 1$.

Case $N \geq 1$: Fix $1 \leq j \leq N$, and suppose that the result holds for $0, 1, \ldots, j-1$. The condition (4.5) is rewritten as

$$\int_{\mathbb{T}} \frac{d\mu(\zeta)}{|1-\zeta|^{N+1}} \leq A. \tag{4.7}$$

Using the formula of f and taking the derivative of both sides gives us

$$f'(z) = \left(\int_{\mathbb{T}} \frac{-2\zeta}{(\zeta - z)^2}\, d\mu(\zeta) \right) f(z). \tag{4.8}$$

Now, take the derivative of both sides $j-1$ times. Leibnitz's formula tells us

$$f^{(j)}(z) = \sum_{k=0}^{j-1} \left(\int_{\mathbb{T}} \frac{-2(k+1)!\zeta}{(\zeta - z)^{k+2}}\, d\mu(\zeta) \right) f^{(j-1-k)}(z). \tag{4.9}$$

On the right side we have $f^{(\ell)}$, where ℓ runs between 0 and $j-1$. Hence, the induction hypothesis applies. To deal with the other term, note that for $z = r < 1$ and also $z = R > 1$, we have

$$\frac{1}{|\zeta - z|} \leq \frac{2}{|\zeta - 1|}, \qquad (\zeta \in \mathbb{T}).$$

Thus, for all $z \in (0, \infty) \setminus \{1\}$ and all k with $0 \leq k \leq j-1 \leq N-1$,

$$\begin{aligned} \left| \frac{-2(k+1)!\zeta}{(\zeta - z)^{k+2}} \right| &\leq \frac{2(k+1)!}{|(\zeta-1)/2|^{k+2}} \\ &\leq \frac{2N!}{|(\zeta-1)/2|^{N+1}} \\ &= \frac{2^{N+2}N!}{|\zeta - 1|^{N+1}}, \qquad (\zeta \in \mathbb{T}). \end{aligned} \tag{4.10}$$

Therefore, by (4.7), (4.10) and the dominated convergence theorem,

$$\lim_{r\to 1^{\pm}} \int_{\mathbb{T}} \frac{-2(k+1)!\zeta}{(\zeta - z)^{k+2}}\, d\mu(\zeta) = \int_{\mathbb{T}} \frac{-2(k+1)!\zeta}{(\zeta - 1)^{k+2}}\, d\mu(\zeta).$$

Note that we again implicitly used the fact $\mu(\{1\}) = 0$. Thus, by induction hypothesis and (4.9), part (i) follows. Moreover, again by the induction hypothesis, assume that the estimation in part (ii) holds for derivatives up to order $j-1$. Then, by (4.9) and (4.10),

$$\begin{aligned} |f^{(j)}(r)| &\leq \sum_{k=0}^{j-1} \left(\int_{\mathbb{T}} \left| \frac{-2(k+1)!\zeta}{(\zeta - r)^{k+2}} \right| d\mu(\zeta) \right) |f^{(j-1-k)}(r)| \\ &\leq \left(\int_{\mathbb{T}} \frac{2^{N+2} N!}{|\zeta - 1|^{N+1}}\, d\mu(\zeta) \right) \sum_{k=0}^{j-1} |f^{(j-1-k)}(r)| \leq j 2^{N+2} N! AC. \end{aligned}$$

Hence, with a bigger constant the result holds for the derivative of order j. We choose the largest constant corresponding to the derivative of order N as the constant C. □

We highlight one property that explicitly mentioned in the proof of Theorem 3.1 for Blaschle products, but it also holds for an arbitrary b. Under the hypothesis of Theorem 4.1, there is a $\delta > 0$ (which depends on b) such that $b^{(j)}(z)$, for $0 \leq j \leq N$, is a continuous function on the ray $[(1-\delta)\zeta_0, (1+\delta)\zeta_0]$.

COROLLARY 4.2. *Let b be a in the closed unit ball of H^∞ with the decomposition* (4.1). *Let $\zeta_0 \in \mathbb{T}$ be such that*

$$\sum_{n=1}^{\infty} \frac{1 - |a_n|}{|\zeta_0 - a_n|^2} + \int_{\mathbb{T}} \frac{d\sigma(\zeta)}{|\zeta_0 - \zeta|^2} + \int_0^{2\pi} \frac{|\log|b(\zeta)||}{|\zeta_0 - \zeta|^2}\, dm(\zeta) < \infty.$$

Then b has derivative in the sense of Carathéodory at ζ_0 and

$$|b'(\zeta_0)| = \sum_{n=1}^{\infty} \frac{1 - |a_n|^2}{|\zeta_0 - a_n|^2} + \int_{\mathbb{T}} \frac{2d\sigma(\zeta)}{|\zeta_0 - \zeta|^2} + \int_0^{2\pi} \frac{2|\log|b(\zeta)||}{|\zeta_0 - \zeta|^2}\, dm(\zeta).$$

PROOF. As we did in (4.3), write $b = Bf$. Corollary 3.2 treats the Blaschke product B and gives a formula (the first term appearing in $|b'(\zeta_0)|$ above). Hence, since on the boundary $|b'| = |B'| + |f'|$, we just need to study f and prove that $|f'(\zeta_0)|$ is precisely the remaining two terms in the formula for $|b'(\zeta_0)|$.

That f has derivative in the sense of Carathéodory at ζ_0 is a direct consequence of Theorem 4.1. To obtain the formula for $|f'(\zeta_0)|$, we use (4.8), i.e.

$$f'(r\zeta_0) = \left(\int_{\mathbb{T}} \frac{-2\zeta}{(\zeta - r\zeta_0)^2}\, d\mu(\zeta) \right) f(r\zeta_0).$$

Now, $r \to 1$ to obtain

$$f'(\zeta_0) = \left(\int_{\mathbb{T}} \frac{-2\zeta}{(\zeta - \zeta_0)^2}\, d\mu(\zeta) \right) f(\zeta_0).$$

The upper bound

$$\frac{1}{|\zeta - r\zeta_0|} \leq \frac{2}{|\zeta - \zeta_0|}, \qquad (\zeta \in \mathbb{T},\ 0 < r < 1).$$

allows the passage of limit inside the integral. Now, note that

$$\frac{-2\zeta}{(\zeta-\zeta_0)^2} = \frac{2\zeta}{(\zeta-\zeta_0)(\bar{\zeta}-\bar{\zeta}_0)\zeta\zeta_0} = \frac{2\bar{\zeta}_0}{|\zeta-\zeta_0|^2}.$$

Hence, we rewrite the formula for $f'(\zeta_0)$ as

$$f'(\zeta_0) = \left(\int_{\mathbb{T}} \frac{2}{|\zeta-\zeta_0|^2}\, d\mu(\zeta)\right) \bar{\zeta}_0 f(\zeta_0).$$

By taking the absolute values of both sides the result follows. □

5. Approximation by Blaschke products

According to (4.3), and arbitrary element of the closed unit ball of H^∞ may be decomposed as $b = Bf$, where B is a Blaschke product and f is a nonvanishing function given by (4.4). Generally speaking, since B is given by a product of some simple fraction of the form $(az+b)/(cz+d)$, it is easy to handle and study its properties. That is why in this section we explore the possibility to approximate f by some Blaschke products. This will enable us to establish certain properties for the family of Blaschke products first, and then extend them to the whole closed unit ball of H^∞.

Given a Blaschke product B with zeros $(a_n)_{n\geq 1}$, we define the measure σ_B on $\overline{\mathbb{D}}$ by

$$\sigma_B = \sum_{n=1}^{\infty} (1-|a_n|)\, \delta_{\{a_n\}},$$

where $\delta_{\{z\}}$ is the Dirac measure anchored at the point z. We consider σ_B as an element of $\mathcal{M}(\overline{\mathbb{D}})$, the space of finite complex Borel measures on $\overline{\mathbb{D}}$. This space is the dual of $\mathbb{C}(\overline{\mathbb{D}})$. Hence, we equip it with the weak-star topology. Since $\mathbb{C}(\overline{\mathbb{D}})$ is separable, this topology is first countable on $\mathcal{M}(\overline{\mathbb{D}})$. More specifically, this means that each measure has a countable local basis. Naively speaking, this implies that we just need to consider sequences of measures to study the properties of this topology.

In the following, we assume that the Blaschke products are normalized so that $B(0) > 0$.

THEOREM 5.1 (Ahern-Clark [**1**]). *Let f be given by (4.4), and let $(B_n)_{n\geq 1}$ be a sequence of Blaschke products. Then B_n converges uniformly to f on compact subsets of $\mathbb{D}$ if and only if $\sigma_{B_n} \to \mu$ in the weak-star topology of $\mathcal{M}(\overline{\mathbb{D}})$.*

PROOF. Assume that $\sigma_{B_n} \to \mu$ in the weak-star topology of $\mathcal{M}(\overline{\mathbb{D}})$, and denote the zeros of B_n by $(a_{nm})_{m\geq 1}$. Since μ is supported on $\mathbb{T}$ the zeros of B_n must tend to $\mathbb{T}$. In fact, fix any $r < 1$ and consider a continuous positive function φ which is identically 1 on $|z| \leq r$, and identically 0 on $|z| > (1+r)/2$. In between, it has a continuous transition from 1 to 0. Since $\sigma_{B_n} \to \mu$ in the weak-star topology, we have

$$\int_{\overline{\mathbb{D}}} \varphi\, d\sigma_{B_n} \longrightarrow \int_{\overline{\mathbb{D}}} \varphi\, d\mu = 0.$$

But,

$$\begin{aligned}\int_{\overline{\mathbb{D}}} \varphi \, d\sigma_{B_n} &\geq \int_{|z|\leq r} \varphi \, d\sigma_{B_n}\\ &= \sum_{|a_{nm}|\leq r} (1-|a_{nm}|)\\ &\geq (1-r)\times \mathrm{Card}\{m : |a_{nm}|\leq r\}.\end{aligned}$$

Therefore, for each $r<1$, there is an $N=N(r)$ such that

$$|a_{nm}| > r, \qquad (n\geq N,\, m\geq 1). \tag{5.1}$$

We now further explore our assumption to show that

$$B_n(0) \longrightarrow f(0). \tag{5.2}$$

Since $\sigma_{B_n}\to\mu$ in the weak-star topology, we have

$$\int_{\overline{\mathbb{D}}} d\sigma_{B_n} \longrightarrow \int_{\overline{\mathbb{D}}} d\mu.$$

By (4.4), $f(0)$ is a positive real number and

$$\int_{\overline{\mathbb{D}}} d\mu = -\log f(0).$$

But, the left side is

$$\int_{\overline{\mathbb{D}}} d\sigma_{B_n} = \sum_{m=1}^{\infty}(1-|a_{nm}|)$$

which is not precisely $-\log B_n(0)$. The actual formula is

$$-\log B_n(0) = -\sum_{m=1}^{\infty} \log|a_{nm}|.$$

However, thanks to (5.1), this difference can be handled. It is elementary to verify that

$$0\leq t-1-\log t\leq (1-t)^2, \qquad (1/2\leq t\leq 1).$$

Hence, for $n\geq N(r)$,

$$(1-|a_{nm}|)\leq -\log|a_{nm}|\leq (1+r)(1-|a_{nm}|), \qquad (m\geq 1).$$

Summing over m gives

$$\int_{\overline{\mathbb{D}}} d\sigma_{B_n} \leq -\log B_n(0)\leq (1+r)\int_{\overline{\mathbb{D}}} d\sigma_{B_n}, \qquad (n\geq N(r)). \tag{5.3}$$

Let $n\to\infty$ to deduce that

$$-\log f(0)\leq \liminf_{n\to\infty} -\log B_n(0)\leq \limsup_{n\to\infty} -\log B_n(0)\leq -(1+r)\log f(0).$$

Now, let $r\to 1$ to conclude that

$$\lim_{n\to\infty} \log B_n(0) = \log f(0).$$

The next step is to show that B_n actually uniformly converges to f on any compact subset of $\mathbb{D}$. In the language of σ_{B_n}, the formula (3.2) is rewritten as

$$\frac{B_n'(z)}{B_n(z)} = \int_{\overline{\mathbb{D}}} \frac{(1+|\zeta|)}{(z-\zeta)(1-\bar{\zeta}z)}\, d\sigma_{B_n}(\zeta).$$

In the first glance, it seems that the function

$$\zeta \longmapsto \frac{(1+|\zeta|)}{(z-\zeta)(1-\bar{\zeta}z)}$$

is not continuous on $\overline{\mathbb{D}}$ and thus we cannot appeal to the weak-star convergence. However, we fix a compact set $|z| \leq r$ and, as we saw above, after a finite number of indices the support of σ_{B_n} is in $|z| > (1+r)/2$. Hence, we can multiply the above function by a transient function which is 1 on $|z| \geq (1+r)/2$ and 0 on $|z| \leq 1$. This operation, on one hand, will not change the value of integrals and, on the other hand, will create a genuine continuous function on $\overline{\mathbb{D}}$. Therefore, we can surely say

$$\begin{aligned}\int_{\overline{\mathbb{D}}} \frac{(1+|\zeta|)}{(z-\zeta)(1-\bar{\zeta}z)}\, d\sigma_{B_n}(\zeta) \quad &\longrightarrow \quad \int_{\overline{\mathbb{D}}} \frac{(1+|\zeta|)}{(z-\zeta)(1-\bar{\zeta}z)}\, d\mu(\zeta)\\ &= \quad \int_{\mathbb{T}} \frac{-2\zeta}{(\zeta-z)^2}\, d\mu(\zeta),\end{aligned}$$

which translates as

$$\frac{B_n'(z)}{B_n(z)} \longrightarrow \frac{f'(z)}{f(z)} \tag{5.4}$$

as $n \to \infty$.

Since $(B_n)_{n\geq 1}$ is uniformly bounded by 1 on $\mathbb{D}$, it is a normal family. Let g be any pointwise limit of a subsequence of $(B_n)_{n\geq 1}$. Then, by (5.2) and (5.4), we must have

$$g(0) = f(0) \qquad \text{and} \qquad \frac{f'(z)}{f(z)} = \frac{f'(z)}{f(z)}, \qquad (z \in \mathbb{D}).$$

Thus, $g = f$, which means that the whole sequence converges uniformly to f on compact sets.

To prove the other way around, assume that B_n converges uniformly to f on compact sets. Thus, $B_n(0) \to f(0)$ and since f has no zeros on $\mathbb{D}$, for each r, (5.1) must hold. Hence, if we let $n \to \infty$ in (5.3), we obtain

$$\limsup_{n\to\infty} \int_{\overline{\mathbb{D}}} d\sigma_{B_n} \leq -\log f(0) \leq (1+r) \liminf_{n\to\infty} \int_{\overline{\mathbb{D}}} d\sigma_{B_n}.$$

Let $r \to 1$ to deduce that

$$\int_{\overline{\mathbb{D}}} d\sigma_{B_n} \longrightarrow -\log f(0).$$

Hence, $(\sigma_{B_n})_{n\geq 1}$ is a bounded sequence in $\mathcal{M}(\overline{\mathbb{D}})$, and any weak-star limit of this sequence must be a positive measure supported on $\mathbb{T}$. But, the sequence has just one weak-star limit, i.e. μ. This is because if ν is any weak-star limit of the sequence, the first part of proof shows that a subsequence of B_n converges to f_ν, where f_ν is given by (4.4) (with μ replaced by ν). Therefore, $f_\nu = f$ on $\mathbb{D}$ and, using the uniqueness theorem for Fourier coefficients of measures, we conclude that $\nu = \mu$. □

To establish our next approximation theorem, we need a result of Frostman which by itself interesting and has numerous other applications. Let Θ be an inner function and, for each $w \in \mathbb{D}$, define

$$\Theta_w(z) = \frac{w - \Theta(z)}{1 - \overline{w}\,\Theta(z)}.$$

The function Θ_w is called a *Frostman shift* of Θ. It is easy to verify that Θ_w is an inner function for each $w \in \mathbb{D}$. However, a lot more is true. Define the *exceptional set of* Θ to be

$$\mathcal{E}(\Theta) = \{\, w \in \mathbb{D} : \Theta_w \text{ is not a Blaschke product} \,\}.$$

Frostman showed that $\mathcal{E}(u)$ is a very small set.

LEMMA 5.2 (Frostman [**17**]). *Let Θ be a non-constant inner function, and let $0 < \rho < 1$. Define*

$$\mathcal{E}_\rho(\Theta) = \{\, \zeta \in \mathbb{T} : \Theta_{\rho\zeta} \textit{ is not a Blaschke product} \}.$$

Then $\mathcal{E}_\rho(\Theta)$ has one dimensional Lebesgue measure zero.

Among numerous applications of Lemma 5.2, we single out the one which states that Blaschke products are uniformly dense in the family of inner functions. A variation of the technic used in the proof of the following result will be exploited in establishing Theorem 5.4.

COROLLARY 5.3 (Frostman [**17**]). *Let Θ be an inner function, and let $\varepsilon > 0$. Then there is a Blaschke product B such that*

$$\|\Theta - B\|_\infty < \varepsilon.$$

PROOF. We have

$$\Theta(z) + \Theta_w(z) = \Theta(z) + \frac{w - \Theta(z)}{1 - \overline{w}\,\Theta(z)} = \frac{w - \bar{w}\Theta^2(z)}{1 - \overline{w}\,\Theta(z)}.$$

Thus,

$$|\Theta(z) + \Theta_w(z)| \le \frac{2|w|}{1 - |w|}.$$

On one hand, this shows that $-\Theta_w \to \Theta$ in the H^∞-norm as $w \to 0$ and, on the other hand, Lemma 5.2 ensures that there are numerous choices of w for which $-\Theta_w$ is a Blaschke product. □

In the following result, we again use $\mathcal{M}(\overline{\mathbb{D}})$, equipped with the weak-star topology. We remind that it is first countable, i.e. each point has a countable local basis of open neighborhood.

THEOREM 5.4 (Ahern-Clark [**1**]). *Let $\lambda \in \overline{\mathbb{D}}$, let $N \ge 1$, and let μ be a positive measure on $\mathbb{T}$ such that*

$$\int_{\mathbb{T}} \frac{d\mu(\zeta)}{|1 - \bar{\lambda}\zeta|^N} < \infty.$$

Then there is a sequence of Blaschke products $(B_n)_{n\ge1}$ such that $\sigma_{B_n} \to \mu$ in the weak-star topology of $\mathcal{M}(\overline{\mathbb{D}})$ and, moreover,

$$\sum_{m=1}^{\infty} \frac{1 - |a_{nm}|^2}{|1 - \bar{\lambda} a_{nm}|^N} \longrightarrow \int_{\mathbb{T}} \frac{2d\mu(\zeta)}{|1 - \bar{\lambda}\zeta|^N}$$

as $n \to \infty$.

PROOF. First, note that the growth restriction on μ implies that μ cannot have a Dirac mass at $1/\bar{\lambda}$. Our strategy is to prove the theorem for discrete measures with finitely many Dirac masses and then appeal to a limiting argument to extend it for the general case.

Assume that $\sigma = \alpha\delta_{\{1\}}$, where $\alpha > 0$. Construct f according to the recipe (4.4), i.e.

$$f(z) = \exp\left(-\alpha\frac{1+z}{1-z}\right).$$

By Lemma 5.2, the function

$$B_c(z) = \gamma_c \frac{f(z) - c}{1 - \bar{c}f(z)}$$

is a Blaschke product for values of c through a sequence which tend to zero and avoids the exceptional set of f. The unimodular constant

$$\gamma_c = \frac{f(0) - \bar{c}}{|f(0) - c|} \frac{|1 - \bar{c}f(0)|}{1 - cf(0)}$$

is added to ensure $B_c(0) > 0$. The precise value of γ_c is not used below. We just need to know that $\gamma_c \to 1$ as $c \to 0$. The formula for B_c implies

$$|f(z) - B_c(z)| \le |1 - \gamma_c| + \frac{2|c|}{1 - |c|}, \qquad (z \in \mathbb{D}).$$

Thus, B_c converges uniformly to f on $\mathbb{D}$ (even uniform convergence on compact sets is enough for us). Therefore, by Theorem 5.1, σ_{B_c} tends to μ in the weak-star topology. The zeros of B_c are

$$\{z : f(z) = c\} = \left\{a_{cm} = \frac{\alpha + \log c + i2\pi m\alpha}{-\alpha + \log c + i2\pi m\alpha} : m \in \mathbb{Z}\right\},$$

which clearly cluster at 1 as $c \to 0$. In this case, $\lambda \ne 1$, and thus the function

$$\zeta \longmapsto \frac{1 + |\zeta|}{|1 - \bar{\lambda}\zeta|^N}$$

can be considered as a continuous function on $\overline{\mathbb{D}}$ when we deal with measures μ and σ_{B_c} (at least for small values of c). Hence,

$$\int_{\overline{\mathbb{D}}} \frac{1 + |\zeta|}{|1 - \bar{\lambda}\zeta|^N}\, d\sigma_{B_c}(\zeta) \longrightarrow \int_{\overline{\mathbb{D}}} \frac{1 + |\zeta|}{|1 - \bar{\lambda}\zeta|^N}\, d\mu(\zeta).$$

But,

$$\int_{\overline{\mathbb{D}}} \frac{1 + |\zeta|}{|1 - \bar{\lambda}\zeta|^N}\, d\sigma_{B_c}(\zeta) = \sum_{m=1}^{\infty} \frac{1 - |a_{cm}|^2}{|1 - \bar{\lambda}a_{cm}|^N}$$

and

$$\int_{\overline{\mathbb{D}}} \frac{1 + |\zeta|}{|1 - \bar{\lambda}\zeta|^N}\, d\mu(\zeta) = \int_{\mathbb{T}} \frac{2}{|1 - \bar{\lambda}\zeta|^N}\, d\mu(\zeta).$$

Therefore, the result follows.

If μ consists of a finite sum of Dirac masses, the result still holds by induction. Now, we turn to the general situation. Assume that μ is an arbitrary positive Borel measure on $\mathbb{T}$, fulfilling the above mentioned growth restriction. Put

$$d\tau(z) = \frac{d\mu(z)}{|1 - \bar{\lambda}z|^N}.$$

Again note that μ cannot have a Dirac mass at $1/\bar{\lambda}$, and this property persists for all measures considered below. The family of discrete measures with finite number of Dirac masses is dense in $\mathcal{M}(\overline{\mathbb{D}})$. Hence, there is a sequence τ_n of such measures such that $\tau_n \to \tau$ in the weak-star topology of $\mathcal{M}(\overline{\mathbb{D}})$. Therefore, for each $f \in \mathcal{C}(\overline{\mathbb{D}})$,

$$\int_{\overline{\mathbb{D}}} f(z)|1-\bar{\lambda}z|^N \, d\tau_n(z) \longrightarrow \int_{\overline{\mathbb{D}}} f(z)|1-\bar{\lambda}z|^N \, d\tau(z) = \int_{\overline{\mathbb{D}}} f(z) d\mu(z).$$

This means that $\sigma_n \to \mu$ in the weak-star topology of $\mathcal{M}(\overline{\mathbb{D}})$, where σ_n is the discrete measure

$$d\sigma_n(z) = |1-\bar{\lambda}z|^N \, d\tau_n(z), \qquad (n \geq 1).$$

We appeal to the first part and find a Blaschke product B_n such that σ_{B_n} is close enough to σ_n in the weak-star topology and also

$$\left| \int_{\overline{\mathbb{D}}} \frac{1+|z|}{|1-\bar{\lambda}z|^N} \, d\sigma_{B_n}(z) - \int_{\mathbb{T}} \frac{2}{|1-\bar{\lambda}\zeta|^N} \, d\sigma_n(\zeta) \right| < \frac{1}{n}.$$

The result thus follows. Our choice of B_n also implies that $\sigma_{B_n} \to \mu$ in the weak-star topology of $\mathcal{M}(\overline{\mathbb{D}})$. □

6. Reproducing kernels for derivatives

Let $\mathcal{H}$ be a reproducing kernel of functions which are analytic on the domain Ω. The kernels of evaluation at point $z \in \Omega$ form a two parameter family of functions $k_z^{\mathcal{H}}(w)$, where z and w run through Ω and $k_z(w)^{\mathcal{H}}$ is analytic with respect to w and conjugate analytic with respect to z. The essential property of $k_z^{\mathcal{H}}(z)$ is

$$f(z) = \langle f, k_z^{\mathcal{H}} \rangle_{\mathcal{H}}, \qquad (f \in \mathcal{H},\, z \in \Omega). \tag{6.1}$$

If we successively take the derivative of f with respect to z, we see that the evaluation functional $f \longmapsto f^{(n)}(z)$ is given by

$$f^{(n)}(z) = \langle f, \partial^n k_z^{\mathcal{H}} / \partial \bar{z}^n \rangle_{\mathcal{H}}, \qquad (f \in \mathcal{H}). \tag{6.2}$$

But, we need to show that $\partial^n k_z^{\mathcal{H}} / \partial \bar{z}^n \in \mathcal{H}$ and also taking the derivative operator inside the inner product is legitimate. We verify this for $n = 1$. For higher derivative, a similar argument works.

For simplicity, write k_z for $k_z^{\mathcal{H}}$. Put $\delta = (1-|z|)/2$. Then, for each $f \in \mathcal{H}$ and each Δ with $0 < |\Delta| < \delta$, we have

$$\left| \left\langle f, \frac{k_{z+\Delta} - k_z}{\bar{\Delta}} \right\rangle_{\mathcal{H}} \right| = \left| \frac{f(z+\Delta) - f(z)}{\Delta} \right| = \left| \frac{1}{\Delta} \int_z^{z+\Delta} f'(\zeta) \, d\zeta \right| \leq C_f,$$

where C_f is the maximum of f' on the disk with center z and radius δ. Therefore, by the uniform boundedness principle, there is a constant C such that

$$\left\| \frac{k_{z+\Delta} - k_z}{\bar{\Delta}} \right\| \leq C, \qquad (0 < |\Delta| < \delta).$$

Let $g \in \mathcal{H}$ be a weak limit of this fraction as $\Delta \to 0$. Then, on one hand, for each $f \in \mathcal{H}$ we have

$$\langle f, g \rangle = \lim_{\Delta \to 0} \left\langle f, \frac{k_{z+\Delta} - k_z}{\bar{\Delta}} \right\rangle = \lim_{\Delta \to 0} \frac{f(z+\Delta) - f(z)}{\Delta} = f'(z).$$

On the other hand,

$$\begin{aligned} g(\zeta) &= \langle g, k_\zeta \rangle = \lim_{\Delta \to 0} \left\langle \frac{k_{z+\Delta} - k_z}{\bar{\Delta}}, k_\zeta \right\rangle \\ &= \lim_{\Delta \to 0} \frac{k_{z+\Delta}(\zeta) - k_z(\zeta)}{\bar{\Delta}} = \frac{\partial k_z}{\partial \bar{z}}(\zeta). \end{aligned}$$

In short, $g = \partial k_z / \partial \bar{z}$.

In the light of relation (6.2), we define the notation

$$k_{z,n}^{\mathcal{H}} = \partial^n k_z^{\mathcal{H}} / \partial \bar{z}^n, \tag{6.3}$$

i.e. the kernel of evaluation functional of the n-th derivative at $z \in \Omega$. The relation (6.2) is rewritten as

$$f^{(n)}(z) = \langle f, k_{z,n}^{\mathcal{H}} \rangle_{\mathcal{H}}, \qquad (f \in \mathcal{H}). \tag{6.4}$$

In the above formula, if we replace f by $k_{z,n}^{\mathcal{H}}$, we obtain

$$(k_{z,n}^{\mathcal{H}})^{(n)}(z) = \|k_{z,n}^{\mathcal{H}}\|_{\mathcal{H}}^2. \tag{6.5}$$

There are some other formulas for $k_{z,n}^{\mathcal{H}}$ and each has its merits and useful in the applications. We treat some of them below. For the space $\mathcal{H}(b)$, instead of $k_{z,n}^{\mathcal{H}(b)}$ we will write $k_{z,n}^b$. Our first formula for $k_{z,n}^b$ is based on the operator X_b.

LEMMA 6.1. *We have*

$$k_{z,n}^b = n!(I - \bar{z}X_b^*)^{-(n+1)} X_b^{*n} k_0^b.$$

PROOF. For each $w, z \in \mathbb{D}$,

$$((I - \bar{w}S)k_w)(z) = k_w(z) - \bar{w}(Sk_w)(z) = \frac{1}{1 - \bar{w}z} - \bar{w}z\frac{1}{1 - \bar{w}z} = 1 = k_0(z).$$

Hence, $(I - \bar{w}S)k_w = k_0$. But, for each $w \in \mathbb{D}$, the operator $I - \bar{w}S$ is invertible. Thus,

$$k_w = (I - \bar{w}S)^{-1} k_0$$

and, for each $f \in H^2(\mathbb{D})$, we have

$$f(w) = \langle (I - wS^*)^{-1} f, k_0 \rangle_2, \qquad (w \in \mathbb{D}).$$

Thus, if $f \in \mathcal{H}(b)$, we can write

$$f(w) = \langle (I - wX_b)^{-1} f, k_0 \rangle_2, \qquad (w \in \mathbb{D}).$$

But, $k_0^b = (I - T_b T_{\bar{b}})k_0$ and

$$\langle (I - wX_b)^{-1} f, k_0 \rangle_2 = \langle (I - wX_b)^{-1} f, (I - T_b T_b^*) k_0 \rangle_b.$$

Hence,

$$f(w) = \langle (I - wX_b)^{-1} f, k_0^b \rangle_b = \langle f, (I - \bar{w}X_b^*)^{-1} k_0^b \rangle_b, \qquad (w \in \mathbb{D}).$$

Since the last relation is valid for every function $f \in \mathcal{H}(b)$, we conclude that $(I - \bar{w}X_b^*)^{-1} k_0^b$ is precisely the reproducing kernel k_w^b. Changing w to z, we obtain

$$k_z^b = (I - \bar{z}X_b^*)^{-1} k_0^b.$$

Hence, using the definition (6.3), we get

$$\begin{aligned} k_{z,n}^b &= \frac{\partial^n k_z^b}{\partial \bar{z}^n} \\ &= \frac{\partial^n}{\partial \bar{z}^n}\Big((I-\bar{z}X_b^*)^{-1}k_0^b\Big) \\ &= n!(I-\bar{z}X_b^*)^{-(n+1)}X_b^{*n}k_0^b. \end{aligned}$$

□

The formula for $k_z^b = k_{z,0}^b$ is

$$k_z^b(w) = \frac{1-\overline{b(z)}\,b(w)}{1-\bar{z}\,w}, \qquad (z, w \in \mathbb{D}).$$

Using Leibnitz' rule, by straightforward computations, we obtain

$$k_{z,n}^b(w) = \frac{\partial^n k_z^b(w)}{\partial \bar{z}^n} = \frac{h_{z,n}^b(w)}{(1-\bar{z}w)^{n+1}}, \tag{6.6}$$

where $h_{z,n}^b$ is the function

$$h_{z,n}^b(w) = n!w^n - b(w)\sum_{j=0}^{n}\binom{n}{j}\overline{b^{(j)}(z)}(n-j)!w^{n-j}(1-\bar{z}w)^j. \tag{6.7}$$

LEMMA 6.2. *Let $z_0 \in \mathbb{D}$ with $b(z_0) \neq 0$. Then*

$$h_{z_0,n}^b(1/\bar{z}_0) = (h_{z_0,n}^b)'(1/\bar{z}_0) = \cdots = (h_{z_0,n}^b)^{(n)}(1/\bar{z}_0) = 0.$$

PROOF. The functional equation (4.2) shows that b is analytic in a neighborhood of the point $1/\bar{z}_0$. (If $b(z_0)=0$, then b has pole at $1/\bar{z}_0$ of the same order as the order of b at z_0.) Therefore, the formula for $k_{z_0}^b(w)$ shows that this kernel is a meromorphic function on $|w|>1$ with poles as described above and a possible pole at $1/\bar{z}_0$. However, again by (4.2), we have $b(z_0)\overline{b(1/\bar{z}_0)} = 1$ and thus the pole is removable. In short, $w \longmapsto k_{z_0}^b(w)$ is analytic at $1/\bar{z}_0$. Therefore, the same is true for the application $w \longmapsto k_{z_0}^b(w)$. Respecting this property, the representation (6.6) implies that $h_{z_0,n}^b$ must have a zero of order $n+1$ at $1/\bar{z}_0$. □

We finish this section by studying $k_{z,n}^B$, where B is a Blaschke product formed with zeros $(a_n)_{n\geq 1}$. It is easy to see that

$$h_j(z) = \left(\prod_{k=1}^{j-1}\frac{a_k - z}{1-\bar{a}_k\,z}\right)\frac{(1-|a_j|^2)^{1/2}}{1-\bar{a}_j\,z}, \qquad (j \geq 1), \tag{6.8}$$

is an orthonormal basis for $K_B = \mathcal{H}(B)$. Sometimes, we will write

$$h_j(z) = (1-|a_j|^2)^{1/2}\,\frac{B_{j-1}(z)}{1-\bar{a}_j\,z}, \qquad (j \geq 1), \tag{6.9}$$

where B_j is the finite product formed with the first j zeros.

LEMMA 6.3. *Let B be a Blaschke product with zeros $(a_n)_{n\geq 1}$. Let $z \in \mathbb{D}$. Then*

$$k_{z,n}^B = \sum_{j=1}^{\infty}\overline{h_j^{(n)}(z)}\,h_j.$$

The series converges in $H^2(\mathbb{D})$-norm.

PROOF. Since $(h_j)_{j\geq 1}$ forms an orthonormal basis for K_B, there are coefficients c_j, $j \geq 1$, such that

$$k_{z,n}^B = \sum_{j=1}^{\infty} c_j\, h_j,$$

where the series converges in $H^2(\mathbb{D})$-norm. Moreover, thanks to orthonormality, c_j is given by

$$c_j = \langle k_{z,n}^B, h_j \rangle_2.$$

But, the formula 6.2 immediately implies $\bar{c}_j = h_j^{(n)}(z)$. □

For a Blaschke product $X_B = S^*|K_B$ and $k_0^B = P_B 1$. Thus, $X_B^* = P_B S = M_B$ is the compressed shift on K_B. Therefore, by Lemma 6.1, we have

$$k_{z,n}^B = n!(I - \bar{z}M_B)^{-(n+1)} M_B^n P_B 1. \tag{6.10}$$

LEMMA 6.4. *Let $z_0 \in \mathbb{D}$, and $N \geq 0$. Let B be a Blaschke product with zeros $(a_n)_{n\geq 1}$. Assume that there are functions $f, g \in H^2(\mathbb{D})$ such that*

$$z^N = (1 - \bar{z}_0 z)^{N+1} f(z) + B(z) g(z), \qquad (z \in \mathbb{D}).$$

Then we have $P_B f = k_{z_0,N}^B / N!$.

PROOF. We write the above equation for f and g as

$$S^N 1 = (1 - \bar{z}_0 S)^{N+1} f + Bg.$$

Since M_B is the compression of S, if we apply P_B to both sides, we obtain

$$M_B^N P_B 1 = (1 - \bar{z}_0 M_B)^{N+1} P_B f.$$

Thus,

$$P_B f = (1 - \bar{z}_0 M_B)^{-N-1} M_B^N P_B 1$$

and the result follows from (6.10). □

7. An interpolation problem

There is a close relation between the existence of derivatives of elements of $\mathcal{H}(b)$ at the boundary and the containment of $X_b^{*N} k_0^b$ to the range of $(I - \overline{\zeta_0} X_b^*)^{N+1}$. This is fully explored in Theorem 8.2. But, to reach that general result, we need to pave the road by studying some special cases. We start doing this by considering Blaschke products. First, a technical lemma.

LEMMA 7.1. *Let $S, (S_n)_{n\geq 1} \in \mathcal{L}(\mathcal{H})$ with the following properties:*

(i) Each S_n is invertible.
(ii) S is injective.
(iii) $S_n \longrightarrow S$ in the norm topology.
(iv) There is a constant M such that

$$\|S_n^{-1} S\| \leq M, \qquad (n \geq 1).$$

Let $y \in \mathcal{H}$. Then $(S_n^{-1} y)_{n\geq 1}$ is a bounded sequence in $\mathcal{H}$ if and only if $y \in \mathbb{R}(S)$. Moreover, if this holds, we actually have $S_n^{-1} y \longrightarrow S^{-1} y$ in the weak topology.

PROOF. Assume that $(S_n^{-1}y)_{n\geq 1}$ is a bounded sequence in $\mathcal{H}$. Hence, it has at least one weak limit point in $\mathcal{H}$. Let $x \in \mathcal{H}$ be a weak limit point of the sequence. Since $S_n \longrightarrow S$ in the norm topology, we surely have (at least for a subsequence) $S_n S_n^{-1} y \longrightarrow Sx$. Therefore, $y = Sx$, i.e. $y \in \mathbb{R}(S)$. But, since S is injective, the above argument shows that the sequence has precisely one weak point (if x' is another weak limit point, we would have $y = Sx = Sx'$). In other words, the whole sequence tends weakly to x.

To prove the other (easy) direction, assume that $y \in \mathbb{R}(S)$, i.e. $y = Sx$ for some $x \in \mathcal{H}$. Then

$$\|S_n^{-1}y\| \leq \|S_n^{-1}Sx\| \leq \|S_n^{-1}S\| \, \|x\| \leq M\|x\|, \qquad (n \geq 1).$$

□

The following corollary is a realization of the preceding lemma. The assumption are adjusted to fit our application in the study of derivatives of $\mathcal{H}(b)$ functions.

COROLLARY 7.2. *Let $T_k \in \mathcal{L}(\mathcal{H})$, $\zeta_k \in \mathbb{T}$ and $\lambda_{k,n} \in \mathbb{D}$, for $n \geq 1$ and $1 \leq k \leq p$, with the following properties:*

(i) Each T_k is a contraction.
(ii) Each $I - \zeta_k T_k$ is one to one.
(iii) $T_k T_{k'} = T_{k'} T_k$ for $k, k' \in \{1, \dots, p\}$.
(iv) For each k, $\lambda_{k,n}$ tends nontangentially to ζ_k as $n \to \infty$.

Let $y \in \mathcal{H}$. Then the sequence

$$\Big((I - \lambda_{1,n}T_1)^{-1} \dots (I - \lambda_{p,n}T_p)^{-1} y \Big)_{n \geq 1}$$

is uniformly bounded if and only if y belongs to the range of the operator $(I - \zeta_1 T_1) \dots (I - \zeta_p T_p)$, in which case,

$$(I - \lambda_{1,n}T_1)^{-1} \dots (I - \lambda_{p,n}T_p)^{-1} y \longrightarrow (I - \zeta_1 T_1)^{-1} \dots (I - \zeta_p T_p)^{-1} y$$

in the weak topology.

PROOF. We apply Lemma 7.1 with

$$S_n = (I - \lambda_{1,n}T_1) \dots (I - \lambda_{p,n}T_p)$$

and

$$S = (I - \zeta_1 T_1) \dots (I - \zeta_p T_p).$$

The only nontrivial property is the boundedness of $S_n^{-1}S$. Since T_k are commuting, it is enough to verify that the sequence

$$\Big((I - \lambda_{k,n}T_k)^{-1} (I - \zeta_k T_k) \Big)_{n \geq 1}$$

is bounded. But,

$$\begin{aligned} \|(I - \lambda_{k,n}T_k)^{-1}(I - \zeta_k T_k)\| &= \|I + (\lambda_{k,n} - \zeta_k)(I - \lambda_{k,n}T_k)^{-1}T_k\| \\ &\leq 1 + |\lambda_{k,n} - \zeta_k| \, \|(I - \lambda_{k,n}T_k)^{-1}\| \\ &\leq 1 + |\lambda_{k,n} - \zeta_k| \, (1 - |\lambda_{k,n}|)^{-1} \\ &\leq 1 + M_k. \end{aligned}$$

The last estimation holds since $\lambda_{k,n}$ tends nontangentially to ζ_k. The result thus follows. □

As a matter of fact, we even need a special case of Corollary 7.2 in which $T_1 = \cdots = T_p$.

COROLLARY 7.3. *Let $T \in \mathcal{L}(\mathcal{H})$ be a contraction, $\zeta \in \mathbb{T}$ and $(\lambda_n)_{n\geq 1} \subset \mathbb{D}$, with the following properties:*

(i) $I - \zeta T$ is one to one.
(ii) λ_n tends nontangentially to ζ as $n \to \infty$.

Let $y \in \mathcal{H}$. Then the sequence

$$\Big((I - \lambda_n T)^{-p} y \Big)_{n\geq 1}$$

is uniformly bounded if and only if y belongs to the range of the operator $(I - \zeta T)^p$, in which case,

$$(I - \lambda_n T)^{-p} y \longrightarrow (I - \zeta T)^{-p} y$$

in the weak topology.

Now we are ready to establish the connection between the existence of boundary derivatives in K_B and an interpolation problem.

THEOREM 7.4. *Let $\zeta \in \mathbb{T}$, and let $N \geq 0$. Let B be a Blaschke product with zeros $(a_n)_{n\geq 1}$ such that*

$$\sum_{n=1}^{\infty} \frac{1 - |a_n|}{|\zeta - a_n|^{2N+2}} \leq A.$$

Then there are functions $f, g \in H^2(\mathbb{D})$ such that

$$z^N = (1 - \bar{\zeta} z)^{N+1} f(z) + B(z) g(z), \qquad (z \in \mathbb{D}),$$

with

$$\|f\|_2 \leq C,$$

where $C = C(N, A)$ is a constant.

PROOF. According to Lemma 6.3,

$$k_{z,N}^B = \sum_{j=1}^{\infty} \overline{h_j^{(N)}(z)}\, h_j.$$

Hence,

$$\|k_{z,n}^B\|^2 = \sum_{j=1}^{\infty} |h_j^{(n)}(z)|^2. \tag{7.1}$$

We rewrite the formula in (6.8) for h_j as

$$h_j(z) = (1 - |a_j|^2)^{1/2}\, \frac{B_{j-1}(z)}{1 - \bar{a}_j\, z}.$$

Hence, by Leibnitz's formula,

$$h_j^{(N)}(z) = (1 - |a_j|^2)^{1/2} \sum_{k=1}^{N} \binom{N}{k} B_{j-1}^{(N-k)}(z)\, \frac{k!(-\bar{a}_j)^k}{(1 - \bar{a}_j\, z)^{k+1}}.$$

Therefore, by Theorem 3.1 and denoting the constant $C(N,A)$ of this theorem by C,

$$\begin{aligned}
|h_j^{(N)}(r\zeta)| &\leq (1-|a_j|^2)^{1/2}\sum_{k=1}^{N}\binom{N}{k}C\frac{k!}{|1-\bar{a}_j\, r\zeta|^{k+1}}\\
&\leq C(1-|a_j|^2)^{1/2}\sum_{k=1}^{N}\binom{N}{k}\frac{2^{k+1}k!}{|1-\bar{a}_j\,\zeta|^{k+1}}\\
&\leq C(1-|a_j|^2)^{1/2}\sum_{k=1}^{N}\binom{N}{k}\frac{2^{N+1}k!}{|1-\bar{a}_j\,\zeta|^{N+1}}\\
&= \left(2^{N+1}C\sum_{k=1}^{N}k!\binom{N}{k}\right)\frac{(1-|a_j|^2)^{1/2}}{|\zeta-a_j|^{N+1}}.\\
&= C'\frac{(1-|a_j|^2)^{1/2}}{|\zeta-a_j|^{N+1}}.
\end{aligned}$$

Considering (7.1), we conclude

$$\|k_{r\zeta,N}^B\|^2 \leq 2AC'^2, \qquad (0<r<1). \tag{7.2}$$

The next step is to appeal to the formula (6.10) and Corollary 7.3. Since $\sigma_p(M_B)\subset\mathbb{D}$, the operator $I-\overline{\zeta}M_B$ is injective. Hence, with $T=M_B$, $p=N+1$ and $y=M_B^N P_B 1$, we see that $M_B^N P_B 1$ belongs to the range of $(I-\overline{\zeta}M_B)^{N+1}$. This means that there is a function $f\in H^2(\mathbb{D})$ such that

$$M_B^N P_B 1 = (I-\overline{\zeta}M_B)^{N+1} f.$$

Since M_B is the compressed shift, we can rewrite the preceding identity as

$$P_B(z^N) = P_B\big((1-\overline{\zeta}z)^{N+1}f\big).$$

Hence, $z^N-(1-\overline{\zeta}z)^{N+1}f\perp K_B$, or equivalently $z^N-(1-\overline{\zeta}z)^{N+1}f\in BH^2(\mathbb{D})$. Therefore, there is $g\in H^2(\mathbb{D})$ such that

$$z^N-(1-\overline{\zeta}z)^{N+1}f = Bg.$$

Finally, Corollary 7.3 also says that

$$f=(I-\overline{\zeta}M_B)^{-N-1}M_B^N P_B 1 = \lim_{r\to 1}(I-\overline{r\zeta}M_B)^{-N-1}M_B^N P_B 1.$$

Hence, by (6.10)

$$f=(I-\overline{\zeta}M_B)^{-N-1}M_B^N P_B 1 = \tfrac{1}{N!}\lim_{r\to 1} k_{r\zeta,N}^B,$$

and, by (7.2), the latter is uniformly bounded by a constant. □

The above result was referees as an interpolation problem since the equation

$$z^n = (1-\bar{\zeta}z)f(z)+B(z)g(z)$$

has a solution if and only if the is a function $f\in H^2(\mathbb{D})$ such that

$$f(a_n)=\frac{a_n}{(1-\bar{\zeta}a_n)^{N+1}}, \qquad (n\geq 1).$$

Since

$$\sum_{n=1}^{\infty} \left| \frac{a_n}{(1-\bar{\zeta} a_n)^{N+1}} \right|^2 (1-|a_n|^2) < \infty,$$

if $(a_n)_{n\geq 1}$ was an interpolation sequence, then the function f trivially exists. The surprising feature of Theorem 7.4 is that it ensures a solution, even with an additional growth restriction, always exists.

Theorem 7.4, in a sense, is reversible. Indeed, this is the version that we need in the proof of Theorem 8.2.

THEOREM 7.5. *Let $N \geq 0$. Let B be a Blaschke product with zeros $(a_n)_{n\geq 1}$. Assume that there are functions $f, g \in H^2(\mathbb{D})$ such that*

$$z^N = (1-\bar{z}_0 z)^{N+1} f(z) + B(z) g(z), \qquad (z \in \mathbb{D}),$$

with

$$\|f\|_2 \leq C \qquad \text{and} \qquad \left(1 - \frac{1}{2C^2}\right)^{1/2} \leq |z_0| < 1,$$

where $C > 1$ is a constant. Then there is a constant $A = A(N, C)$ such that

$$\sum_{n=1}^{\infty} \frac{1-|a_n|}{|1-\bar{a}_n z_0|^{2N+2}} \leq A.$$

PROOF. Since we appeal to induction, the function f and g that appear in the N-th step will be denoted by f_N and g_N. Note that, by Lemma 6.4,

$$P_B f_N = \frac{k^B_{z_0,N}}{N!}. \tag{7.3}$$

Case $N = 0$: By Lemma 6.3, the

$$k^B_{z_0} = \sum_{j=1}^{\infty} \overline{h_j(z_0)}\, h_j.$$

Hence, by (7.3), our condition $\|f_0\|_2 \leq C$ translates as

$$\sum_{j=1}^{\infty} |h_j(z_0)|^2 \leq C^2.$$

We use (6.9), to reqrite this estimation as

$$\sum_{j=1}^{\infty} |B_{j-1}(z_0)|^2 \, \frac{1-|a_j|^2}{|1-\bar{a}_j\, z_0|^2} \leq C^2. \tag{7.4}$$

We just need to get rid of $|B_{j-1}(z_0)|^2$ to establish the result. To do so, just note that since B_j is a subproduct of B, we have

$$P_{B_j} k^B_{z_0}(z) = k^{B_j}_{z_0}(z) = \frac{1-\overline{B_j(z_0)} B_j(z)}{1-\bar{z}_0 z}.$$

Hence,

$$\frac{1-|B_j(z_0)|^2}{1-|z_0|^2} = k^{B_j}_{z_0}(z_0) = \|k^{B_j}_{z_0}\|^2 \leq \|k^{B_j}_{z}\|^2 \leq C^2.$$

The restriction $1-1/2C^2 \le |z_0|^2 < 1$ now implies $|B_j(z_0)|^2 \ge 1/2$. Therefore, from (7.4), we conclude

$$\sum_{j=1}^{\infty} \frac{1-|a_j|^2}{|1-\bar{a}_j\, z_0|^2} \le 2C^2.$$

This settles the case $N=0$.

Case $N \ge 1$: Assume that the result holds for $N-1$. Our assumption is that there are functions $f_N, g_N \in H^2(\mathbb{D})$ such that

$$z^N = (1-\bar{z}_0 z)^{N+1} f_N(z) + B(z) g_N(z), \qquad (z \in \mathbb{D}), \tag{7.5}$$

with $\|f_N\|_2 \le C$. Write

$$1-(1-\bar{z}_0 z)^N = -\sum_{k=1}^{N} \binom{N}{k} (-\bar{z}_0)^k z^k.$$

Multiply by z^{N-1} to get

$$z^{N-1} = (1-\bar{z}_0 z)^N z^{N-1} - \left(\sum_{k=1}^{N} \binom{N}{k} (-\bar{z}_0)^k z^{k-1} \right) z^N.$$

Plugging (7.5) gives

$$z^{N-1} = (1-\bar{z}_0 z)^N f_{N-1}(z) + B(z) g_{N-1}(z), \qquad (z \in \mathbb{D}),$$

where

$$f_{N-1}(z) = z^{N-1} - \left(\sum_{k=1}^{N} \binom{N}{k} (-\bar{z}_0)^k z^{k-1} \right) (1-\bar{z}_0 z) f_N(z).$$

Hence,

$$\|f_{N-1}\|_2 \le 1 + 2^{N+1} C.$$

This means that all required conditions are fulfilled and we can apply the induction for $N-1$. Thus, there is a constant A such that

$$\sum_{n=1}^{\infty} \frac{1-|a_n|}{|1-\bar{a}_n z_0|^{2N}} \le A. \tag{7.6}$$

Theorem 3.1 now ensures that $B_j^{(k)}(z_0)$, $0 \le k \le 2N-1$, exist and are uniformly bounded by a constant A', where B_j is any subproduct of B.

If we take N times the derivative of both sides in (6.9), we obtain

$$h_j^{(N)}(z) = (1-|a_j|^2)^{1/2} \sum_{k=0}^{N} \binom{N}{k} B_{j-1}^{(k)}(z) \frac{(N-k)!(-\bar{a}_j)^{N-k}}{(1-\bar{a}_j\, z)^{N-k+1}}, \qquad (j \ge 1).$$

We rewrite this as

$$\begin{aligned} &(1-|a_j|^2)^{1/2}\, B_{j-1}(z) \frac{N!(-\bar{a}_j)^N}{(1-\bar{a}_j\, z)^{N+1}} \\ &= h_j^{(N)}(z) - (1-|a_j|^2)^{\frac{1}{2}} \sum_{k=1}^{N} \binom{N}{k} B_{j-1}^{(k)}(z) \frac{(N-k)!(-\bar{a}_j)^{N-k}}{(1-\bar{a}_j\, z)^{N-k+1}}. \end{aligned} \tag{7.7}$$

As we saw above, for $1 \le k \le N$,

$$\begin{aligned}\left| \binom{N}{k} B_{j-1}^{(k)}(z_0) \, \frac{(N-k)!(-\bar{a}_j)^{N-k}}{(1-\bar{a}_j\, z)^{N-k+1}} \right| &\le \frac{A' N!}{|1-\bar{a}_j\, z_0|^{N-k+1}} \\ &\le \frac{A' N! 2^N}{|1-\bar{a}_j\, z_0|^{N}}.\end{aligned}$$

Thus, the right side of (7.7) is majorized by

$$|h_j^{(N)}(z)| + A' N N! 2^N \frac{(1-|a_j|^2)^{1/2}}{|1-\bar{a}_j\, z_0|^N}.$$

The left side of (7.7) is minorized by

$$\frac{N!}{2^{N+1}} \frac{(1-|a_j|^2)^{1/2}}{|1-\bar{a}_j\, z_0|^{N+1}}$$

for zeros $|a_j| \ge 1/2$. Hence, for such j, we have

$$\frac{(1-|a_j|^2)^{1/2}}{|1-\bar{a}_j\, z_0|^{N+1}} \le 2^{N+1}|h_j^{(N)}(z_0)| + A' N 2^{2N+1} \frac{(1-|a_j|^2)^{1/2}}{|1-\bar{a}_j\, z_0|^N}.$$

Hence, by Minkowski's inequality, Lemmas 6.3 and 6.4, and (7.6),

$$\begin{aligned}&\left(\sum_{|a_j|\ge 1/2} \frac{1-|a_j|^2}{|1-\bar{a}_j\, z_0|^{2N+2}} \right)^{1/2} \\ \le\;& 2^{N+1} \left(\sum_{j=1}^{\infty} |h_j^{(N)}(z_0)|^2 \right)^{1/2} + A' N 2^{2N+1} \left(\sum_{j=1}^{\infty} \frac{1-|a_j|^2}{|1-\bar{a}_j\, z_0|^{2N}} \right)^{1/2} \\ \le\;& 2^{N+1} \|k_{z_0,N}^B\| + A' A N 2^{2N+1} \\ \le\;& 2^{N+1} N! \|P_B f_N\| + A' A N 2^{2N+1} \\ \le\;& 2^{N+1} N! C + A' A N 2^{2N+1}.\end{aligned}$$

For zeros with $|a_j| < 1/2$, we have

$$\sum_{|a_j|<1/2} \frac{1-|a_j|^2}{|1-\bar{a}_j\, z_0|^{2N+2}} \le 4 \sum_{|a_j|<1/2} \frac{1-|a_j|^2}{|1-\bar{a}_j\, z_0|^{2N}} \le 4A.$$

Hence, the result follows. □

8. Derivatives of $\mathcal{H}(b)$ functions

There is a close connection between the analytic continuation of b across a subarc of $\mathbb{T}$ on one hand, and the analytic continuation of all function of $\mathcal{H}(b)$ across the same subarc on the other hand. In this section, we treat a similar result. While we are studying the derivative of elements in $\mathcal{H}(b)$, we are content with the existence of nontangential boundary values.

We begin with a simple lemma which is a simple exercise from calculus and is interesting in its own rights. We do not it in such a generality since in our application even the derivative of order $n+m+1$ exists at all points. However, the proof for the general case is essentially the same.

LEMMA 8.1. *Let I be an open interval, and let $a, b \in I$. Suppose that the function $h : I \longrightarrow \mathbb{C}$ satisfies the following properties:*

(i) h has $n+m$ continuous derivatives on the I.
(ii) $h^{(n+m+1)}$ is continuous and bounded on $I \setminus \{a\}$.
(iii) $h(b) = h'(b) = \cdots = h^{(n-1)}(b) = 0$.

Put

$$k(x) = \frac{h(x)}{(x-b)^n}, \qquad (x \in I).$$

Then k is $m+1$ times differentiable on I and, moreover,

$$k^{(m+1)}(x) = \int_0^1 \cdots \int_0^1 h^{(m+n+1)}\big(b + t_1 \cdots t_n(x-b)\big)\, v(t)\, dt_1 \cdots dt_n,$$

where $v(t) = t_1^{p_1} \cdots t_n^{p_n}$ is some monomial.

PROOF. Since $h(b) = 0$, the fundamental theorem of calculus says

$$\begin{aligned} h(x) &= \int_0^1 \frac{d}{dt_1}\Big(h\big(b + t_1(x-b)\big)\Big)\, dt_1 \\ &= (x-b) \int_0^1 h'\big(b + t_1(x-b)\big)\, dt_1. \end{aligned}$$

Applying the same result to the function $x \longmapsto h'\big(b + t_1(x-b)\big)$ gives

$$h'\big(b + t_1(x-b)\big) = t_1(x-b) \int_0^1 h''\big(b + t_1 t_2(x-b)\big)\, dt_2.$$

Therefore,

$$h(x) = (x-b)^2 \int_0^1 \int_0^1 t_1\, h''\big(b + t_1 t_2(x-b)\big)\, dt_1 dt_2.$$

Continuing this process n times gives

$$k(x) = \int_0^1 \cdots \int_0^1 t_1^{n-1} t_2^{n-2} \cdots t_{n-1}\, h^{(n)}\big(b + t_1 \cdots t_n(x-b)\big)\, dt_1 \cdots dt_n.$$

Write $m(t) = t_1^{n-1} t_2^{n-2} \cdots t_{n-1}$.

Since h has $n+m$ continuous derivatives on the I, and $h^{(n+m+1)}$ is continuous and bounded on $I \setminus \{a\}$, the function k has $m+1$ continuous derivatives on the I and

$$\begin{aligned} k^{(m+1)}(x) &= \int_0^1 \cdots \int_0^1 m(t)\, \frac{\partial^{m+1}}{\partial x^{m+1}} h^{(n)}\big(b + t_1 \cdots t_n(x-b)\big)\, dt_1 \cdots dt_n \\ &= \int_0^1 \cdots \int_0^1 v(t)\, h^{(m+n+1)}\big(b + t_1 \cdots t_n(x-b)\big)\, dt_1 \cdots dt_n, \end{aligned}$$

where $v(t) = t_1^{m+n} t_2^{m+n-1} \cdots t_{n-1}^{m+2} t_n^{m+1}$. □

The following result gives a criterion for the existence of the derivatives for functions of $\mathcal{H}(b)$ and it generalizes the Ahern-Clark result.

THEOREM 8.2 (Fricain–Mashreghi [**14**]). *Let b be a point in the unit ball of $H^\infty(\mathbb{D})$ with the canonical factorization* (4.1), *let $\zeta_0 \in \mathbb{T}$ and let N be a nonnegative integer. Then the following are equivalent.*

(i) For every $f \in \mathcal{H}(b)$, the functions $f(z), f'(z), \ldots, f^{(N)}(z)$ have finite limits as z tends radially to ζ_0.

(*ii*) *For every* $f \in \mathcal{H}(b)$, *the function* $|f^{(N)}(z)|$ *remains bounded as* z *tends radially to* ζ_0.

(*iii*) $\|k^b_{z,N}\|_b$ *is bounded on the ray* $z \in [0, \zeta_0]$.

(*iv*) $X_b^{*N} k_0^b$ *belongs to the range of* $(I - \overline{\zeta_0} X_b^*)^{N+1}$.

(*v*) *We have*

$$\sum_n \frac{1 - |a_n|^2}{|\zeta_0 - a_n|^{2N+2}} + \int_{\mathbb{T}} \frac{d\mu(\zeta)}{|\zeta_0 - \zeta|^{2N+2}} + \int_{\mathbb{T}} \frac{|\log|b(\zeta)||}{|\zeta_0 - \zeta|^{2N+2}} \, dm(\zeta) < \infty.$$

PROOF. (*i*) $\Longrightarrow$ (*ii*): This is trivial.

(*ii*) $\Longrightarrow$ (*iii*): In the light of representation (6.4), this implication follows from the principle of uniform boundedness.

(*iii*) $\Longrightarrow$ (*iv*): By Lemma 6.1,

$$k^b_{z,N} = N!(I - \bar{z} X_b^*)^{-(N+1)} X_b^{*N} k_0^b. \tag{8.1}$$

Since $\sigma_p(X_b^*) \subset \mathbb{D}$, the operator $I - \overline{\zeta_0} X_b^*$ is injective. By assumption, $(I - \overline{z_n} X_b^*)^{-(N+1)} X_b^{*N} k_0^b$ is uniformly bounded for any sequence $z_n \in \mathbb{D}$ tending radially to ζ_0. Now, we apply Corollary 7.3 with $T = X_b^*$, $p = N+1$ and $y = X_b^{*N} k_0^b$ to conclude that $X_b^{*N} k_0^b$ belongs to the range of $(I - \overline{\zeta_0} X_b^*)^{N+1}$.

(*iv*) $\Longrightarrow$ (*i*): Using once more Corollary 7.3, we see that

$$(I - \overline{z_n} X_b^*)^{-(N+1)} X_b^{*N} k_0^b \longrightarrow (I - \overline{\zeta_0} X_b^*)^{-(N+1)} X_b^{*N} k_0^b$$

in the weak topology, for any sequence $z_n \in \mathbb{D}$ tending radially to ζ. But, (8.1) says that the left side is precisely $k^b_{z_n,N}$. Hence, in the light of (6.4), for every function f in $\mathcal{H}(b)$, The N-th derivative $f^{(N)}(z)$ has a finite limit as z tends radially to ζ_0.

The rest is by induction. We have

$$I - (I - \overline{\zeta_0} X_b^*)^N = -\sum_{k=1}^{N} \binom{N}{k} (-\zeta_0)^k X_b^{*k}.$$

Multiply both sides by $X_b^{*(N-1)} k_0^b$ to get

$$X_b^{*(N-1)} k_0^b = (I - \overline{\zeta_0} X_b^*)^N X_b^{*(N-1)} k_0^b - \sum_{k=1}^{N} \binom{N}{k} (-\zeta_0)^k X_b^{*(k-1)} X_b^{*N} k_0^b.$$

Hence, $X_b^{*(N-1)} k_0^b$ belongs to the range of $(I - \overline{\zeta_0} X_b^*)^N$. The above argument applies with N replaced by $N-1$. We continue this process N times. Therefore, for every function f in $\mathcal{H}(b)$, $f^{(j)}(z)$, $0 \le j \le N$, has a finite limit as z tends radially to ζ_0.

(*v*) $\Longrightarrow$ (*iii*): Without loss of generality we assume that $\zeta_0 = 1$. By Theorem 4.1, the condition (*v*) implies that

$$\lim_{r \to 1^-} b^{(j)}(r) \quad \text{and} \quad \lim_{R \to 1^+} b^{(j)}(R)$$

exist and are equal for $0 \le j \le 2N+1$. Moreover, since b can have only a finite number of real zeros, we can take $\delta > 0$ such that the interval $[1-\delta, 1)$ is free of zeros of b. Therefore, b has $2N+1$ continuous bounded derivatives on $[1-\delta, 1+\delta]$. Now, fix r in the interval $(1-\delta, 1)$.

We remind that by (6.6) and (6.7), $k^b_{r,N}(x) = h^b_{r,N}(x)/(1 - rx)^{N+1}$, where

$$h^b_{r,N}(x) = N! x^N - b(x) \sum_{j=0}^{N} \binom{N}{j} \overline{b^{(j)}(r)} (N-j)! x^{N-j} (1 - rx)^j.$$

Hence, $h^b_{r,N}$ has $2N+1$ continuous bounded derivatives on $(1-\delta, 1+\delta)$. Moreover, by Lemma 6.2, we have

$$h^b_{r,N}(1/r) = (h^b_{r,N})'(1/r) = \cdots = (h^b_{r,n})^{(N)}(1/r) = 0.$$

We now apply Lemma 8.1 with $I = (1-\delta, 1+\delta)$, $a = 1$, $b = 1/r$, $n = N+1$, $m = N$, and $h = h^b_{r,N}$. Note that

$$\frac{h(x)}{(x-b)^n} = \frac{h^b_{r,N}(x)}{(x-1/r)^{N+1}} = (-r)^{N+1} k^b_{N,r}(x).$$

Thus, lemma says that $(-r)^{N+1}(k^b_{N,r})^{(N)}(x)$ is equal to

$$\int_0^1 \cdots \int_0^1 (h^b_{r,N})^{(2N+1)}\left(\frac{1}{r} + t_1 \cdots t_{N+1}(x - \frac{1}{r})\right) v(t)\, dt_1 \ldots dt_{N+1}.$$

Since there is an M such that

$$|(h^b_{r,N})^{(2N+1)}(s)| \le M, \qquad (1-\delta < s < 1+\delta),$$

we deduce that

$$|(k^b_{r,N})^{(N)}(x)| \le M\delta^{-N-1}, \qquad (1-\delta < x < 1).$$

In particular, $(k^b_{r,N})^{(N)}(r)$ is bounded as $r \to 1^-$. But, according to (6.5),

$$\|k^b_{z,N}\|_b^2 = (k^b_{z,N})^{(N)}(z).$$

Thus, $\|k^b_{r,N}\|_b$ remains bounded as $r \to 1^-$.

$(iii) \implies (v)$: Again, without loss of generality, assume that $\zeta_0 = 1$. Fix $r \in (0,1)$. Considering the canonical factorization of b, since b is in the unit ball of H^∞, we have

$$\sum_n \frac{1-|a_n|^2}{|1-a_n r|^{2N+2}} + \int_{\mathbb{T}} \frac{d\mu(\zeta)}{|r-\zeta|^{2N+2}} + \int_{\mathbb{T}} \frac{|\log|b(\zeta)||}{|r-\zeta|^{2N+2}}\, dm(\zeta) < \infty.$$

For simplicity of formulas, denote the left side by Δ_r. According to Theorem 5.4, there is a sequence $(B_j)_{j\ge1}$ of Blaschke products, with zeros $(a_{jk})_{k\ge1}$, converging uniformly to b on compact subsets of $\mathbb{D}$ and such that

$$\sum_{k=1}^{\infty} \frac{1-|a_{jk}|^2}{|1-ra_{jk}|^{2N+2}} \longrightarrow \Delta_r$$

as $j \to \infty$. Hence, the formulas (6.6) and (6.7) show that $k^{B_j}_{w,N}$ tends to $k^b_{w,N}$ uniformly on compact subsets of $\mathbb{D}$. In particular, we must have

$$\lim_{j\to\infty} (k^{B_j}_{w,N})^{(N)}(w) = (k^b_{w,N})^{(N)}(w).$$

In the light of (6.5), we rewrite this identity as

$$\lim_{j\to\infty} \|k^{B_j}_{w,N}\|_2 = \|k^b_{w,N}\|_b.$$

The assumption (iii) implies that there is a $C>0$ such that

$$\|k^b_{r,N}\|_b \le C, \qquad (0<r<1).$$

Therefore, there is an index j_r such that

$$\|k^{B_j}_{r,N}\|_2 \le C+1, \qquad (j \ge j_r).$$

The formulas (6.6) and (6.7) also show that

$$(1 - rz)^{N+1} k^{B_j}_{r,N}(z) = N! z^N - B_j(z) g_j(z),$$

where $g_j \in H^2(\mathbb{D})$. Hence, it follows from Theorem 7.5 that there is a constant $A = A(C, N)$ (independent of r) such that

$$\sum_k \frac{1 - |a_{jk}|^2}{|1 - r a_{jk}|^{2N+2}} \leq A, \qquad (j \geq j_r),$$

Letting $j \to \infty$, we obtain $\Delta_r \leq A$ for all $r \in (0, 1)$. Finally, we let $r \to 1^-$ to get the desired condition (v). □

We also mention that Sarason has obtained another criterion in terms of the Clark measure σ_λ associated with b.

THEOREM 8.3 (Sarason, [**26**]). *Let ζ_0 be a point of $\mathbb{T}$ and let ℓ be a nonnegative integer. The following conditions are equivalent.*

(i) *Each function in $\mathcal{H}(b)$ and all its derivatives up to order ℓ have nontangential limits at ζ_0.*

(ii) *There is a point $\lambda \in \mathbb{T}$ such that*

$$\int_{\mathbb{T}} |e^{i\theta} - \zeta_0|^{-2\ell - 2} \, d\sigma_\lambda(e^{i\theta}) < +\infty. \tag{8.2}$$

(iii) *The last inequality holds for all $\lambda \in \mathbb{T} \setminus \{b(\zeta_0)\}$.*

(iv) *There is a point $\lambda \in \mathbb{T}$ such that μ_λ has a point mass at ζ_0 and*

$$\int_{\mathbb{T} \setminus \{z_0\}} |e^{i\theta} - \zeta_0|^{-2\ell} \, d\sigma_\lambda(e^{i\theta}) < \infty.$$

Recently, Bolotnikov and Kheifets [**5**] gave a third criterion (in some sense more algebraic) in terms of the Schwarz-Pick matrix. Recall that if b is a function in the unit ball of H^∞, then the matrix $\mathbf{P}^\omega_\ell(z)$, which will be refered to as to a Schwarz-Pick matrix and defined by

$$\mathbf{P}^b_\ell(z) := \left[\frac{1}{i! j!} \frac{\partial^{i+j}}{\partial z^i \partial \bar{z}^j} \frac{1 - |b(z)|^2}{1 - |z|^2} \right]^\ell_{i,j=0},$$

is positive semidefinite for every $\ell \geq 0$ and $z \in \mathbb{D}$. We extend this notion to boundary points as follows: given a point $\zeta_0 \in \mathbb{T}$, the boundary Schwarz-Pick matrix is

$$\mathbf{P}^b_\ell(\zeta_0) = \lim_{\substack{z \longrightarrow \zeta_0 \\ \sphericalangle}} \mathbf{P}^b_\ell(z) \qquad (\ell \geq 0),$$

provided this non tangential limit exists.

THEOREM 8.4. *Let b be a point in the unit ball of H^∞, let $\zeta_0 \in \mathbb{T}$ and let ℓ be a nonnegative integer. Assume that the boundary Schwarz-Pick matrix $\mathbf{P}^b_\ell(\zeta_0)$ exists. Then each function in $\mathcal{H}(b)$ and all its derivatives up to order ℓ have nontangential limits at ζ_0.*

Further it is shown in [**5**] that the boundary Schwarz-Pick matrix $\mathbf{P}^b_\ell(\zeta_0)$ exists if and only if

$$\lim_{\substack{z \longrightarrow \zeta_0 \\ \sphericalangle}} d_{b,\ell}(z) < +\infty, \tag{8.3}$$

where

$$d_{b,\ell}(z) := \frac{1}{(\ell!)^2} \frac{\partial^{2\ell}}{\partial z^\ell \partial \bar{z}^\ell} \frac{1-|b(z)|^2}{1-|z|^2}.$$

We should mention that it is not clear to show direct connections between conditions (8.2), (8.3) and condition (v) of Theorem 8.2.

9. Passage to the upper half plane $\mathbb{C}_+$

Let $\mathbb{C}_+$ denote the upper half plane in the complex plane and let $H^2(\mathbb{C}_+)$ denote the usual Hardy space consisting of analytic functions f on $\mathbb{C}_+$ which satisfy the growth restriction

$$\|f\|_2 = \sup_{y>0} \left(\int_{\mathbb{R}} |f(x+iy)|^2 \, dx \right)^{1/2} < \infty.$$

For each function $f \in H^2(\mathbb{C}_+)$ and for almost all $x_0 \in \mathbb{R}$,

$$f^*(x_0) = \lim_{t \to 0^+} f(x_0 + it)$$

exists. Moreover, we have $f^* \in L^2(\mathbb{R})$ and $\mathcal{F}f^* = 0$ on $(-\infty, 0)$, where $\mathcal{F}$ is the Fourier–Plancherel transformation, and $\|f^*\|_2 = \|f\|_2$.

A particularly interesting class of subspaces of $H^2(\mathbb{C}_+)$ is the $\mathcal{H}(b)$ classes of the upper half plane. The definitions are similar to those for the open unit disc. However, for the sake of completeness, we mention them below. We also mention some facts without proof.

For $\varphi \in L^\infty(\mathbb{R})$, let T_φ stand for the Toeplitz operator defined on $H^2(\mathbb{C}_+)$ by

$$T_\varphi(f) = P_+(\varphi f), \qquad (f \in H^2(\mathbb{C}_+)),$$

where P_+ is the orthogonal projection of $L^2(\mathbb{R})$ onto $H^2(\mathbb{C}_+)$. Then, for $b \in H^\infty$, with $\|b\|_\infty \le 1$, the de Branges–Rovnyak space $\mathcal{H}(b)$ consists of those $H^2(\mathbb{C}_+)$ functions which are in the range of the operator $(I - T_b T_{\bar{b}})^{1/2}$. As before, $\mathcal{H}(b)$ is a Hilbert space when equipped with the inner product

$$\langle (I - T_b T_{\bar{b}})^{1/2} f, (I - T_b T_{\bar{b}})^{1/2} g \rangle_b = \langle f, g \rangle_2,$$

where $f, g \in H^2(\mathbb{C}_+) \ominus \ker\, (I - T_\varphi T_{\bar{b}})^{1/2}$. For each $w \in \mathbb{C}_+$, the function

$$k_w^b(z) = \frac{i}{2\pi} \frac{1 - \overline{b(w)} b(z)}{z - \overline{w}}, \qquad (z \in \mathbb{C}_+),$$

is the reproducing kernel of $\mathcal{H}(b)$, that is

$$f(w) = \langle f, k_w^b \rangle_b, \qquad (f \in \mathcal{H}(b)). \tag{9.1}$$

In particular, with $f = k_w^b(z)$, we obtain

$$\|k_w^b\|_b^2 = k_w^b(w) = \frac{1}{2\pi} \frac{1 - |b(w)|^2}{2\Im w}, \qquad (f \in \mathcal{H}(b)). \tag{9.2}$$

Let $\rho(t) = 1 - |b(t)|^2$, $t \in \mathbb{R}$, and let $L^2(\rho)$ stand for the usual Hilbert space of measurable functions $f : \mathbb{R} \longrightarrow \mathbb{C}$ with $\|f\|_\rho < \infty$, where

$$\|f\|_\rho^2 = \int_{\mathbb{R}} |f(t)|^2 \rho(t) \, dt.$$

For each $w \in \mathbb{C}_+$, the Cauchy kernel k_w belongs to $L^2(\rho)$. Hence, we define $H^2(\rho)$ to be the span in $L^2(\rho)$ of the functions k_w ($w \in \mathbb{C}_+$). If g is a function in $L^2(\rho)$, then $g\rho$ is in $L^2(\mathbb{R})$, being the product of $g\rho^{1/2} \in L^2(\mathbb{R})$ and the bounded function $\rho^{1/2}$. Thus, we define the operator $C_\rho : L^2(\rho) \longrightarrow H^2(\mathbb{C}_+)$ by

$$C_\rho(g) = P_+(g\rho).$$

Then C_ρ is a partial isometry from $L^2(\rho)$ onto $\mathcal{H}(\bar{b})$ whose initial space equals to $H^2(\rho)$ and it is an isometry if and only if b is an extreme point of the unit ball of $H^\infty(\mathbb{C}_+)$.

Write $k_w^b = k_w - \overline{b(w)} b k_w$. Since $T_{\bar{b}} k_w = \overline{b(w)} k_w$, we obtain

$$\begin{aligned} T_{\bar{b}} k_w^b &= \overline{b(w)} \left(k_w - P_+(|b|^2 k_w)\right) \\ &= \overline{b(w)} P_+\left((1-|b|^2)k_w\right) \\ &= \overline{b(w)} P_+\left(\rho k_w\right) \\ &= \overline{b(w)}\, C_\rho(k_w). \end{aligned} \tag{9.3}$$

In Section 8, we have studied the boundary behavior of the derivatives of functions in $\mathcal{H}(b)$ spaces of the open unit disc $\mathbb{D}$. We mention some parts of Theorems 4.1 and 8.2, modified for the upper half plane $\mathbb{C}_+$, that are needed below.

THEOREM 9.1. *Let b be in the unit ball of $H^\infty(\mathbb{C}_+)$ and let $b = BI_\mu O_b$ be its canonical factorization, where*

$$B(z) = \prod_k e^{i\alpha_k} \frac{z - z_k}{z - \bar{z}_k}$$

is a Blaschke product, the singular inner function I_μ is given by

$$I_\mu(z) = \exp\left(iaz - \frac{i}{\pi}\int_{\mathbb{R}}\left(\frac{1}{z-t} + \frac{t}{t^2+1}\right)d\mu(t)\right)$$

with a positive singular measure μ and $a \geq 0$, and O_b is the outer function

$$O_b(z) = \exp\left(\frac{i}{\pi}\int_{\mathbb{R}}\left(\frac{1}{z-t} + \frac{t}{t^2+1}\right)\log|b(t)|\,dt\right).$$

Put

$$S_n(x) = \sum_{k=1}^{\infty} \frac{\Im z_k}{|x - z_k|^n} + \int_{\mathbb{R}} \frac{d\mu(t)}{|x-t|^n} + \int_{\mathbb{R}} \frac{\big|\log|b(t)|\big|}{|x-t|^n}\,dt. \tag{9.4}$$

Then, for $x \in \mathbb{R}$ and for $n \geq 0$, the following hold.

(i) If $S_{n+1}(x) < \infty$, then the limits

$$b^{(j)}(x) = \lim_{t\to 0^+} b^{(j)}(x+it), \qquad (0 \leq j \leq n),$$

exist.

(ii) For every function $f \in \mathcal{H}(b)$, the limits

$$f^{(j)}(x) = \lim_{t\to 0^+} f^{(j)}(x+it), \qquad (0 \leq j \leq n),$$

exist if and only if $S_{2n+2}(x) < \infty$.

Theorem 9.1 suggests to define

$$E_n(b) = \{x \in \mathbb{R} : S_n(x) < \infty\}.$$

The upper half plane version of Corollary 4.2 says that, in $x \in E_2(b)$, then the modulus of the angular derivative of b at a point x is given by

$$|b'(x)| = a + \sum_k \frac{2\Im z_k}{|x - z_k|^2} + \frac{1}{\pi}\int_{\mathbb{R}} \frac{d\mu(t)}{|x-t|^2} + \frac{1}{\pi}\int_{\mathbb{R}} \frac{\big|\log|b(t)|\big|}{|x-t|^2}\, dt. \tag{9.5}$$

10. Integral representations for derivatives

In this section, our goal is to prove an integral representation for the derivatives of elements of $\mathcal{H}(b)$. we start by finding such a formula at a point w in the upper half plane. Since w is away from the boundary, the representation is easy to establish. In order to get an integral representation for the nth derivative of f at point w for functions in the de-Branges-Rovnyak spaces, we need to introduce the following kernels

$$k_{w,n}^b(z) = -\frac{n!}{2\pi i} \cdot \frac{1 - b(z)\displaystyle\sum_{p=0}^{n} \frac{\overline{b^{(p)}(w)}}{p!}(z-\bar{w})^p}{(z-\bar{w})^{n+1}}, \qquad (z \in \mathbb{C}_+), \tag{10.1}$$

and

$$k_{w,n}^\rho(t) = -\frac{n!}{2\pi i} \cdot \frac{\displaystyle\sum_{p=0}^{n} \frac{\overline{b^{(p)}(w)}}{p!}(t-\bar{w})^p}{(t-\bar{w})^{n+1}}, \qquad (t \in \mathbb{R}). \tag{10.2}$$

For $n = 0$, we see that $k_{w,0}^b = k_w^b$ and $k_{w,0}^\rho = \overline{b(w)}k_w$. As a matter of fact, (10.1) is the upper half plane version of (6.6), and we have

$$f^{(n)}(w) = \langle f, k_{w,n}^b \rangle_b, \qquad (f \in \mathcal{H}(b)).$$

In the following lemma, we obtain a more friendly representation for $f^{(n)}(w)$.

LEMMA 10.1. *Let b be a point in the unit ball of $H^\infty(\mathbb{C}_+)$, let $f \in \mathcal{H}(b)$ and let $g \in H^2(\rho)$ be such that $T_{\bar{b}}f = C_\rho(g)$. Then, for all $w \in \mathbb{C}_+$ and for any integer $n \geq 0$, we have $k_{w,n}^b \in \mathcal{H}(b)$ and $k_{w,n}^\rho \in H^2(\rho)$ and*

$$f^{(n)}(w) = \int_{\mathbb{R}} f(t)\overline{k_{w,n}^b(t)}\, dt + \int_{\mathbb{R}} g(t)\rho(t)\overline{k_{w,n}^\rho(t)}\, dt. \tag{10.3}$$

PROOF. According to (9.1), we have

$$f(w) = \langle f, k_w^b \rangle_b = \langle f, k_w^b \rangle_2 + \langle T_{\bar{b}}f, T_{\bar{b}}k_w^b \rangle_{\bar{b}}.$$

Hence, by (9.3),

$$f(w) = \langle f, k_w^b \rangle_2 + b(w)\langle C_\rho(g), C_\rho(k_w) \rangle_{\bar{b}}.$$

Since C_ρ is a partial isometry from $L^2(\rho)$ onto $\mathcal{H}(\bar{b})$, with initial space equals to $H^2(\rho)$, we conclude that

$$f(w) = \langle f, k_w^b \rangle_2 + b(w)\langle g, k_w \rangle_\rho = \langle f, k_w^b \rangle_2 + \langle \rho g, k_w^\rho \rangle_2.$$

We rewrite this identity as

$$f(w) = \langle f, k_{w,0}^b \rangle_2 + \langle \rho g, k_{w,0}^\rho \rangle_2,$$

which is precisely the representation (10.3) for $n = 0$.

Now straightforward computations show that

$$\frac{\partial^n k^b_{w,0}}{\partial \bar{w}^n} = k^b_{w,n} \qquad \text{and} \qquad \frac{\partial^n k^\rho_{w,0}}{\partial \bar{w}^n} = k^\rho_{w,n}.$$

Since $k^b_{w,0} \in \mathcal{H}(b)$ and $k^\rho_{w,0} \in H^2(\rho)$, as we justified similarly in Section 6, we have $k^b_{w,n} \in \mathcal{H}(b)$ and $k^\rho_{w,n} \in H^2(\rho)$, $n \geq 0$. The representation (10.3) follows now by induction and by differentiating under the integral sign. □

The next step is to show that (10.3) is still valid at the boundary points x_0 which satisfy $S_{2n+2}(x_0) < \infty$. To do so, we need the boundary analogues of the kernels (10.1) and (10.2). In fact, both formulas make sense if we simply replace w by x_0 and assume that $S_{n+1}(x_0) < \infty$. However, we see that, under the stronger condition $S_{2n+2}(x_0) < \infty$, $k^b_{x_0,n}$ is actually the kernel function in $\mathcal{H}(b)$ for the functional of the n-th derivative at x_0.

LEMMA 10.2. *Let b be a point in the unit ball of $H^\infty(\mathbb{C}_+)$, let $n \geq 0$, and let $x_0 \in \mathbb{R}$. Assume that x_0 satisfies the condition $S_{2n+2}(x_0) < \infty$. Then $k^b_{x_0,n} \in \mathcal{H}(b)$ and, for every function $f \in \mathcal{H}(b)$, we have*

$$f^{(n)}(x_0) = \langle f, k^b_{x_0,n} \rangle_b. \tag{10.4}$$

PROOF. According to Theorem 9.1, the condition $S_{2n+2}(x_0) < \infty$ guarantees that, for every function $f \in \mathcal{H}(b)$, $f^{(n)}(w)$ tends to $f^{(n)}(x_0)$, as w tends radially to x_0. Therefore, an application of the uniform boundedness principle shows that the functional $f \longmapsto f^{(n)}(x_0)$ is bounded on $\mathcal{H}(b)$. Hence, by Riesz' theorem, there exists $\varphi_{x_0,n} \in \mathcal{H}(b)$ such that

$$f^{(n)}(x_0) = \langle f, \varphi_{x_0,n} \rangle_b, \qquad (f \in \mathcal{H}(b)).$$

Now, the formula

$$f^{(n)}(w) = \langle f, k^b_{w,n} \rangle_b, \qquad (f \in \mathcal{H}(b)),$$

implies that $k^b_{w,n}$ tends weakly to $\varphi_{x_0,n}$, as w tends radially to x_0. Thus, for $z \in \mathbb{C}_+$, we can write

$$\begin{aligned}
\varphi_{x_0,n}(z) &= \langle \varphi_{x_0,n}, k^b_z \rangle_b \\
&= \lim_{t \to 0^+} \langle k^b_{x_0+it,n}, k^b_z \rangle_b \\
&= \lim_{t \to 0^+} k^b_{x_0+it,n}(z) \\
&= \lim_{t \to 0^+} -\frac{n!}{2\pi i} \frac{1 - b(z) \sum_{p=0}^{n} \frac{\overline{b^{(p)}(x_0+it)}}{p!} (z - x_0 + it)^p}{(z - x_0 + it)^{n+1}} \\
&= -\frac{n!}{2\pi i} \frac{1 - b(z) \sum_{p=0}^{n} \frac{\overline{b^{(p)}(x_0)}}{p!} (z - x_0)^p}{(z - x_0)^{n+1}} = k^b_{x_0,n}(z).
\end{aligned}$$

Hence, $k^b_{x_0,n} \in \mathcal{H}(b)$ and, for every function $f \in \mathcal{H}(b)$, (10.4) holds. □

The next result gives a (standard) Taylor formula at a point on the boundary.

LEMMA 10.3. *Let h be a holomorphic function in the upper-half plane $\mathbb{C}_+$, let $n \geq 0$, and let $x_0 \in \mathbb{R}$. Assume that $h^{(n)}$ has a radial limit at x_0. Then*

$h, h', \ldots, h^{(n-1)}$ have radial limits at x_0 and

$$h(w) = \sum_{p=0}^{n} \frac{h^{(p)}(x_0)}{p!}(w-x_0)^p + (w-x_0)^n \varepsilon(w), \qquad (w \in \mathbb{C}_+),$$

with $\lim_{t \to 0^+} \varepsilon(x_0 + it) = 0$.

PROOF. The proof is by induction.

Case $n = 0$: This is trivial.

Case $n \geq 1$: Assumes that the property is true for $n-1$. Applying the induction hypothesis to $g = h'$, we see that $g = h', g' = h^{(2)}, \ldots, g^{(n-1)} = h^{(n)}$ have a radial limit at x_0 and

$$h'(w) = \sum_{p=0}^{n-1} \frac{h^{(p+1)}(x_0)}{p!}(w-x_0)^p + (w-x_0)^{n-1} \varepsilon_1(w),$$

with $\lim_{t \to 0^+} \varepsilon_1(x_0 + it) = 0$.

Since h' has a radial limit at x_0, we can define $h(x_0)$ by

$$h(x_0) = h(x_0 + it) - \int_{[x_0, x_0+it]} h'(u)\, du.$$

By Cauchy's theorem, $h(x_0)$ is well-defined and its value does not depend on t. However, this freedom in choosing t shows that $h(x_0) = \lim_{t \to 0} h(x_0 + it)$. Then another application of Cauchy's theorem reveals that we can even write

$$h(w) = h(x_0) + \int_{\Gamma_w} h'(u)\, du,$$

for all $w \in \mathbb{C}_+$. The path Γ_w is from x_0 to w. To have the uniform continuity of h on the path, we assume that in the beginning Γ_w is a vertical segment starting at x_0 (the hight of segment is not important), and then it goes to w via a rectifiable path in $\mathbb{C}_+$. Hence, we have

$$\begin{aligned} h(w) &= h(x_0) + \int_{\Gamma_w} \left(\sum_{p=0}^{n-1} \frac{h^{(p+1)}(x_0)}{p!}(u-x_0)^p + (u-x_0)^{n-1}\varepsilon_1(u) \right) du \\ &= \sum_{p=0}^{n} \frac{h^{(p)}(x_0)}{p!}(w-x_0)^p + \int_{\Gamma_w} (u-x_0)^{n-1}\varepsilon_1(u)\, du. \end{aligned}$$

The natural choice for ε is

$$\varepsilon(w) = \frac{1}{(w-x_0)^n} \int_{\Gamma_w} (u-x_0)^{n-1}\varepsilon_1(u)\, du, \qquad (w \in \mathbb{C}_+).$$

Thus,

$$\varepsilon(x_0 + it) = \frac{1}{(it)^n} \int_{s=0}^{t} (is)^{n-1} \varepsilon_1(x_0 + is)\, ds.$$

Therefore, based on the induction hypothesis on ε_1, we have

$$\begin{aligned} |\varepsilon(x_0 + it)| &\leq \frac{1}{t^n} \int_{s=0}^{t} s^{n-1} |\varepsilon_1(x_0 + is)|\, ds \\ &\leq \frac{1}{t} \int_{s=0}^{t} |\varepsilon_1(x_0 + is)|\, ds \longrightarrow 0 \end{aligned}$$

as $t \to 0^+$. □

If x_0 satisfies the condition $S_{2n+2}(x_0) < \infty$ we also have $k^\rho_{x_0,n} \in L^2(\rho)$. Indeed, according to (10.2), it suffices to prove that $(t - x_0)^{-j} \in L^2(\rho)$, for $1 \le j \le n+1$. Since $\rho \le 1$, it is enough to verify this fact in a neighborhood of x_0, say $I_{x_0} = [x_0 - 1, x_0 + 1]$. But according to the condition $S_{2n+2}(x_0) < \infty$, we have

$$\int_{I_{x_0}} \frac{1 - |b(t)|^2}{|t - x_0|^{2j}}\, dt \le 2 \int_{I_{x_0}} \frac{|\log|b(t)||}{|t - x_0|^{2j}}\, dt \le 2 \int_{I_{x_0}} \frac{|\log|b(t)||}{|t - x_0|^{2n+2}}\, dt < \infty.$$

THEOREM 10.4 (Fricain–Mashreghi [**15**]). *Let b be a point in the unit ball of $H^\infty(\mathbb{C}_+)$, let $n \ge 0$, let $f \in \mathcal{H}(b)$, and let $g \in H^2(\rho)$ be such that $T_{\overline{b}} f = C_\rho(g)$. Then, for every point $x_0 \in \mathbb{R}$ satisfying the condition $S_{2n+2}(x_0) < \infty$, we have*

$$f^{(n)}(x_0) = \int_{\mathbb{R}} f(t)\overline{k^b_{x_0,n}(t)}\, dt + \int_{\mathbb{R}} g(t)\rho(t)\overline{k^\rho_{x_0,n}(t)}\, dt. \tag{10.5}$$

PROOF. Recall that the condition $S_{2n+2}(x_0) < \infty$ guarantees that $b^{(j)}(x_0)$ exists for $0 \le j \le 2n+1$. Moreover, Lemma 10.2 implies that $k^b_{x_0,p} \in \mathcal{H}(b)$, for $0 \le p \le n$.

Put

$$h_{x_0,n}(z) = \frac{b(z) - \sum_{p=0}^n \frac{b^{(p)}(x_0)}{p!}(z - x_0)^p}{(z - x_0)^{n+1}}, \qquad (z \in \mathbb{C}_+).$$

Let us verify that $h_{x_0,n}$ satisfies

$$h_{x_0,n} = 2\pi i \sum_{p=0}^n \frac{b^{(n-p)}(x_0)}{(n-p)!p!} k^b_{x_0,p}. \tag{10.6}$$

To simplify a little bit the next computations, we put $a_p := \frac{b^{(p)}(x_0)}{p!}$, $0 \le p \le n$. According to (10.1), we have

$$\begin{aligned}
&2\pi i \sum_{p=0}^n a_{n-p} \frac{k^b_{x_0,p}(z)}{p!} \\
&= \sum_{p=0}^n a_{n-p} \left(b(z) \frac{\sum_{j=0}^p \overline{a_j}(z - x_0)^j - 1}{(z - x_0)^{p+1}} \right) \\
&= \frac{1}{(z - x_0)^{n+1}} \left[\sum_{p=0}^n a_{n-p}(z - x_0)^{n-p} \left(b(z) \sum_{j=0}^p \overline{a_j}(z - x_0)^j - 1 \right) \right] \\
&= \frac{1}{(z - x_0)^{n+1}} \left[b(z) \left(\sum_{p=0}^n \sum_{j=0}^p a_{n-p}\overline{a_j}(z - x_0)^{n-p+j} \right) - \sum_{k=0}^n a_k (z - x_0)^k \right].
\end{aligned}$$

Therefore, we see that (10.6) is equivalent to

$$\sum_{p=0}^n \sum_{j=0}^p a_{n-p}\overline{a_j}(z - x_0)^{n-p+j} = 1. \tag{10.7}$$

But, putting $j = \ell - n + p$, we obtain

$$\begin{aligned}\sum_{p=0}^{n}\sum_{j=0}^{p} a_{n-p}\overline{a_j}(z-x_0)^{n-p+j} &= \sum_{\ell=0}^{n}\left(\sum_{p=n-\ell}^{n} a_{n-p}\overline{a_{\ell-n+p}}\right)(z-x_0)^{\ell}\\ &= \sum_{\ell=0}^{n}\left(\sum_{q=0}^{\ell} a_{\ell-q}\overline{a_q}\right)(z-x_0)^{\ell}.\end{aligned}$$

Consequently, (10.7) is equivalent to

$$|a_0|^2 = |b(x_0)|^2 = 1 \qquad (\text{for } \ell = 0), \tag{10.8}$$

and

$$\sum_{q=0}^{\ell} a_{\ell-q}\overline{a_q} = 0, \qquad (1 \le \ell \le n). \tag{10.9}$$

To establish (10.8) and (10.8), put

$$\varphi(z) = 1 - b(z)\sum_{p=0}^{n}\overline{a_p}(z-x_0)^p, \qquad (z \in \mathbb{C}_+).$$

Then φ is holomorphic in $\mathbb{C}_+$ and φ and its derivatives up to order $2n+1$ have radial limits at x_0. An application of Lemma 10.3 shows that we can write

$$\varphi(z) = \sum_{p=0}^{n}\frac{\varphi^{(p)}(x_0)}{p!}(z-x_0)^p + o((z-x_0)^n),$$

as z tends radially to x_0. Assume that there exists $p \in \{0,\dots,n\}$ such that $\varphi^{(p)}(x_0) \ne 0$ and put

$$p_0 = \min\{0 \le p \le n : \varphi^{(p)}(x_0) \ne 0\}.$$

Hence, as $t \to 0^+$,

$$|k^b_{x_0,n}(x_0+it)| \sim \frac{1}{2\pi}\frac{|\varphi^{(p_0)}(x_0)|}{p_0!}t^{p_0-(n+1)},$$

which implies that $\lim_{t\to 0^+}|k^b_{x_0,n}(x_0+it)| = \infty$. This is a contradiction with the fact that $k^b_{x_0,n}$ belongs to $\mathcal{H}(b)$ and has a finite radial limit at x_0. Therefore, we necessarily have $\varphi^{(p)}(x_0) = 0$, $0 \le p \le n$. But,

$$\varphi(x_0) = 1 - b(x_0)\overline{b(x_0)} = 1 - |b(x_0)|^2,$$

and if we use Leibnitz rule to compute the derivative of φ, for $1 \le \ell \le n$, we get

$$\varphi^{(\ell)}(x_0) = -\sum_{p=0}^{\ell}\overline{a_p}\binom{\ell}{p}p!\,b^{(\ell-p)}(x_0) = -\ell!\sum_{p=0}^{\ell}\overline{a_p}a_{\ell-p}.$$

Thus, we established (10.9) and (10.8), which in return proves (10.6). According to Lemma 10.2, (10.6) implies $h_{x_0,n} \in \mathcal{H}(b)$.

Now, for almost all $t \in \mathbb{R}$, we have

$$\begin{aligned}
&\overline{b(t)}\, k^b_{x_0,n}(t)\\
&= -\frac{n!}{2\pi i}\cdot\frac{\overline{b(t)}-|b(t)|^2\sum_{p=0}^n \overline{a_p}(t-x_0)^p}{(t-x_0)^{n+1}}\\
&= -\frac{n!}{2\pi i}\cdot(1-|b(t)|^2)\frac{\sum_{p=0}^n \overline{a_p}(t-x_0)^p}{(t-x_0)^{n+1}} - \frac{n!}{2\pi i}\cdot\frac{\overline{b(t)}-\sum_{p=0}^n \overline{a_p}(t-x_0)^p}{(t-x_0)^{n+1}}\\
&= \rho(t)k^\rho_{x_0,n}(t) - \frac{n!}{2\pi i}\cdot\overline{h_{x_0,n}(t)}.
\end{aligned}$$

Since $h_{x_0,n} \in \mathcal{H}(b) \subset H^2(\mathbb{C}_+)$, we get that $P_+(\bar{b}k^b_{x_0,n}) = P_+(\rho k^\rho_{x_0,n})$, which can be written as $T_{\bar{b}}k^b_{x_0,n} = C_\rho k^\rho_{x_0,n}$. It follows from Lemma 10.2 that

$$\begin{aligned}
f^{(n)}(x_0) &= \langle f, k^b_{x_0,n}\rangle_b\\
&= \langle f, k^b_{x_0,n}\rangle_2 + \langle T_{\bar{b}}f, T_{\bar{b}}k^b_{x_0,n}\rangle_{\bar{b}}\\
&= \langle f, k^b_{x_0,n}\rangle_2 + \langle g, k^\rho_{x_0,n}\rangle_\rho\\
&= \int_{\mathbb{R}} f(t)\overline{k^b_{x_0,n}(t)}\,dt + \int_{\mathbb{R}} g(t)\rho(t)\overline{k^\rho_{x_0,n}(t)}\,dt,
\end{aligned}$$

which proves the relation (10.5). □

If b is inner, then it is clear that the second integral in (10.5) is zero and we obtain the easier formula

$$f^{(n)}(w) = \int_{\mathbb{R}} f(t)\,\overline{k^b_{w,n}(t)}\,dt. \tag{10.10}$$

We now provide a variation of Theorem 10.4 which is more suitable for obtaining Bernstein-type inequalities. To do so, we need the new kernel

$$\mathfrak{K}^\rho_{z_0,n}(t) = \overline{b(z_0)}\,\frac{\sum_{p=0}^n \binom{n+1}{p+1}(-1)^p\,\overline{b^p(z_0)}\,b^p(t)}{(t-\overline{z_0})^{n+1}}, \qquad (t \in \mathbb{R}), \tag{10.11}$$

which is well-defined for all $z_0 \in \mathbb{C}_+$. It also makes sense whenever $z_0 = x_0 \in E_1(b)$. We highlight that b^p is the p-power of b and should not be mistaken with the p-th derivative.

COROLLARY 10.5 (Baranov–Fricain–Mashreghi [**6**]). *Let b be in the unit ball of $H^\infty(\mathbb{C}_+)$, let $z_0 \in \mathbb{C}_+ \cup E_{2n+2}(b)$, and let $n \geq 0$. Then $(k^b_{z_0})^{n+1} \in H^2(\mathbb{C}_+)$ and $\mathfrak{K}^\rho_{z_0,n} \in L^2(\rho)$. Moreover, for every function $f \in \mathcal{H}(b)$, we have*

$$f^{(n)}(z_0) = \frac{n!}{2\pi i}\left(\int_{\mathbb{R}} f(t)\overline{(k^b_{z_0})^{n+1}(t)}\,dt + \int_{\mathbb{R}} g(t)\rho(t)\overline{\mathfrak{K}^\rho_{z_0,n}(t)}\,dt\right), \tag{10.12}$$

where $g \in H^2(\rho)$ is such that $T_{\bar{b}}f = C_\rho g$.

PROOF. *Step 1*: To show that the first integral in (10.12) is well-defined, we show that $(k^b_{z_0})^{n+1} \in H^2(\mathbb{C}_+)$.

Let $a_p = b^{(p)}(z_0)/p!$. Fix $0 \leq j \leq n$. Then we rewrite (10.1) as

$$-\frac{2\pi i}{j!}k^b_{z_0,j}(z) = \frac{1-\overline{b(z_0)}b(z)}{(z-\overline{z_0})^{j+1}} - b(z)\sum_{p=1}^{j}\frac{\overline{a_p}}{(z-\overline{z_0})^{j+1-p}}.$$

Hence, multiplying by $(1-\overline{b(z_0)}b(z))^j$ and rearranging the terms, we obtain

$$\begin{aligned}\left(k_{z_0}^b(z)\right)^{j+1} &= (1-\overline{b(z_0)}b(z))^j\left(-\frac{2\pi i}{j!}k_{z_0,j}^b(z)\right)\\ &+ b(z)\sum_{p=1}^{j}\overline{a_{j+1-p}}(1-\overline{b(z_0)}b(z))^{j-p}\left(k_{z_0}^b(z)\right)^p. \end{aligned}\tag{10.13}$$

Since $z_0 \in \mathbb{C}_+ \cup E_{2n+2}(b)$, according to Lemmas 10.1 and 10.2, the functions $k_{z_0}^b$ and $k_{z_0,j}^b$ $(1 \le j \le n)$ belong to $\mathcal{H}(b)$. Hence, using the recurrence relation (10.13) and that $1-\overline{b(z_0)}b(z) \in H^\infty(\mathbb{C}_+)$, we see immediately by induction that $(k_{z_0}^b)^{n+1} \in H^2(\mathbb{C}_+)$.

Step 2: To show that the second integral in (10.12) is well-defined, we show that $\mathfrak{K}_{z_0,n}^\rho \in L^2(\rho)$.

We have $= (t-\overline{z_0})^{-(n+1)}\varphi(t)$, with

$$|\mathfrak{K}_{z_0,n}^\rho(t)| \le \frac{2^{n+1}}{|t-z_0|^{n+1}}, \qquad (t \in \mathbb{R}).$$

Hence, it is sufficient to prove that $(t-z_0)^{-(n+1)} \in L^2(\rho)$. If $z_0 \in \mathbb{C}_+$, this fact is trivial and if $z_0 \in E_{2n+2}(b)$, the inequality $1-x \le |\log x|$, $x \in [0,1]$, implies

$$\int_{\mathbb{R}} \frac{\rho(t)}{|t-z_0|^{2n+2}}\,dt \le \int_{\mathbb{R}} \frac{1-|b(t)|^2}{|t-z_0|^{2n+2}}\,dt \le 2\int_{\mathbb{R}} \frac{|\log|b(t)||}{|t-z_0|^{2n+2}}\,dt < \infty,$$

which is the required result.

Step 3: It remains to prove that (10.12) holds.

Let ψ be any element of $H^2(\mathbb{C}_+)$. According to (10.5), we have

$$\begin{aligned} f^{(n)}(z_0) &= \langle f, k_{z_0,n}^b\rangle_2 + \langle \rho g, k_{z_0,n}^\rho\rangle_2\\ &= \langle f, k_{z_0,n}^b - b\psi\rangle_2 + \langle \bar b f, \psi\rangle_2 + \langle \rho g, k_{z_0,n}^\rho\rangle_2. \end{aligned}$$

But we have $T_{\bar b} f = C_\rho g$, which means that $\bar b f - \rho g \perp H^2(\mathbb{C}_+)$. Since $\psi \in H^2(\mathbb{C}_+)$, it follows that $\langle \bar b f, \psi\rangle_2 = \langle \rho g, \psi\rangle_2$. Hence, the identity

$$f^{(n)}(z_0) = \langle f, k_{z_0,n}^b - b\psi\rangle_2 + \langle \rho g, k_{z_0,n}^\rho + \psi\rangle_2 \tag{10.14}$$

holds for each $\psi \in H^2(\mathbb{C}_+)$. A very specific ψ gives us the required representation. To find the appropriate ψ note that, on one hand, we have

$$\begin{aligned} &-\frac{2\pi i}{n!}k_{z_0,n}^b(t) - (k_{z_0}^b)^{n+1}(t)\\ &= \frac{1-b(t)\sum_{p=0}^n \overline{a_p}(t-\overline{z_0})^p - (1-\overline{b(z_0)}b(t))^{n+1}}{(t-\overline{z_0})^{n+1}}\\ &= \frac{1-(1-\overline{b(z_0)}b(t))^{n+1}}{(t-\overline{z_0})^{n+1}} - b(t)\frac{\sum_{p=0}^n \overline{a_p}(t-\overline{z_0})^p}{(t-\overline{z_0})^{n+1}}\\ &= b(t)\psi(t), \end{aligned}$$

where

$$\psi(t) = \frac{\sum_{p=1}^{n+1}(-1)^{p+1}\binom{n+1}{p}(\overline{b(z_0)})^p(b(t))^{p-1}}{(t-\overline{z_0})^{n+1}} - \frac{\sum_{p=0}^n \overline{a_p}(t-\overline{z_0})^p}{(t-\overline{z_0})^{n+1}}.$$

On the other hand, we easily see that

$$\begin{aligned}
&-\frac{2\pi i}{n!} k^{\rho}_{z_0,n}(t) + \psi(t) \\
&= \frac{\sum_{p=1}^{n+1}(-1)^{p+1}\binom{n+1}{p}(\overline{b(z_0)})^p (b(t))^{p-1}}{(t-\overline{z_0})^{n+1}} \\
&= \overline{b(z_0)}\frac{\sum_{p=0}^{n}(-1)^{p}\binom{n+1}{p+1}(\overline{b(z_0)})^p (b(t))^{p}}{(t-\overline{z_0})^{n+1}} = \mathfrak{K}^{\rho}_{z_0,n}(t).
\end{aligned}$$

Therefore, (10.12) follows immediately from (10.14).

□

References

[1] P. R. Ahern and D. N. Clark, *Radial limits and invariant subspaces*, Amer. J. Math. **92** (1970), 332–342. MR0262511 (41 #7117)

[2] P. R. Ahern and D. N. Clark, *Radial nth derivatives of Blaschke products*, Math. Scand. **28** (1971), 189–201. MR0318495 (47 #7042)

[3] Daniel Alpay, *Algorithme de Schur, espaces à noyau reproduisant et théorie des systèmes* (French, with English and French summaries), Panoramas et Synthèses [Panoramas and Syntheses], vol. 6, Société Mathématique de France, Paris, 1998. MR1638044 (99g:47016)

[4] J. Ball, V. Bolotnikov, *de Branges–Rovnyak spaces: basics and theory*, submitted, ArXiv:1405.2980.

[5] Vladimir Bolotnikov and Alexander Kheifets, *A higher order analogue of the Carathéodory-Julia theorem*, J. Funct. Anal. **237** (2006), no. 1, 350–371, DOI 10.1016/j.jfa.2006.03.016. MR2239269 (2007d:46023)

[6] Anton Baranov, Emmanuel Fricain, and Javad Mashreghi, *Weighted norm inequalities for de Branges-Rovnyak spaces and their applications*, Amer. J. Math. **132** (2010), no. 1, 125–155, DOI 10.1353/ajm.0.0094. MR2597508 (2011a:46047)

[7] G. T. Cargo, *Angular and tangential limits of Blaschke products and their successive derivatives*, Canad. J. Math. **14** (1962), 334–348. MR0136743 (25 #204)

[8] William S. Cohn, *Radial limits and star invariant subspaces of bounded mean oscillation*, Amer. J. Math. **108** (1986), no. 3, 719–749, DOI 10.2307/2374661. MR844637 (87j:30076)

[9] Louis de Branges, *A proof of the Bieberbach conjecture*, Acta Math. **154** (1985), no. 1-2, 137–152, DOI 10.1007/BF02392821. MR772434 (86h:30026)

[10] Louis de Branges and James Rovnyak, *Canonical models in quantum scattering theory*, Perturbation Theory and its Applications in Quantum Mechanics (Proc. Adv. Sem. Math. Res. Center, U.S. Army, Theoret. Chem. Inst., Univ. of Wisconsin, Madison, Wis., 1965), Wiley, New York, 1966, pp. 295–392. MR0244795 (39 #6109)

[11] Louis de Branges and James Rovnyak, *Square summable power series*, Holt, Rinehart and Winston, New York-Toronto, Ont.-London, 1966. MR0215065 (35 #5909)

[12] B. Mark Davis and John E. McCarthy, *Multipliers of de Branges spaces*, Michigan Math. J. **38** (1991), no. 2, 225–240, DOI 10.1307/mmj/1029004330. MR1098860 (92k:46034)

[13] P. Fatou, *Séries trigonométriques et séries de Taylor* (French), Acta Math. **30** (1906), no. 1, 335–400, DOI 10.1007/BF02418579. MR1555035

[14] Emmanuel Fricain and Javad Mashreghi, *Boundary behavior of functions in the de Branges-Rovnyak spaces*, Complex Anal. Oper. Theory **2** (2008), no. 1, 87–97, DOI 10.1007/s11785-007-0028-8. MR2390675 (2009a:46054)

[15] Emmanuel Fricain and Javad Mashreghi, *Integral representation of the n-th derivative in de Branges-Rovnyak spaces and the norm convergence of its reproducing kernel* (English, with English and French summaries), Ann. Inst. Fourier (Grenoble) **58** (2008), no. 6, 2113–2135. MR2473631 (2009k:46050)

[16] E. Fricain and J. Mashreghi, *An Introduction to $\mathcal{H}(b)$ spaces, Vol.I & II*, New math. Monographs 20 and 21, Cambridge University Press, 2015.

[17] Otto Frostman, *Sur les produits de Blaschke* (French), Kungl. Fysiografiska Sällskapets i Lund Förhandlingar [Proc. Roy. Physiog. Soc. Lund] **12** (1942), no. 15, 169–182. MR0012127 (6,262e)

[18] Henry Helson, *Lectures on invariant subspaces*, Academic Press, New York-London, 1964. MR0171178 (30 #1409)
[19] Michael T. Jury, *Reproducing kernels, de Branges-Rovnyak spaces, and norms of weighted composition operators*, Proc. Amer. Math. Soc. **135** (2007), no. 11, 3669–3675 (electronic), DOI 10.1090/S0002-9939-07-08931-9. MR2336583 (2008h:47050)
[20] Benjamin A. Lotto and Donald Sarason, *Multipliers of de Branges-Rovnyak spaces. II*, Harmonic analysis and hypergroups (Delhi, 1995), Trends Math., Birkhäuser Boston, Boston, MA, 1998, pp. 51–58. MR1614726 (99c:46022)
[21] B. A. Lotto and Donald Sarason, *Multipliers of de Branges-Rovnyak spaces*, Indiana Univ. Math. J. **42** (1993), no. 3, 907–920, DOI 10.1512/iumj.1993.42.42042. MR1254125 (95a:46039)
[22] B. A. Lotto and D. Sarason, *Multiplicative structure of de Branges's spaces*, Rev. Mat. Iberoamericana **7** (1991), no. 2, 183–220, DOI 10.4171/RMI/110. MR1133377 (92k:46035)
[23] Benjamin A. Lotto, *Inner multipliers of de Branges's spaces*, Integral Equations Operator Theory **13** (1990), no. 2, 216–230, DOI 10.1007/BF01193757. MR1038151 (91g:46032)
[24] R.E. Paley, N. Wiener, *Fourier Transforms in the Complex Domain*, Amer. Math. Soc. Colloq. Publ., Vol. 19, Providence, 1934.
[25] D. Sarason, *Kernel of Toeplitz operators*, Oper. Theory Adv. Appl., **71** (1994), 153–164.
[26] Donald Sarason, *Sub-Hardy Hilbert spaces in the unit disk*, University of Arkansas Lecture Notes in the Mathematical Sciences, 10, John Wiley & Sons, Inc., New York, 1994. A Wiley-Interscience Publication. MR1289670 (96k:46039)
[27] Donald Sarason, *Exposed points in H^1. II*, Topics in operator theory: Ernst D. Hellinger memorial volume, Oper. Theory Adv. Appl., vol. 48, Birkhäuser, Basel, 1990, pp. 333–347. MR1207406 (94a:46031)
[28] Donald Sarason, *Exposed points in H^1. I*, The Gohberg anniversary collection, Vol. II (Calgary, AB, 1988), Oper. Theory Adv. Appl., vol. 41, Birkhäuser, Basel, 1989, pp. 485–496. MR1038352 (91h:46043)
[29] Donald Sarason, *Nearly invariant subspaces of the backward shift*, Contributions to operator theory and its applications (Mesa, AZ, 1987), Oper. Theory Adv. Appl., vol. 35, Birkhäuser, Basel, 1988, pp. 481–493. MR1017680 (90m:47012)
[30] Jonathan E. Shapiro, *Relative angular derivatives*, J. Operator Theory **46** (2001), no. 2, 265–280. MR1870407 (2002m:46042)
[31] Jonathan E. Shapiro, *More relative angular derivatives*, J. Operator Theory **49** (2003), no. 1, 85–97. MR1978323 (2004c:46042)
[32] Fernando Daniel Suárez, *Multipliers of de Branges-Rovnyak spaces in H^2*, Rev. Mat. Iberoamericana **11** (1995), no. 2, 375–415, DOI 10.4171/RMI/177. MR1344898 (97g:46031)

Laboratoire Paul Painlevé, UFR de Mathématiques, Bâtiment M2, Université des Sciences et Technologies Lille 1, 59 655 Villeneuve d'Ascq Cédex, France

E-mail address: `emmanuel.fricain@math.univ-lille1.fr`

Département de mathématiques et de statistique, Université Laval, Québec, QC, Canada G1K 7P4

E-mail address: `Javad.Mashreghi@mat.ulaval.ca`

Contemporary Mathematics
Volume **638**, 2015
http://dx.doi.org/10.1090/conm/638/12809

Interpolation and moment in weighted Hardy spaces

André Boivin and Changzhong Zhu

ABSTRACT. We study interpolation and moment problems in weighted Hardy spaces for the upper half-plane with the weight function satisfying Muckenhoupt's (A^q) condition.

1. Introduction

In [**2**], we studied approximation properties (expansion, interpolation and moment) in the weighted Hardy spaces $H^p_w(\mathbb{D})$ $(p > 1)$ for the unit disc $\mathbb{D}$ with the weight function w satisfying Muckenhoupt's (A^q) condition (for the definition, see Section 1.1 below). In this paper, we study the corresponding approximation properties in the (A^q)-weighted Hardy spaces for the upper half-plane, which we denote by H^p_{+w}.

Due to the existence of the (A^q) weight function w, it does not seem possible, by simply using conformal mappings between the upper half-plane and the unit disc, to obtain directly the corresponding results in the spaces H^p_{+w} from known results in the spaces $H^p_w(\mathbb{D})$.

Moreover, some close connections are known between the spaces $H^p_w(\mathbb{D})$ $(p > 1)$ and the un-weighted (i.e. $w \equiv 1$) Hardy spaces. Two important such connections are: $H^p_w(\mathbb{D}) \subset H^{p_0}(\mathbb{D})$ for some $1 < p_0 < p$; and $f \in H^p_w(\mathbb{D})$ if and only if $fW_p \in H^p(\mathbb{D})$ where W_p is an outer function related with w and p (see [**2**, Lemma 2.1 and Lemma 2.3] and [**6**]). When we studied approximation properties in $H^p_w(\mathbb{D})$ in [**2**], one of our basic ideas was to use these connections to transfer related problems from the weighted case to the un-weighted case, and then apply related well-known results in classical Hardy spaces.

In the upper-half plane, there seem to be no such obvious connections between the weighted and un-weighted Hardy spaces, and though some approximation properties are known in the corresponding un-weighted setting (see, for examples, [**5**] and [**9**]), we were unable to use the idea mentioned above.

We will use instead some important known results for the spaces H^p_{+w}, obtained in for examples, [**6**] and [**12**], which are closely related with our study and we will

2010 *Mathematics Subject Classification.* Primary 30D55.
Key words and phrases. Weighted Hardy spaces, Muckenhoupt condition.
Research supported by NSERC (Canada).

prove approximation properties (expansion, interpolation and moment) in these spaces similar to those obtained in [**2**].

Let us first, in the remaining part of this section, recall some basic definitions and state some of these properties of the spaces H^p_{+w} referred above. Next, in Section 2, we will obtain expansions with respect to two systems of analytic functions in H^p_{+w}, and then, in Sections 3 and 4, show the connections between these expansions and interpolation and moment problems.

1.1. The (A^q) weights. Let $w(x)$ be a non-negative, locally integrable function defined on the real line $\mathbb{R} := (-\infty, +\infty)$. For $1 < q < \infty$, we say $w(x)$ satisfies the (Muckenhoupt's) condition (A^q) (see [**10**], [**13**, Chapter IX], [**7**]), or simply write $w \in (A^q)$, if we have

$$\sup_I \left(\frac{1}{|I|}\int_I w(x)dx\right)\left(\frac{1}{|I|}\int_I w(x)^{-\frac{1}{q-1}}dx\right)^{q-1} < \infty,$$

where $I \subset \mathbb{R}$ denotes an arbitrary interval and $|I|$ denotes its length.

We say that $w \in (A^\infty)$ if there is an $\alpha > 0$ such that for any interval $I \subset \mathbb{R}$ and any measurable subset E of I,

$$\frac{W(E)}{W(I)} \le C\left(\frac{|E|}{|I|}\right)^\alpha,$$

where C is a constant independent of I and E and, for $F \subset \mathbb{R}$ measurable, $W(F) := \int_F w(x)dx$.

For example (see [**13**, p. 236]), if $1 < q < \infty$ and $-1 < \beta < q-1$, then $w(x) := |x|^\beta \in (A^q)$. It is known that (i) $w(x) \in (A^{p'})$ if and only if $[w(x)]^{1-p} \in (A^p)$ with $\frac{1}{p} + \frac{1}{p'} = 1$; (ii) if $w(x) \in (A^q)$ and $q' > q$, then $w(x) \in (A^{q'})$; (iii) if $w(x) \in (A^q)$ with $1 < q \le \infty$, then $w(x) \in (A^{q'})$ for some $1 < q' < q$. Denote by q_w the *critical exponent* for $w(x)$, that is, the infimum of all the q's such that $w(x)$ satisfies the condition (A^q). We have $q_w \ge 1$.

NOTE. If we assume $q_w < p < \infty$, then it follows that $w \in (A^p)$, a fact that we will use frequently.

It is known (see [**8**, p. 232]) that if $1 < p < \infty$ and $w(x) \in (A^p)$, then for each interval $I \subset \mathbb{R}$,

$$\int_{x \notin I} \frac{|I|^{p-1}w(x)}{|x-a_I|^p}dx \le \frac{K}{|I|}\int_I w(x)dx, \tag{1.1}$$

where a_I denotes the center of I, and K is a constant independent of I.

From (1.1), we can deduce another inequality which is useful for our study.

LEMMA 1.1. *If $w \in A^q$ with $q_w < p < \infty$, then for $z = x + iy$ with $y > 0$,*

$$\int_{-\infty}^{\infty} \frac{w(t)dt}{|t-z|^p} \le \frac{C}{y^p}\int_x^{x+y} w(t)dt, \tag{1.2}$$

where C is a constant independent of z. Moreover (1.2) remains true if z is replaced by $\overline{z}$,

PROOF. Take $I = [x, x+y]$ for $y > 0$, then by (1.1),

$$\left[\int_{-\infty}^{x} + \int_{x+y}^{\infty}\right]\frac{w(t)dt}{|t-(x+\frac{y}{2})|^p} \le \frac{K}{y^p}\int_x^{x+y} w(t)dt.$$

For any point $z = x + iy$ with $y > 0$, and any real $t \notin (x, x+y)$, we have $2|t - (x \pm iy)| > |t - (x + \frac{y}{2})|$, and for any $t \in (x, x+y)$, we have $|t - (x \pm iy)| > y$, so

$$\begin{aligned} \int_{-\infty}^{\infty} \frac{w(t)dt}{|t-z|^p} &= \left[\int_{-\infty}^{x} + \int_{x+y}^{\infty}\right] \frac{w(t)dt}{|t-z|^p} + \int_{x}^{x+y} \frac{w(t)dt}{|t-z|^p} \\ &\leq \frac{2^p K}{y^p} \int_{x}^{x+y} w(t)dt + \frac{1}{y^p} \int_{x}^{x+y} w(t)dt \\ &= \frac{2^p K + 1}{y^p} \int_{x}^{x+y} w(t)dt. \end{aligned}$$

□

1.2. The spaces H^p_{+w}. Let $w \in (A^q)$ and assume that $q_w < p < \infty$. The weighted Hardy space H^p_{+w} is the collection of functions $f(z)$ which are analytic in the upper half-plane $R^2_+ = \{z \in \mathbb{C} : \mathrm{Im}(z) > 0\}$ and satisfy

$$\|f(z)\|^p_{H^p_{+w}} = \sup_{y>0} \int_{-\infty}^{+\infty} |f(x+iy)|^p w(x) dx < +\infty \qquad (z = x + iy).$$

The weighted Hardy space H^p_{-w} for the lower half-plane is defined similarly by replacing "upper half-plane" $(y > 0)$ by "lower half-plane" $(y < 0)$. Clearly, if $w(x) \equiv 1$, H^p_{+w} and H^p_{-w} are the classical Hardy spaces H^p_+ (for the upper half-plane) and H^p_- (for the lower half-plane), respectively.

The space $L^p_w(\mathbb{R})$ is the collection of measurable functions $f(x)$ on $\mathbb{R}$ which satisfy

$$\|f(x)\|^p_{L^p_w(\mathbb{R})} = \int_{-\infty}^{+\infty} |f(x)|^p w(x) dx < +\infty.$$

Now we list some known properties which are closely related with our study of the approximation properties in the spaces.

LEMMA 1.2. (see [**6**, Theorem II.1.1]) *Let $w \in (A^q)$ and assume that $q_w < p < \infty$ and $f(z) \in H^p_{+w}$, then $f(z)$ has non-tangential limits almost everywhere (a.e.) in $\mathbb{R}$. The limit function $f(x)$ belongs to $L^p_w(\mathbb{R})$ and is called the boundary function of $f(z)$. Moreover, for any $f(z) \in H^p_{+w}$,*

$$\|f(x)\|_{L^p_w(\mathbb{R})} \leq \|f(z)\|_{H^p_{+w}} \leq C_p \|f(x)\|_{L^p_w(\mathbb{R})}, \tag{1.3}$$

where C_p is a constant which depends only on p.

LEMMA 1.3. (see [**12**]) *(i) If $w \in (A^q)$ with $q_w < p < \infty$, and $f(x) \in L^p_w(\mathbb{R})$, then the integral*

$$F(z) = \frac{1}{2\pi i} \int_{-\infty}^{+\infty} \frac{f(x)}{x - z} dx \tag{1.4}$$

defines two functions: for $\mathrm{Im}(z) > 0$, $F(z) = F^+(z) \in H^p_{+w}$; for $\mathrm{Im}(z) < 0$, $F(z) = F^-(z) \in H^p_{-w}$. If $F^+(x)$ and $F^-(x)$ are the boundary functions of $F^+(z)$ and $F^-(z)$ respectively, then we have

$$f(x) = F^+(x) - F^-(x) \quad \textit{a.e. on } \mathbb{R}; \tag{1.5}$$

and

$$\|F^{\pm}(x)\|_{L^p_w(\mathbb{R})} \leq C_p \|f(x)\|_{L^p_w(\mathbb{R})}, \tag{1.6}$$

where C_p is a constant dependent only of p.

(ii) If $w \in (A^q)$ with $q_w < p < \infty$, and $f(z) \in H^p_{+w}$, then

$$\frac{1}{2\pi i}\int_{-\infty}^{+\infty} \frac{f(x)}{x-z}dx = \begin{cases} f(z), & \mathrm{Im}(z) > 0, \\ 0, & \mathrm{Im}(z) < 0. \end{cases} \tag{1.7}$$

Let p' denote the conjugate exponent of p, that is $1/p + 1/p' = 1$. If $f \in L^p(\mathbb{R}, w(x)dx)$, $1 < p < \infty$, then $f = \varphi w^{-1/p}$, where $\varphi \in L^p(\mathbb{R})$ and $\|\varphi\|_p = \|f\|_{p,w}$. Therefore to each (linear, continuous) functional ℓ on $L^p(\mathbb{R}, w(x)dx)$ corresponds a (unique) function $\psi \in L^{p'}(\mathbb{R})$ such that for any $f \in L^p(\mathbb{R}, w(x)dx)$

$$\ell(f) = (\varphi, \psi) = (fw^{1/p}, \psi) = (f, w^{1/p}\psi) = (f, g),$$

where $g = w^{1/p}\psi \in L^{p'}(\mathbb{R}, w^{1-p'}(x)dx)$, and every functional on $L^p(\mathbb{R}, w(x)dx)$ is obtained in this way. In our setting, using Lemma 1.3, this has the following easy consequence which we state for convenience and further reference.

LEMMA 1.4. *Let $w \in (A^q)$ and assume that $q_w < p < \infty$. Any linear functional ℓ_+ on H^p_{+w} has the representation*

$$\ell_+(f) = \int_{-\infty}^{+\infty} f(x)\overline{g(x)}dx, \qquad f \in H^p_{+w}, \quad g \in H^{p'}_{+w^{1-p'}}, \quad \frac{1}{p} + \frac{1}{p'} = 1, \tag{1.8}$$

and its norm is comparable to that of g, that is,

$$\|\ell_+\| \le C_0 \|g\|_{H^{p'}_{+w^{1-p'}}} \le C_1 \|\ell_+\| \tag{1.9}$$

(see [**6**, p. 27] and [**12**, Theorem 2.2]).

LEMMA 1.5. *If $w \in (A^q)$ with $q_w < p < \infty$, and $f(z) \in H^p_{+w}$, then for each compact subset K in the upper half-plane,*

$$|f(z)| \le C_{K,p}\|f\|_{L^p_w(\mathbb{R})}, \qquad z \in K,$$

where $C_{K,p}$ is a constant only dependent of K and p.

PROOF. By Lemma 1.3, and using the Hölder inequality, we have

$$|f(z)| \le \frac{1}{2\pi}\left(\int_{-\infty}^{\infty} \frac{(w(x))^{-\frac{p'}{p}}dx}{|x-z|^{p'}}\right)^{1/p'} \|f\|_{L^p_w(\mathbb{R})}, \qquad z \in K,$$

where $\frac{1}{p} + \frac{1}{p'} = 1$. But noting that $-\frac{p'}{p} = 1 - p'$ and $(w(x))^{1-p'} \in A^{p'}$, we obtain from Lemma 1.1 that

$$\frac{1}{2\pi}\left(\int_{-\infty}^{\infty} \frac{(w(x))^{-\frac{p'}{p}}dx}{|x-z|^{p'}}\right)^{\frac{1}{p'}} \le C(K,p), \qquad z \in K,$$

where $C(K,p)$ is a constant only dependent of K and p. The lemma is proved. □

REMARK 1.6. From Lemma 1.5, it follows that convergence in H^p_{+w} (or equivalently in $L^p_w(\mathbb{R})$, using Lemma 1.2) implies uniform convergence on each compact subset of the upper half-plane.

2. Expansions in H^p_{+w}

A system of functions $\{\phi_k(z)\}$ in H^p_{+w} $(k = 1, 2, \ldots)$ is called complete in H^p_{+w} if for any function $f(z) \in H^p_{+w}$, its boundary function $f(x)$ on $\mathbb{R}$ can be approximated arbitrarily well in $L^p_w(\mathbb{R})$ by functions in the linear span of the system $\{\phi_k(x)\}$, where $\phi_k(x)$ are the boundary functions of $\phi_k(z)$ on $\mathbb{R}$. Otherwise, the system is called incomplete in H^p_{+w}. The completeness of a system in H^p_{-w} is defined similarly.

Let $\{a_k\}$ $(k = 1, 2, \ldots)$ be a sequence of complex numbers in R^2_+ (i.e. $\mathrm{Im}(a_k) > 0$ for all k) satisfying

$$\sum_{k=1}^{\infty} \frac{\mathrm{Im}(a_k)}{1 + |a_k|^2} < +\infty. \tag{2.1}$$

It is well-known (see [**5**], [**7**] and [**4**]) that the Blaschke product

$$B(z) = \prod_{k=1}^{\infty} \Big[\frac{z - a_k}{z - \overline{a}_k} \cdot \frac{|1 + a_k^2|}{1 + a_k^2}\Big] \tag{2.2}$$

and the finite Blaschke products

$$B_n(z) = \prod_{k=1}^{n} \Big[\frac{z - a_k}{z - \overline{a}_k} \cdot \frac{|1 + a_k^2|}{1 + a_k^2}\Big] \qquad (n = 1, 2, \ldots) \tag{2.3}$$

define functions that are analytic in R^2_+. Moreover

$$\overline{B_n(\overline{z})} = \frac{1}{B_n(z)}, \tag{2.4}$$

and as $n \to \infty$, $B_n(z) \to B(z)$ uniformly on each compact subset in R^2_+; we also have $|B_n(z)| \le 1$ for $z \in R^2_+$ and $|B_n(x)| = 1$ for $x \in \mathbb{R}$, and thus $|B(z)| \le 1$ for $z \in R^2_+$, and $|B(x)| = 1$ a.e. on $\mathbb{R}$.

LEMMA 2.1. *Let $w \in (A^q)$ and assume that $q_w < p < \infty$ and $h(x) \in L^p_w(\mathbb{R})$, then as $n \to \infty$,*

$$\|h(x)[B_n(x) - B(x)]\|_{L^p_w(\mathbb{R})} \to 0. \tag{2.5}$$

Indeed, since $h(x)[w(x)]^{1/p} \in L^p(\mathbb{R})$, by [**5**, Lemma 1.1], (2.5) follows immediately.

For a sequence $\{a_k\}$ $(k = 1, 2, \ldots)$ located in R^2_+ satisfying (2.1), and with the additional property $a_k \neq a_j$ for $k \neq j$, consider the following two systems of functions:

$$e_k(z) = \frac{1}{2\pi i} \cdot \frac{1}{z - \overline{a}_k}, \qquad k = 1, 2, \ldots, \tag{2.6}$$

$$\phi_k(z) = -\frac{B(z)}{(z - a_k)B'(a_k)}, \qquad k = 1, 2, \ldots. \tag{2.7}$$

Clearly we have $\phi_k(a_j) = 0$ for $k \neq j$, and $\phi_k(a_k) = -1$, i.e. $\phi_k(a_j) = -\delta_{kj}$. For $z \neq a_k$ $(k = 1, 2, \ldots)$, recalling that $a_k \neq a_j$ for $k \neq j$, it follows from the residue theorem that $\phi_k(z)$ has the following integral expression:

$$\phi_k(z) = \frac{B(z)}{2\pi i} \int_{c_k} \frac{d\xi}{B(\xi)(\xi - z)}, \quad k = 1, 2, \ldots, \tag{2.8}$$

where $c_k \subset R^2_+$ is a sufficiently small circle with centre at a_k such that the closed disc bounded by c_k does not contain any other a_j $(j \neq k)$, and does not contain z.

The systems (2.6) and (2.7) are bi-orthogonal on $\mathbb{R}$ (see [**9**]), that is

$$\int_{-\infty}^{+\infty} e_k(x)\overline{\phi_j(x)}dx = \int_{-\infty}^{+\infty} \overline{e_k(x)}\phi_j(x)dx = \delta_{kj},$$

where $\delta_{kj} = 1$ if $k = j$; $\delta_{kj} = 0$ if $k \neq j$.

Let $w \in (A^q)$ and assume that $q_w < p < \infty$. By [**12**, Section 3], it is known that the functions $e_k(z)$ and $\phi_k(z)$ $(k = 1, 2, \ldots)$ all belong to H^p_{+w}, and that the two systems they formed are incomplete in H^p_{+w}. Denote by $E_{p,w}$ and $\Psi_{p,w}$ the closed linear spans in H^p_{+w} of $\{e_k(z)\}$ and $\{\phi_k(z)\}$, respectively. We have $E_{p,w} = \Psi_{p,w}$ (which, by incompleteness, is a proper subspace of H^p_{+w}). Indeed, define the space $H^p_w(\{a_k\}) \subset H^p_{+w}$ consisting of all functions $f(z) \in H^p_{+w}$ satisfying $f(x)/B(x) = \lim_{y\to 0^-} F(x + iy)$ for some $F(z) \in H^p_{-w}$. By [**12**], under the conditions (2.1) and $q_w < p < \infty$, both the systems $\{e_k(z)\}$ and $\{\phi_k(z)\}$ belong to $H^p_w(\{a_k\})$ and both are dense in $H^p_w(\{a_k\})$. It thus follows that $E_{p,w} = \Psi_{p,w}$.

LEMMA 2.2. *Let $w \in (A^q)$ and assume that $q_w < p < \infty$. If $f(z) \in E_{p,w}$ (i.e. $\Psi_{p,w}$), then*

$$\frac{1}{2\pi i}\int_{-\infty}^{\infty} \frac{f(x)dx}{B(x)(x-z)} \equiv 0, \qquad \text{Im}(z) > 0, \tag{2.9}$$

where $B(z)$ is the Blaschke product of $\{a_k\}$ defined by (2.2).

PROOF. Since $f(z) \in E_{p,w}$, there are linear combinations

$$P_n(f, z) = \sum_{k=1}^{n} b_k^{(n)}(f) \cdot e_k(z), \quad n = 1, 2, \ldots,$$

such that

$$\lim_{n\to\infty} \|f - P_n\|_{L^p_w(\mathbb{R})} = 0.$$

Since for $k = 1, 2, \ldots$, we have

$$\frac{e_k(z)}{B(z)} = \frac{1}{2\pi i} \cdot \frac{1}{(z - \bar{a}_k)} \cdot \frac{1}{B(z)} \in H^p_{-w},$$

then, by Lemma 1.3,

$$\int_{-\infty}^{\infty} \frac{e_k(x)}{B(x)} \cdot \frac{dx}{x-z} \equiv 0, \quad \text{Im}(z) > 0.$$

Thus, for $n = 1, 2, \ldots$,

$$\int_{-\infty}^{\infty} \frac{P_n(f, x)}{B(x)} \cdot \frac{dx}{x-z} \equiv 0, \quad \text{Im}(z) > 0.$$

Since $|B(x)| = 1$ a.e. on $(-\infty, \infty)$, by Hölder's inequality, we have

$$\Big|\int_{-\infty}^{\infty} \frac{f(x)}{B(x)} \cdot \frac{dx}{x-z}\Big| = \Big|\int_{-\infty}^{\infty} \frac{f(x) - P_n(f, x)}{B(x)} \cdot \frac{dx}{x-z}\Big|$$

$$\leq \|f - P_n\|_{L^p_w(\mathbb{R})} \cdot \left\{\int_{-\infty}^{\infty} \frac{(w(x))^{1-p'}dx}{|x-z|^{p'}}\right\}^{1/p'}, \quad \text{Im}(z) > 0, \quad n = 1, 2, \ldots,$$

where p' satisfies $1/p + 1/p' = 1$. Applying Lemma 1.1 and taking the limit produces the required result. □

Next, we produce expansions for the functions f in H^p_{+w} $(q_w < p < \infty)$ with respect to the systems $\{\phi_k(z)\}$ and $\{e_k(z)\}$, respectively. Since the proofs are formally almost exactly the same as in [**1**, Section 4] or [**2**] (except for the use of the residue theorem, and Lemmas 1.3 and 2.1 above), we omit them here.

THEOREM 2.3. *Assume that* $\{a_k\} \subset R^2_+$ *satisfies* (2.1), *and* $w \in (A^q)$ *with* $q_w < p < \infty$. *If* $f(z) \in H^p_{+w}$, *then*

$$f(z) = P_{\Phi_{p,w}} f(z) + H_f(z), \qquad \mathrm{Im}(z) > 0, \tag{2.10}$$

where

$$H_f(z) = \frac{B(z)}{2\pi i} \int_{-\infty}^{+\infty} \frac{f(x)dx}{B(x)(x-z)}, \qquad \mathrm{Im}(z) > 0, \tag{2.11}$$

and

$$P_{\Phi_{p,w}} f(z) = \lim_{n\to\infty} \sum_{k=1}^{n} \alpha(f, e_k) \cdot \frac{B(a_k)}{B_n(a_k)} \cdot \phi_k(z), \qquad \mathrm{Im}(z) > 0, \tag{2.12}$$

where the limit is in the sense of both $L^p_w(\mathbb{R})$ *convergence and uniform convergence on each compact subset in the upper half-plane, and*

$$\alpha(f, e_k) = \int_{-\infty}^{+\infty} f(x)\overline{e_k(x)}dx = -f(a_k), \qquad k = 1, 2, \ldots \tag{2.13}$$

where the last equality holds by Lemma 1.3 (ii).

Note that since $|B(z)| \le 1$ for $\mathrm{Im}(z) > 0$, $|B(x)| = 1$ a.e. in $\mathbb{R}$, and $f(x) \in L^p_w(\mathbb{R})$, hence we have $f(x)/B(x) \in L^p_w(\mathbb{R})$, and thus by Lemma 1.3(i), $H_f(z) \in H^p_{+w}$.

THEOREM 2.4. *Assume that* $\{a_k\} \subset R^2_+$ *satisfies* (2.1), *and* $w \in (A^q)$ *with* $q_w < p < \infty$. *If* $f(z) \in H^p_{+w}$, *then*

$$f(z) = P_{E_{p,w}} f(z) + H_f(z), \qquad \mathrm{Im}(z) > 0, \tag{2.14}$$

where $H_f(z)$ *is defined by* (2.11), *and*

$$P_{E_{p,w}} f(z) = \lim_{n\to\infty} \sum_{k=1}^{n} \alpha(f, \phi_k) \cdot \overline{\left(\frac{B(a_k)}{B_n(a_k)}\right)} \cdot e_k(z), \qquad \mathrm{Im}(z) > 0, \tag{2.15}$$

where the limit is in the same sense as in Theorem 2.3, *and*

$$\alpha(f, \phi_k) = \int_{-\infty}^{+\infty} f(x)\overline{\phi_k(x)}dx, \qquad k = 1, 2, \ldots. \tag{2.16}$$

By Theorems 2.3 and 2.4, we see that

$$P_{E_{p,w}} f(z) = P_{\Phi_{p,w}} f(z) = f(z) - H_f(z), \qquad \mathrm{Im}(z) > 0,$$

which is the orthogonal projection of $f(z)$ onto $E_{p,w}$ (i.e. $\Phi_{p,w}$). Indeed, noting that $H_f(z) \in H^p_{+w}$, by Lemma 1.3,

$$\int_{-\infty}^{\infty} H_f(x)\overline{e_k(x)}dx = -H_f(a_k) = 0, \qquad k = 1, 2, \ldots,$$

hence

$$\int_{-\infty}^{\infty} [f(x) - P_{E_{p,w}} f(x)]\overline{e_k(x)}dx = 0, \qquad k = 1, 2, \ldots.$$

It should be pointed out that for the classical (i.e. un-weighted) case, the corresponding results were obtained in [**5**] where it is not assumed that $a_k \neq a_j$ for $k \neq j$, i.e. a value a_j can appear repeatedly in the sequence $\{a_k\}$, and this is reflected in the definition of the functions $e_k(z)$ and $\phi_k(z)$ in [**5**] as they must then take a more general form.

Corollary 2.5. *Let $w \in (A^q)$ with $q_w < p < \infty$. If $f(z) \in H^p_{+w}$, then $f(z) \in E_{p,w}$ if and only if $H_f(z) \equiv 0$.*

Proof. Combine Theorem 2.4 with Lemma 2.2. □

Corollary 2.6. *Let $w \in (A^q)$ with $q_w < p < \infty$, and let $f(z) \in H^p_{+w}$. If $\alpha(f, \phi_k) = 0$ then $f(a_k) = 0$ $(k = 1, 2, \ldots)$.*

Proof. If $\alpha(f, \phi_k) = 0$ $(k = 1, 2, \ldots)$, by Theorem 2.4, $f(z) = H_f(z)$ for $\mathrm{Im}(z) > 0$. Thus, $f(a_k) = H_f(a_k) = 0$ $(k = 1, 2, \ldots)$. □

Corollary 2.7. *Let $w \in (A^q)$ with $q_w < p < \infty$. If $f(z) \in \Phi_{p,w}$ (or equivalently, $f(z) \in E_{p,w}$), and $f(a_k) = 0$ $(k = 1, 2, \ldots)$, then $f(z) \equiv 0$ for $\mathrm{Im}(z) > 0$.*

Proof. If $f(z) \in \Phi_{p,w}$, or equivalently $f(z) \in E_{p,w}$, then by Lemma 2.2, $H_f(z) \equiv 0$. And since $f(a_k) = -\alpha(f, e_k) = 0$, then by Theorem 2.3, we have $f(z) \equiv 0$ for $\mathrm{Im}(z) > 0$. □

Corollary 2.7, as a uniqueness theorem, will play important role in the next two sections.

3. Interpolation and expansion

We say that a sequence $\{a_k\}$ in the upper half-plane satisfies the Carleson condition, or we say it is uniformly separated (see [**11**, p. 3]), if

$$\inf_k \prod_{\substack{j=1\\ j\neq k}} \left| \frac{a_j - a_k}{a_j - \bar{a}_k} \right| \geq \delta > 0. \tag{3.1}$$

It is known (see [**5**, § 3.1]) that if $\mathrm{Im}(a_k) > 0$ $(k = 1, 2, \ldots)$ and $\{a_k\}$ satisfies the Carleson condition (3.1), then $\{a_k\}$ satisfies the Blaschke condition (2.1).

In this section, using the results obtained above and the relationship between interpolation and approximation, we will give an expansion for functions $f(z) \in H^p_{+w}$ with respect to the system $\{\phi_k(z)\}$ under the Carleson condition.

Assume that $w(x) \in (A^\infty)$, $a_k = x_k + iy_k$ $(y_k > 0)$ with $a_k \neq a_j$ for $k \neq j$, and let

$$W_k = \int_{x_k}^{x_k+y_k} w(x)dx, \qquad k = 1, 2, \ldots.$$

For $1 < p < \infty$, define an operator T_p mapping H^p_{+w} into a sequence space by

$$(T_p f)_k = W_k^{\frac{1}{p}} f(a_k), \quad f(z) \in H^p_{+w}, \qquad k = 1, 2, \ldots. \tag{3.2}$$

By [**12**, Theorem 5.2], we have

Lemma 3.1. *If $\{a_k\}$ satisfies the Carleson condition (3.1), then the operator T_p is bounded from H^p_{+w} into l^p.*

Now we study an interpolation problem: Given two sequences of complex numbers $\{a_k\}$ and $\{b_k\}$, find conditions on the sequences that will guarantee the existence of a function $I \in H^p_{+w}$ with the property that

$$I(a_k) = b_k \quad (k = 1, 2, \ldots). \tag{3.3}$$

THEOREM 3.2. *Assume that* $a_k = x_k + iy_k$ *with* $y_k > 0$ $(k = 1, 2, \ldots)$. *Let* $w \in (A^q)$ *with* $q_w < p < \infty$, *and let* p' *denote the conjugate exponent of* p, *that is* $\frac{1}{p} + \frac{1}{p'} = 1$, *and set*[1]

$$V_k = \int_{x_k}^{x_k+y_k} [w(x)]^{1-p'} dx, \qquad k = 1, 2, \ldots.$$

If $\{a_k\}$ *satisfies the Carleson condition* (3.1) *and* $\{b_k\}$ *satisfies*

$$\sum_{k=1}^{\infty} |b_k|^p (\mathrm{Im}(a_k))^p V_k^{-\frac{p}{p'}} < +\infty, \tag{3.4}$$

then the series

$$\sum_{n=1}^{\infty} (-b_n) \cdot \phi_n(z) \tag{3.5}$$

converges in H^p_{+w}, *and thus defines a function* $I \in \Phi_{p,w} \subset H^p_{+w}$ *satisfying*

$$I(a_k) = b_k, \qquad k = 1, 2, \ldots.$$

PROOF. First we note that since

$$B'(a_k) = \frac{-i}{2\mathrm{Im}(a_k)} \cdot \frac{|1 + a_k^2|}{1 + a_k^2} \cdot \prod_{\substack{n=1 \\ n \neq k}}^{\infty} \left[\frac{a_k - a_n}{a_k - \bar{a}_n} \cdot \frac{|1 + a_n^2|}{1 + a_n^2} \right], \tag{3.6}$$

and $\{a_k\}$ satisfies the Carleson condition, we have

$$|B'(a_k)| \geq \frac{\delta}{2\mathrm{Im}(a_k)}, \qquad k = 1, 2, \ldots. \tag{3.7}$$

Also note that

$$\phi_k(a_j) = -\frac{B(a_j)}{B'(a_k)(a_j - a_k)} = -\delta_{kj}. \tag{3.8}$$

Now we prove that the series (3.5) converges in H^p_{+w}. We follow the proof for the classical H^p_+ spaces in [**5**, Theorem 5.1]. For any positive integer n, by the

---

[1]Note: As mentioned before, by [**13**, p. 250-251], if $w(x) \in (A^p), 1 < p < \infty$, then $w(x) \in (A^\infty)$. Since $w(x) \in (A^p)$, we have $[w(x)]^{1-p'} \in (A^{p'})$, hence $v(x) = [w(x)]^{1-p'} \in (A^\infty)$.

Hahn-Banach theorem and Lemmas 1.4 and 1.2, we have

$$\begin{aligned}\|S_n(x)\|_{L^p_w(\mathbb{R})} &:= \left\|\sum_{k=1}^n b_k\cdot\phi_k(x)\right\|_{L^p_w(\mathbb{R})}\\ &\le C_0 \sup_{\substack{g(x)\in L^{p'}_{w^{1-p'}}(\mathbb{R})\\ \|g(x)\|_{L^{p'}_{w^{1-p'}}(\mathbb{R})}\le 1}} \left|\int_{-\infty}^{+\infty}\left(\sum_{k=1}^n b_k\phi_k(x)\right)\overline{g(x)}dx\right|\\ &= C_0 \sup_{\substack{g(x)\in L^{p'}_{w^{1-p'}}(\mathbb{R})\\ \|g(x)\|_{L^{p'}_{w^{1-p'}}(\mathbb{R})}\le 1}} \left|\sum_{k=1}^n b_k\left(\int_{-\infty}^{+\infty}\overline{g(x)}\phi_k(x)dx\right)\right|,\end{aligned}$$

where C_0 is a constant depending only on p. We note that for $g(x)\in L^{p'}_{w^{1-p'}}(\mathbb{R})$ with $\|g(x)\|_{L^{p'}_{w^{1-p'}}(\mathbb{R})}\le 1$,

$$\frac{1}{2\pi i}\int_{-\infty}^{+\infty}\overline{g(x)}\phi_k(x)dx = \frac{-1}{B'(a_k)}\frac{1}{2\pi i}\int_{-\infty}^{+\infty}\frac{\overline{g(x)}B(x)}{x-a_k}dx = \frac{-1}{B'(a_k)}G(a_k).$$

By Lemma 1.3,

$$G(z) := \frac{1}{2\pi i}\int_{-\infty}^{\infty}\frac{\overline{g(x)}B(x)}{x-z}dx\in H^{p'}_{+w^{1-p'}},$$

and

$$\begin{aligned}\|G(x)\|_{L^{p'}_{w^{1-p'}}(\mathbb{R})} &\le C_1\|\overline{g(x)}B(x)\|_{L^{p'}_{w^{1-p'}}(\mathbb{R})}\\ &= C_1\|g(x)\|_{L^{p'}_{w^{1-p'}}(\mathbb{R})}\le C_1,\end{aligned}$$

where C_1 is a constant independent of g. Thus, noting (3.7), we obtain

$$\left\|\sum_{k=1}^n \frac{b_k}{\mu_k}\cdot\phi_k(x)\right\|_{L^p_w(\mathbb{R})} \le C_0 \sup_{\substack{G(z)\in H^{p'}_{+w^{1-p'}}\\ \|G(x)\|_{L^{p'}_{w^{1-p'}}(\mathbb{R})}\le C_1}} \sum_{k=1}^n\left|b_k\frac{2\mathrm{Im}(a_k)}{\delta}G(a_k)\right|. \tag{3.9}$$

By Hölder's inequality,
(3.10)

$$\sum_{k=1}^n |b_k\mathrm{Im}(a_k)G(a_k)| \le \left(\sum_{k=1}^n V_k|G(a_k)|^{p'}\right)^{1/p'}\left(\sum_{k=1}^n |b_k|^p\left(\mathrm{Im}(a_k)\right)^p V_k^{-\frac{p}{p'}}\right)^{1/p}.$$

Consider the operator $T_{p'}$ which maps $H^{p'}_{+w^{1-p'}}$ into a sequence space defined by

$$(T_{p'}G)_k = V_k^{\frac{1}{p'}}G(a_k),\quad G(z)\in H^{p'}_{+w^{1-p'}},\quad k=1,2,\ldots.$$

By Lemma 3.1, $T_{p'}$ is bounded from $H^{p'}_{+w^{1-p'}}$ into $l^{p'}$. So,

$$\left(\sum_{k=1}^\infty V_k|G(a_k)|^{p'}\right)^{1/p'} < C_2\|G\|_{L^{p'}_{w^{1-p'}}(\mathbb{R})} < C_1C_2 = C, \tag{3.11}$$

where C is a constant independent of G and g.

Thus, by (3.4), (3.9), (3.10) and (3.11), and noting that $S_n(z) \in \Phi_{p,w} \subset H^p_{+w}$, it follows that the series in (3.5) converges in H^p_{+w} to a function $-I(z) \in \Phi_{p,w}$, hence $I(z) \in \Phi_{p,w} \subset H^p_{+w}$. Also, by Lemma 1.5, it is convergent on each compact subset of the upper half-plane, so is pointwise convergent in $\operatorname{Im}(z) > 0$. Noting (3.8), it is easy to see that

$$I(a_k) = \sum_{j=1}^{\infty}(-b_j)\cdot\phi_j(a_k) = b_k, \quad k = 1, 2, \ldots.$$

The proof is complete. □

Note that when $w \equiv 1$, this is Theorem 5.1 in [**5**].

As an application of this result, based on Theorem 2.3, under the Carleson condition, we can obtain an expansion with respect to the system $\{\phi_k(z)\}$ for any $f \in H^p_{+w}$. We first give an inequality:

LEMMA 3.3. *If $w \in (A^p)$ with $1 < p < \infty$ and $a_k = x_k + iy_k$ with $y_k > 0$, then*

$$y_k^p V_k^{-\frac{p}{p'}} \le W_k, \tag{3.12}$$

i.e.

$$y_k^p\Big(\int_{x_k}^{x_k+y_k}[w(x)]^{1-p'}dx\Big)^{-\frac{p}{p'}} \le \int_{x_k}^{x_k+y_k} w(x)dx. \tag{3.13}$$

where p' denote the conjugate exponent of p.

PROOF. Apply Hölder's inequality to

$$\int_{x_k}^{x_k+y_k}[w(x)]^{\frac{1}{p}}[w(x)]^{-\frac{1}{p}}dx,$$

and note that $-p'/p = 1 - p'$. □

THEOREM 3.4. *Let $w \in (A^q)$ and assume that $q_w < p < \infty$ and $f(z) \in H^p_{+w}$. If $\{a_k\}$ satisfies the Carleson condition (3.1), then*

$$f(z) = \sum_{k=1}^{\infty}\alpha(f, e_k)\phi_k(z) + H_f(z), \quad \operatorname{Im}(z) > 0, \tag{3.14}$$

where $\alpha(f, e_k) = -f(a_k)$ $(k = 1, 2, \ldots)$, and $H_f(z)$ is as in Theorem 2.3, and the convergence of the series is as in Theorem 2.3.

PROOF. By Lemma 3.1, we have

$$\left(\sum_{k=1}^{\infty} W_k|f(a_k)|^p\right)^{1/p} \le C\|f\|_{H^p_{+w}},$$

where C is a constant independent of f, and by Lemma 3.3, it follows that

$$\left(\sum_{k=1}^{\infty}|f(a_k)|^p y_k^p V_k^{-\frac{p}{p'}}\right)^{1/p} \le \left(\sum_{k=1}^{\infty}|f(a_k)|^p W_k\right)^{1/p} \le C\|f\|_{H^p_{+w}},$$

where p' satisfies $\frac{1}{p} + \frac{1}{p'} = 1$. Thus, by Theorem 3.2, $I(z) = \sum_{k=1}^{\infty}[-f(a_k)]\cdot \phi_k(z) \in \Phi_{p,w} \subset H^p_{+w}$ with $I(a_k) = f(a_k)$ $(k = 1, 2, \ldots)$. Since under Carleson's condition (3.1), the sequence $\{a_k\}$ also satisfies the Blaschke condition (2.1), then by Theorem 2.3, $f(z) - H_f(z) \in \Phi_{p,w}$. Now consider the function $g(z) = f(z) -$

$H_f(z) - I(z)$. Clearly, $g(z) \in \Phi_{p,w}$ and $g(a_k) = f(a_k) - H_f(a_k) - I(a_k) = 0$ $(k = 1, 2, \ldots)$. Hence, by Corollary 2.7, $g(z) \equiv 0$, getting $f(z) = I(z) + H_f(z)$ for $\operatorname{Im}(z) > 0$. That is (3.14) and the theorem is proved. □

THEOREM 3.5. *Under the assumptions of Theorem 3.2, if $\{a_k\}$ satisfies Carleson's condition (3.1) and $\{b_k\}$ satisfies (3.4), then all the solutions of the interpolation problem (3.3) can be expressed as*

$$f(z) = f_1(z) + f_2(z), \qquad \operatorname{Im}(z) > 0,$$

where

$$f_1(z) = \sum_{k=1}^{\infty} (-b_k)\phi_k(z) \in \Phi_{p,w},$$

and $f_2(z) = B(z)g(z)$ with $g(z) \in H^p_{+w}$.

PROOF. If $f(z) \in H^p_{+w}$ is a solution of the interpolation problem, then $f(a_k) = -\alpha(f, e_k) = b_k$ $(k = 1, 2, \ldots)$. Since $\{a_k\}$ satisfies the Carleson condition, by Theorem 3.4,

$$f(z) = \sum_{k=1}^{\infty} \alpha(f, e_k)\phi_k(z) + H_f(z), \qquad \operatorname{Im}(z) > 0,$$

hence

$$f(z) = \sum_{k=1}^{\infty} (-b_k)\phi_k(z) + H_f(z) = f_1(z) + f_2(z), \qquad \operatorname{Im}(z) > 0,$$

where

$$f_1(z) = \sum_{k=1}^{\infty} (-b_k)\phi_k(z) \in \Phi_{p,w},$$

and

$$f_2(z) = H_f(z) = B(z)g(z)$$

with

$$g(z) = \frac{1}{2\pi i} \int_{-\infty}^{+\infty} \frac{f(x)}{B(x)(x - z)}\, dx, \qquad \operatorname{Im}(z) > 0.$$

By Lemma 1.3, $g(z) \in H^p_{+w}$.

Conversely, if $f(z) = f_1(z) + f_2(z)$ where

$$f_1(z) = \sum_{k=1}^{\infty} (-b_k)\phi_k(z),$$

and $f_2(z) = B(z)g(z)$ with $g(z) \in H^p_{+w}$, then clearly, $f(a_k) = b_k$ $(k = 1, 2, \ldots)$. The proof is complete. □

Note that if we restrict the solution of the interpolation problem (3.3) to the subspace $\Phi_{p,w}$, then the solution is unique. Indeed, if $I_1(z), I_2(z) \in \Phi_{p,w}$ and $I_1(a_k) = f(a_k) = I_2(a_k)$, then $I(z) = I_1(z) - I_2(z) \in \Phi_{p,w}$ and $I(a_k) = 0$ $(k = 1, 2, \ldots)$, hence $I(z) \equiv 0$ by Corollary 2.7.

THEOREM 3.6. *Under the assumptions of Theorem 3.2, if $\{a_k\}$ satisfies the Carleson's condition (3.1), then for any $f(z) \in \Phi_{p,w}$,*

$$\|f(z)\|_{H^p_{+w}} \le C \left(\sum_{k=1}^{\infty} |f(a_k)|^p \left(\operatorname{Im}(a_k)\right)^p V_k^{-\frac{p}{p'}} \right)^{1/p}. \tag{3.15}$$

where C is a constant independent of f.

PROOF. By Lemmas 3.1 and 3.3, the linear operator

$$T \,:\, f \mapsto \{f(a_k)\mathrm{Im}(a_k)V_k^{-1/p'}\} \tag{3.16}$$

is bounded from $\Phi_{p,w}$ onto l^p. It is onto since for any $\{x_k\} \in l^p$, letting

$$b_k = \frac{x_k V_k^{1/p'}}{\mathrm{Im}(a_k)}, \qquad k = 1, 2, \ldots,$$

then

$$\sum_{k=1}^{\infty} |b_k|^p (\mathrm{Im}(a_k))^p V_k^{-\frac{p}{p'}} = \sum_{k=1}^{\infty} |x_k|^p < \infty,$$

and thus there is an $f(z) \in \Phi_{p,w}$ such that $f(a_k) = b_k$, i.e. $T(f) = \{x_k\}$.

By the uniqueness of the solution of the interpolation problem in $\Phi_{p,w}$, we have *ker* $T = \{0\}$. Thus by Banach's theorem on inverse operators or the open-mapping theorem (see for example, [**3**, p. 80]), the inverse T^{-1} exists, and is continuous, hence bounded, and the inequality (3.15) follows. □

4. Moment and expansion

In this section, under Carleson's condition, we give another expansion for $f(z) \in H^p_{+w}$. First, let us consider a moment problem.

THEOREM 4.1. *Assume that $a_k = x_k + iy_k$ with $y_k > 0$ $(k = 1, 2, \ldots)$. Let $w \in (A^q)$ with $q_w < p < \infty$, and let p' denote the conjugate exponent of p, that is $\frac{1}{p} + \frac{1}{p'} = 1$, and set*

$$V_k = \int_{x_k}^{x_k+y_k} [w(x)]^{1-p'}\,dx, \qquad k = 1, 2, \ldots.$$

If $\{a_k\}$ satisfies Carleson's condition (3.1) and $\{b_k\}$ satisfies

$$\sum_{k=1}^{\infty} |b_k|^p V_k^{-\frac{p}{p'}} < +\infty, \tag{4.1}$$

then

$$h(z) = \sum_{k=1}^{\infty} b_k e_k(z) \in E_{p,w}, \qquad \mathrm{Im}(z) > 0, \tag{4.2}$$

and

$$\alpha(h, \phi_k) = b_k, \qquad k = 1, 2, \ldots. \tag{4.3}$$

where the series is both H^p_{+w} convergent and uniformly convergent on each compact subset in the upper half-plane $\mathrm{Im}(z) > 0$.

PROOF. We need to prove that the series (4.2) converges in H^p_{+w}. For any positive integer n, let $S_n(z) = \sum_{k=1}^{n} b_k e_k(z)$. By the Hahn-Banach theorem and

Lemmas 1.4 and 1.2, we have

$$\begin{aligned} \|S_n(x)\|_{L^p_w(\mathbb{R})} &= \left\| \sum_{k=1}^n b_k e_k(x) \right\|_{L^p_w(\mathbb{R})} \\ &\leq C_0 \sup_{\substack{g(x)\in L^{p'}_{w^{1-p'}}(\mathbb{R}) \\ \|g(x)\|_{L^{p'}_{w^{1-p'}}(\mathbb{R})}\leq 1}} \left| \int_{-\infty}^{+\infty} \left(\sum_{k=1}^n b_k e_k(x) \right) \overline{g(x)} dx \right| \\ &= C_0 \sup_{\substack{g(x)\in L^{p'}_{w^{1-p'}}(\mathbb{R}) \\ \|g(x)\|_{L^{p'}_{w^{1-p'}}(\mathbb{R})}\leq 1}} \left| \sum_{k=1}^n \overline{b_k} \left(\int_{-\infty}^{+\infty} g(x)\overline{e_k(x)} dx \right) \right) \end{aligned}$$

where C_0 is a constant depending only on p. We note that for the above $g(x) \in L^{p'}_{w^{1-p'}}(\mathbb{R})$ with $\|g(x)\|_{L^{p'}_{w^{1-p'}}(\mathbb{R})} \leq 1$,

$$\int_{-\infty}^{\infty} g(x)\overline{e_k(x)} dx = -\frac{1}{2\pi i} \int_{-\infty}^{+\infty} \frac{g(x)}{x - a_k} dx = G(a_k).$$

By Lemma 1.3,

$$G(z) := -\frac{1}{2\pi i} \int_{-\infty}^{+\infty} \frac{g(x)}{x - z} dx \in H^{p'}_{+w^{1-p'}},$$

and

$$\|G(x)\|_{L^{p'}_{w^{1-p'}}(\mathbb{R})} \leq C_1 \|g(x)\|_{L^{p'}_{w^{1-p'}}(\mathbb{R})} \leq C_1,$$

where C_1 is a constant independent of g. Thus we have

$$\left\| \sum_{k=1}^n b_k e_k(x) \right\|_{L^p_w(\mathbb{R})} \leq C_0 \sup_{\substack{G(z)\in H^{p'}_{+w^{1-p'}} \\ \|G(x)\|_{L^{p'}_{w^{1-p'}}(\mathbb{R})}\leq C_1}} \sum_{k=1}^n \left| \overline{b_k} G(a_k) \right| . \tag{4.4}$$

By Hölder's inequality,

$$\sum_{k=1}^n \left| \overline{b_k} G(a_k) \right| \leq \left(\sum_{k=1}^n V_k |G(a_k)|^{p'} \right)^{1/p'} \left(\sum_{k=1}^n |b_k|^p V_k^{-\frac{p}{p'}} \right)^{1/p} . \tag{4.5}$$

Consider the operator $T_{p'}$ which maps $H^{p'}_{+w^{1-p'}}$ into a sequence space defined by

$$(T_{p'}G)_k = V_k^{\frac{1}{p'}} G(a_k), \quad G(z) \in H^{p'}_{+w^{1-p'}}, \quad k = 1, 2, \ldots .$$

By Lemma 3.1, $T_{p'}$ is bounded from $H^{p'}_{+w^{1-p'}}$ into $l^{p'}$. So,

$$\left(\sum_{k=1}^\infty V_k |G(a_k)|^{p'} \right)^{1/p'} < C_2 \|G\|_{L^{p'}_{w^{1-p'}}(\mathbb{R})} < C_1 C_2 = C, \tag{4.6}$$

where C is a constant independent of G and g.

Thus, by (4.1), (4.4), (4.5) and (4.6), and noting that $S_n(z) \in E_{p,w} \subset H^p_{+w}$, it follows that series (4.2) converges in H^p_{+w} to a function $h(z) \in E_{p,w} \subset H^p_{+w}$.

Then, since $\phi_k(x) \in L^{p'}_{w^{1-p'}}(\mathbb{R})$, and $h \to \alpha(h, \phi_k)$ is a continuous linear functional on $L^p_w(\mathbb{R})$, the biorthogonality of these two systems produces (4.3):

$$\alpha(h, \phi_k) = \alpha\Big(\sum_{j=1}^{\infty} b_j e_j, \phi_k\Big) = \sum_{j=1}^{\infty} b_j \alpha(e_j, \phi_k) = b_k, \qquad k = 1, 2, \ldots.$$

The proof is complete. □

As an application of Theorem 4.1, based on Theorem 2.4, under the Carleson condition, we can obtain an expansion with respect to the system $\{e_k(z)\}$ for any $f \in H^p_{+w}$.

THEOREM 4.2. *Let $w \in (A^q)$ with $q_w < p < \infty$, and assume that $f(z) \in H^p_{+w}$ and $\{a_k\}$ satisfies Carleson's condition (3.1). Then*

$$f(z) = \sum_{k=1}^{\infty} \alpha(f, \phi_k) e_k(z) + H_f(z), \qquad \mathrm{Im}(z) > 0, \tag{4.7}$$

where

$$\alpha(f, \phi_k) = \int_{-\infty}^{\infty} f(x)\overline{\phi_k(x)}dx \quad (k = 1, 2, \ldots),$$

$H_f(z)$ is as in Theorem 2.3, and the convergence of the series is as in Theorem 2.4.

PROOF. First, we claim that

$$\sum_{k=1}^{\infty} |\alpha(f, \phi_k)|^p V_k^{-\frac{p}{p'}} < \infty, \tag{4.8}$$

where V_k is defined in Theorem 3.2. Note that, by (3.7),

$$\frac{1}{|B'(a_k)|} \le \frac{2y_k}{\delta}, \quad k = 1, 2, \ldots$$

where $a_k = x_k + iy_k$ $(y_k > 0)$, and $\delta > 0$ is the constant in the Carleson condition. We have

$$|\alpha(f, \phi_k)| = |\overline{\alpha(f, \phi_k)}| \le \frac{4\pi y_k}{\delta} \left| \frac{1}{2\pi i} \int_{-\infty}^{\infty} \frac{\overline{f(x)}B(x)}{x - a_k} dx \right|.$$

Since $\overline{f(x)}B(x) \in L^p_w(\mathbb{R})$, by Lemma 1.3, $\overline{f(x)}B(x) = F^+(x) - F^-(x)$ a.e. on $\mathbb{R}$, where

$$\frac{1}{2\pi i} \int_{-\infty}^{\infty} \frac{\overline{f(x)}B(x)}{x - z} dx = \begin{cases} F^+(z) \in H^p_{+w}, & \mathrm{Im}(z) > 0; \\ F^-(z) \in H^p_{-w}, & \mathrm{Im}(z) < 0, \end{cases}$$

and

$$\|F^{\pm}\|_{H^p_{\pm w}} \le C\|\overline{f}B\|_{L^p_w(\mathbb{R})} = C\|f\|_{L^p_w(\mathbb{R})}.$$

By Lemma 1.3 (ii),

$$\frac{1}{2\pi i} \int_{-\infty}^{\infty} \frac{F^-(x)}{x - a_k} dx = 0.$$

Thus,

$$|\alpha(f, \phi_k)| \le \frac{4\pi y_k}{\delta} \left| \frac{1}{2\pi i} \int_{-\infty}^{\infty} \frac{F^+(x) - F^-(x)}{x - a_k} dx \right| = \frac{4\pi y_k}{\delta} |F^+(a_k)|,$$

and

$$|\alpha(f, \phi_k)|^p \le C_1 y_k^p |F^+(a_k)|^p,$$

where $C_1 = \left(\frac{4\pi}{\delta}\right)^p$. By Lemma 3.3, and Lemma 3.1,

$$\begin{aligned}\left(\sum_{k=1}^{\infty} y_k^p |F^+(a_k)|^p V_k^{-\frac{p}{p'}}\right)^{1/p} &\leq \left(\sum_{k=1}^{\infty} |F^+(a_k)|^p W_k\right)^{1/p} \\ &\leq C_1 \|F^+\|_{H^p_{+w}} \leq C_2 \|f\|_{L^p_w(\mathbb{R})} < \infty,\end{aligned}$$

as claimed in (4.8). Thus, by Theorem 4.1,

$$h(z) = \sum_{k=1}^{\infty} \alpha(f, \phi_k) e_k(z) \in E_{p,w}, \quad \mathrm{Im}(z) > 0,$$

and

$$\alpha(h, \phi_k) = \alpha(f, \phi_k), \quad k = 1, 2, \ldots.$$

By Theorem 2.4, $f(z) - H_f(z) \in E_{p,w}$. Let

$$g(z) = f(z) - H_f(z) - h(z).$$

We have $g(z) \in E_{p,w} = \Phi_{p,w}$, and

$$\alpha(g, \phi_k) = \alpha(f, \phi_k) - \alpha(H_f, \phi_k) - \alpha(h, \phi_k), \quad k = 1, 2, \ldots.$$

Since $\alpha(H_f, \phi_k) = 0$ $(k = 1, 2, \ldots)$ by Corollary 2.6, we have $\alpha(g, e_k) = 0$, and hence $g(a_k) = 0$ $(k = 1, 2, \ldots)$, and thus, by Corollary 2.7, we get $g(z) \equiv 0$ for $\mathrm{Im}(z) > 0$. This gives (4.7). □

References

[1] André Boivin and Changzhong Zhu, *Bi-orthogonal expansions in the space* $L^2(0, \infty)$, Blaschke products and their applications, Fields Inst. Commun., vol. 65, Springer, New York, 2013, pp. 99–112, DOI 10.1007/978-1-4614-5341-3_4. MR3052288

[2] André Boivin, Paul M. Gauthier, and Changzhong Zhu, *Weighted Hardy spaces for the unit disc: approximation properties*, Complex and harmonic analysis, DEStech Publ., Inc., Lancaster, PA, 2007, pp. 129–155. MR2387285 (2009d:30086)

[3] Béla Bollobás, *Linear analysis*, Cambridge Mathematical Textbooks, Cambridge University Press, Cambridge, 1990. An introductory course. MR1087297 (92a:46001)

[4] Peter L. Duren, *Theory of* H^p *spaces*, Pure and Applied Mathematics, Vol. 38, Academic Press, New York-London, 1970. MR0268655 (42 #3552)

[5] M. M. Džrbašjan, *The basis property of some biorthogonal systems, and the solution of a multiple interpolation problem in* H^p *classes in a half plane* (Russian), Izv. Akad. Nauk SSSR Ser. Mat. **42** (1978), no. 6, 1322–1384, 1439–1440. MR522941 (81d:30008)

[6] José García-Cuerva, *Weighted* H^p *spaces*, Dissertationes Math. (Rozprawy Mat.) **162** (1979), 63. MR549091 (82a:42018)

[7] John B. Garnett, *Bounded analytic functions*, Pure and Applied Mathematics, vol. 96, Academic Press, Inc. [Harcourt Brace Jovanovich, Publishers], New York-London, 1981. MR628971 (83g:30037)

[8] Richard Hunt, Benjamin Muckenhoupt, and Richard Wheeden, *Weighted norm inequalities for the conjugate function and Hilbert transform*, Trans. Amer. Math. Soc. **176** (1973), 227–251. MR0312139 (47 #701)

[9] M. S. Martirosyan, *On a representation in an incomplete system of rational functions* (Russian, with English and Russian summaries), Izv. Nats. Akad. Nauk Armenii Mat. **32** (1997), no. 6, 30–38; English transl., J. Contemp. Math. Anal. **32** (1997), no. 6, 26–34 (1998). MR1717670 (2000k:30057)

[10] Benjamin Muckenhoupt, *Weighted norm inequalities for the Hardy maximal function*, Trans. Amer. Math. Soc. **165** (1972), 207–226. MR0293384 (45 #2461)

[11] Kristian Seip, *Interpolation and sampling in spaces of analytic functions*, University Lecture Series, vol. 33, American Mathematical Society, Providence, RI, 2004. MR2040080 (2005c:30038)

[12] Xie Chang Shen and Ming Wen, *Biorthogonal bases and multiple interpolation in weighted Hardy spaces*, Approx. Theory Appl. **6** (1990), no. 1, 1–22. MR1070611 (92f:42025)
[13] Alberto Torchinsky, *Real-variable methods in harmonic analysis*, Pure and Applied Mathematics, vol. 123, Academic Press, Inc., Orlando, FL, 1986. MR869816 (88e:42001)

Department of Mathematics, University of Western Ontario, London, Ont., Canada N6A 5B7
E-mail address: boivin@uwo.ca

Department of Mathematics, University of Western Ontario, London, Ont., Canada N6A 5B7
E-mail address: changzhong_zhu@yahoo.com

Contemporary Mathematics
Volume **638**, 2015
http://dx.doi.org/10.1090/conm/638/12811

Model spaces: A survey

Stephan Ramon Garcia and William T. Ross

ABSTRACT. This is an introduction, aimed at graduate students, to the subject of model subspaces of the Hardy space.

1. Introduction

In the summer of 2013, the authors gave a series of lectures and minicourses in Montréal, Lens, and Helsinki on the topic of model spaces. In preparing for these lectures, we discovered the need for an easy introduction to model spaces suitable for the graduate students who formed the intended audience for our lectures. The standard texts on the subject [**14, 27, 69–71, 78**] are thorough and encyclopedic, but are sufficiently intimidating that the beginner might find this whole beautiful subject out of reach. The purpose of this survey is to give the novice a friendly, albeit incomplete, introduction to model spaces.

Model spaces are Hilbert spaces of the form $(uH^2)^\perp$, where u is an inner function, H^2 is the classical Hardy space on the open unit disk $\mathbb{D}$, and $\perp$ denotes the orthogonal complement in H^2. On a functional analysis level, model spaces are the orthogonal complements of the nontrivial invariant subspaces of the unilateral shift $Sf = zf$ on H^2. These subspaces were characterized as uH^2 by Beurling in his famous 1949 paper [**16**]. As such, the spaces $(uH^2)^\perp$ are the invariant subspaces of the backward shift operator $S^*f = (f - f(0))/z$ on H^2. However, unlike the spaces uH^2 which are simple to understand (i.e., all H^2 multiples of the inner function u), the model spaces $(uH^2)^\perp$ are much more troublesome. For instance, it is not immediately clear which functions actually belong $(uH^2)^\perp$ or what properties these functions have.

A major breakthrough in the study of model spaces occurred in 1970, with the publication of the seminal paper of Douglas, Shapiro, and Shields [**31**]. Extending some partial results of earlier authors, they showed that functions in $(uH^2)^\perp$ have analytic continuations in the same neighborhoods of points on the unit circle as does u. However, a generic inner function u need not have an analytic continuation across any point of the unit circle $\mathbb{T}$ and thus a new type of continuation was needed. Fortunately this type of continuation, called a *pseudocontinuation*, was discovered and formalized in two earlier papers [**87, 88**] of Shapiro and indeed turned

2010 *Mathematics Subject Classification.* Primary 47A15, 30D55.
First author partially supported by National Science Foundation Grant DMS-1265973.

out to be the determining characterization of functions in $(uH^2)^\perp$. In fact, a notion of continuation more general than analytic continuation was already being discussed, in various forms (rational approximation for example), often never actually being called a "continuation." For example an earlier paper of Tumarkin **[98]** discussed functions in $(uH^2)^\perp$ via controlled approximation by rational functions.

From these beginnings, the function theoretic aspects of model spaces have truly grown. We now understand much of the subtle relationship between the boundary behavior of functions in $(uH^2)^\perp$ and the angular derivative of u through the papers of Ahern and Clark **[3, 4]**.

On an operator theory level, work of Sz.-Nagy and Foias showed that the compression of the unilateral shift S to a model space $(uH^2)^\perp$ is a model for a certain class of contractions **[95]**. This seminal work has been studied by many people and continues to have relevance in operator theory **[77]**. Clark later examined unitary rank-one perturbations of this compressed shift and was able to come up with an exact spectral realization of this unitary operator using a family of measures that now bear his name **[29]**. These measures enjoy truly fascinating properties that have been harnessed by Aleksandrov **[6, 9]**, Poltoratski **[75]**, Sarason **[74, 84, 85]**, and others to explore fine boundary properties of the inner function u as well as completeness of families of reproducing kernels in $(uH^2)^\perp$. These measures have even appeared in a somewhat different form in mathematical physics **[92, 93]**.

Model spaces have also been helpful is examining nearly invariant subspaces (subspaces that contain f/z whenever they contain a function f satisfying $f(0) = 0$) **[82]**. These nearly invariant subspaces have been useful in developing characterizations of the invariant subspaces for the shift $f \mapsto zf$ on Hardy spaces of planar domains with holes or slits **[10, 11, 57]**.

Though model spaces have many connections to old ideas in analysis (analytic continuation, factorization, pseudocontinuation, etc.), they continue to be relevant. Recent work of Makarov and Poltoratski **[65]** show that a spectral realization of certain selfadjoint Schrödinger operators can be realized through model spaces and Clark measures. Model spaces also make important connections to the subject of complex symmetric operators **[40–44]**, a certain class of Hilbert space operators that are frequently modeled by compressions of Toeplitz operators to model spaces **[28, 51, 94]**.

As the subject of model spaces is quite vast, we again emphasize that these notes are not meant to be an encyclopedic treatment. These notes are instead meant to give the beginning student a reason to want to study this material and to provide the means for them to take their first few steps into this rich and fertile territory.

2. Preliminaries

Before proceeding, we start with a brief review of Hardy space theory. The material presented in this section is now considered classical and can be found in many standard texts **[33, 53, 58, 60]**. Relatively new texts that might be more suitable for a student who is new to Hardy spaces are **[67, 68]**.

2.1. Lebesgue spaces. Let m denote normalized Lebesgue measure on the unit circle $\mathbb{T}$ (i.e., $m = d\theta/2\pi$) and let $L^2 := L^2(\mathbb{T}, m)$ denote the space of m-measurable (i.e., Lebesgue measurable) functions $f : \mathbb{T} \to \mathbb{C}$ such that

$$\|f\| := \left(\int_{\mathbb{T}} |f(\zeta)|^2 dm(\zeta) \right)^{\frac{1}{2}}$$

is finite. As such, L^2 is a Hilbert space endowed with the inner product

$$\langle f, g \rangle := \int_{\mathbb{T}} f(\zeta)\overline{g(\zeta)}\, dm(\zeta).$$

A simple calculation using the fact that $m(\mathbb{T}) = 1$ shows that the family of functions $\{\zeta \mapsto \zeta^n : n \in \mathbb{Z}\}$ is an orthonormal basis for L^2. The coefficients

$$\widehat{f}(n) := \langle f, \zeta^n \rangle = \int_{\mathbb{T}} f(\zeta)\overline{\zeta}^n dm(\zeta)$$

of a function f in L^2 with respect to this basis are called the (complex) *Fourier coefficients* of f. In light of Parseval's Identity

$$\|f\|^2 = \sum_{n \in \mathbb{Z}} |\widehat{f}(n)|^2,$$

we see that the L^2 norm of f coincides with the norm of the sequence $\{\widehat{f}(n) : n \in \mathbb{Z}\}$ of Fourier coefficients in the space $\ell^2(\mathbb{Z})$ of all square-summable sequences on $\mathbb{Z}$. We therefore identify the Hilbert spaces L^2 and $\ell^2(\mathbb{Z})$ via $f \leftrightarrow \{\widehat{f}(n) : n \in \mathbb{Z}\}$.

We also require the space $L^\infty := L^\infty(\mathbb{T})$ of all essentially bounded functions on $\mathbb{T}$ which, when equipped with the norm

$$\|f\|_\infty := \text{ess-sup}_{\zeta \in \mathbb{T}} |f(\zeta)|,$$

becomes a Banach algebra. We also remark that for any φ in L^∞, the multiplication operator $f \mapsto \varphi f$ on L^2 is bounded and has operator norm equal to $\|\varphi\|_\infty$.

2.2. Hardy spaces. For an analytic function f on $\mathbb{D}$ the integral means

$$\int_{\mathbb{T}} |f(r\zeta)|^2\, dm(\zeta)$$

are increasing as a function of r on $(0, 1)$. Indeed, if $f(z) = \sum_{n=0}^{\infty} a_n z^n$, then

$$\int_{\mathbb{T}} |f(r\zeta)|^2 dm(\zeta) = \sum_{n=0}^{\infty} |a_n|^2 r^{2n},$$

which is clearly increasing in r. This leads us to define the *Hardy space* H^2 as those f for which

$$\|f\| := \lim_{r \to 1^-} \left(\int_{\mathbb{T}} |f(r\zeta)|^2\, dm(\zeta) \right)^{\frac{1}{2}} \tag{2.1}$$

is finite. It is no accident that we use $\|\cdot\|$ to denote both the norm in L^2 and in H^2. Indeed, classical work of Fatou and Riesz show that the *radial limit*[1]

$$f(\zeta) := \lim_{r \to 1^-} f(r\zeta) \tag{2.2}$$

[1]It turns out that each f in H^2 has a finite *non-tangential limit* at ζ for almost every ζ. By this we mean that the limit of $f(z)$ exists as z approaches ζ in every *Stolz region* $\{z \in \mathbb{D} : |z - \zeta| < \alpha(1 - |z|)\}$, $\alpha > 1$.

exists for m-almost every ζ in $\mathbb{T}$ and it turns out that the H^2 norm of f and the L^2 norm of its boundary function, defined in (2.2), coincide. In this manner, we often regard H^2 as a closed subspace of L^2. As such, H^2 inherits the structure of a Hilbert space from L^2.

Let H^∞ denote the space of all bounded analytic functions on $\mathbb{D}$, endowed with the supremum norm

$$\|f\|_\infty := \sup_{z\in\mathbb{D}} |f(z)|.$$

In a similar manner, the radial boundary function of an H^∞ function belongs to L^∞ and one has a corresponding maximum modulus type result

$$\sup_{z\in\mathbb{D}} |f(z)| = \text{ess-sup}_{\zeta\in\mathbb{T}} |f(\zeta)|,$$

that allows us to view H^∞ as a Banach subalgebra of L^∞. Moreover, H^∞ happens to be the multiplier algebra for H^2, meaning that the operator of multiplication by an analytic function φ on H^2 (i.e., $f \mapsto \varphi f$) is bounded if and only if φ belongs to H^∞. The norm of this multiplication operator is precisely $\|\varphi\|_\infty$.

The inner product on the Hardy space H^2 is given by

$$\langle f, g\rangle = \int_{\mathbb{T}} f(\zeta)\overline{g(\zeta)}\,dm(\zeta) = \sum_{n=0}^{\infty} a_n\overline{b_n},$$

where $f(z) = \sum_{n=0}^\infty a_n z^n$ and $g(z) = \sum_{n=0}^\infty b_n z^n$ denote typical elements of H^2. In other words, we have a natural identification of H^2 with the sequence space $\ell^2(\mathbb{N})$, where each f in H^2 is identified with its sequence of Taylor coefficients $\{a_n\}_{n\geqslant 0}$.

Of great importance is the manner in which H^2 sits inside of L^2. If $f(z) = \sum_{n=0}^\infty a_n z^n$ belongs to H^2, then the almost everywhere defined boundary function f has an associated Fourier series

$$f \sim \sum_{n=0}^{\infty} a_n\zeta^n$$

which belongs to the first component in the direct sum

$$L^2 = H^2 \oplus \overline{zH^2},$$

where

$$\overline{zH^2} = \{\overline{zh} : h \in H^2\}.$$

In terms of Fourier coefficients,

$$H^2 = \bigvee\{z^n : n \geqslant 0\}, \qquad \overline{zH^2} = \bigvee\{z^n : n \leqslant -1\},$$

where $\bigvee$ denotes the closed linear span in L^2. In particular, note that

$$a_n = \begin{cases} \widehat{f}(n) & \text{if } n \geqslant 0,\\ 0 & \text{if } n < 0,\end{cases}$$

and that the polynomials in z are dense in H^2. These results are summarized in the following diagram:

$$\begin{array}{ccc} H^2 & \xleftrightarrow{\text{identified}} & \ell^2(\mathbb{N}) \\ \cap & & \cap \\ L^2 & \xleftrightarrow{\text{identified}} & \ell^2(\mathbb{Z}) \end{array}$$

2.3. The Cauchy-Szegő kernel. In light of the inequality

$$|f(\lambda)| \leqslant \sum_{n=0}^{\infty} |a_n||\lambda|^n \leqslant \left(\sum_{n=0}^{\infty} |a_n|^2\right)^{\frac{1}{2}} \left(\sum_{n=0}^{\infty} |\lambda|^{2n}\right)^{\frac{1}{2}} = \frac{\|f\|}{\sqrt{1-|\lambda|^2}}, \tag{2.3}$$

which holds for all λ in $\mathbb{D}$ and all f in H^2, it follows that for fixed $\lambda \in \mathbb{D}$ the point evaluation functionals $f \mapsto f(\lambda)$ are bounded on H^2 and hence, by the Riesz Representation Theorem, must be of the form

$$f(\lambda) = \langle f, c_\lambda \rangle \tag{2.4}$$

for some c_λ in H^2. In fact, it is not hard to show that

$$c_\lambda(z) = \frac{1}{1-\overline{\lambda}z}, \quad \lambda \in \mathbb{D}, \tag{2.5}$$

which is called the *Cauchy-Szegő kernel* or, perhaps more frequently, the *Cauchy kernel*. In more general terms, one says that the Cauchy-Szegő kernel is the *reproducing kernel* for H^2. Of great interest to us are other reproducing kernels, which satisfy equations analogous to (2.4) on various subspaces of H^2 (see Section 5).

Before proceeding, we should also remark that the reproducing formula (2.4) is simply a restatement of the identity

$$f(\lambda) = \sum_{n=0}^{\infty} a_n \lambda^n = \big\langle (a_0, a_1, a_2, \ldots), (1, \overline{\lambda}, \overline{\lambda}^2, \ldots) \big\rangle_{\ell^2(\mathbb{N})}.$$

When written as a contour integral, (2.4) reduces to the Cauchy Integral Formula

$$f(\lambda) = \int_{\mathbb{T}} \frac{f(\zeta)}{1-\lambda\overline{\zeta}}\, dm(\zeta)$$

for H^2 functions (recall from (2.2) that for each f in H^2, the almost everywhere defined boundary function $\zeta \mapsto f(\zeta)$ belongs to L^2, allowing the preceding integral to be well-defined). Along similar lines, for fixed $n \geqslant 0$ we have

$$f^{(n)}(\lambda) = \langle f, c_\lambda^{(n)} \rangle,$$

where

$$c_\lambda^{(n)}(z) = \frac{n!\overline{\lambda}^n}{(1-\overline{\lambda}z)^{n+1}}$$

is the nth derivative of c_λ with respect to the variable z.

PROPOSITION 2.6. *The set* $\{c_\lambda : \lambda \in \mathbb{D}\}$ *is linearly independent.*

PROOF. If $\lambda_1, \lambda_2, \ldots, \lambda_n$ are distinct elements of $\mathbb{D}$ and

$$\sum_{i=1}^{n} \alpha_i c_{\lambda_i} = 0,$$

then

$$\sum_{i=1}^{n} \overline{\alpha_i} f(\lambda_i) = 0$$

holds for all f in H^2. The Lagrange Interpolation Theorem provides us with a polynomial $p(z)$ satisfying $p(\lambda_i) = \alpha_i$ so that $\sum_{i=1}^{n} |\alpha_i|^2 = 0$. Thus $\alpha_i = 0$ for $i = 1, 2, \ldots, n$. □

2.4. Canonical factorization. Although H^2 is a linear space, it is its multiplicative structure that reveals its true function-theoretic depth. We recall here the main ingredients necessary to describe the canonical factorization of H^2 functions.

DEFINITION 2.7. An *inner function* is a bounded analytic function u on $\mathbb{D}$ such that $|u(\zeta)| = 1$ for almost every ζ in $\mathbb{T}$.[2]

The simplest nontrivial example of an inner function is a Möbius transformation of the form

$$e^{i\theta}\frac{w-z}{1-\overline{w}z},$$

where $|w| < 1$ and $0 \leqslant \theta < 2\pi$, which is easily seen to be an automorphism of $\mathbb{D}$ mapping $\mathbb{T}$ onto $\mathbb{T}$. More generally, if $\{z_n\}_{n\geqslant 1}$ is a sequence of points in $\mathbb{D}\backslash\{0\}$, repeated according to multiplicity, then the *Blaschke condition*

$$\sum_{n=1}^{\infty}(1-|z_n|) < \infty \tag{2.8}$$

is necessary and sufficient for the convergence (uniformly on compact subsets of $\mathbb{D}$) of the corresponding *Blaschke product*

$$B(z) := z^m \prod_{n=1}^{\infty} \frac{|z_n|}{z_n}\frac{z_n - z}{1-\overline{z_n}z}, \tag{2.9}$$

where m denotes a nonnegative integer. With some work, one can show that every Blaschke product is an inner function [**33**, Thm. 2.4] (the only thing left to check is that the boundary function is unimodular almost everywhere). The importance of these functions stems from the fact that the Blaschke condition (2.8) completely characterizes the zero sets for H^2 functions.

THEOREM 2.10. *A sequence $\{z_n\}_{n\geqslant 1} \subset \mathbb{D}$, repeated according to multiplicity, is the zero set of a nonconstant H^2 function if and only if it satisfies the Blaschke condition* (2.8).

Other examples of inner functions are furnished by the following construction. For a positive, finite, singular (with respect to m) Borel measure μ on $\mathbb{T}$, we claim that the analytic function

$$S_\mu(z) := \exp\left(-\int \frac{\zeta+z}{\zeta-z}d\mu(\zeta)\right), \quad z \in \mathbb{D}, \tag{2.11}$$

is inner. Such a function is known as a *singular inner function*. First notice that for any $z \in \mathbb{D}$,

$$\begin{aligned} |S_\mu(z)| &= \exp\left[\Re\left(-\int \frac{\zeta+z}{\zeta-z}d\mu(\zeta)\right)\right] \\ &= \exp\left(-\int_{\mathbb{T}} \frac{1-|z|^2}{|\zeta-z|^2}d\mu(\zeta)\right) \\ &\leqslant 1 \end{aligned}$$

[2]The reader is reminded that whenever we use the term "boundary function" or write $f(\zeta)$ for $f \in H^2$ and $\zeta \in \mathbb{T}$, we are referring to the almost everywhere defined radial (non-tangential) limit in (2.2).

since both the measure μ and the *Poisson kernel*

$$(\zeta \in \mathbb{T}, z \in \mathbb{D}) \qquad P_z(\zeta) := \frac{1-|z|^2}{|\zeta - z|^2},$$

are nonnegative. Since S_μ belongs to H^∞, it follows that S_μ has nontangential boundary values m-a.e. on $\mathbb{T}$. To show that these boundary values are almost everywhere unimodular, we require some basic facts from the theory of harmonic functions. Let

$$(w \in \mathbb{T}) \qquad (D\mu)(w) := \lim_{t \to 0^+} \frac{\mu\big((e^{-it}w, e^{it}w)\big)}{2t}$$

denote the *symmetric derivative* of μ on $\mathbb{T}$, where $(e^{-it}w, e^{it}w)$ denotes the circular arc subtended by the points $e^{-it}w$ and $e^{it}w$. We also have the identity

$$\lim_{r \to 1^-} \int_{\mathbb{T}} P_{rw}(\zeta)\,d\mu(\zeta) = (D\mu)(w). \qquad m\text{-a.e. } w \in \mathbb{T}. \tag{2.12}$$

Since μ is singular, it follows that $D\mu = 0$ holds m-almost everywhere and so it now follows from the identity

$$|S_\mu(z)| = \exp\left(-\int_{\mathbb{T}} P_z(\zeta)\,d\mu(\zeta)\right)$$

that S has unimodular boundary values m-almost everywhere. For instance, if $\mu = \delta_1$ denotes the point mass at $\zeta = 1$, then

$$S_{\delta_1}(z) = \exp\left(\frac{z+1}{z-1}\right).$$

This type of inner function is often called an *atomic inner function.*

THEOREM 2.13. *Every inner function u can be factored uniquely as*

$$u = e^{i\gamma} B_\Lambda S_\mu,$$

where $\gamma \in [0, 2\pi)$, Λ is a Blaschke sequence, and μ is a positive singular measure on $\mathbb{T}$. Conversely, any such product is inner.

The reader is warned that, as an abuse of language, one frequently permits a function of the form $e^{i\gamma} B_\Lambda$ (resp. $e^{i\gamma} S_\mu$) to be called a Blaschke product (resp. a singular inner function). For the sake of convenience, we adopt this common practice here in these notes.

DEFINITION 2.14. Let u_1, u_2 be inner functions and $f \in H^2$.

(i) We say that u_1 *divides* u_2, written $u_1 | u_2$, if $u_2/u_1 \in H^\infty$.

(ii) We say that u_1 *divides* f, written $u_1 | f$, if $f/u_1 \in H^2$.

(iii) We say that u_1 and u_2 are *relatively prime* if the only inner divisors of both u_1 and u_2 are constant functions of unit modulus.

For example, if u_1 and u_2 are Blaschke products with simple zeros, then u_1 divides u_2 if and only if the zero set of u_1 is contained in the zero set of u_2. Moreover, u_1 is relatively prime to u_2 if and only if u_1 and u_2 have no common zeros.

DEFINITION 2.15. An *outer function* is an analytic function F on $\mathbb{D}$ of the form

$$F(z) = e^{i\gamma} \exp\left(\int_{\mathbb{T}} \frac{\zeta + z}{\zeta - z} \varphi(\zeta)\,dm(\zeta)\right), \tag{2.16}$$

where γ is a real constant and φ is a real-valued function in L^1.

The significance of the somewhat unwieldy formula in (2.16) lies in the fact that

$$\log|F(z)| = \int_{\mathbb{T}} P_z(\zeta)\varphi(\zeta)dm(\zeta),$$

from which it follows that $\log|F(z)|$ equals the harmonic extension to $\mathbb{D}$ of the boundary function $\varphi : \mathbb{T} \to \mathbb{R}$ so that $\varphi = \log|F|$ a.e. on $\mathbb{T}$. In particular, an outer function is determined up to a unimodular constant factor by its modulus on $\mathbb{T}$. On the other hand, it is possible to show that every f in H^2 that does not vanish identically satisfies

$$\int_{\mathbb{T}} \log|f(\zeta)|\, dm(\zeta) > -\infty$$

and hence one can form the outer function (2.16) corresponding to the boundary data $\varphi = \log|f|$ [**33**, Thm. 2.2]. Putting this all together, we can now state the canonical factorization for H^2 functions [**33**, Thm. 2.8].

THEOREM 2.17. *Every function f in $H^2\backslash\{0\}$ has a unique factorization of the form*

$$f = BSF \tag{2.18}$$

where B is a Blaschke product, S is a singular inner function, and F is an outer function in H^2. Conversely, any product of the form (2.18) *belongs to H^2.*

2.5. Bounded type. In order to adequately discuss the cyclic vectors for the backward shift operator, we need two more additional classes of meromorphic functions on $\mathbb{D}$.

DEFINITION 2.19. Let f be a meromorphic function on $\mathbb{D}$.

(i) We say f is of *bounded type* if f can be written as quotient of two bounded analytic functions. The set of functions of bounded type is denoted by N (often called the Nevanlinna class).
(ii) We say f is in the *Smirnov class* if is the quotient of two bounded analytic functions where the denominator is an outer function. The Smirnov class is denoted by N^+. Notice in this case that N^+ is a space of analytic functions on $\mathbb{D}$ since the denominator is outer and outer functions have no zeros on $\mathbb{D}$.

REMARK 2.20. It is known that $H^2(\mathbb{D}) \subset N^+$ and that every $f \in N$ has finite non-tangential limits almost everywhere [**33**, Ch. 2]. Furthermore, if $f \in N^+$ and the boundary function belongs to L^2, then $f \in H^2$. This fact is no longer true for $f \in N$.[3]

2.6. Beurling's Theorem. Of supreme importance in the world of operator-related function theory are the shift operators. Chief among these is the *unilateral shift* $S : H^2 \to H^2$ defined by

$$[Sf](z) = zf(z),$$

[3]Even if f is analytic on $\mathbb{D}$, this is no longer true. Just consider the function $f(z) = \exp(-\frac{z+1}{z-1})$, which is the reciprocal of the atomic inner function defined earlier. This function belongs to N, is analytic on $\mathbb{D}$, has unimodular boundary values, but does not belong to H^2 since it does not satisfy the growth estimate (2.3).

or, in terms of Taylor coefficients, by

$$S(a_0, a_1, \ldots) = (0, a_0, a_1, \ldots).$$

Because of its ubiquity in the realm of operator theory, one often refers to S as *the* shift operator. One easily sees that S is an isometry that is not unitary. The adjoint S^* of the unilateral shift is the *backward shift* $S^* : H^2 \to H^2$ given by

$$[S^* f](z) = \frac{f(z) - f(0)}{z},$$

or, in terms of Taylor coefficients, by

$$S^*(a_0, a_1, \ldots) = (a_1, a_2, \ldots).$$

If u is inner, then the operator $f \mapsto uf$ is an isometry on H^2 and thus uH^2 is a subspace (i.e., a closed linear manifold) of H^2. Moreover, assuming that u is a nonconstant inner function, uH^2 is a nontrivial invariant subspace for the operator S. A celebrated theorem of Beurling **[16]** (also see any of the standard texts mentioned above for a complete proof and further discussions and generalizations) says that these are all of them.

THEOREM 2.21 (Beurling's Theorem). *The nontrivial invariant subspaces of H^2 for the unilateral shift S are precisely the subspaces*

$$uH^2 := \{uh : h \in H^2\},$$

where u is an inner function. Moreover, f is cyclic for S, i.e.,

$$\bigvee\{f, Sf, S^2 f, \ldots\} = \bigvee\{qf : q \textit{ is a polynomial}\} = H^2,$$

if and only if f is an outer function.

From an operator theoretic perspective, Beurling's Theorem is notable for providing an explicit description of the lattice of invariant subspaces for the unilateral shift operator. Indeed,

$$u_1 H^2 \subseteq u_2 H^2 \iff u_1/u_2 \in H^\infty. \tag{2.22}$$

For our purposes, however, it is the invariant subspaces for the backward shift that are of greatest importance. These are the so-called *model spaces* that are the primary focus of our investigations.

3. Model spaces

The details of much of the following material can be found in the original sources, which we attempt to quote whenever possible, as well as the texts **[26, 27, 69, 70, 77, 78]**. On the other hand, many of the results discussed below are part of the folklore on the subject and occasionally proper references cannot be readily identified.

3.1. Basic properties. We are now ready to introduce our primary object of study.

DEFINITION 3.1. If u is an inner function, then the corresponding *model space* is

$$\mathcal{K}_u := (uH^2)^\perp = H^2 \ominus uH^2. \tag{3.2}$$

The definition above is somewhat unenlightening since the function theoretic properties of the model space $\mathcal{K}_u$ do not immediately present themselves. A more direct description of $\mathcal{K}_u$ via the boundary values of these functions comes from the following result.

PROPOSITION 3.3. *For inner u, the model space $\mathcal{K}_u$ is the set of functions f in H^2 such that $f = \overline{gz}u$ almost everywhere on $\mathbb{T}$ for some g in H^2. In other words,*

$$\mathcal{K}_u = H^2 \cap u\overline{zH^2},$$

where the right hand side is regarded as a set of functions on $\mathbb{T}$.

PROOF. For each f in H^2, we see that

$$\langle f, uh\rangle = 0,\ \forall h \in H^2 \quad \iff \quad \langle \overline{u}f, h\rangle = 0,\ \forall h \in H^2 \quad \iff \quad \overline{u}f \in \overline{zH^2}.$$

Since $u\overline{u} = 1$ almost everywhere on $\mathbb{T}$, we see that an H^2 function f belongs to the orthocomplement of uH^2 if and only if f belongs to $u\overline{zH^2}$ (i.e., $f = \overline{gz}u$ for some g in H^2).[4] □

Just as the Beurling-type subspaces uH^2 constitute the nontrivial invariant subspaces for the unilateral shift on H^2, the subspaces $\mathcal{K}_u$ play an analogous role for the backward shift.

COROLLARY 3.4. *The model spaces $\mathcal{K}_u$, where u is inner, are precisely the proper invariant subspaces of H^2 for the backward shift*

$$f(z) \mapsto \frac{f(z) - f(0)}{z}, \qquad (a_0, a_1, a_2, \ldots) \mapsto (a_1, a_2, a_3, \ldots). \tag{3.5}$$

The following corollary is an immediate consequence of (2.22) (recall what it means for one inner function to *divide* another from Definition 2.14).

COROLLARY 3.6. *If u_1, u_2 are inner functions, then*

$$u_1 | u_2 \iff \mathcal{K}_{u_1} \subseteq \mathcal{K}_{u_2}.$$

REMARK 3.7. We pause here to mention the mildly surprising fact, not widely known, that if u is not a finite Blaschke product, then $\mathcal{K}_u$ contains a linearly ordered chain of S^*-invariant subspaces of uncountable length. This is obvious if u has a singular inner factor S_μ since $\mathcal{K}_{S_{\alpha\mu}} \subseteq \mathcal{K}_u$ for $0 \leqslant \alpha \leqslant 1$. On the other hand, if u is an infinite Blaschke product, then things are not so clear. Suppose that $u = \prod_{n=1}^{\infty} b_n$ where each b_n is a Blaschke factor. Let $I \subseteq \mathbb{R}$ be a nonempty interval and let $\tau : \mathbb{Q} \cap I \to \mathbb{N}$ be a bijection. For each α in I, define $I_\alpha = \{q \in \mathbb{Q} \cap I : q < \alpha\}$ and notice that each α corresponds to a distinct Blaschke product $u_\alpha = \prod_{n \in \tau(I_\alpha)} b_n$ that divides u. This yields the desired chain of S^*-invariant subspaces.

Building upon the proof of Proposition 3.3, we have the following important result [**31**, Thm. 3.1.5].

PROPOSITION 3.8. *A function f in H^2 is noncyclic for S^* if and only if there exists a function g in H^2 and an inner function u such that*

$$f = \overline{gz}u \tag{3.9}$$

[4] In fact, g actually belongs to $\mathcal{K}_u$ as well (see Section 7).

almost everywhere on $\mathbb{T}$. *If* u *and the inner factor of* g *are relatively prime, then the* S^*-*invariant subspace of* H^2 *generated by* f *is* $\mathcal{K}_u$ *itself:*

$$\bigvee\{f, S^*f, S^{*2}f, \dots\} = \mathcal{K}_u.$$

REMARK 3.10. We will see another description of $\mathcal{K}_u$ as well as the noncyclic vectors for S^* when we discuss pseudocontinuations in Section 6.

Being invariant under the backward shift operator, it is not surprising that the spaces $\mathcal{K}_u$ are also invariant under certain functions of the backward shift. In what follows, we let P denote the *Riesz projection*

$$P(\dots, a_{-1}, a_0, a_1, a_2, \dots) = (a_0, a_1, a_2, \dots), \tag{3.11}$$

the orthogonal projection that returns the "analytic part" of a Fourier series in L^2. For instance, $P(1 + 2\cos\theta) = P(e^{-i\theta} + 1 + e^{i\theta}) = 1 + e^{i\theta} = 1 + \zeta$. We also remark that, as an orthogonal projection, the operator P is self-adjoint and hence satisfies $\langle Pf, g\rangle = \langle f, Pg\rangle$ for all f, g in L^2.

DEFINITION 3.12. For φ in $L^\infty(\mathbb{T})$ the *Toeplitz operator* $T_\varphi : H^2 \to H^2$ with *symbol* φ is defined by

$$T_\varphi(f) = P(\varphi f).$$

The study of Toeplitz operators is a vast subject and we refer the reader to the texts **[32, 67]** for the basics and to **[18]** for an encyclopedia on the subject. Note that when φ belongs to H^∞, the Toeplitz operator $T_\varphi f = \varphi f$ is just a multiplication operator. A simple calculation shows that T_z and $T_{\overline{z}}$ are precisely the forward and backward shift operators on H^2. The operator $T_z : H^2 \to H^2$ enjoys an H^∞-functional calculus given by $\varphi(T_z) = T_\varphi$ for φ in H^∞ **[95]**. Being $T_{\overline{z}}$-invariant already, it should come as little surprise that the spaces $\mathcal{K}_u$ are also invariant under conjugate-analytic Toeplitz operators (i.e., Toeplitz operators of the form $T_{\overline{\varphi}}$ where φ belongs to H^∞).

PROPOSITION 3.13. *If* $\varphi \in H^\infty$ *and* u *is inner then* $T_{\overline{\varphi}}\mathcal{K}_u \subseteq \mathcal{K}_u$.

PROOF. For each f in $\mathcal{K}_u$ we have

$$\langle T_{\overline{\varphi}} f, uh\rangle = \langle P(\overline{\varphi} f), uh\rangle = \langle \overline{\varphi} f, P(uh)\rangle = \langle \overline{\varphi} f, uh\rangle = \langle f, u(\varphi h)\rangle = 0. \qquad \square$$

Proposition 3.13 shows that the spaces $\mathcal{K}_u$ enjoy the so-called *F-property*. In general, a set $\mathscr{C}$ of functions contained in H^2 has the *F-property* if whenever θ divides f, then $f/\theta \in \mathscr{C}$. Good sources for this are **[90, 91]**.

COROLLARY 3.14. *If* f *belongs to* $\mathcal{K}_u$ *and* θ *is an inner function that divides* f, *i.e.,* f/θ *belongs to* H^2, *then* f/θ *also belongs to* $\mathcal{K}_u$. *In particular, the outer factor of any function in* $\mathcal{K}_u$ *also belongs to* $\mathcal{K}_u$.

PROOF. Simply observe that $T_{\overline{\theta}} f = P(\overline{\theta} f) = P(f/\theta) = f/\theta$ since f/θ is in H^2. By the preceding proposition, f/θ belongs to $\mathcal{K}_u$. $\square$

3.2. Finite dimensional model spaces. The simplest examples of model spaces are those corresponding to finite Blaschke products

$$u(z) = \prod_{j=1}^{n} \frac{z - \lambda_j}{1 - \overline{\lambda_j} z}. \qquad (\lambda_j \in \mathbb{D})$$

Indeed, these are the only model spaces that are finite-dimensional and whose elements can be completely characterized in an explicit fashion. To do this, one should first notice that the Cauchy-Szegő kernel c_λ (from (2.5)) belongs to $\mathcal{K}_u$ whenever λ is a zero of u. Indeed, if $u(\lambda) = 0$, then clearly $\langle uh, c_\lambda \rangle = u(\lambda)h(\lambda) = 0$ for all h in H^2.

PROPOSITION 3.15. *If u is a finite Blaschke product with simple zeros $\lambda_1, \lambda_2, \ldots, \lambda_n$, then*

$$\mathcal{K}_u = \left\{ \frac{a_0 + a_1 z + \cdots + a_{n-1} z^{n-1}}{(1 - \overline{\lambda_1} z)(1 - \overline{\lambda_2} z) \cdots (1 - \overline{\lambda_n} z)} : a_0, a_1, \ldots, a_{n-1} \in \mathbb{C} \right\}. \tag{3.16}$$

In particular, $\mathcal{K}_{z^n}$ is the space of all polynomials of degree at most $n-1$.

PROOF. Since $\langle uh, c_{\lambda_i} \rangle = u(\lambda_i)h(\lambda_i) = 0$ for all h in H^2, it follows that

$$\operatorname{span}\{c_{\lambda_1}, c_{\lambda_2}, \ldots, c_{\lambda_n}\} \subseteq \mathcal{K}_u.$$

If $f(\lambda_i) = \langle f, c_{\lambda_i} \rangle = 0$ for all i, then $u|f$ and hence f belongs to uH^2. Thus

$$\operatorname{span}\{c_{\lambda_1}, c_{\lambda_2}, \ldots, c_{\lambda_n}\}^\perp \subseteq \mathcal{K}_u^\perp.$$

Hence, $\mathcal{K}_u = \operatorname{span}\{c_{\lambda_1}, c_{\lambda_2}, \ldots, c_{\lambda_n}\}$ and the result follows by simple algebra. To be more specific, any linear combination of the Cauchy kernels $\{c_{\lambda_j} : 1 \leqslant j \leqslant n\}$ can be expressed as a rational function of the type prescribed in (3.16). Conversely, any expression of the type encountered in (3.16) can be decomposed, via partial fractions, into a linear combination of the functions $c_{\lambda_1}, c_{\lambda_2}, \ldots, c_{\lambda_n}$. □

If λ has multiplicity m as a zero of u, then one must also include the functions

$$c_\lambda, c_\lambda', c_\lambda'', \ldots, c_\lambda^{(m-1)}$$

in place of c_λ in the preceding proof. Along similar lines, the proof of Proposition 3.15 and the preceding comment provides us with the following useful fact.

PROPOSITION 3.17. *Suppose that u is the finite Blaschke product with distinct zeros $\lambda_1, \lambda_2, \ldots, \lambda_n$ with respective multiplicities $m_1, m_2, \ldots, m_n$, then*

$$\mathcal{K}_u = \operatorname{span}\{c_{\lambda_i}^{(\ell_i - 1)} : 1 \leqslant i \leqslant n,\ 1 \leqslant \ell_i \leqslant m_i\}.$$

In fact, the preceding observation makes it clear why $\dim \mathcal{K}_u < \infty$ occurs if and only if u is a finite Blaschke product. If u has a factor that is an infinite Blaschke product, then Proposition 2.6 ensures that $\mathcal{K}_u$ contains an infinite, linearly independent set (namely the Cauchy kernels corresponding to the distinct zeros of u). On the other hand, if u is a singular inner function, then $u^{1/n}$ is an inner function that divides u whence $\mathcal{K}_{u^{1/n}} \subseteq \mathcal{K}_u$ for $n \geqslant 1$. We may find an infinite orthonormal sequence in $\mathcal{K}_u$ by selecting unit vectors f_n in $\mathcal{K}_{u^{1/n}} \ominus \mathcal{K}_{u^{1/(n+1)}}$, from which it follows that $\dim \mathcal{K}_u = \infty$. For the general case we point out the following decomposition that is interesting in its own right.

PROPOSITION 3.18. *If $\{u_j\}_{j \geqslant 1}$ is a possibly finite sequence of inner functions such that $u = \prod_{j \geqslant 1} u_j$ exists, then*

$$\mathcal{K}_u = \mathcal{K}_{u_1} \oplus \bigoplus_{n \geqslant 2} \Big(\prod_{j=1}^{n-1} u_j \Big) \mathcal{K}_{u_n}.$$

In particular, if u and v are inner functions, then

$$\mathcal{K}_{uv} = \mathcal{K}_u \oplus u\mathcal{K}_v. \tag{3.19}$$

We refer the reader to [**3**, **62**] for further details, although we provide a proof of the special case (3.19) in Subsection 5.1.

3.3. Three unitary operators. The following three transformations from one model space to another are often useful.

PROPOSITION 3.20. *Suppose u is a fixed inner function.*

(i) *If $w \in \mathbb{D}$, then*

$$f \mapsto \frac{\sqrt{1-|w|^2}}{1-\overline{w}u} f$$

defines a unitary operator from $\mathcal{K}_u$ onto $\mathcal{K}_{\frac{u-w}{1-\overline{w}u}}$.

(ii) *If φ is a disk automorphism, i.e., $\varphi(z) = \zeta(z-a)(1-\overline{a}z)^{-1}$ for some $a \in \mathbb{D}$ and $\zeta \in \mathbb{T}$, then*

$$f \mapsto \sqrt{\varphi'}(f \circ \varphi)$$

defines a unitary operator from $\mathcal{K}_u$ onto $\mathcal{K}_{u\circ\varphi}$.

(iii) *If $u^{\#}(z) := \overline{u(\overline{z})}$, then, in terms of boundary functions, the map*

$$f(\zeta) \mapsto \overline{\zeta} f(\overline{\zeta}) u^{\#}(\zeta)$$

is a unitary operator from $\mathcal{K}_u$ into $\mathcal{K}_{u^{\#}}$.

The first unitary operator is due to Crofoot [**30**] (a more detailed discussion of these so-called *Crofoot transforms* can be found in [**85**, Sec. 13]). When dealing with a model space $\mathcal{K}_u$ where u has a nontrivial Blaschke factor, one can often make the simplifying assumption that $u(0) = 0$. Also of great importance is the fact that Crofoot transforms intertwine the conjugations (see Section 7) on the corresponding models spaces [**85**, Lem. 3.1]. The second unitary operator is clearly unitary on H^2 (change of variables formula). Showing that its restriction to $\mathcal{K}_u$ has the correct range is a little tricky and the proof of this can be found in [**28**, Prop. 4.1]. The third unitary operator depends on a discussion on conjugations that we take up in more detail in Section 7. Indeed, on the face of it, the map does not even seem to take analytic functions to analytic functions. However, when one thinks of model space functions in terms of their boundary values as in Proposition 3.3, everything works out. The proof can be found in [**28**, Lemma 4.3].

4. Model operators

One of the main reasons that model spaces are worthy of study in their own right stems from the so-called *model theory* developed by Sz.-Nagy and Foiaş, which shows that a wide range of Hilbert space operators can be realized concretely as restrictions of the backward shift operator to model spaces. These ideas have since been generalized in many directions (e.g., de Branges-Rovnyak spaces, vector-valued Hardy spaces, etc.) and we make no attempt to provide an encyclopedic account of the subject, referring the reader instead to the influential texts [**14**, **69**, **71**, **77**, **95**].

4.1. Contractions. In the following, we let $\mathcal{H}$ denote a separable complex Hilbert space. If T is an arbitrary bounded operator on $\mathcal{H}$, then we may assume that $\|T\| \leqslant 1$ (i.e., T is a *contraction*). As such, T enjoys a decomposition of the form $T = K \oplus U$ (see [**95**, p. 8] for more details) where U is a unitary operator and K is a *completely nonunitary (CNU) contraction* (i.e., there does not exist a reducing subspace for K upon which K is unitary).

Since the structure and behavior of unitary operators is well-understood, via the spectral theorem, the study of arbitrary bounded Hilbert space operators can therefore be focused on CNU contractions. With a few additional hypotheses, one can obtain a concrete *functional model* for such operators. In light of the fact that the following theorem justifies, to an extent, the further study of model subspaces, we feel obliged to provide a complete proof. Moreover, the proof itself is surprisingly simple and is worthy of admiration for its own sake.

THEOREM 4.1 (Sz.-Nagy-Foiaş). *If T is a contraction on a Hilbert space that satisfies*

(i) $\|T^n\mathbf{x}\| \to 0$ *for all* $\mathbf{x} \in \mathcal{H}$,

(ii) $\operatorname{rank}(I - T^*T) = \operatorname{rank}(I - TT^*) = 1$,

then there exists an inner function u such that T is unitarily equivalent to $S^*|\mathcal{K}_u$, where S^* is the backward shift operator on H^2.

PROOF. We let $\cong$ denote the unitary equivalence of Hilbert spaces or their operators. Since the *defect operator* $D = \sqrt{I - T^*T}$ has rank 1, we see that $\operatorname{ran} D \cong \mathbb{C}$ so that

$$\widetilde{\mathcal{H}} := \bigoplus_{n=1}^{\infty} \operatorname{ran} D \cong H^2. \tag{4.2}$$

It follows that for each $n \in \mathbb{N}$ we have

$$\begin{aligned}
\sum_{j=0}^{n} \|DT^j\mathbf{x}\|^2 &= \sum_{j=0}^{n} \langle (I - T^*T)^{\frac{1}{2}} T^j\mathbf{x}, (I - T^*T)^{\frac{1}{2}} T^j\mathbf{x} \rangle \\
&= \sum_{j=0}^{n} \langle (I - T^*T) T^j\mathbf{x}, T^j\mathbf{x} \rangle \\
&= \sum_{j=0}^{n} \Big(\langle T^j\mathbf{x}, T^j\mathbf{x} \rangle - \langle T^*TT^j\mathbf{x}, T^j\mathbf{x} \rangle \Big) \\
&= \sum_{j=0}^{n} \left(\|T^j\mathbf{x}\|^2 - \|T^{j+1}\mathbf{x}\|^2 \right) \\
&= \|\mathbf{x}\|^2 - \|T^{n+1}\mathbf{x}\|^2.
\end{aligned}$$

Since, by hypothesis, $\|T^n\mathbf{x}\| \to 0$ for each $\mathbf{x} \in \mathcal{H}$, we conclude that

$$\sum_{j=0}^{\infty} \|DT^j\mathbf{x}\|^2 = \|\mathbf{x}\|^2 \qquad (\mathbf{x} \in \mathcal{H})$$

and hence the operator $\Phi : \mathcal{H} \to H^2$ defined by

$$\Phi\mathbf{x} = (D\mathbf{x}, DT\mathbf{x}, DT^2\mathbf{x}, DT^3\mathbf{x}, \ldots)$$

is an isometric embedding of $\mathcal{H}$ into H^2 (here we have identified a function in H^2 with its sequence of Taylor coefficients). Since Φ is an isometry, its image

$$\Phi\mathcal{H} = (D\mathcal{H}, DT\mathcal{H}, DT^2\mathcal{H}, \ldots)$$

is closed in H^2 and clearly S^*-invariant. Therefore, by Corollary 3.4, $\operatorname{ran} \Phi = \mathcal{K}_u$ for some u (the possibility that $\operatorname{ran} \Phi = H^2$ is ruled out because the following argument would show that $T \cong S^*$, violating the assumption that $\operatorname{rank}(I - TT^*) = 1$). Now observe that

$$(\mathbf{x} \in \mathcal{H}) \qquad \Phi T\mathbf{x} = (DT\mathbf{x}, DT^2\mathbf{x}, DT^3\mathbf{x}, \ldots) = S^*\Phi\mathbf{x}.$$

Letting $U : \mathcal{H} \to \mathcal{K}_u$ denote the unitary operator obtained from Φ by reducing its codomain from H^2 to $\mathcal{K}_u$, it follows that

$$UT = (S^*|\mathcal{K}_u)U.$$

Thus T is unitarily equivalent to the restriction of S^* to $\mathcal{K}_u$. □

Implicit in Theorem 4.1 is the fact that $S^*|\mathcal{K}_u$ satisfies conditions (i) and (ii), as can be shown through some routine computations.

The case of higher defect indices (i.e., $\operatorname{rank}(I - T^*T) = \operatorname{rank}(I - TT^*) \geqslant n$, $n > 1$), is treated by moving to vector-valued Hardy spaces $H^2(\mathcal{E})$ and operator-valued inner functions. However, in making such a move one sacrifices a large variety of tools and techniques inherited from classical function theory (making the theory of model spaces more difficult and less interesting from our perspective). For instance, the multiplication of operator-valued inner functions is no longer commutative and the corresponding factorization theory is more complicated.

4.2. Spectrum of an inner function. In light of the Sz.-Nagy-Foiaş Theorem, the restriction of the backward shift to the spaces $\mathcal{K}_u$ is of premier importance. However, it turns out that the *compressed shift*, the compression $S_u : \mathcal{K}_u \to \mathcal{K}_u$ of the unilateral shift to $\mathcal{K}_u$, is more prevalent in the literature. Here

$$S_u f = P_u(zf),$$

where P_u denotes the orthogonal projection from L^2 onto $\mathcal{K}_u$ (see Subsection 5.1). In reality, the distinction alluded to above is artificial since the operator S_u is unitarily equivalent to the restriction of the backward shift to the space $\mathcal{K}_{u^\#}$ where $u^\#(z) = \overline{u(\overline{z})}$. In fact the unitary operator which intertwines S_u and $S^*|\mathcal{K}_{u^\#}$ is the one given in Proposition 3.20. See [**28**, Lemma 4.5] for more details on this.

There happens to be two convenient ways to describe the spectrum $\sigma(S_u)$ of S_u. In fact, one can show that $\sigma(S_u)$ coincides with the so-called *spectrum* of the inner function u, as defined below (this result is sometimes referred to as the Livšic-Möller Theorem).

DEFINITION 4.3. The *spectrum* of an inner function is the set

$$\sigma(u) := \left\{ \lambda \in \mathbb{D}^- : \liminf_{z \to \lambda} |u(z)| = 0 \right\}. \tag{4.4}$$

It follows from the preceding definition, for instance, that every zero of u in $\mathbb{D}$ belongs to $\sigma(u)$. Moreover, any limit point of zeros of u must also lie in $\sigma(u)$. On the other hand, if u is the singular inner function associated to the singular measure μ, then the nontangential limit of u is zero μ-almost everywhere [**53**, Thm. 6.2]. Even further, if $\lambda \in \mathbb{D}$ and $u(\lambda) = 0$ then $c_\lambda(z) = (1 - \overline{\lambda}z)^{-1}$ belongs to $\mathcal{K}_u$ and satisfies $S^*c_\lambda = \overline{\lambda}c_\lambda$. In other words, $\overline{\lambda}$ belongs to $\sigma_p(S^*|\mathcal{K}_u)$, the point spectrum (i.e., set of eigenvalues) of $S^*|\mathcal{K}_u$. These observations suggest the following important theorem that provides us with a convenient description of the spectrum of an inner function in terms of its canonical factorization.

THEOREM 4.5. *If $u = B_\Lambda S_\mu$, where B_Λ is a Blaschke product with zero sequence Λ and S_μ is a singular inner function with corresponding singular measure μ, then*

(i) $\sigma(S_u) = \sigma(u) = \Lambda^- \cup \operatorname{supp} \mu$,

(ii) $\sigma_p(S^*|\mathcal{K}_u) = \{\overline{\lambda} : \lambda \in \Lambda\}$,

(iii) $\sigma_p(S_u) = \Lambda$.

5. Reproducing kernels

As a closed subspace of the reproducing kernel Hilbert space[5] H^2, each model space $\mathcal{K}_u$ itself possesses a reproducing kernel. In this section, we identify these kernels and explore their basic properties.

5.1. Basic properties. Recalling that the Cauchy kernels $c_\lambda = (1 - \overline{\lambda} z)^{-1}$ are the reproducing kernels for the Hardy space (see Subsection 2.3), let us attempt to compute the corresponding reproducing kernels for $\mathcal{K}_u$. We first observe that if $f = uh$, with $h \in H^2$, then

$$f(\lambda) = u(\lambda)h(\lambda) = u(\lambda)\langle h, c_\lambda \rangle = u(\lambda)\langle f\overline{u}, c_\lambda \rangle = \langle f, \overline{u(\lambda)} u c_\lambda \rangle,$$

from which it follows that the reproducing kernel for uH^2 is given by

$$\overline{u(\lambda)} u(z) c_\lambda(z).$$

If f belongs to $\mathcal{K}_u$, then it follows that

$$\begin{aligned} f(\lambda) &= \langle f, c_\lambda \rangle \\ &= \langle f, c_\lambda \rangle - u(\lambda)\langle f, u c_\lambda \rangle \\ &= \big\langle f, (1 - \overline{u(\lambda)} u) c_\lambda \big\rangle. \end{aligned}$$

Moreover, the function $(1 - \overline{u(\lambda)} u) c_\lambda$ belongs to $\mathcal{K}_u$ since

$$\begin{aligned} \big\langle uh, (1 - \overline{u(\lambda)} u) c_\lambda \big\rangle &= u(\lambda)h(\lambda) - u(\lambda)\langle uh, u c_\lambda \rangle \\ &= u(\lambda)h(\lambda) - u(\lambda)\langle h, c_\lambda \rangle \\ &= u(\lambda)h(\lambda) - u(\lambda)h(\lambda) \\ &= 0 \end{aligned}$$

for all h in H^2. Putting this all together we see that

$$(f \in \mathcal{K}_u) \qquad f(\lambda) = \langle f, k_\lambda \rangle,$$

where

$$k_\lambda(z) = \frac{1 - \overline{u(\lambda)} u(z)}{1 - \overline{\lambda} z}. \tag{5.1}$$

The function k_λ is called the *reproducing kernel* for $\mathcal{K}_u$. When dealing with more than one model space at a time, the notation k_λ^u is often used to denote the reproducing kernel for $\mathcal{K}_u$.

[5] A Hilbert space $\mathcal{H}$ of analytic functions on a domain $\Omega \subset \mathbb{C}$ is called a *reproducing kernel Hilbert space* if for each $\lambda \in \Omega$ there is a function $K_\lambda \in \mathcal{H}$ for which $f(\lambda) = \langle f, K_\lambda \rangle_{\mathcal{H}}$ for all $f \in \mathcal{H}$. All of the classical Hilbert spaces of analytic functions on the disk (e.g., Hardy, Bergman, Dirichlet, etc.) are reproducing kernel Hilbert spaces and understanding the properties of the kernels often yields valuable information about the functions in $\mathcal{H}$. A few good sources for this are **[2, 12, 72]**.

From the general theory of reproducing kernels [**2, 12, 72**], it follows that if $e_1, e_2, \ldots$ is any orthonormal basis for $\mathcal{K}_u$, then

$$k_\lambda(z) = \sum_{n\geqslant 1} \overline{e_n(\lambda)} e_n(z), \tag{5.2}$$

the sum converging in the norm of $\mathcal{K}_u$ [**2**]. The following example illustrates both expressions (5.1) and (5.2).

EXAMPLE 5.3. If $u = z^n$, then $\mathcal{K}_u = \operatorname{span}\{1, z, z^2, \ldots, z^{n-1}\}$ and

$$k_\lambda(z) = \frac{1 - \overline{\lambda}^n z^n}{1 - \overline{\lambda} z} = 1 + \overline{\lambda} z + \overline{\lambda}^2 z^2 + \cdots + \overline{\lambda}^{n-1} z^{n-1}.$$

We are now in a position to provide an elegant derivation of the decomposition (3.19). Indeed, for inner functions u and v, divide the trivial equality

$$1 - \overline{u(\lambda)v(\lambda)}u(z)v(z) = \big(1 - \overline{u(\lambda)}u(z)\big) + \overline{u(\lambda)}u(z)\big(1 - \overline{v(\lambda)}v(z)\big)$$

by $1 - \overline{\lambda} z$ to get the identity

$$k_\lambda^{uv} = k_\lambda^u + \overline{u(\lambda)}u(z)k_\lambda^v.$$

However, the preceding is simply a restatement, in terms of reproducing kernels, of the fact that $\mathcal{K}_{uv} = \mathcal{K}_u \oplus u\mathcal{K}_v$.

Since

$$|k_\lambda(z)| \leqslant \frac{2}{1 - |\lambda|} \tag{5.4}$$

for each z in $\mathbb{D}$, it follows that each k_λ belongs to $\mathcal{K}_u \cap H^\infty$. Moreover, the kernel functions k_λ are among the few readily identifiable functions that belong to $\mathcal{K}_u$. As such, they provide invaluable insight into the structure and properties of model spaces. In fact, kernel functions often wind up being "test functions" for various statements about model spaces. As an example of what we mean by this, suppose that the quantity

$$\|k_\lambda\|^2 = \frac{1 - |u(\lambda)|^2}{1 - |\lambda|^2}$$

remains bounded as $\lambda \to \zeta \in \mathbb{T}$ along some path $\Gamma \subset \mathbb{D}$. Then, since

$$|f(\lambda)| = |\langle f, k_\lambda\rangle| \leqslant \|f\|\|k_\lambda\|,$$

we see that every f in $\mathcal{K}_u$ is bounded along Γ. We will see more of these types of results in Section 6.

The orthogonal projection P_u from L^2 onto $\mathcal{K}_u$ arises frequently in the study of Hankel operators, model operators and truncated Toeplitz operators. This important operator can be expressed in a simple manner using reproducing kernels.

PROPOSITION 5.5. *For each f in L^2 and λ in $\mathbb{D}$,*

$$(P_u f)(\lambda) = \langle f, k_\lambda\rangle. \tag{5.6}$$

PROOF. Using the selfadjointness of P_u we get

$$\langle f, k_\lambda\rangle = \langle f, P_u k_\lambda\rangle = \langle P_u f, k_\lambda\rangle = (P_u f)(\lambda). \qquad \square$$

Rewriting (5.6) as an integral we obtain

$$(P_u f)(\lambda) = \int_{\mathbb{T}} f(\zeta) \frac{1 - u(\lambda)\overline{u(\zeta)}}{1 - \lambda\overline{\zeta}}\, dm(\zeta),$$

which can be used to estimate or control the behavior of $P_u f$. In particular, the preceding formula highlights the dependence of $P_u f$ on the behavior of u itself.

5.2. Density results. As mentioned earlier, in general, the model spaces $\mathcal{K}_u$ contain few readily identifiable functions. One therefore relies heavily upon the kernel functions k_λ in the study of model spaces since they are among the few explicitly describable functions contained in $\mathcal{K}_u$. As the following proposition shows, one can always find a collection of kernels whose span is dense in the whole space.

PROPOSITION 5.7. *If Λ is a subset of $\mathbb{D}$ such that either (i) Λ has an accumulation point in $\mathbb{D}$, or (ii) $\sum_{\lambda \in \Lambda} (1 - |\lambda|)$ diverges, then for any inner function u*

$$\bigvee \{k_\lambda : \lambda \in \Lambda\} = \mathcal{K}_u.$$

PROOF. The containment $\subseteq$ is obvious. If $f \perp k_\lambda$ for all $\lambda \in \Lambda$, then f vanishes on Λ. (i) If Λ has an accumulation point in $\mathbb{D}$, then the Identity Theorem implies that $f \equiv 0$. (ii) If $\sum_{\lambda \in \Lambda}(1 - |\lambda|)$ diverges, then $f \equiv 0$ since the zero set of a nonzero H^2 function must satisfy the Blaschke condition (2.8). □

A somewhat more general result is provided by [**45**, Thm. 1]:

PROPOSITION 5.8. *If u is a nonconstant inner function, then there exists a subset $E \subset \mathbb{D}$ of area measure zero such that for each w in $\mathbb{D} \backslash E$, the inverse image $u^{-1}(\{w\})$ is nonempty, $u'(\lambda) \neq 0$ for all λ in $u^{-1}(\{w\})$, and*

$$\bigvee \{k_\lambda : \lambda \in u^{-1}(\{w\})\} = \mathcal{K}_u.$$

REMARK 5.9. We will take up the general question of whether a sequence of kernel functions has a dense linear span in $\mathcal{K}_u$ when we discuss completeness problems in Section 9.

Since each kernel function k_λ belongs to H^∞ by (5.4), an immediate consequence of Proposition 5.7 is the following useful result.

PROPOSITION 5.10. *The linear manifold $\mathcal{K}_u \cap H^\infty$ is dense in $\mathcal{K}_u$.*

A second proof can be obtained by noting that

$$\mathcal{K}_u = \bigvee \{S^{*n} u : n \geqslant 1\}$$

and that each of the backward shifts $S^{*n}u$ of u belongs to H^∞. In certain sophisticated applications one requires a dense subset of $\mathcal{K}_u$ consisting of functions having a certain degree of smoothness on $\mathbb{D}^-$. This next result is a restatement of Proposition 3.15 for infinite Blaschke products.

PROPOSITION 5.11. *If u is a Blaschke product with simple zeros $\lambda_1, \lambda_2, \ldots$, then*

$$\bigvee \{k_{\lambda_1}, k_{\lambda_2}, \ldots\} = \mathcal{K}_u.$$

Thus $\mathcal{K}_u$ contains a dense subset whose elements are continuous on $\mathbb{D}^-$ and whose boundary functions are infinitely differentiable on $\mathbb{T}$.

PROOF. Indeed, suppose that f is a function in $\mathcal{K}_u$ that satisfies $\langle f, k_{\lambda_n}\rangle = 0$ for all n. This implies that $f(\lambda_n) = 0$ for all n, whence $u|f$ so that f belongs to uH^2. In other words, f must be identically zero. Since

$$k_{\lambda_n}(z) = \frac{1-\overline{u(\lambda_n)}u(z)}{1-\overline{\lambda_n}z} = \frac{1}{1-\overline{\lambda_n}z} = c_{\lambda_n}(z), \tag{5.12}$$

the second statement follows immediately. □

If the zeros of u are not simple, then the preceding statement is true if one also includes the appropriate derivatives of the kernel functions in the spanning set. When u is not a Blaschke product, finding a dense set of functions in $\mathcal{K}_u$, each of which is continuous on $\mathbb{D}^-$, is much more difficult. A deep result in this direction is due to A.B. Aleksandrov [7], who proved the following astonishing result (see [26] for a discussion of this).

THEOREM 5.13 (Aleksandrov). *For an inner function u, the set $\mathcal{K}_u \cap \mathcal{A}$ is dense in $\mathcal{K}_u$. Here $\mathcal{A}$ denotes the disk algebra, the Banach algebra of all H^∞ functions that are continuous on $\mathbb{D}^-$.*

This theorem is remarkable due to the fact that $\mathcal{K}_u$ often does not contain a single readily identifiable function that is continuous on $\mathbb{D}^-$. For example, if u is the singular inner function

$$u(z) = \exp\left(\frac{z+1}{z-1}\right),$$

then it is not at all obvious that $\mathcal{K}_u$ contains any functions that are continuous on $\mathbb{D}^-$, let alone a dense set of them. See [34] for some related results concerning when $\mathcal{K}_u$ contains smoother functions than those in $\mathcal{A}$.

5.3. Cauchy-Szegő bases. If u is an inner function and $u(\lambda) = 0$, then (5.12) tells us that $k_\lambda(z) = c_\lambda(z) = (1-\overline{\lambda}z)^{-1}$. Thus the kernel functions corresponding to zeros of u are extremely simple functions to work with since they do not depend explicitly upon u. In certain situations, these functions can be used to construct bases for models spaces (in a sense to be made precise shortly). However, no two Cauchy kernels are orthogonal, so one cannot hope to obtain an orthonormal basis of kernel functions.[6] We therefore need a somewhat more flexible definition [25].

DEFINITION 5.14. A linearly independent sequence $\mathbf{x}_n$ in a Hilbert space $\mathcal{H}$ is called a *Riesz basis* for $\mathcal{H}$ if $\bigvee\{\mathbf{x}_1, \mathbf{x}_2, \ldots\} = \mathcal{H}$ and there exist constants $M_1, M_2 > 0$ such that

$$M_1 \sum_{i=1}^n |a_i|^2 \leqslant \|\sum_{i=1}^n a_i \mathbf{x}_i\|^2 \leqslant M_2 \sum_{i=1}^n |a_i|^2$$

for all finite numerical sequences $a_1, a_2, \ldots, a_n$.

The following seminal result of L. Carleson tells us when the normalized reproducing kernels corresponding to the zero set of a Blaschke product forms a Riesz basis for $\mathcal{K}_u$ [69, p. 133]

[6]One can sometimes obtain orthonormal bases consisting of *boundary kernels* (see Section 8).

THEOREM 5.15. *If u is a Blaschke product with simple zeros $\lambda_1, \lambda_2, \ldots$, then the normalized kernels*

$$\frac{k_{\lambda_n}}{\|k_{\lambda_n}\|} = \frac{\sqrt{1-|\lambda_n|^2}}{1-\overline{\lambda_n}z}$$

form a Riesz basis for $\mathcal{K}_u$ if and only if there exists a $\delta > 0$ such that

$$(i = 1, 2, \ldots) \qquad \delta < \prod_{\substack{j=1 \\ j \neq i}}^{\infty} \left| \frac{\lambda_i - \lambda_j}{1 - \overline{\lambda_j}\lambda_i} \right|.$$

There is a generalization of this result where one can somewhat relax the condition that the λ_n are the zeros of u **[59]**.

5.4. Takenaka-Malmquist-Walsh bases. Unlike the Hardy space H^2 itself, the model spaces $\mathcal{K}_u$ do not come pre-equipped with a canonical orthonormal basis. It turns out that an orthonormal basis for $\mathcal{K}_u$, where u is a Blaschke product, can be obtained by orthogonalizing the kernel functions corresponding to the zeros of u.

For w in $\mathbb{D}$, we let

$$b_w(z) = \frac{z-w}{1-\overline{w}z}. \tag{5.16}$$

If u is a Blaschke product with simple zeros $\lambda_1, \lambda_2, \ldots$, then observe that

$$\langle b_{\lambda_1} k_{\lambda_2}, k_{\lambda_1} \rangle = b_{\lambda_1}(\lambda_1) k_{\lambda_2}(\lambda_1) = 0.$$

Similarly, we have

$$\langle b_{\lambda_1} b_{\lambda_2} k_{\lambda_3}, k_{\lambda_1} \rangle = \langle b_{\lambda_1} b_{\lambda_2} k_{\lambda_3}, k_{\lambda_2} \rangle = 0. \tag{5.17}$$

This process suggests the following iterative definition: If u is a Blaschke product with zeros $\lambda_1, \lambda_2, \ldots$, let

$$v_1(z) := \frac{\sqrt{1-|\lambda_1|^2}}{1-\overline{\lambda_1}z},$$

$$(k \geqslant 2) \qquad v_k(z) = \Big(\prod_{i=1}^{k-1} b_{\lambda_i} \Big) \frac{\sqrt{1-|\lambda_k|^2}}{1-\overline{\lambda_k}z}.$$

One can show that $\{v_n : n \geqslant 1\}$ is an orthonormal basis for $\mathcal{K}_u$ that is the Gram-Schmidt orthonormalization of the kernels $k_{\lambda_1}, k_{\lambda_2}, \ldots$.

The terminology here is not completely standard. It seems that these bases first appeared in Takenaka's 1925 paper involving finite Blaschke products **[96]**. The classic text **[69]** considers the general case where u is a potentially infinite Blaschke product, referring to this result as the Malmquist-Walsh Lemma. In light of Takenaka's early contribution to the subject, the authors chose in **[50]** to refer to such a basis for $\mathcal{K}_u$ as a *Takenaka-Malmquist-Walsh (TMW) basis*.

Another important family of orthonormal bases of $\mathcal{K}_u$ is (sometimes) provided by the Aleksandrov-Clark spectral theory of rank-one unitary perturbations of model operators. The so-called *modified Aleksandrov-Clark bases* are particularly important in the study of finite-dimensional model spaces. To discuss these bases, we first require a few words about boundary behavior and angular derivatives.

6. Boundary behavior

6.1. Pseudocontinuations. Functions in model spaces enjoy certain "continuation" properties across $\mathbb{T}$. These types of continuation properties appear in many places [**78**] and in many different settings but they all go under the broad heading of "generalized analytic continuation." The type of generalized analytic continuation that is relevant to model spaces is called *pseudocontinuation* and it was first explored by H. S. Shapiro [**87, 88**]. In what follows, we let

$$\mathbb{D}_e := \{|z| > 1\} \cup \{\infty\}$$

denote the *extended exterior disk*, the complement of the closed unit disk in the extended complex plane.

DEFINITION 6.1. Let f and $\widetilde{f}$ be meromorphic functions on $\mathbb{D}$ and $\mathbb{D}_e$, respectively. If the nontangential limiting values of f (from $\mathbb{D}$) agree with the nontangential limiting values of $\widetilde{f}$ (from $\mathbb{D}_e$) almost everywhere on $\mathbb{T}$, then we say that f and $\widetilde{f}$ are *pseudocontinuations* of one another.

REMARK 6.2. Pseudocontinuations are unique in the sense that if F and G are meromorphic functions on $\mathbb{D}_e$ that are both pesudocontinuations of f, then $F = G$. This follows from the Privalov uniqueness theorem [**60**, p. 62], which states that if f and g are meromorphic on $\mathbb{D}$ with equal non-tangential limits on any set of positive Lebesgue measure on $\mathbb{T}$, then $f = g$. This is why the definition of pseudocontinuation is stated in terms of nontangential limits as opposed to radial limits, where Privalov's theorem is no longer true. However, in the context we will apply this definition, all the functions involved will be of *bounded type*, that is to say the quotient of two bounded analytic functions (recall Definition 2.19), where the nontangential limits exist almost everywhere.

By Privalov's uniqueness theorem again, we can show that pseudocontinuation is compatible with analytic continuation in that if f (on $\mathbb{D}$) has a pseudocontinuation F (on $\mathbb{D}_e$) and f also has an analytic continuation F_1 across some neighborhood of a point on $\mathbb{T}$, then $F_1 = F$.

EXAMPLE 6.3. Let us provide a few instructive examples (see [**78**] for further details).

(i) Any inner function u has a pseudocontinuation to $\mathbb{D}_e$ defined by

$$\widetilde{u}(z) := \frac{1}{\overline{u(1/\overline{z})}}. \tag{6.4}$$

In fact, this pseucontinuation is one of bounded type (written PCBT), being a quotient of two bounded analytic functions on $\mathbb{D}_e$. Pseudocontinuations of bounded type will play an important role momentarily.

(ii) If f is a rational function whose poles lie in $\mathbb{D}_e$, then f is PCBT.

(iii) Since pseudocontinuations must be compatible with analytic continuations (see the previous remark), functions with isolated branch points on $\mathbb{T}$, such as $\sqrt{1-z}$, do not possess pseudocontinuations.

(iv) The function $f(z) = \exp z$ is not pseudocontinuable. Although it is analytically continuable to $\mathbb{C}$, it is not meromorphic on $\mathbb{D}_e$ due to its essential singularity at ∞.

(v) The classical gap theorems of Hadamard [**54**] and Fabry [**37**] show that analytic functions on $\mathbb{D}$ with lacunary power series do not have analytic continuations across any point of the unit circle. Such functions do not have pseudocontinuations either [**1**, **8**, **88**].

The following seminal result demonstrates the direct connection between membership in a model space and pseudocontinuability. In what follows,

$$H^2(\mathbb{D}_e) := \{f(1/z) : f \in H^2\} \tag{6.5}$$

denotes the Hardy space of the extended exterior disk.

THEOREM 6.6 (Douglas, Shapiro, Shields [**31**]). *A function $f \in H^2$ belongs to $\mathcal{K}_u$ if and only if f/u has a pseudocontinuation $F_u \in H^2(\mathbb{D}_e)$ satisfying $F_u(\infty) = 0$.*

PROOF. Suppose that f belongs to $\mathcal{K}_u$. By Proposition 3.3, $f = \overline{g\zeta u}$ almost everywhere on $\mathbb{T}$. Define the function F_u on $\mathbb{D}_e$ by

$$F_u(z) := \frac{1}{z}\overline{g\Big(\frac{1}{\overline{z}}\Big)} \tag{6.7}$$

and note by (6.5) that F_u is in $H^2(\mathbb{D}_e)$ and $F_u(\infty) = 0$. Moreover, by the identity $f = \overline{g\zeta u}$ almost everywhere on $\mathbb{T}$ we see that $F_u(\zeta) = f(\zeta)/u(\zeta)$ for almost every $\zeta \in \mathbb{T}$, i.e., F_u is a pseudocontinuation of f/u. This proves one direction.

For the other direction, suppose that f/u has a pseudocontinuation $F_u \in H^2(\mathbb{D}_e)$ with $F_u(\infty) = 0$. Then, again, by (6.5) we have

$$F_u(z) = \frac{1}{z}h\Big(\frac{1}{z}\Big)$$

for some $h \in H^2$ and $f(\zeta)/u(\zeta) = F_u(\zeta)$ for almost every ζ in $\mathbb{T}$. Define a function g on $\mathbb{D}$ by $g(z) := \overline{h(\overline{z})}$ and observe that $g \in H^2$. From here we see that $f(\zeta) = u(\zeta)\overline{\zeta g(\zeta)}$ for almost every ζ in $\mathbb{T}$ from which it follows, via Proposition 3.3, that f belongs to $\mathcal{K}_u$. □

With a little more work, one can also determine the *cyclic vectors* for S^*, those f in H^2 for which

$$\bigvee\{S^{*n}f : n = 0, 1, 2, \cdots\} = H^2.$$

COROLLARY 6.8. *A function f in H^2 is not cyclic for S^* if and only if f is PCBT.*

PROOF. We will prove only one direction. Indeed, suppose f is non-cyclic for S^*. Then f must belong to some model space $\mathcal{K}_u$. By the previous theorem, f/u has a pseudocontinuation F_u that belongs to $H^2(\mathbb{D}_e)$. But since u already has a natural pseudocontinuation $\widetilde{u}$ given by (6.4), we see that f has a pseudocontinuation $\widetilde{u}F_u$ that is of bounded type[7]. □

Along with Example 6.3 this corollary shows that inner functions and rational functions are noncyclic for S^* whereas functions like $\sqrt{1-z}$ and e^z are cyclic.

In principle the cyclic vector problem for S^* is solved (i.e., f is noncyclic for S^* if and only if f is PCBT). However, this solution is not as explicit as the solution to the cyclic vector problem for S (i.e., f is cyclic for S if and only f is outer). Outer functions are, in a sense, readily identifiable. On the other hand, PCBT functions

[7]We are using the well-known fact here that F_u (or any function in $H^2(\mathbb{D}_e)$) is of bounded type (see Remark 2.20).

are not so easily recognized. We refer the reader to some papers which partially characterize the noncyclic vectors for S^* by means of growth of Taylor coefficients **[61, 86]**, the modulus of the function **[31, 61]**, and gaps in the Taylor coefficients **[1, 8, 31]**.

6.2. Analytic continuation. Recall from (4.4) and Theorem 4.5 that if $u = B_\Lambda S_\mu$, where B_Λ is a Blaschke product with zero set Λ and S_μ is a singular inner function with corresponding singular measure μ, then the *spectrum* of u is the set

$$\sigma(u) = \Lambda^- \cup \operatorname{supp} \mu.$$

The relevance of $\sigma(u)$ to the function theoretic properties of $\mathcal{K}_u$ lies in the following observation **[69**, p. 65], **[27**, p. 84].

PROPOSITION 6.9. *Every f in $\mathcal{K}_u$ can be analytically continued to*

$$\widehat{\mathbb{C}} \backslash \{1/\overline{z} : z \in \sigma(u)\},$$

where $\widehat{\mathbb{C}} = \mathbb{C} \cup \{\infty\}$ denotes the extended complex plane.

PROOF. A sketch of the proof goes as follows. By **[53**, Sect. 2.6] u has an analytic continuation across $\mathbb{T} \setminus \sigma(u)$. By Theorem 6.6, f/u has a pseudocontinuation F_u in $H^2(\mathbb{D}_e)$ for any $f \in \mathcal{K}_u$. Let J be an open arc whose closure is contained in $\mathbb{T} \setminus \sigma(u)$. We know that u has an analytic continuation across J. We also know that the integral means

$$\int_J \left|\frac{f}{u}(r\zeta)\right| dm(\zeta)$$

and

$$\int_J |F_u(s\zeta)| dm(\zeta)$$

are uniformly bounded in $0 < r < 1$ and $s > 1$. One can now take a triangle Δ cut by J and show (by cutting Δ with $(1-\epsilon)J$ and $(1+\epsilon)J$ and using the boundedness of the above integral means) that the contour integral of the function defined by f/u on $\mathbb{D}$ and F_u on $\mathbb{D}_e$ integrates to zero on Δ. Now apply Morera's theorem[8]. □

6.3. Nontangential limits. Functions in H^2 possess nontangential limits almost everywhere on $\mathbb{T}$. However, one can easily see that for each fixed ζ in $\mathbb{T}$ there exists an H^2 function that does not have a finite nontangential limit at ζ (e.g., $f(z) = \log(\zeta - z)$). In sharp contrast to this, for a given model space $\mathcal{K}_u$, there might exist points ζ in $\mathbb{T}$ such that each function in $\mathcal{K}_u$ possesses a nontangential limit at ζ. We begin with a definition.

DEFINITION 6.10. If u is inner and $\zeta \in \mathbb{T}$, then u has an *angular derivative in the sense of Carathéodory* (ADC) *at* ζ if the nontangential limits of u and u' exist at ζ and $|u(\zeta)| = 1$.

A number of equivalent conditions for the existence of an angular derivative (called the Julia-Carathéodory theorem) can be found in **[89**, p. 57]. The precise relationship between angular derivatives and model spaces is contained in the following important result.

[8] There is a precise version of this type of Morera's theorem found in **[53]**.

THEOREM 6.11 (Ahern-Clark [**3,4**]). *For an inner function $u = B_\Lambda S_\mu$, where B_Λ is a Blaschke product with zeros $\Lambda = \{\lambda_n\}_{n=1}^\infty$, repeated according to multiplicity, S_μ is a singular inner function with corresponding singular measure μ, and $\zeta \in \mathbb{T}$, the following are equivalent:*

(i) *Every $f \in \mathcal{K}_u$ has a nontangential limit at ζ.*

(ii) *For every $f \in \mathcal{K}_u$, $f(\lambda)$ is bounded as $\lambda \to \zeta$ nontangentially.*

(iii) *u has an ADC at ζ.*

(iv) *the nontangential limit of u at ζ exists and* $k_\zeta(z) = \dfrac{1-\overline{u(\zeta)}u(z)}{1-\overline{\zeta}z} \in H^2$,

(v) $\displaystyle\sum_{n\geqslant 1} \frac{1-|\lambda_n|^2}{|\zeta-\lambda_n|^2} + \int_{\mathbb{T}} \frac{d\mu(\xi)}{|\xi-\zeta|^2} < \infty.$

It is worth comparing condition (v) of the preceding theorem to an old result of Frostman [**39**] that says that an inner function u (and all its divisors) has a nontangential limits of modulus one at ζ in $\mathbb{T}$ whenever

$$\sum_{n\geqslant 1} \frac{1-|\lambda_n|}{|\zeta-\lambda_n|} + \int_{\mathbb{T}} \frac{d\mu(\xi)}{|\xi-\zeta|} < \infty.$$

We should also mention that there are inner functions (in fact Blaschke products) u for which $|u'(\zeta)| = \infty$ at every point of $\mathbb{T}$ [**39**].

There are also results about *tangential* limits of functions from model spaces [**15, 20**]. When condition (v) fails then there are functions in $\mathcal{K}_u$ that do not have nontangential limits at ζ. In particular, there are functions f in $\mathcal{K}_u$ for which $|f(r\zeta)|$ is unbounded as $r \to 1^-$. However, in [**55**] there are results that give bounds (both upper and lower) on the growth of $|f(r\zeta)|$ as a function of r.

7. Conjugation

Each model space $\mathcal{K}_u$ comes equipped with a conjugation, a certain type of conjugate-linear operator that generalizes complex conjugation on $\mathbb{C}$. Not only does this conjugation cast a new light on pseudocontinuations, it also interacts with a number of important linear operators that act upon model spaces. Most of the material in this section can be found in [**40, 47**].

7.1. Basic properties. We say that a conjugate-linear function $C : \mathcal{H} \to \mathcal{H}$ on a complex Hilbert space $\mathcal{H}$ is a *conjugation* if C is isometric (i.e., $\|Cx\| = \|x\|$ for all x in $\mathcal{H}$) and involutive (i.e., $C^2 = I$). In light of the polarization identity, it turns out that the isometric condition is equivalent to the assertion that $\langle Cx, Cy\rangle = \langle y, x\rangle$ for all x, y in $\mathcal{H}$.

The structure of conjugations is remarkably simple. If C is a conjugation on $\mathcal{H}$, then there exists an orthonormal basis $\{e_n\}$ of $\mathcal{H}$ such that $Ce_n = e_n$ for all n [**41**, Lem. 1]. We refer to such a basis as a *C-real* orthonormal basis of $\mathcal{H}$. As a consequence, any conjugation is unitarily equivalent to the *canonical conjugation*

$$(z_1, z_2, \ldots) \mapsto (\overline{z_1}, \overline{z_2}, \ldots)$$

on an ℓ^2-space of the appropriate dimension.

In order to understand the natural conjugation on $\mathcal{K}_u$, we must first remind the reader that $\mathcal{K}_u = H^2 \cap u\overline{zH^2}$, as a space of functions on $\mathbb{T}$ (see Proposition 3.3).

PROPOSITION 7.1. *The conjugate-linear operator $C : \mathcal{K}_u \to \mathcal{K}_u$, defined in terms of boundary functions on $\mathbb{T}$ by $Cf = \overline{fz}u$, is a conjugation. In particular, $|f| = |Cf|$ almost everywhere on $\mathbb{T}$ so that f and Cf share the same outer factor.*

PROOF. Since $|u| = 1$ almost everywhere on $\mathbb{T}$, it follows that C is conjugate-linear, isometric, and involutive. We need only prove that C maps $\mathcal{K}_u$ into $\mathcal{K}_u$. Since f is orthogonal to uH^2, it follows that

$$\begin{aligned}
\langle Cf, \overline{zh}\rangle &= \langle \overline{fz}u, \overline{zh}\rangle \\
&= \int_{\mathbb{T}} \overline{f(\zeta)\zeta} u(\zeta)\zeta h(\zeta)\, dm(\zeta) \\
&= \int_{\mathbb{T}} u(\zeta)h(\zeta)\overline{f(\zeta)}\, dm(\zeta) \\
&= \langle uh, f\rangle \\
&= 0
\end{aligned}$$

for all h in H^2. In other words, Cf belongs to H^2. Similarly,

$$\langle Cf, uh\rangle = \langle \overline{fz}u, uh\rangle = \langle \overline{fz}, h\rangle = 0,$$

from which it follows that Cf belongs to $\mathcal{K}_u$. □

EXAMPLE 7.2. If $u(z) = z^n$, then $\mathcal{K}_u = \bigvee\{1, z, \dots, z^{n-1}\}$ and the conjugation C assumes the form

$$C(a_0 + a_1 z + \cdots a_{n-1}z^{n-1}) = \overline{a_{n-1}} + \overline{a_{n-2}}z + \cdots + \overline{a_0}z^{n-1}.$$

Now suppose that u is a finite Blaschke product having zeros $\lambda_1, \lambda_2, \dots, \lambda_n$, repeated according to multiplicity. Referring back to Proposition 3.15 for an explicit description of $\mathcal{K}_u$, one can show that

$$C\left(\frac{a_0 + a_1 z + \cdots + a_{n-1}z^{n-1}}{\prod_{i=1}^n (1 - \overline{\lambda_i}z)}\right) = \frac{\overline{a_{n-1}} + \overline{a_{n-2}}z + \cdots + \overline{a_0}z^{n-1}}{\prod_{i=1}^n (1 - \overline{\lambda_i}z)}.$$

In other words, each element of $\mathcal{K}_u$ can be represented as a rational function whose denominator is $\prod_{i=1}^n (1 - \overline{\lambda_i}z)$ and whose numerator is a polynomial of degree $\leqslant n-1$. The conjugation C acts on $\mathcal{K}_u$ by reversing the order of the polynomial in the numerator and conjugating its coefficients.

EXAMPLE 7.3. A straightforward computation reveals that C sends reproducing kernels to difference quotients:

$$\begin{aligned}
[Ck_\lambda](z) &= \overline{\left(\frac{1 - \overline{u(\lambda)}u(z)}{1 - \overline{\lambda}z}\right)}\overline{z}u(z) \\
&= \frac{1 - u(\lambda)\overline{u(z)}}{1 - \lambda\overline{z}} \cdot \frac{u(z)}{z} \\
&= \frac{u(z) - u(\lambda)}{z - \lambda},
\end{aligned}$$

where, as usual, we consider all of the functions involved as functions on $\mathbb{T}$ (i.e., so that $z\overline{z} = 1$). For each λ, the kernel function k_λ is an outer function since it is the quotient of the two outer functions $1 - \overline{u(\lambda)}u(z)$ and $1 - \overline{\lambda}z$ [**33**, Ex. 1, Ch. 3]. In light of Proposition 7.1, we expect that the difference quotient Ck_λ is simply an inner multiple of k_λ. This is indeed the case as the following computation shows.

$$\begin{aligned}
[Ck_\lambda](z) &= \frac{u(z) - u(\lambda)}{z - \lambda} \\
&= \frac{u(z) - u(\lambda)}{1 - \overline{u(\lambda)}u(z)} \cdot \frac{1 - \overline{\lambda}z}{z - \lambda} \cdot \frac{1 - \overline{u(\lambda)}u(z)}{1 - \overline{\lambda}z} \\
&= \frac{b_{u(\lambda)}(u(z))}{b_\lambda(z)} k_\lambda(z),
\end{aligned}$$

where $b_w(z) = (z - w)/(1 - \overline{w}z)$ is the disk automorphism (5.16).

EXAMPLE 7.4. If ζ is a point on $\mathbb{T}$ for which u has an ADC (see Subsection 6.3), then the boundary kernel k_ζ, as defined in Theorem 6.11, belongs to $\mathcal{K}_u$. In this case, something remarkable occurs. Since ζ and $u(\zeta)$ are of unit modulus, it follows that

$$\begin{aligned}
[Ck_\zeta](z) &= \frac{u(z) - u(\zeta)}{z - \zeta} \\
&= \frac{u(\zeta)}{\zeta} \cdot \frac{1 - \overline{u(\zeta)}u(z)}{1 - \overline{\zeta}z} \\
&= \overline{\zeta}u(\zeta)k_\zeta(z).
\end{aligned}$$

For either branch of the square root, it follows that the outer function

$$(\overline{\zeta}u(\zeta))^{\frac{1}{2}} k_\zeta$$

belongs to $\mathcal{K}_u$ and is fixed by C.

Under certain circumstances, it is possible to construct C-real orthonormal bases for $\mathcal{K}_u$ using boundary kernels. This is relatively straightforward for finite dimensional model spaces, as the following example demonstrates (see Section 8 for a discussion of the general setting).

EXAMPLE 7.5. Let u be a finite Blaschke product with n zeros repeated according to multiplicity. For the sake of simplicity, suppose that $u(0) = 0$. For each fixed α in $\mathbb{T}$, the equation $u(\zeta) = \alpha$ has precisely n distinct solutions $\zeta_1, \zeta_2, \ldots, \zeta_n$ on $\mathbb{T}$. The functions

$$e_j(z) = \frac{e^{\frac{i}{2}(\arg\alpha - \arg\zeta_j)}}{\sqrt{|u'(\zeta_j)|}} \frac{1 - \overline{\alpha}u(z)}{1 - \overline{\zeta_j}z}$$

for $j = 1, 2, \ldots, n$ form a C-real orthonormal basis of $\mathcal{K}_u$.

7.2. Associated inner functions. Fix an inner function u and let $f = I_1 F$ denote the inner-outer factorization of a nonzero function f in $\mathcal{K}_u$. Proposition 7.1 ensures that $g = Cf$ has the same outer factor as f whence we may write $g = I_2 F$ for some inner function I_2. Since $g = \overline{fzu}$ almost everywhere on $\mathbb{T}$ it follows that $I_2 F = \overline{I_1 F zu}$ whence

$$\mathcal{I}_F = \frac{\overline{Fzu}}{F}, \tag{7.6}$$

in which the inner function $\mathcal{I}_F = I_1 I_2$ is called the *associated inner function* of F with respect to u **[47]**.[9] In light of (7.6), we see that $\mathcal{I}_F$ is uniquely determined by u and F.

On the other hand, if I_1 and I_2 are inner functions such that $I_1 I_2 = \mathcal{I}_F$, then (7.6) implies that

$$I_1 F = \overline{I_2 F z} u$$

so that the functions $f = I_1 F$ and $g = I_2 F$ satisfy $f = \overline{gz}u$ almost everywhere on $\mathbb{T}$. By Proposition 3.3, it follows that f and g belong to $\mathcal{K}_u$ and satisfy $Cf = g$. Putting this all together, we obtain the following result.

PROPOSITION 7.7. *Let $F \in \mathcal{K}_u$ be outer. The set of all functions in $\mathcal{K}_u$ having outer factor F is precisely*

$$\{\theta F : \theta \text{ inner and } \theta | \mathcal{I}_F\}. \tag{7.8}$$

We define a partial ordering on (7.8) by declaring that $\theta_1 F \leq \theta_2 F$ if and only if $\theta_1 | \theta_2$. With respect to this ordering, F and $\mathcal{I}_F F$ are the unique minimal and maximal elements, respectively. Moreover, C restricts to an order-reversing bijection from (7.8) to itself.

7.3. Generators of $\mathcal{K}_u$. We say that a function f in $\mathcal{K}_u$ *generates* $\mathcal{K}_u$ if

$$\bigvee \{f, S^* f, S^{*2} f, \ldots\} = \mathcal{K}_u. \tag{7.9}$$

Recall that Proposition 3.8 tells us that if $f = \overline{gz}u$ belongs to $\mathcal{K}_u$, then (7.9) holds if and only if u and the inner factor of g are relatively prime. Recognizing that $g = Cf$ in this setting immediately yields **[47**, Prop. 4.3].

PROPOSITION 7.10. *If f belongs to $\mathcal{K}_u$ and the inner factor of Cf is relatively prime to u, then f generates $\mathcal{K}_u$.*

EXAMPLE 7.11. Each difference quotient $Ck_\lambda = (u(z) - u(\lambda))/(z - \lambda)$ generates $\mathcal{K}_u$ since its conjugate $C(Ck_\lambda) = k_\lambda$ is outer. If u is a singular inner function (i.e., u is not divisible by any Blaschke product), then Frostman's Theorem **[33**, Ex. 8, Ch. 2], **[39]** asserts that k_λ generates $\mathcal{K}_u$ for almost every (with respect to area measure) λ in $\mathbb{D}$. Indeed, Example 7.3 tells us that the inner factor of Ck_λ is precisely $b_{u(\lambda)}(u(z)))/b_\lambda(z)$, which is a Blaschke product whenever $u(\lambda)$ is nonzero and does not lie in the exceptional set for u.

COROLLARY 7.12. *If f is an outer function in $\mathcal{K}_u$, then Cf generates $\mathcal{K}_u$. In particular, any self-conjugate outer function in $\mathcal{K}_u$ generates $\mathcal{K}_u$.*

EXAMPLE 7.13. Any boundary kernel k_ζ in $\mathcal{K}_u$ generates $\mathcal{K}_u$. Indeed, Example 7.4 shows that any such function is a constant multiple of a self-conjugate outer function.

7.4. Quaternionic structure of 2×2 inner functions. In at least one instance, it turns out that conjugations on model spaces can yield structural insights into the nature of higher order inner functions. The following result can be found in **[48]**, although it is stated in **[41]** without proof (note that the reverse implication is straightforward), along with several other related results.

[9] We should note that the influential article [**31**, Rem. 3.1.6] of Douglas, Shapiro, and Shields refers to u itself as the associated inner function of f.

THEOREM 7.14. *If u is a nonconstant inner function, then the function $\Theta : \mathbb{D} \to M_2(\mathbb{C})$ defined by*

$$\Theta = \begin{bmatrix} a & -b \\ Cb & Ca \end{bmatrix} \tag{7.15}$$

is unitary almost everywhere on $\mathbb{T}$ and satisfies $\det \Theta = u$ *if and only if*

(i) *a, b, c, d belong to $\mathcal{K}_{zu}$.*

(ii) *$|a|^2 + |b|^2 = 1$ almost everywhere on $\mathbb{T}$.*

(iii) *$Ca = d$ and $Cb = c$.*

Here $C : \mathcal{K}_{zu} \to \mathcal{K}_{zu}$ is the conjugation $Cf = \overline{f}u$ on $\mathcal{K}_{zu}$.

We remark that (7.15) is entirely analogous to the representation of quaternions of unit modulus using 2×2 complex matrices. This representation was exploited in **[24]** to study the characteristic function of a complex symmetric contraction.

7.5. Cartesian decomposition. Each f in $\mathcal{K}_u$ enjoys a *Cartesian decomposition* $f = a + ib$ where the functions

$$a = \tfrac{1}{2}(f + Cf), \qquad b = \tfrac{1}{2i}(f - Cf),$$

are both fixed by C. With respect to this decomposition the conjugation C on $\mathcal{K}_u$ assumes the form $Cf = a - ib$. To describe the functions that belong to $\mathcal{K}_u$, it suffices to describe those functions that are *C-real* (i.e., such that $Cf = f$).

If $f = Cf$, then

$$f = \overline{fzu}.$$

Since u has unit modulus a.e. on $\mathbb{T}$, by replacing u with a suitable unimodular constant multiple of u (which does not change the model space $\mathcal{K}_u$), we may assume that $u(\zeta) = \zeta$ for some ζ in $\mathbb{T}$.

If u has an ADC at ζ, then Theorem 6.11 tells us that the boundary kernel k_ζ belongs to $\mathcal{K}_u$. In the event that k_ζ fails to belong to $\mathcal{K}_u$ (i.e., k_ζ is not in H^2), we still can assert that k_ζ resides in the *Smirnov class* N^+ (quotients of bounded analytic functions with outer denominator, see Definition 2.19) **[33**, Sect. 2.5**]**. A little arithmetic tells us that

$$\begin{aligned}
k_\zeta / \overline{k_\zeta} &= \frac{1 - \overline{\zeta} u}{1 - \overline{\zeta} z} \cdot \frac{1 - \zeta \overline{z}}{1 - \zeta \overline{u}} \\
&= \frac{1 - \overline{\zeta} u}{1 - \zeta \overline{u}} \cdot \frac{1 - \zeta \overline{z}}{1 - \overline{\zeta} z} \\
&= \overline{\zeta} u \left(\frac{\zeta \overline{u} - 1}{1 - \zeta \overline{u}} \right) \cdot \zeta \overline{z} \left(\frac{\overline{\zeta} z - 1}{1 - \overline{\zeta} z} \right) \\
&= \zeta \overline{\zeta} \overline{z} u \\
&= \overline{z} u,
\end{aligned}$$

which, when substituted into the equation $f = \overline{fzu}$, reveals that

$$f / k_\zeta = \overline{f / k_\zeta}.$$

Thus a function f in $\mathcal{K}_u$, where u has been normalized so that $u(\zeta) = \zeta$, satisfies $f = Cf$ if and only if f / k_ζ belongs to N^+ and is real almost everywhere on $\mathbb{T}$.

DEFINITION 7.16. A function f belonging to the Smirnov class N^+ is called a *real Smirnov function* if its boundary function is real valued almost everywhere on $\mathbb{T}$. Denote the set of all real Smirnov functions by $\mathcal{R}^+$.

The following elegant theorem of Helson **[56]** provides an explicit formula for real Smirnov functions.

THEOREM 7.17. *If $f \in \mathcal{R}^+$, then there exist inner functions ψ_1, ψ_2 such that*

$$f = i\frac{\psi_1 + \psi_2}{\psi_1 - \psi_2} \tag{7.18}$$

and $\psi_1 - \psi_2$ is outer.

PROOF. The function

$$\tau(z) = i\frac{1+z}{1-z}$$

maps $\mathbb{D}$ onto the upper half-plane. Thus $\tau^{-1} \circ f$ is of bounded characteristic (i.e., a quotient of H^∞ functions) and is unimodular almost everywhere on $\mathbb{T}$. Thus $\tau^{-1} \circ f = \psi_1/\psi_2$ is a quotient of inner functions and hence f has the desired form. □

Since each f in $\mathcal{K}_u$ can be expressed in the form $(\alpha + i\beta)k_\zeta$ where α and β belong to $\mathcal{R}^+$, it follows from Helson's Theorem that any function in $\mathcal{K}_u$ can be written in a simple algebraic manner using only a handful of inner functions. In particular, pseudocontinuations arise directly through the mechanism of Schwarz reflection (i.e., through pseudocontinuations of inner functions).

Unfortunately, the Helson representation (7.18) can be difficult to compute with. For instance, it is a somewhat messy calculation to find a Helson representation for the *Köbe function*

$$k(z) = \frac{z}{(1-z)^2} = z + 2z^2 + 3z^3 + \cdots,$$

which maps $\mathbb{D}$ bijectively onto $\mathbb{C}\backslash(-\infty, -\frac{1}{4}]$. We therefore propose a more constructive description of the functions in $\mathbb{R}^+$, which originates in **[46, 52]**. First note that if $f = I_f F$ is the inner-outer factorization of a function $\mathcal{R}^+$, then

$$I_f F = \left[\frac{-4I_f}{(1-I_f)^2}\right] \cdot \left[\frac{(1-I_f)^2 F}{-4}\right],$$

where the first term is in $\mathcal{R}^+$ and has the same inner factor as f and where the second is *outer* and in $\mathcal{R}^+$. Thus to describe functions in $\mathcal{R}^+$, it suffices to describe *real outer (RO) functions*. An infinite product expansion, in terms of Cayley-like transforms of inner functions, is obtained in **[52]**. We refer the reader there for further details.

8. Aleksandrov-Clark measures

The study of Aleksandrov-Clark measures dates back to the seminal work of Clark **[29]** and it continued with deep work of Aleksandrov **[5, 6, 9]** and Poltoratski **[75]**. Since then, Aleksandrov-Clark measures have appeared in the study of spectral theory, composition operators, completeness problems, and mathematical physics. Several sources for this important topic, including references to their applications and the connections mentioned above, are **[26, 74, 79]**.

8.1. Clark measures. If μ is a probability measure on $\mathbb{T}$ then the function

$$F_\mu(z) := \int_{\mathbb{T}} \frac{\zeta + z}{\zeta - z}\, d\mu(\zeta)$$

is analytic on $\mathbb{D}$ and satisfies $F_\mu(0) = 1$. Furthermore, a simple computation shows that

$$\Re F_\mu(z) = \int_{\mathbb{T}} \frac{1 - |z|^2}{|\zeta - z|^2}\, d\mu(\zeta)$$

is positive on $\mathbb{D}$. The following classical result of Herglotz asserts that this process can be reversed [**33**, p. 10].

THEOREM 8.1 (Herglotz). *If F is analytic on $\mathbb{D}$, $F(0) = 1$, and $\Re F > 0$ on $\mathbb{D}$, then there is a unique probability measure μ on $\mathbb{T}$ such that $F = F_\mu$.*

If u is inner and $u(0) = 0$, then for each α in $\mathbb{T}$ the function

$$\frac{1 + \overline{\alpha} u(z)}{1 - \overline{\alpha} u(z)}$$

satisfies the hypothesis of Herglotz's theorem. Indeed, this function is clearly analytic on $\mathbb{D}$, evaluates to 1 at $z = 0$, and satisfies

$$\Re\left(\frac{1 + \overline{\alpha} u(z)}{1 - \overline{\alpha} u(z)}\right) = \frac{1 - |u(z)|^2}{|\alpha - u(z)|^2} > 0.$$

By Herglotz' Theorem, there is a unique probability measure σ_α on $\mathbb{T}$ such that

$$\frac{1 + \overline{\alpha} u(z)}{1 - \overline{\alpha} u(z)} = \int \frac{\zeta + z}{\zeta - z}\, d\sigma_\alpha(\zeta). \tag{8.2}$$

The resulting family of measures

$$\{\sigma_\alpha : \alpha \in \mathbb{T}\}$$

are called the *Clark measures corresponding to u* (see Subsection 8.3 for their generalization, the *Aleksandrov-Clark measures*). The following proposition summarizes some of their basic properties (see [**26**] for details).

PROPOSITION 8.3. *For an inner function u satisfying $u(0) = 0$, the corresponding family of Clark measures $\{\sigma_\alpha : \alpha \in \mathbb{T}\}$ satisfy the following.*

(i) *$\sigma_\alpha \perp m$ for all α.*

(ii) *$\sigma_\alpha \perp \sigma_\beta$ for $\alpha \neq \beta$.*

(iii) *σ_α has a point mass at ζ if and only if $u(\zeta) = \alpha$ and $|u'(\zeta)| < \infty$. Furthermore*

$$\sigma_\alpha(\{\zeta\}) = \frac{1}{|u'(\zeta)|}.$$

(iv) *A carrier*[10] *for σ_α is the set*

$$\left\{\zeta \in \mathbb{T} : \lim_{r \to 1^-} u(r\zeta) = \alpha\right\}.$$

---

[10]By a *carrier* for σ_α we mean a Borel set $C \subset \mathbb{T}$ for which $\sigma_\alpha(A \cap C) = \sigma_\alpha(A)$ for all Borel subsets $A \subset \mathbb{T}$. Carriers are not unique and a carrier is often different from a support. For example, if a measure consists of a dense set of point masses on $\mathbb{T}$, then a carrier would consist of just these point masses, whereas the support would be $\mathbb{T}$.

This process can also be reversed. Starting with a singular probability measure μ on $\mathbb{T}$ (e.g., $\mu = \delta_1$) one forms the *Herglotz integral*

$$H_\mu(z) := \int_{\mathbb{T}} \frac{\zeta + z}{\zeta - z} d\mu(\zeta), \quad z \in \mathbb{D}.$$

This function has positive real part on $\mathbb{D}$ and satisfies $H_\mu(0) = 1$. Thus

$$(z \in \mathbb{D}) \qquad u_\mu(z) := \frac{H_\mu(z) - 1}{H_\mu(z) + 1}$$

satisfies

$$\Re\left(\frac{1 + u_\mu(z)}{1 - u_\mu(z)}\right) = \frac{1 - |u_\mu(z)|^2}{|1 - u_\mu(z)|^2} = \Re H_\mu(z) = \int_{\mathbb{T}} P_z(\zeta) d\mu(\zeta).$$

Using the fact that

$$\lim_{r \to 1^-} \int_{\mathbb{T}} P_{rw}(\zeta) d\mu(\zeta) = 0$$

for m-almost every $w \in \mathbb{T}$ (since $\mu \perp m$ – see (2.12)) we see that u_μ is inner and μ is the Clark measure with $\alpha = 1$ corresponding to u_μ.

8.2. Clark unitary operator. For an inner function u with $u(0) = 0$, recall that the compressed shift is the operator $S_u : \mathcal{K}_u \to \mathcal{K}_u$ defined by $S_u f = P_u(zf)$. For each $\alpha \in \mathbb{T}$ define

$$U_\alpha : \mathcal{K}_u \to \mathcal{K}_u, \quad U_\alpha := S_u + \alpha\Big(1 \otimes \frac{u}{z}\Big), \tag{8.4}$$

where $1 \otimes \frac{u}{z}$ denotes the rank one operator defined by $(1 \otimes \frac{u}{z})f = \langle f, \frac{u}{z}\rangle 1$ (note that since $u(0) = 0$ the constant function 1 belongs to $\mathcal{K}_u$). The main theorem here is the following (see **[26]** for details).

THEOREM 8.5 (Clark **[29]**). *For an inner function u with $u(0) = 0$, the operator* (8.4) *is a cyclic unitary operator whose spectral measure is carried by the Borel set*

$$\Big\{\zeta \in \mathbb{T} : \lim_{r \to 1^-} u(r\zeta) = \alpha\Big\}.$$

The eigenvalues of U_α are the points ζ in $\mathbb{T}$ so that $u(\zeta) = \alpha$ and $|u'(\zeta)| < \infty$. The corresponding eigenvectors are the boundary kernels k_ζ.

EXAMPLE 8.6. Consider the model space $\mathcal{K}_{z^3} = \bigvee\{1, z, z^2\}$. Letting $S_3 := S_{z^3}$, we see that

$$S_3 1 = z, \quad S_3 z = z^2, \quad S_3 z^2 = 0,$$

so that the matrix representation of S_3 with respect to the basis $\{1, z, z^2\}$ is

$$\begin{bmatrix} 0 & 0 & 0 \\ 1 & 0 & 0 \\ 0 & 1 & 0 \end{bmatrix}.$$

The Clark unitary operator in this case is

$$U_\alpha = S_3 + \alpha(1 \otimes z^2).$$

Similarly, since

$$U_\alpha 1 = z, \quad U_\alpha z = z^2, \quad U_\alpha z^2 = \alpha,$$

the matrix representation of U_α with respect to this basis is

$$\begin{bmatrix} 0 & 0 & \alpha \\ 1 & 0 & 0 \\ 0 & 1 & 0 \end{bmatrix}.$$

The unitarity of U_α follows easily from the fact that $|\alpha| = 1$. The eigenvalues of U_α are precisely the solutions to $u(z) = \alpha$ (i.e., the three cube roots of α) and corresponding eigenvectors are the boundary kernels

$$k_\beta(z) = \frac{1 - \overline{\beta}^3 z^3}{1 - \overline{\beta} z} = 1 + \overline{\beta} z + \overline{\beta}^2 z^2,$$

where $\beta^3 = \alpha$. Each eigenvector k_β has norm $\sqrt{3}$, which is precisely the square root of the modulus of the angular derivative (namely $3z^2$) of z^3 at the point β in $\mathbb{T}$. Moreover, the normalized eigenvectors of U_α form an orthonormal basis for $\mathcal{K}_{z^3}$ (as expected since U_α is a unitary operator on a finite dimensional space).

The Clark theory also gives us a concrete spectral representation for U_α via the Clark measure σ_α.

THEOREM 8.7 (Clark [29]). *For an inner function u with $u(0) = 0$, let σ_α be the unique finite positive Borel measure on $\mathbb{T}$ satisfying*

$$\frac{1 + \overline{\alpha} u(z)}{1 - \overline{\alpha} u(z)} = \int \frac{\zeta + z}{\zeta - z}\, d\sigma_\alpha(\zeta).$$

The operator

$$(V_\alpha f)(z) = (1 - \overline{\alpha} u(z)) \int \frac{f(\zeta)}{1 - \overline{\zeta} z}\, d\sigma_\alpha(\zeta)$$

is a unitary operator from $L^2(\sigma_\alpha)$ onto $\mathcal{K}_u$. Furthermore, if

$$Z_\alpha : L^2(\sigma_\alpha) \to L^2(\sigma_\alpha), \quad (Z_\alpha f)(\zeta) = \zeta f(\zeta),$$

then $V_\alpha Z_\alpha = U_\alpha V_\alpha$.

EXAMPLE 8.8. Returning to Example 8.6. Let $\zeta_1, \zeta_2, \zeta_3$ be the three solutions to $u(z) = \alpha$ (i.e., to $z^3 = \alpha$). Then the discrete measure

$$d\sigma_\alpha = \tfrac{1}{3}\delta_{\zeta 1} + \tfrac{1}{3}\delta_{\zeta_2} + \tfrac{1}{3}\delta_{\zeta_3}$$

satisfies

$$\frac{1 + \overline{\alpha} z^3}{1 - \overline{\alpha} z^3} = \int \frac{\zeta + z}{\zeta - z}\, d\sigma_\alpha(\zeta).$$

From the theorem above,

$$V_\alpha : L^2(\sigma_\alpha) \to \mathcal{K}_{z^3}, \quad (V_\alpha f)(z) = (1 - \overline{\alpha} z^3) \int \frac{f(\zeta)}{1 - \overline{\zeta} z}\, d\sigma_\alpha(\zeta).$$

To see this worked out, note that if

$$f = c_1 \chi_{\zeta_1} + c_2 \chi_{\zeta_2} + c_3 \chi_{\zeta_3},$$

a typical element of $L^2(\sigma_\alpha)$, then

$$(Vf)(z) = (1 - \overline{\alpha} z^3) \left(c_1 \frac{1/3}{1 - \overline{\zeta_1} z} + c_2 \frac{1/3}{1 - \overline{\zeta_2} z} + c_3 \frac{1/3}{1 - \overline{\zeta_3} z} \right).$$

Since $\zeta_j^3 = \alpha$, one can verify that the expression above is a polynomial of degree at most 2, which are precisely the elements of $\mathcal{K}_{z^3}$. Furthermore, $\{\sqrt{3}\chi_{\zeta_1}, \sqrt{3}\chi_{\zeta_2}, \sqrt{3}\chi_{\zeta_3}\}$ is an orthonormal basis for $L^2(\sigma_\alpha)$ and then

$$(V_\alpha \sqrt{3}\chi_{\zeta_j})(z) = \frac{\sqrt{3}}{3} k_{\zeta_j}(z)$$

is an orthonormal basis for $\mathcal{K}_{z^3}$. We can also directly verify that V_α intertwines U_α with Z_α.

The adjoint of the unitary operator V_α is of particular interest. We know that $V_\alpha f$ belongs to $\mathcal{K}_u$ for every f in $L^2(\sigma_\alpha)$ and, as such, $V_\alpha f$ has a radial limit almost everywhere with respect to Lebesgue measure m. The following theorem of Poltoratski **[75]** says much more.

THEOREM 8.9 (Poltoratski **[75]**). *For $f \in L^2(\sigma_\alpha)$,*

$$\lim_{r\to 1^-} (V_\alpha f)(r\zeta) = f(\zeta)$$

for σ_α-almost every $\zeta \in \mathbb{T}$.

The significance of this result is that it says that functions in model spaces (i.e., the functions $V_\alpha f$ for $f \in L^2(\sigma_\alpha)$ in the previous theorem) have nontangential limits on finer sets than generic functions from H^2. We have seen a result along these lines already in Theorem 6.11. The previous theorem says a bit more in that all functions from $\mathcal{K}_u$ have non-tangential limits σ_α almost everywhere. This is especially notable since $\sigma_\alpha \perp m$ and a classical result **[63, 64]** says that if E is any closed subset of $\mathbb{T}$ with Lebesgue measure zero, then there exists an f in H^2 (which can be taken to be inner) whose radial limits do not exists on E. For example, if σ_α has an isolated point pass at ζ, then every function in $\mathcal{K}_u$ has a nontangential limit at ζ. In this particular case, the result should not be surprising since, by Proposition 8.3, u will have a finite angular derivative at ζ and so by the Ahern-Clark result (Theorem 6.11) every function in $\mathcal{K}_u$ has a nontangential limit at ζ.

Here is one more fascinating and useful gem about Clark measures due to Aleksandrov **[5]** (see also **[26]**).

THEOREM 8.10 (Aleksandrov **[5]**). *Let u be an inner function with associated family of Clark measures $\{\sigma_\alpha : \alpha \in \mathbb{T}\}$. If $f \in C(\mathbb{T})$, then*

$$\int_{\mathbb{T}} \left(\int_{\mathbb{T}} f(\zeta) d\sigma_\alpha(\zeta) \right) dm(\alpha) = \int_{\mathbb{T}} f(\xi)\, dm(\xi).$$

PROOF. For a fixed $z \in \mathbb{D}$ notice that if $P_z(\zeta)$ is the Poisson kernel, then taking real parts of both sides of (8.2) we get

$$\begin{aligned}
\int_{\mathbb{T}} \left(\int_{\mathbb{T}} P_z(\zeta) d\sigma_\alpha(\zeta) \right) dm(\alpha) &= \int_{\mathbb{T}} \left(\frac{1 - |u(z)|^2}{|\alpha - u(z)|^2} \right) dm(\alpha) \\
&= \int_{\mathbb{T}} P_{u(z)}(\alpha) dm(\alpha) \\
&= 1 \\
&= \int_{\mathbb{T}} P_z(\zeta) dm(\zeta).
\end{aligned}$$

Thus the theorem works for finite linear combinations of Poisson kernels. Using the fact that finite linear combinations of Poisson kernels are dense in $C(\mathbb{T})$ one can use a limiting argument to get the general result[11]. □

REMARK 8.11. Aleksandrov **[5]** showed that $C(\mathbb{T})$ can be replaced by L^1 in the preceding disintegration theorem. There are a few technical issues to work out here. For example, the inner integrals

$$\int f(\zeta) d\sigma_\alpha(\zeta)$$

in the disintegration formula do not seem to be well defined for L^1 functions since indeed the measure σ_α can contain point masses on $\mathbb{T}$ while L^1 functions are defined merely m-almost everywhere. However, amazingly, the function

$$\alpha \mapsto \int f(\zeta) d\sigma_\alpha(\zeta)$$

is defined for m-almost every α and is integrable. An argument with the monotone class theorem is used to prove this more general result. See also **[26]** for the details.

8.3. Aleksandrov-Clark measures. The alert reader will have noticed that the title of this section was Aleksandrov-Clark measures but we seem to have only discussed *Clark* measures. As it turns out, one can still examine the equation

$$\frac{1+\overline{\alpha}u(z)}{1-\overline{\alpha}u(z)} = \int \frac{\zeta+z}{\zeta-z} d\sigma_\alpha(\zeta)$$

but where u belongs to H^∞ and $\|u\|_\infty \leqslant 1$, but u is not necessarily inner. The measures σ_α will no longer be singular but they do form an interesting class of measures explored by Aleksandrov and others (see **[26]** for references). One can identify all the parts of these measures (absolutely continuous, singular continuous, point masses, etc.) as well as develop a decomposition theorem as before. Versions of these measures appear in mathematical physics through some papers of B. Simon and T. Wolff **[92, 93]**.

9. Completeness problems

In Subsection 5.3, we discussed circumstances under which a sequence $\{k_\lambda\}_{\lambda\in\Lambda}$ of kernel functions forms a Riesz basis for $\mathcal{K}_u$ (see Theorem 5.15). More generally, one can ask the following question: If Λ is a sequence in $\mathbb{D}$, when does

$$K_\Lambda := \bigvee_{\lambda\in\Lambda} k_\lambda = \mathcal{K}_u?$$

Such problems are known as *completeness problems*.

By considering orthogonal complements and the reproducing property of k_λ, we see that $K_\Lambda = \mathcal{K}_u$ precisely when there are no nonzero functions in $\mathcal{K}_u$ that vanish on Λ. We have already seen that $K_\Lambda = \mathcal{K}_u$ if Λ is not a Blaschke sequence or if Λ has an accumulation point in $\mathbb{D}$ (Proposition 5.7). A little further thought shows that the same holds if Λ has an accumulation point in $\mathbb{T}\setminus\sigma(u)$ since otherwise there would exist a nonzero function in $\mathcal{K}_u$ whose analytic continuation vanishes on a set having a limit point in its domain (Proposition 6.9). For general

[11] See **[26**, Ch. 9] for the details on the density of the linear span of the Poisson kernels in $C(\mathbb{T})$ as well as the completion of this limiting argument

Blaschke sequences, the situation is more complicated. However, we can rephrase the completeness problem in terms of Toeplitz kernels[12].

PROPOSITION 9.1. *If Λ is a Blaschke sequence in $\mathbb{D}$ and B is the Blaschke product corresponding to Λ, then for any inner function u we have*

$$K_\Lambda \neq \mathcal{K}_u \quad \iff \quad \ker T_{\overline{u}B} \neq \{0\}.$$

PROOF. Note that if $K_\Lambda \neq \mathcal{K}_u$, then there is a nonzero function f in $\mathcal{K}_u$ that vanishes on Λ with the appropriate multiplicities. This happens precisely when $f = gB$ for some g in H^2 that is not identically zero. But since f belongs to $\mathcal{K}_u$ we know that $\overline{u}f$ is contained in $\overline{zH^2}$. This means that $\overline{u}Bg$ belongs to $\overline{zH^2}$. Now apply the Riesz projection (of L^2 onto H^2) to $\overline{u}Bg$ to see that $T_{\overline{u}B}g = P(\overline{u}Bg) = 0$. For the other direction, simply reverse the argument. □

Clark's approach in [29] to the completeness problem is based on the following related approximation problem of Paley-Wiener. Suppose that $\{x_n\}_{n\geqslant 1}$ is an orthonormal basis for a Hilbert space $\mathcal{H}$ and $\{y_n\}_{n\geqslant 1}$ is a sequence in $\mathcal{H}$. If these sequences are *close* in some way, does $\{y_n\}_{n\geqslant 1}$ span $\mathcal{H}$? There are various results which say this is often the case (see [76, Sec. 86] or [69] for a survey of these results) with various interpretations of the term *close*. We will see a specific example of this below.

Clark's approach to the completeness problem $K_\Lambda = \mathcal{K}_u$ was to find an orthonormal basis for $\mathcal{K}_u$ consisting of (normalized) boundary kernel functions. One way to accomplish this is to find a unitary operator U on $\mathcal{K}_u$ whose spectrum consists only of eigenvalues (pure point spectrum) and such that the eigenvectors of U are boundary kernel functions. But indeed we have already seen this can be done. From Clark's theorem (Theorem 8.5) we see that U_α is unitary and, under the right spectral circumstances (i.e., the corresponding Clark measure σ_α is purely atomic), its normalized eigenvectors

$$\left\{ \frac{k_\zeta}{\sqrt{|u'(\zeta)|}} : u(\zeta) = \alpha, |u'(\zeta)| < \infty \right\}$$

form an orthonormal basis for $\mathcal{K}_u$. From here one can apply a Paley-Wiener type result to obtain completeness results for the family $\{k_\lambda : \lambda \in \Lambda\}$. Let us give two illustrative examples from the original Clark paper [29].

EXAMPLE 9.2. Suppose that u is an inner function and $\alpha \in \mathbb{T}$ such that the corresponding Clark measure σ_α is discrete, i.e.,

$$\sigma_\alpha = \sum_{n=1}^{\infty} \frac{1}{|u'(\zeta_n)|} \delta_{\zeta_n}.$$

Note the use of Proposition 8.3 here where $\{\zeta_n\}_{n\geqslant 1}$ are the solutions to $u = \alpha$. Since σ_α is a discrete measure and U_α is unitary, the normalized eigenvectors

$$h_{\zeta_n} := \frac{k_{\zeta_n}}{\sqrt{|u'(\zeta_n)|}}$$

[12] Kernels of Toeplitz operators have been studied before and we refer the reader to the papers [35, 47, 83] for further details.

are an orthonormal basis for $\mathcal{K}_u$. For our given sequence $\Lambda = \{\lambda_n\}_{n \geqslant 1} \subset \mathbb{D}$, let

$$h_{\lambda_n} := \frac{k_{\lambda_n}}{\|k_{\lambda_n}\|}$$

be the normalized kernel functions. If the sequence $\{\lambda_n\}_{n\geqslant 1}$ satisfies

$$\sum_{n=1}^{\infty} \|h_{\zeta_n} - h_{\lambda_n}\|^2 < 1,$$

then a generalization of a theorem of Paley and Wiener (see **[76**, Sec. 86]) says that $\{h_{\lambda_n}\}_{n\geqslant 1}$ forms an unconditional basis for $\mathcal{K}_u$, meaning that every f in $\mathcal{K}_u$ has an expansion $f = \sum_n a_n h_{\lambda_n}$.

This next example brings in some earlier work of Sarason **[80]**.

EXAMPLE 9.3. Consider the model space $\mathcal{K}_u$ with inner function

$$u(z) := \exp\left(\frac{z+1}{z-1}\right).$$

The Clark measure σ_1 corresponding to u has the set $\{u = 1\}$ as its carrier, which turns out to be the discrete set

$$(n \in \mathbb{Z}) \qquad \zeta_n := \frac{1 + 2\pi i n}{1 - 2\pi i n}.$$

In **[80]** it is shown that the operator

$$[Wf](z) := \frac{\sqrt{2}}{1-z} \int_0^1 f(t) u^t(z)\, dt$$

defines a unitary operator which maps $L^2[0,1]$ onto $\mathcal{K}_u$. In our current setting, the real usefulness of this operator comes from the formula

$$W(e^{i\gamma t}) = q k_\lambda, \quad \gamma := i\frac{1+\overline{\lambda}}{1-\overline{\lambda}},$$

where $|\lambda| \leqslant 1$ and q is a constant. In particular, this means that W maps each orthonormal basis element $e^{2\pi i n t}$ for $L^2[0,1]$ to a constant times k_{ζ_n}. So, for a given a sequence $\{\lambda_n\}_{n\in\mathbb{Z}} \subset \mathbb{D}$ satisfying

$$\max_{n\in\mathbb{Z}} |\gamma_n - 2\pi n| < \frac{\log 2}{\pi}, \qquad \gamma_n = i\frac{1+\overline{\lambda_n}}{1-\overline{\lambda_n}},$$

a theorem of Paley and Wiener (see **[76**, Sec. 86]) says that $\{e^{i\gamma_n t}\}_{n\in\mathbb{Z}}$ forms an unconditional basis for $L^2[0,1]$. Under this criterion, we get that the normalized kernels $\{k_{\lambda_n}/\|k_{\lambda_n}\| : n \in \mathbb{Z}\}$ form an unconditional basis for $\mathcal{K}_u$.

We will consider other unitary operators that transfer orthonormal bases of L^2 to kernel functions in the next section.

10. Model spaces for the upper-half plane

10.1. The Hardy space again. Let

$$\mathbb{C}_+ := \{z \in \mathbb{C} : \Im z > 0\}$$

denote the upper half plane and let $\mathscr{H}^2$ denote the *Hardy space of the upper half plane*. These are the analytic functions f on $\mathbb{C}_+$ for which

$$\sup_{y>0} \int_{-\infty}^{\infty} |f(x+iy)|^2 \, dx < \infty.$$

The reader will immediately notice the analogue of the "bounded integral mean" condition (2.1) from the Hardy space H^2 on the unit disk. As it turns out, the majority of the theory from H^2 (on the disk) transfers over *mutatis mutandis* to $\mathscr{H}^2$. For example, every f in $\mathscr{H}^2$ has an almost everywhere well-defined "radial" boundary function

$$f(x) := \lim_{y \to 0^+} f(x+iy)$$

and this function belongs to $L^2(dx) := L^2(\mathbb{R}, dx)$. Moreover,

$$\int_{-\infty}^{\infty} |f(x)|^2 dx = \sup_{y>0} \int_{-\infty}^{\infty} |f(x+iy)|^2 dx.$$

This allows us to endow $\mathscr{H}^2$ with an inner product

$$\langle f, g \rangle := \int_{-\infty}^{\infty} f(x)\overline{g(x)} dx,$$

where, in a manner that is analogous to the H^2 case, $f(x)$ and $g(x)$ are the almost everywhere defined boundary functions of f, g in $\mathscr{H}^2$. Still further, equating H^2 with the functions in $L^2(\mathbb{T}, m)$ whose negatively indexed Fourier coefficients vanish, we have the corresponding characterization

$$\mathscr{H}^2 := \{\widehat{f} : f \in L^2(0, \infty)\},$$

where

$$\widehat{f}(t) = \int_{-\infty}^{\infty} f(x) e^{-2\pi i x t} \, dx$$

is the *Fourier transform* of f.

The reason that most of the Hardy space theory on the disk can be imported into $\mathscr{H}^2$ is because of the fact that the operator

$$\mathcal{U} : L^2(m) \to L^2(dx), \quad (\mathcal{U}f)(x) := \frac{1}{\sqrt{\pi}(x+i)} f(w(x)),$$

where

$$w(z) := \frac{z-i}{z+i}$$

is a conformal map from $\mathbb{C}_+$ onto $\mathbb{D}$, is a unitary map from $L^2(\mathbb{T}, m)$ onto $L^2(\mathbb{R}, dx)$. In particular, $\mathcal{U}$ maps H^2 unitarily onto $\mathscr{H}^2$ and maps $L^2(\mathbb{T}, m)$ isometrically onto $L^2(\mathbb{R}, dx)$.

10.2. Inner functions. We say that an analytic function Θ on $\mathbb{C}_+$ is *inner* if $|\Theta(z)| \leqslant 1$ for all $z \in \mathbb{C}_+$ and if the almost everywhere defined boundary function $\Theta(x)$ is unimodular almost everywhere on $\mathbb{R}$. The two most basic types of inner functions on $\mathbb{C}_+$ are

$$(c > 0) \qquad S^c(z) := e^{icz}$$

and

$$(\lambda \in \mathbb{C}_+) \qquad b_\lambda(z) := \frac{z - \lambda}{z - \overline{\lambda}}.$$

The first class of examples are the most basic type of *singular inner function* on $\mathbb{C}_+$ and the second type is the most basic type of *Blaschke product*.

Further examples of singular inner functions are given by

$$S_{\mu,c}(z) := e^{icz} \exp\left(-\frac{1}{\pi i}\int_{-\infty}^{\infty}\left(\frac{1}{x-z}-\frac{x}{1+x^2}\right)d\mu(x)\right),$$

where $c > 0$ and μ is a positive measure on $\mathbb{R}$ that is singular with respect to Lebesgue measure and is *Poisson finite*, in the sense that

$$\int_{-\infty}^{\infty}\frac{1}{1+x^2}d\mu(x) < \infty.$$

The general Blaschke product is formed by specifying its zeros

$$\Lambda = \{\lambda_n\}_{n\geqslant 1} \subset \mathbb{C}_+\backslash\{i\}$$

and forming the product

$$B_\Lambda(z) = b_i(z)^m \prod_{n\geqslant 1} \epsilon_n b_{\lambda_n}(z),$$

where the constants

$$\epsilon_n := \frac{|\lambda_n^2+1|}{\lambda_n^2+1}$$

are chosen so that $\epsilon_n b_{\lambda_n}(i) > 0$. It is well known **[53]** that the product above converges uniformly on compact subsets of $\mathbb{C}_+$ to an inner function if and only if the zeros $\{\lambda_n\}_{n\geqslant 1}$ satisfy the *Blaschke condition*

$$\sum_{n\geqslant 1}\frac{\Im\lambda_n}{1+|\lambda_n|^2} < \infty.$$

Every inner function Θ can be factored uniquely as

$$(\gamma \in [0,2\pi)) \qquad \Theta = e^{i\gamma}B_\Lambda S_{\mu,c}$$

10.3. Model spaces. For an inner function Θ define a *model space* $\mathscr{K}_\Theta$ on the upper-half plane as

$$\mathscr{K}_\Theta := \mathscr{H}^2 \ominus \Theta\mathscr{H}^2.$$

Much of the theory for the model spaces $\mathcal{K}_u$ on the disk carries over to the upper-half plane. For instance, via Proposition 3.3, we can regard $\mathscr{K}_\Theta$ as a space of boundary functions on $\mathbb{R}$ by noting that

$$\mathscr{K}_\Theta := \mathscr{H}^2 \cap \Theta\overline{\mathscr{H}^2}. \tag{10.1}$$

An inner function Θ has an analytic continuation across $\mathbb{R}\setminus\sigma(\Theta)$, where

$$\sigma(\Theta) := \left\{z \in \mathbb{C}_+^- : \liminf_{\lambda\to z}|\Theta(\lambda)| = 0\right\}$$

is the *spectrum* of Θ, which turns to be the union of the closure of the zeros of Θ along with the support of the singular measure for Θ. Every f in $\mathscr{K}_\Theta$ has an analytic continuation across $\mathbb{R}\backslash\sigma(\Theta)$. Using the identity (10.1), we see that for each f in $\mathscr{K}_\Theta$ there is a corresponding g in $\mathscr{H}^2$ for which $f = \overline{g}\Theta$ almost everywhere on $\mathbb{R}$. As was done with (6.7) this allows us to obtain a formula

$$(z \in \mathbb{C}_-) \qquad \widetilde{f}(z) := \frac{\overline{g(\overline{z})}}{\overline{\Theta(\overline{z})}}$$

for a pseudocontinuation of f (to the lower half-plane $\mathbb{C}_-$) of bounded type.

Of particular importance here are the reproducing kernel functions

$$(\lambda, z \in \mathbb{C}_+) \qquad K_\lambda(z) = \frac{i}{2\pi} \frac{1 - \overline{\Theta(\lambda)}\Theta(z)}{z - \overline{\lambda}}$$

These belong to $\mathscr{K}_\Theta$ and satisfy

$$(\lambda \in \mathbb{C}_+) \qquad f(\lambda) = \langle f, K_\lambda \rangle$$

10.4. Clark theory again. As to be expected, there is a Clark theory for $\mathscr{K}_\Theta$. Indeed, for an inner function Θ, the function

$$m = i\frac{1+\Theta}{1-\Theta}$$

is analytic on $\mathbb{C}_+$ and satisfies $\Im m \geqslant 0$ there (i.e., m is a *Herglotz function*). In this setting, the Herglotz representation theorem guarantees the existence of parameters $b \geqslant 0, c \in \mathbb{R}$, and a positive Poisson finite measure μ_Θ on $\mathbb{R}$ so that

$$m(\lambda) = b\lambda + c + \frac{1}{\pi}\int \left(\frac{1}{x-\lambda} - \frac{x}{1+x^2} \right) d\mu_\Theta(x).$$

Using Poltoratski's result (Theorem 8.9) we can define the operator

$$Q_\Theta : \mathscr{K}_\Theta \to L^2(\mu_\Theta), \quad Qf = f|C_\Theta, \tag{10.2}$$

where C_Θ is a carrier for Θ. This operator turns out to be unitary.

As a final remark, the alert reader is probably getting *déjà vu* when looking at the function m and its associated inner function Θ. The current discussion is probably reminding the reader of our earlier treatment of Clark measures. Indeed, if one begins with an inner Θ and looks at the family of inner functions

$$\{\overline{\alpha}\Theta : \alpha \in \mathbb{T}\},$$

there is an associated family of Poisson finite measures

$$\{\mu_{\overline{\alpha}\Theta} : \alpha \in \mathbb{T}\},$$

the Clark measures associated with Θ. Many of the properties of Clark measures that hold for inner u on $\mathbb{D}$ have direct analogs for inner functions Θ on $\mathbb{C}_+$.

10.5. Weyl-Titchmarsh inner functions. In this section we point out a relationship between Schrödinger operators and model spaces. In order not to get too deep into technical details, we will be a little vague about proper definitions. The reader looking for precision should consult **[65, 97]**.

For a real potential q in $L^2(a,b)$, define the *Schrödinger operator*

$$u \mapsto u'' + qu.$$

If one imposes the boundary condition

$$u(b)\cos\varphi + u'(b)\sin\varphi = 0,$$

in which $\varphi \in [0, 2\pi)$, this operator becomes a densely defined, self-adjoint operator on $L^2(a,b)$.

For each λ in $\mathbb{C}$ the differential equation $u'' + qu = \lambda u$ has a non-trivial solution u_λ and one can form the *Weyl-Titchmarsh* function (at a) by

$$m(\lambda) := \frac{u_\lambda'(a)}{u_\lambda(a)}.$$

One can show that the function Θ defined on $\mathbb{C}_+$ by

$$\Theta := \frac{m-i}{m+i}$$

is inner.

For each $\lambda \in \mathbb{C}_+$ define

$$w_\lambda(x) := \frac{u_\lambda(x)}{u'_\lambda(a) + iu_\lambda(a)}$$

and notice that w_λ is also a solution to the Schrödinger equation $u'' + qu = \lambda u$, the so called *normalized solution*. Now define the following transform W on $L^2(a,b)$ (sometimes called the *modified Fourier transform*) by

$$[Wf](\lambda) := \int_a^b f(x) w_\lambda(x) dx.$$

Schrödinger operators and model spaces are related by means of the following theorem from **[65]**.

THEOREM 10.3 (Makarov-Poltoratski). *The modified Fourier transform W is, up to a factor of $\sqrt{\pi}$, a unitary operator from $L^2(a,b)$ onto $\mathcal{K}_\Theta$. Furthermore, $W\overline{w_\lambda} = \pi K_\lambda$.*

With the preceding theorem one can recast the completeness problems for solutions of the Schrödinger equation in $L^2(a,b)$ in terms of the completeness problem for reproducing kernels in $\mathcal{K}_\Theta$. Secondly, the unitary operator $WQ_\Theta : L^2(a,b) \to L^2(\mu_\Theta)$ yields a spectral representation of the Schrödinger operator $u \mapsto u'' + qu$ in that it is unitarily equivalent to the densely defined operator multiplication by the independent variable on $L^2(\mu_\Theta)$.

11. Generalizations of model spaces

11.1. Model spaces in H^p. One can also consider model subspaces of H^p for $p \neq 2$. Recall that if $0 < p < \infty$, then H^p is the space of analytic functions f on $\mathbb{D}$ for which

$$\|f\|_p := \lim_{r \to 1^-} \left(\int_{\mathbb{T}} |f(r\zeta)|^2 dm(\zeta) \right)^{1/p}$$

is finite. If $p \geqslant 1$, then H^p is a Banach space while if $0 < p < 1$, then H^p is a topological vector space. Almost all of the basic function-theoretic results that hold for H^2 functions carry over to H^p (e.g., each H^p has an almost everywhere defined nontangential boundary function that belongs to L^p).

For $p \neq 2$, the spaces H^p are not Hilbert spaces, although they do have readily identifiable duals:

- For $1 < p < \infty$, the dual space $(H^p)^*$ can be identified with H^q, where q is the Hölder conjugate index to p (i.e., $\frac{1}{p} + \frac{1}{q} = 1$).
- For $p = 1$, the dual space $(H^1)^*$ can be identified with the space of all analytic functions of bounded mean oscillation (BMOA),
- For $0 < p < 1$, $(H^p)^*$ can be identified with certain classes of smooth functions on $\mathbb{D}^-$ (Lipschitz or Zygmund classes).

In each of these cases, the dual pairing is given by the *Cauchy pairing*

$$\lim_{r \to 1^-} \int_{\mathbb{T}} f(r\zeta)\overline{g(r\zeta)}\, dm(\zeta) \tag{11.1}$$

where $f \in H^p$ and $g \in (H^p)^*$. Standard references are [**33, 53**].

The unilateral shift $S : H^p \to H^p$ defined as usual by $Sf = zf$ is continuous on H^p and, for $1 \leqslant p < \infty$, the invariant subspaces of S are still the Beurling-type invariant subspaces uH^p for some inner u. The invariant subspaces of the backward shift[13]

$$Bf = \frac{f - f(0)}{z}$$

on H^p are known to be

$$(1 \leqslant p < \infty) \qquad \mathcal{K}_u^p := H^p \cap u\overline{zH^p}$$

For $1 < p < \infty$, the proof relies on using the duality relationship between $(H^p)^*$ and H^q via the integral pairing (11.1) to compute the annihilator of uH^q. For $p = 1$, the backward shift invariant subspaces are still the same (i.e., $\mathcal{K}_u^1 = H^1 \cap u\overline{zH^1}$), except that the proof has to deal with some complications from the more difficult dual space of H^1. When $0 < p < 1$, the backward shift invariant subspaces are too complicated to describe in this survey. A complete account of these results can be found in [**27**]. Most of the function-theoretic results of this survey for model spaces $\mathcal{K}_u$ in H^2 can be restated appropriately for $\mathcal{K}_u^p$ spaces – usually with nearly the same proof.

11.2. de Branges-Rovnyak spaces. There is an important generalization of model spaces, the *de Branges-Rovnyak spaces*, that play an increasingly important role in analysis. Unlike the model spaces $\mathcal{K}_u$, which are parameterized by inner functions u, these spaces are parameterized by functions u in H_1^∞, the unit ball in H^∞ (i.e., $u \in H^\infty, \|u\|_\infty \leqslant 1$). These spaces often have similar properties as model spaces but they also have many differences. We will not go into great detail here but just point out the existence of these types of spaces and their very basic properties. The standard reference for this subject is Sarason's book [**84**] from 1994, although a new book by Fricain and Mashreghi should appear soon [**38**].

For $u \in H_1^\infty$ (not necessarily inner!), define the kernel function

$$(\lambda, z \in \mathbb{D}) \qquad k_\lambda(z) := \frac{1 - \overline{u(\lambda)}u(z)}{1 - \overline{\lambda}z}.$$

The author will recognize this kernel, when u is an inner function, as the reproducing kernel for the model space $\mathcal{K}_u$, but here u is not necessarily inner. We construct a reproducing kernel Hilbert space $\mathscr{H}(u)$ of analytic functions on $\mathbb{D}$ from this kernel in the following way. First we notice that this kernel is positive in the sense that

$$\sum_{1 \leqslant l,j \leqslant n} c_j \overline{c_l} k_{\lambda_j}(\lambda_l) \geqslant 0$$

for every set of constants $c_1, \ldots, c_n$ and points $\lambda_1, \ldots, \lambda_n$ in $\mathbb{D}$. We initially populate our space $\mathscr{H}(u)$ with finite linear combinations of kernel functions and define

[13]Much of the literature here abuses notation and often uses the notation $S^*f = (f - f(0))/z$ for the backward shit on H^p even though, technically speaking, the adjoint S^* is defined on $(H^p)^*$ and not H^p.

a norm on these functions in a way that makes the k_λ the reproducing kernels for $\mathscr{H}(u)$. We do this by defining

$$\left\|\sum_{j=1}^{n} c_j k_{\lambda_j}\right\|^2 = \left\langle \sum_{j=1}^{n} c_j k_{\lambda_j}, \sum_{l=1}^{n} c_l k_{\lambda_l} \right\rangle := \sum_{1 \leqslant l,j \leqslant n} c_j \overline{c_l} k_{\lambda_j}(\lambda_l).$$

One can check that this is a norm and that the corresponding inner product on the vector space of finite linear combinations of kernel functions, make this vector space a pre-Hilbert space. The de Branges-Rovnyak space $\mathscr{H}(u)$ is the closure of this pre-Hilbert space under this norm. A more standard way of defining $\mathscr{H}(u)$ in terms of Toeplitz operators is described in [**84**].

Using the positivity of the kernel, one can prove that $\mathscr{H}(u)$ is reproducing kernel Hilbert space that is contractively contained in H^2. When $\|u\|_\infty < 1$, then $\mathscr{H}(u) = H^2$ with an equivalent norm. On the other hand when u is inner, then $\mathscr{H}(u) = \mathcal{K}_u$ with equality of norms.

Recall that the compressed shift S_u on $\mathcal{K}_u$ serves as a model for certain types of contraction operators. It turns out that $\mathscr{H}(u)$ spaces can be used to model other types of contractions where the condition (i) of Theorem 4.1 is somewhat relaxed.

12. Truncated Toeplitz operators

We do not attempt to give a complete overview of the rapidly developing theory of truncated Toeplitz operators. The recent survey article [**51**] and the seminal article [**85**] of Sarason should be consulted for a more thorough treatment. We content ourselves here with a brief summary of the principal definitions and basic results.

Recall from Subsection 3.1 that the Toeplitz operator $T_\varphi : H^2 \to H^2$ with symbol φ in L^∞ is defined by $P(\varphi f)$, where P is the orthogonal projection from L^2 onto H^2. One can show that T_φ is a bounded operator on H^2 that satisfies $T_\varphi^* = T_{\overline{\varphi}}$ and $\|T_\varphi\| = \|\varphi\|_\infty$. The (j,k) entry of the matrix representation of T_φ with respect to the orthonormal normal basis $\{1, z, z^2, \ldots\}$ of H^2 is $\widehat{\varphi}(j-k)$, which yields an infinite Toeplitz matrix (constant along the diagonals).

An old result of Brown and Halmos [**19**] gives a convenient algebraic characterization of Toeplitz operators: a bounded operator T on H^2 is a Toeplitz operator if and only

$$T = STS^*, \tag{12.1}$$

where S is the unilateral shift on H^2. Many other things are known about Toeplitz operators: their spectrum, essential spectrum, commutator ideals, algebras generated by certain collections of Toeplitz operators, etc. These topics are all thoroughly discussed in [**18, 32, 67**].

It is well-known that the commutant

$$\{S\}' := \{A \in \mathcal{B}(H^2) : SA = AS\}$$

of the unilateral shift S is equal to the set

$$\{T_\varphi : \varphi \in H^\infty\}$$

of all *analytic* Toeplitz operators. This naturally leads one to consider the commutant $\{S_u\}'$ of the compressed shift S_u. The answer, due to Sarason [**81**], prompts us to consider the following.

DEFINITION 12.2. If u is inner and $\varphi \in L^\infty$, then the *truncated Toeplitz operator* (TTO) with symbol φ is the operator $A^u_\varphi : \mathcal{K}_u \to \mathcal{K}_u$ defined by $A^u_\varphi f = P_u(\varphi f)$, where P_u is the orthogonal projection from L^2 onto $\mathcal{K}_u$.

THEOREM 12.3 (Sarason). *For inner u,*

$$\{S_u\}' = \{A^u_\varphi : \varphi \in H^\infty\}.$$

This theorem, which initiated the general study of *commutant lifting theorems*, gives an initial impetus for the study of truncated Toeplitz operators. However, to obtain the full class of truncated Toeplitz operators, one needs to consider symbols in L^2, as opposed to L^∞. For φ in L^2, one can define an operator A^u_φ on $\mathcal{K}_u \cap H^\infty$ (which is dense in $\mathcal{K}_u$ via Proposition 5.10) by

$$A^u_\varphi f := P_u(\varphi f).$$

If A^u_φ can be extended to a bounded operator on all of $\mathcal{K}_u$, then we say A^u_φ is a bounded truncated Toeplitz operator. When there is no chance of confusion, we write A_φ instead of A^u_φ. We let $\mathcal{T}_u$ denote the set of all bounded truncated Toeplitz operators on $\mathcal{K}_u$, remarking that $\mathcal{T}_u$ is a weakly closed linear space.

There are some superficial similarities between truncated Toeplitz operators and Toeplitz operators. For instance, one has the adjoint formula $A^*_\varphi = A_{\overline{\varphi}}$ and an analogue of the Brown-Halmos result (12.1): a bounded operator A on $\mathcal{K}_u$ is a truncated Toeplitz operator if and only if

$$A_z A A^*_z = A + R,$$

where R is a certain operator on $\mathcal{K}_u$ of rank $\leqslant 2$. However, the differences are greater than the similarities. For instance, the symbol of a Toeplitz operator is unique, yet for truncated Toeplitz operators $A_\varphi = A_\psi$ if and and only if

$$\varphi - \psi \in uH^2 + \overline{uH^2}.$$

In particular, the inequality $\|A_\varphi\| \leqslant \|\varphi\|_\infty$ is frequently strict. While there are no compact Toeplitz operators, many compact TTOs exist (e.g., any TTO on $\mathcal{K}_u$ if $\dim \mathcal{K}_u < \infty$). Finally, not every bounded truncated Toeplitz operator can be represented with a bounded symbol **[13]**.

Recall from Section 7 that $Cf = \overline{fzu}$ defines a conjugation on $\mathcal{K}_u$. It turns out that each truncated Toeplitz operator is *complex symmetric*, in the sense that $A_\varphi = CA^*_\varphi C$. More generally, we say that a bounded operator T on a complex Hilbert space $\mathcal{H}$ is a complex symmetric operator (CSO) if there exists a conjugation C on $\mathcal{H}$ so that $T = CT^*C$ **[40–44]**. The class of complex symmetric operators is surprisingly large and, somewhat surprisingly, many CSOs can be shown to be unitarily equivalent to truncated Toeplitz operators **[28, 49, 51, 94]**. Clarifying the precise relationship between CSOs and TTOs is an ongoing effort, chronicled to some extent in **[51]**.

Finally, we mention the following beautiful connection between Clark's unitary operators (Section 8) and truncated Toeplitz operators.

THEOREM 12.4. *If $\varphi \in L^\infty$ and u is inner, then*

$$A^u_\varphi = \int \varphi(U_\alpha)\, dm(\alpha),$$

in the sense that

$$\langle A^u_\varphi f, g\rangle = \int \langle \varphi(U_\alpha) f, g\rangle \, dm(\alpha),$$

for f, g *in* $\mathcal{K}_u$.

13. Things we did not mention

In this final section we point the reader in the direction of other aspects of model spaces that we do not have the time to fully explore.

13.1. Carleson measures. Carleson in **[21, 22]** (see also **[53]**) characterized those measures μ on $\mathbb{D}$ for which the inclusion operator from H^2 to $L^2(\mu)$ is continuous, i.e.,

$$(f \in H^2) \qquad \int_{\mathbb{D}} |f|^2 d\mu \leqslant C\|f\|^2$$

where $C > 0$ is a positive number independent of f. This result has been generalized to many other Hilbert spaces of analytic functions. What are the Carleson measures for the model spaces? A discussion of these measures along with historical references can be found in **[17]**.

13.2. Vector-valued model spaces. One can form the vector-valued Hardy space H^2 of $\mathbb{C}^n$-valued analytic functions on $\mathbb{D}$ for which the integral means

$$\int_{\mathbb{T}} \|f(r\zeta)\|^2_{\mathbb{C}^n} dm(\zeta)$$

are bounded as $r \to 1^-$. For a matrix valued inner function U on $\mathbb{D}$ (i.e., a matrix-valued analytic function $U(z)$ on $\mathbb{D}$ that satisfies $\|U\| \leqslant 1$ on $\mathbb{D}$ and which is unitary valued a.e. on $\mathbb{T}$), we can define the $\mathbb{C}^n$-valued model space $H^2 \ominus UH^2$. The compression of the shift to these model spaces are the model operators for contractions via a similar Sz.-Nagy-Foiaş theory for higher defect indices. Some well-known sources for all this operator theory are **[14, 69, 71, 77]**. There is even a Clark theory for these vector-valued model spaces including a version of the Aleksandrov disintegration theorem (Theorem 8.10)**[36, 66]**.

13.3. Hankel operators. There are strong connections between model spaces and Hankel operators. The standard texts on the subject are **[70, 71, 73]**.

13.4. Extremal problems. There is a nice connection between model spaces, truncated Toeplitz operators, and classical extremal problems. For a rational function ψ whose poles are in $\mathbb{D}$, a classical problem in complex analysis is to compute the quantity

$$\sup_{f \in H^1_1} \left| \frac{1}{2\pi i} \oint_{\mathbb{T}} \psi(\zeta) f(\zeta) d\zeta \right|, \tag{13.1}$$

along with the functions f for which the above supremum is achieved. In the above definition, H^1_1 denotes the unit ball in the Hardy space H^1. These types of problems and extensions of them, where ψ belongs to L^∞ and is not assumed to be rational, have been studied since the early twentieth century. We refer the reader to **[50]** for a survey of these classical results. In that same paper is the following result:

THEOREM 13.2. *If ψ is a rational function whose poles are contained in $\mathbb{D}$ then*

$$\sup_{f\in H_1^1}\left|\frac{1}{2\pi i}\oint_{\mathbb{T}}\psi(\zeta)f(\zeta)d\zeta\right| = \sup_{g\in H_1^2}\left|\frac{1}{2\pi i}\oint_{\mathbb{T}}\psi(\zeta)g^2(\zeta)d\zeta\right|. \tag{13.3}$$

Furthermore, if u is the finite Blaschke product whose zeros are precisely those poles with corresponding multiplicities and if $\varphi := u\psi$, then

$$\sup_{f\in H_1^1}\left|\frac{1}{2\pi i}\oint_{\mathbb{T}}\psi(\zeta)f(\zeta)d\zeta\right| = \|A_\varphi^u\|,$$

where A_φ^u is the analytic truncated Toeplitz operator on $\mathcal{K}_u$ with symbol φ.

In fact, the linear and quadratic supremums in (13.3) are the same for general ψ in $L^\infty(\mathbb{T})$ **[23]**.

References

[1] Evgeny V. Abakumov, *Cyclicity and approximation by lacunary power series*, Michigan Math. J. **42** (1995), no. 2, 277–299, DOI 10.1307/mmj/1029005229. MR1342491 (96f:47056)

[2] Jim Agler and John E. McCarthy, *Pick interpolation and Hilbert function spaces*, Graduate Studies in Mathematics, vol. 44, American Mathematical Society, Providence, RI, 2002. MR1882259 (2003b:47001)

[3] P. R. Ahern and D. N. Clark, *On functions orthogonal to invariant subspaces*, Acta Math. **124** (1970), 191–204. MR0264385 (41 #8981a)

[4] P. R. Ahern and D. N. Clark, *Radial limits and invariant subspaces*, Amer. J. Math. **92** (1970), 332–342. MR0262511 (41 #7117)

[5] A. B. Aleksandrov, *Multiplicity of boundary values of inner functions* (Russian, with English and Armenian summaries), Izv. Akad. Nauk Armyan. SSR Ser. Mat. **22** (1987), no. 5, 490–503, 515. MR931885 (89e:30058)

[6] A. B. Aleksandrov, *Inner functions and related spaces of pseudocontinuable functions* (Russian, with English summary), Zap. Nauchn. Sem. Leningrad. Otdel. Mat. Inst. Steklov. (LOMI) **170** (1989), no. Issled. Linein. Oper. Teorii Funktsii. 17, 7–33, 321, DOI 10.1007/BF01099304; English transl., J. Soviet Math. **63** (1993), no. 2, 115–129. MR1039571 (91c:30063)

[7] A. B. Aleksandrov, *On the existence of angular boundary values of pseudocontinuable functions* (Russian, with English and Russian summaries), Zap. Nauchn. Sem. S.-Peterburg. Otdel. Mat. Inst. Steklov. (POMI) **222** (1995), no. Issled. po Linein. Oper. i Teor. Funktsii. 23, 5–17, 307, DOI 10.1007/BF02355824; English transl., J. Math. Sci. (New York) **87** (1997), no. 5, 3781–3787. MR1359992 (97a:30046)

[8] A. B. Aleksandrov, *Gap series and pseudocontinuations. An arithmetic approach*, Algebra i Analiz **9** (1997), no. 1, 3–31. MR1458416 (98d:30006)

[9] A. B. Aleksandrov, *Embedding theorems for coinvariant subspaces of the shift operator. II* (Russian, with English and Russian summaries), Zap. Nauchn. Sem. S.-Peterburg. Otdel. Mat. Inst. Steklov. (POMI) **262** (1999), no. Issled. po Linein. Oper. i Teor. Funkts. 27, 5–48, 231, DOI 10.1023/A:1015379002290; English transl., J. Math. Sci. (New York) **110** (2002), no. 5, 2907–2929. MR1734326 (2001g:46047)

[10] Alexandru Aleman, Nathan S. Feldman, and William T. Ross, *The Hardy space of a slit domain*, Frontiers in Mathematics, Birkhäuser Verlag, Basel, 2009. MR2548414 (2011m:30095)

[11] Alexandru Aleman and Stefan Richter, *Simply invariant subspaces of H^2 of some multiply connected regions*, Integral Equations Operator Theory **24** (1996), no. 2, 127–155, DOI 10.1007/BF01193457. MR1371943 (99b:47010a)

[12] N. Aronszajn, *Theory of reproducing kernels*, Trans. Amer. Math. Soc. **68** (1950), 337–404. MR0051437 (14,479c)

[13] Anton Baranov, Isabelle Chalendar, Emmanuel Fricain, Javad Mashreghi, and Dan Timotin, *Bounded symbols and reproducing kernel thesis for truncated Toeplitz operators*, J. Funct. Anal. **259** (2010), no. 10, 2673–2701, DOI 10.1016/j.jfa.2010.05.005. MR2679022 (2011h:47051)

[14] Hari Bercovici, *Operator theory and arithmetic in H^∞*, Mathematical Surveys and Monographs, vol. 26, American Mathematical Society, Providence, RI, 1988. MR954383 (90e:47001)

[15] Robert D. Berman and William S. Cohn, *Tangential limits of Blaschke products and functions of bounded mean oscillation*, Illinois J. Math. **31** (1987), no. 2, 218–239. MR882111 (88g:30036)

[16] Arne Beurling, *On two problems concerning linear transformations in Hilbert space*, Acta Math. **81** (1948), 17. MR0027954 (10,381e)

[17] A. Blandigneres, E. Fricain, F. Gaunard, A. Hartmann, and W. T. Ross, *Reverse Carleson embeddings for model spaces*, To appear, J. London. Math. Soc.

[18] Albrecht Böttcher and Bernd Silbermann, *Analysis of Toeplitz operators*, 2nd ed., Springer Monographs in Mathematics, Springer-Verlag, Berlin, 2006. Prepared jointly with Alexei Karlovich. MR2223704 (2007k:47001)

[19] Arlen Brown and P. R. Halmos, *Algebraic properties of Toeplitz operators*, J. Reine Angew. Math. **213** (1963/1964), 89–102. MR0160136 (28 #3350)

[20] G. T. Cargo, *Angular and tangential limits of Blaschke products and their successive derivatives*, Canad. J. Math. **14** (1962), 334–348. MR0136743 (25 #204)

[21] Lennart Carleson, *An interpolation problem for bounded analytic functions*, Amer. J. Math. **80** (1958), 921–930. MR0117349 (22 #8129)

[22] Lennart Carleson, *Interpolations by bounded analytic functions and the corona problem*, Ann. of Math. (2) **76** (1962), 547–559. MR0141789 (25 #5186)

[23] I. Chalendar, E. Fricain, and D. Timotin, *On an extremal problem of Garcia and Ross*, Oper. Matrices **3** (2009), no. 4, 541–546, DOI 10.7153/oam-03-31. MR2597679 (2011b:30130)

[24] Nicolas Chevrot, Emmanuel Fricain, and Dan Timotin, *The characteristic function of a complex symmetric contraction*, Proc. Amer. Math. Soc. **135** (2007), no. 9, 2877–2886 (electronic), DOI 10.1090/S0002-9939-07-08803-X. MR2317964 (2008c:47025)

[25] Ole Christensen, *An introduction to frames and Riesz bases*, Applied and Numerical Harmonic Analysis, Birkhäuser Boston, Inc., Boston, MA, 2003. MR1946982 (2003k:42001)

[26] Joseph A. Cima, Alec L. Matheson, and William T. Ross, *The Cauchy transform*, Mathematical Surveys and Monographs, vol. 125, American Mathematical Society, Providence, RI, 2006. MR2215991 (2006m:30003)

[27] Joseph A. Cima and William T. Ross, *The backward shift on the Hardy space*, Mathematical Surveys and Monographs, vol. 79, American Mathematical Society, Providence, RI, 2000. MR1761913 (2002f:47068)

[28] Joseph A. Cima, Stephan Ramon Garcia, William T. Ross, and Warren R. Wogen, *Truncated Toeplitz operators: spatial isomorphism, unitary equivalence, and similarity*, Indiana Univ. Math. J. **59** (2010), no. 2, 595–620, DOI 10.1512/iumj.2010.59.4097. MR2648079 (2011i:47035)

[29] Douglas N. Clark, *One dimensional perturbations of restricted shifts*, J. Analyse Math. **25** (1972), 169–191. MR0301534 (46 #692)

[30] R. Bruce Crofoot, *Multipliers between invariant subspaces of the backward shift*, Pacific J. Math. **166** (1994), no. 2, 225–246. MR1313454 (95k:47009)

[31] R. G. Douglas, H. S. Shapiro, and A. L. Shields, *Cyclic vectors and invariant subspaces for the backward shift operator.* (English, with French summary), Ann. Inst. Fourier (Grenoble) **20** (1970), no. fasc. 1, 37–76. MR0270196 (42 #5088)

[32] Ronald G. Douglas, *Banach algebra techniques in operator theory*, Academic Press, New York-London, 1972. Pure and Applied Mathematics, Vol. 49. MR0361893 (50 #14335)

[33] Peter L. Duren, *Theory of H^p spaces*, Pure and Applied Mathematics, Vol. 38, Academic Press, New York-London, 1970. MR0268655 (42 #3552)

[34] Konstantin Dyakonov and Dmitry Khavinson, *Smooth functions in star-invariant subspaces*, Recent advances in operator-related function theory, Contemp. Math., vol. 393, Amer. Math. Soc., Providence, RI, 2006, pp. 59–66, DOI 10.1090/conm/393/07371. MR2198372 (2007b:30069)

[35] Konstantin M. Dyakonov, *Kernels of Toeplitz operators via Bourgain's factorization theorem*, J. Funct. Anal. **170** (2000), no. 1, 93–106, DOI 10.1006/jfan.1999.3499. MR1736197 (2000m:47036)

[36] Sam Elliott, *A matrix-valued Aleksandrov disintegration theorem*, Complex Anal. Oper. Theory **4** (2010), no. 2, 145–157, DOI 10.1007/s11785-009-0007-3. MR2652521 (2011g:46076)

[37] E. Fabry, *Sur les séries de Taylor qui ont une infinité de points singuliers*, Acta Math. **22** (1898–1899), 65–88.

[38] E. Fricain and J. Mashreghi, *Theory of $H(b)$ spaces, Vol I & II*, New monograph series, vol. 20, Cambridge U. Press, 2014.

[39] Otto Frostman, *Sur les produits de Blaschke* (French), Kungl. Fysiografiska Sällskapets i Lund Förhandlingar [Proc. Roy. Physiog. Soc. Lund] **12** (1942), no. 15, 169–182. MR0012127 (6,262e)

[40] Stephan Ramon Garcia, *Conjugation and Clark operators*, Recent advances in operator-related function theory, Contemp. Math., vol. 393, Amer. Math. Soc., Providence, RI, 2006, pp. 67–111, DOI 10.1090/conm/393/07372. MR2198373 (2007b:47073)

[41] Stephan Ramon Garcia and Mihai Putinar, *Complex symmetric operators and applications*, Trans. Amer. Math. Soc. **358** (2006), no. 3, 1285–1315 (electronic), DOI 10.1090/S0002-9947-05-03742-6. MR2187654 (2006j:47036)

[42] Stephan Ramon Garcia and Mihai Putinar, *Complex symmetric operators and applications. II*, Trans. Amer. Math. Soc. **359** (2007), no. 8, 3913–3931 (electronic), DOI 10.1090/S0002-9947-07-04213-4. MR2302518 (2008b:47005)

[43] Stephan Ramon Garcia and Warren R. Wogen, *Complex symmetric partial isometries*, J. Funct. Anal. **257** (2009), no. 4, 1251–1260, DOI 10.1016/j.jfa.2009.04.005. MR2535469 (2011g:47005)

[44] Stephan Ramon Garcia and Warren R. Wogen, *Some new classes of complex symmetric operators*, Trans. Amer. Math. Soc. **362** (2010), no. 11, 6065–6077, DOI 10.1090/S0002-9947-2010-05068-8. MR2661508 (2011g:47086)

[45] Stephan R. Garcia and Mihai Putinar, *Interpolation and complex symmetry*, Tohoku Math. J. (2) **60** (2008), no. 3, 423–440, DOI 10.2748/tmj/1223057737. MR2453732 (2009k:47048)

[46] Stephan Ramon Garcia, *A $*$-closed subalgebra of the Smirnov class*, Proc. Amer. Math. Soc. **133** (2005), no. 7, 2051–2059 (electronic), DOI 10.1090/S0002-9939-05-07735-X. MR2137871 (2005m:30040)

[47] Stephan Ramon Garcia, *Conjugation, the backward shift, and Toeplitz kernels*, J. Operator Theory **54** (2005), no. 2, 239–250. MR2186351 (2006g:30055)

[48] Stephan Ramon Garcia, *Inner matrices and Darlington synthesis*, Methods Funct. Anal. Topology **11** (2005), no. 1, 37–47. MR2130935 (2005m:30039)

[49] Stephan Ramon Garcia, Daniel E. Poore, and William T. Ross, *Unitary equivalence to a truncated Toeplitz operator: analytic symbols*, Proc. Amer. Math. Soc. **140** (2012), no. 4, 1281–1295, DOI 10.1090/S0002-9939-2011-11060-8. MR2869112 (2012m:47004)

[50] Stephan Ramon Garcia and William T. Ross, *A non-linear extremal problem on the Hardy space*, Comput. Methods Funct. Theory **9** (2009), no. 2, 485–524, DOI 10.1007/BF03321742. MR2572653 (2010k:47062)

[51] Stephan Ramon Garcia and William T. Ross, *Recent progress on truncated Toeplitz operators*, Blaschke products and their applications, Fields Inst. Commun., vol. 65, Springer, New York, 2013, pp. 275–319. MR3052299

[52] Stephan Ramon Garcia and Donald Sarason, *Real outer functions*, Indiana Univ. Math. J. **52** (2003), no. 6, 1397–1412, DOI 10.1512/iumj.2003.52.2511. MR2021044 (2004k:30129)

[53] John B. Garnett, *Bounded analytic functions*, 1st ed., Graduate Texts in Mathematics, vol. 236, Springer, New York, 2007. MR2261424 (2007e:30049)

[54] J. Hadamard, *Essai sur l'étude des fonctions données par leur développement de Taylor*, J. Math. **8** (1892), 101–186.

[55] Andreas Hartmann and William T. Ross, *Bad boundary behavior in star invariant subspaces II*, Ann. Acad. Sci. Fenn. Math. **37** (2012), no. 2, 467–478, DOI 10.5186/aasfm.2012.3744. MR2987079

[56] Henry Helson, *Large analytic functions. II*, Analysis and partial differential equations, Lecture Notes in Pure and Appl. Math., vol. 122, Dekker, New York, 1990, pp. 217–220. MR1044789 (92c:30039)

[57] D. Hitt, *Invariant subspaces of $\mathcal{H}^2$ of an annulus*, Pacific J. Math. **134** (1988), no. 1, 101–120. MR953502 (90a:46059)

[58] Kenneth Hoffman, *Banach spaces of analytic functions*, Prentice-Hall Series in Modern Analysis, Prentice-Hall, Inc., Englewood Cliffs, N. J., 1962. MR0133008 (24 #A2844)

[59] S. V. Hruščëv, N. K. Nikol'skiĭ, and B. S. Pavlov, *Unconditional bases of exponentials and of reproducing kernels*, Complex analysis and spectral theory (Leningrad, 1979/1980), Lecture Notes in Math., vol. 864, Springer, Berlin-New York, 1981, pp. 214–335. MR643384 (84k:46019)

[60] Paul Koosis, *Introduction to H_p spaces*, 2nd ed., Cambridge Tracts in Mathematics, vol. 115, Cambridge University Press, Cambridge, 1998. With two appendices by V. P. Havin [Viktor Petrovich Khavin]. MR1669574 (2000b:30052)

[61] T. L. Kriete III, *On the Fourier coefficients of outer functions*, Indiana Univ. Math. J. **20** (1970/71), 147–155. MR0412429 (54 #555)

[62] T. L. Kriete III, *A generalized Paley-Wiener theorem*, J. Math. Anal. Appl. **36** (1971), 529–555. MR0288275 (44 #5473)

[63] A. J. Lohwater and G. Piranian, *The boundary behavior of functions analytic in a disk*, Ann. Acad. Sci. Fenn. Ser. A. I. **1957** (1957), no. 239, 17. MR0091342 (19,950c)

[64] N. Lusin and J. Priwaloff, *Sur l'unicité et la multiplicité des fonctions analytiques* (French), Ann. Sci. École Norm. Sup. (3) **42** (1925), 143–191. MR1509265

[65] N. Makarov and A. Poltoratski, *Meromorphic inner functions, Toeplitz kernels and the uncertainty principle*, Perspectives in analysis, Math. Phys. Stud., vol. 27, Springer, Berlin, 2005, pp. 185–252, DOI 10.1007/3-540-30434-7_10. MR2215727 (2007b:47079)

[66] R. T. W. Martin, *Unitary perturbations of compressed n-dimensional shifts*, Complex Anal. Oper. Theory **7** (2013), no. 4, 765–799, DOI 10.1007/s11785-012-0236-8. MR3079829

[67] Rubén A. Martínez-Avendaño and Peter Rosenthal, *An introduction to operators on the Hardy-Hilbert space*, Graduate Texts in Mathematics, vol. 237, Springer, New York, 2007. MR2270722 (2007k:47051)

[68] Javad Mashreghi, *Representation theorems in Hardy spaces*, London Mathematical Society Student Texts, vol. 74, Cambridge University Press, Cambridge, 2009. MR2500010 (2011e:30001)

[69] N. K. Nikol'skiĭ, *Treatise on the shift operator*, Grundlehren der Mathematischen Wissenschaften [Fundamental Principles of Mathematical Sciences], vol. 273, Springer-Verlag, Berlin, 1986. Spectral function theory; With an appendix by S. V. Hruščev [S. V. Khrushchëv] and V. V. Peller; Translated from the Russian by Jaak Peetre. MR827223 (87i:47042)

[70] Nikolai K. Nikolski, *Operators, functions, and systems: an easy reading. Vol. 1*, Mathematical Surveys and Monographs, vol. 92, American Mathematical Society, Providence, RI, 2002. Hardy, Hankel, and Toeplitz; Translated from the French by Andreas Hartmann. MR1864396 (2003i:47001a)

[71] Nikolai K. Nikolski, *Operators, functions, and systems: an easy reading. Vol. 2*, Mathematical Surveys and Monographs, vol. 93, American Mathematical Society, Providence, RI, 2002. Model operators and systems; Translated from the French by Andreas Hartmann and revised by the author. MR1892647 (2003i:47001b)

[72] V. Paulsen, *An introduction to the theory of reproducing kernel Hilbert spaces*, www.math.uh.edu/ vern/rkhs.pdf, 2009.

[73] Vladimir V. Peller, *Hankel operators and their applications*, Springer Monographs in Mathematics, Springer-Verlag, New York, 2003. MR1949210 (2004e:47040)

[74] Alexei Poltoratski and Donald Sarason, *Aleksandrov-Clark measures*, Recent advances in operator-related function theory, Contemp. Math., vol. 393, Amer. Math. Soc., Providence, RI, 2006, pp. 1–14, DOI 10.1090/conm/393/07366. MR2198367 (2006i:30048)

[75] A. G. Poltoratskiĭ, *Boundary behavior of pseudocontinuable functions* (Russian, with Russian summary), Algebra i Analiz **5** (1993), no. 2, 189–210; English transl., St. Petersburg Math. J. **5** (1994), no. 2, 389–406. MR1223178 (94k:30090)

[76] Frigyes Riesz and Béla Sz.-Nagy, *Functional analysis*, Frederick Ungar Publishing Co., New York, 1955. Translated by Leo F. Boron. MR0071727 (17,175i)

[77] Marvin Rosenblum and James Rovnyak, *Hardy classes and operator theory*, Oxford Mathematical Monographs, The Clarendon Press, Oxford University Press, New York, 1985. Oxford Science Publications. MR822228 (87e:47001)

[78] William T. Ross and Harold S. Shapiro, *Generalized analytic continuation*, University Lecture Series, vol. 25, American Mathematical Society, Providence, RI, 2002. MR1895624 (2003h:30003)

[79] Eero Saksman, *An elementary introduction to Clark measures*, Topics in complex analysis and operator theory, Univ. Málaga, Málaga, 2007, pp. 85–136. MR2394657 (2009g:47041)

[80] Donald Sarason, *A remark on the Volterra operator*, J. Math. Anal. Appl. **12** (1965), 244–246. MR0192355 (33 #580)

[81] Donald Sarason, *Generalized interpolation in H^∞*, Trans. Amer. Math. Soc. **127** (1967), 179–203. MR0208383 (34 #8193)

[82] Donald Sarason, *Nearly invariant subspaces of the backward shift*, Contributions to operator theory and its applications (Mesa, AZ, 1987), Oper. Theory Adv. Appl., vol. 35, Birkhäuser, Basel, 1988, pp. 481–493. MR1017680 (90m:47012)

[83] Donald Sarason, *Kernels of Toeplitz operators*, Toeplitz operators and related topics (Santa Cruz, CA, 1992), Oper. Theory Adv. Appl., vol. 71, Birkhäuser, Basel, 1994, pp. 153–164. MR1300218 (95k:47039)

[84] Donald Sarason, *Sub-Hardy Hilbert spaces in the unit disk*, University of Arkansas Lecture Notes in the Mathematical Sciences, 10, John Wiley & Sons, Inc., New York, 1994. A Wiley-Interscience Publication. MR1289670 (96k:46039)

[85] Donald Sarason, *Algebraic properties of truncated Toeplitz operators*, Oper. Matrices **1** (2007), no. 4, 491–526, DOI 10.7153/oam-01-29. MR2363975 (2008i:47060)

[86] Harold S. Shapiro, *Weakly invertible elements in certain function spaces, and generators in $\updownarrow_1$*, Michigan Math. J. **11** (1964), 161–165. MR0166343 (29 #3620)
[87] Harold S. Shapiro, *Generalized analytic continuation*, Symposia on Theoretical Physics and Mathematics, Vol. 8 (Symposium, Madras, 1967), Plenum, New York, 1968, pp. 151–163. MR0241614 (39 #2953)
[88] H. S. Shapiro, *Functions nowhere continuable in a generalized sense*, Publ. Ramanujan Inst. No. **1** (1968/1969), 179–182. MR0267080 (42 #1982)
[89] Joel H. Shapiro, *Composition operators and classical function theory*, Universitext: Tracts in Mathematics, Springer-Verlag, New York, 1993. MR1237406 (94k:47049)
[90] N. A. Širokov, *Ideals and factorization in algebras of analytic functions that are smooth up to the boundary* (Russian), Trudy Mat. Inst. Steklov. **130** (1978), 196–222, 223. Spectral theory of functions and operators. MR505686 (80j:46085)
[91] N. A. Shirokov, *Division and multiplication by inner functions in spaces of analytic functions smooth up to the boundary*, Complex analysis and spectral theory (Leningrad, 1979/1980), Lecture Notes in Math., vol. 864, Springer, Berlin-New York, 1981, pp. 413–439. MR643387 (83c:30028)
[92] Barry Simon, *Spectral analysis of rank one perturbations and applications*, Mathematical quantum theory. II. Schrödinger operators (Vancouver, BC, 1993), CRM Proc. Lecture Notes, vol. 8, Amer. Math. Soc., Providence, RI, 1995, pp. 109–149. MR1332038 (97c:47008)
[93] Barry Simon and Tom Wolff, *Singular continuous spectrum under rank one perturbations and localization for random Hamiltonians*, Comm. Pure Appl. Math. **39** (1986), no. 1, 75–90, DOI 10.1002/cpa.3160390105. MR820340 (87k:47032)
[94] E. Strouse, D. Timotin, and M. Zarrabi, *Unitary equivalence to truncated Toeplitz operators*, Indiana Univ. Math. J. **61** (2012), no. 2, 525–538, DOI 10.1512/iumj.2012.61.4562. MR3043586
[95] Béla Sz.-Nagy, Ciprian Foias, Hari Bercovici, and László Kérchy, *Harmonic analysis of operators on Hilbert space*, second ed., Universitext, Springer, New York, 2010. MR2760647 (2012b:47001)
[96] S. Takenaka, *On the orthonormal functions and a new formula of interpolation*, Jap. J. Math. **2** (1925), 129 – 145.
[97] Gerald Teschl, *Mathematical methods in quantum mechanics*, Graduate Studies in Mathematics, vol. 99, American Mathematical Society, Providence, RI, 2009. With applications to Schrödinger operators. MR2499016 (2010h:81002)
[98] G. C. Tumarkin, *Description of a class of functions admitting an approximation by fractions with preassigned poles* (Russian, with Armenian and English summaries), Izv. Akad. Nauk Armjan. SSR Ser. Mat. **1** (1966), no. 2, 89–105. MR0206304 (34 #6123)

Department of Mathematics, Pomona College, Claremont, California 91711
E-mail address: Stephan.Garcia@pomona.edu
URL: http://pages.pomona.edu/~sg064747

Department of Mathematics and Computer Science, University of Richmond, Richmond, Virginia 23173
E-mail address: wross@richmond.edu
URL: http://facultystaff.richmond.edu/~wross

Contemporary Mathematics
Volume **638**, 2015
http://dx.doi.org/10.1090/conm/638/12812

Note on a Julia operator related to model spaces

Dan Timotin

ABSTRACT. The properties of a certain Julia-type operator are shown to provide a unitary approach to several results in model spaces for contractions on a Hilbert space.

1. Introduction

There are several equivalent functional models for completely nonunitary contractions on a Hilbert space, most notably the Sz-Nagy–Foias model [**9**] and the de Branges–Rovnyak model [**1**]. The first is more extensively used by operator theorists; the second has regained some popularity in the last years. In both cases a complete unitary invariant is given by a contractive operator valued analytic function defined in the unit disc $\mathbb{D}$, traditionally denoted by Θ in the first case and by B in the second; this function is the basic ingredient in the construction of both models.

In the context of the Sz.-Nagy–Foias model, it is known (and easy to prove) that if a contraction T is associated to Θ, then T^* is associated to $\tilde{\Theta}$, where $\tilde{\Theta}(z) = \Theta(\bar{z})^*$. The translation of this fact and its consequences for the de Branges–Rovnyak model is less transparent. The purpose of this note is to make explicit this connection as well as some other related results. It is worth to emphasize that the central role is played by a certain Julia-type operator, whose properties are used in several directions. The results contained are mostly known to specialists, but they are not readily available in the literature, so it seemed interesting to gather them in a short presentation.

One should note that, geometrically, the simplest way to understand the role of the Julia operator is by using the so called coordinate–free model introduced by Nikolski and Vasyunin [**6, 7**]. However, this is not appropriate for concrete calculations, and so is less popular. The development in this paper is mainly based on the Sz.-Nagy–Foias model.

2. Preliminaries

2.1. Basic notations and facts. We will deal with functions defined on the unit circle $\mathbb{T} = \{z \in \mathbb{C} : |z| = 1\}$ or the unit disc $\mathbb{D} = \{z \in \mathbb{C} : |z| < 1\}$ and whose values are either vectors in a Hilbert space or bounded operators (actually

2010 *Mathematics Subject Classification.* Primary 47A45, 47B32.
Key words and phrases. Hilbert space, contraction, Julia operator, model space.

contractions) between two Hilbert spaces. Usually ζ will denote the variable in $\mathbb{T}$ and z the variable in $\mathbb{D}$.

If $\mathcal{E}$ is a separable Hilbert space, then $L^2(\mathcal{E})$ is the Hilbert space of Lebesgue square integrable measurable functions defined on $\mathbb{T}$ and taking values in $\mathcal{E}$, while $H^2(\mathcal{E})$ is its subspace consisting of functions with null negative Fourier coefficients; also, $H^2_-(\mathcal{E}) = L^2(\mathcal{E}) \ominus H^2(\mathcal{E})$. The space $H^2(\mathcal{E})$ can alternately be viewed as a space of analytic functions defined in $\mathbb{D}$; radial convergence to the boundary values takes place almost everywhere. The operator of multiplication by the variable ζ on any $L^2(\mathcal{E})$ will be denoted by same letter Z, and its restriction to $H^2(\mathcal{E})$ by S.

If we are now given two separable Hilbert spaces $\mathcal{E}, \mathcal{E}_*$, we denote by $\mathcal{L}(\mathcal{E}, \mathcal{E}_*)$ the space of bounded operators from $\mathcal{E}$ to $\mathcal{E}_*$. Then $L^\infty(\mathcal{E}, \mathcal{E}_*)$ is the space of measurable essentially bounded functions on $\mathbb{T}$ with values in $\mathcal{L}(\mathcal{E}, \mathcal{E}_*)$. This Banach space has as a closed space $H^\infty(\mathcal{E}, \mathcal{E}_*)$, characterized again by null negative Fourier coefficients, and this last space may also be considered as a space of functions analytic in $\mathbb{D}$ with values in $\mathcal{L}(\mathcal{E}, \mathcal{E}_*)$; radial convergence to the boundary values takes place in the strong operator topology.

For an operator valued function F we denote $\tilde{F}(\zeta) = F(\bar{\zeta})^*$, while for a vector valued function f we denote $\hat{f}(\zeta) = \bar{\zeta} f(\bar{\zeta})$. The map $f \mapsto \hat{f}$ is unitary on each space $L^2(\mathcal{E})$, and it interchanges $H^2(\mathcal{E})$ and $H^2_-(\mathcal{E})$.

A function $F \in L^\infty(\mathcal{E}, \mathcal{E}_*)$ defines a multiplication operator $M_F \in \mathcal{L}(L^2(\mathcal{E}), L^2(\mathcal{E}_*))$ by the formula

$$(M_F f)(\zeta) = F(\zeta) f(\zeta).$$

We have $\|M_F\| = \|F\|_\infty$ and $M_F^* = M_{F^*}$. Occasionally we will drop the distinction in notation and write Ff for the function $M_F f$.

If $T \in \mathcal{L}(H_1, H_2)$ is a contraction between two Hilbert spaces, we define $D_T = (I - T^*T)^{1/2}$ and $\mathcal{D}_T = \operatorname{clos}(D_T H_1)$. The *Julia operator* $J(T) : H_1 \oplus \mathcal{D}_{T^*} \to H_2 \oplus \mathcal{D}_T$ is

$$J(T) = \begin{pmatrix} T & D_{T^*} \\ D_T & -T^* \end{pmatrix}.$$

It is a unitary operator, and $J(T)^* = J(T^*)$.

In particular, if $F \in L^\infty(\mathcal{E}, \mathcal{E}_*)$ and $\|F\|_\infty \le 1$, then

$$D_{M_F} = M_{\Delta_F},$$

where $\Delta_F \in L^\infty(\mathcal{E}, \mathcal{E}_*)$ is defined by $\Delta_F(\zeta) = (I_{\mathcal{E}} - F(\zeta)^* F(\zeta))^{1/2}$. The Julia operator

$$J(M_F) : L^2(\mathcal{E}) \oplus \operatorname{clos}(\Delta_{F^*} L^2(\mathcal{E}_*)) \to L^2(\mathcal{E}_*) \oplus \operatorname{clos}(\Delta_F L^2(\mathcal{E}))$$

has the formula

$$J(M_F) = \begin{pmatrix} M_F & M_{\Delta_{F^*}} \\ M_{\Delta_F} & -M_{F^*} \end{pmatrix}. \tag{2.1}$$

2.2. The Nagy–Foias model. Suppose that $\Theta \in H^\infty(\mathcal{E}, \mathcal{E}_*)$ and $\|\Theta\|_\infty \le 1$. We will also assume that Θ is *pure*, that is, $\|\Theta(0)x\| < \|x\|$ for all $x \in \mathcal{E}$. (Actually, this technical condition is irrelevant to us, but it is standard in model theory.) We denote in this section $\Delta = \Delta_\Theta$; further on we will have to keep the subscript for clarity.

The corresponding Nagy–Foias model space is

$$\mathbf{K}_\Theta = \big(H^2(\mathcal{E}_*) \oplus \operatorname{clos}(\Delta L^2(\mathcal{E}))\big) \ominus \{\Theta h \oplus \Delta h : h \in H^2(\mathcal{E})\} \tag{2.2}$$

and the model operator $\mathbf{T}_\Theta \in \mathcal{L}(\mathbf{K}_\Theta)$ is

$$\mathbf{T}_\Theta = P_{\mathbf{K}_\Theta}(S \oplus Z)|\mathbf{K}_\Theta. \tag{2.3}$$

Actually $\mathbf{K}_\Theta$ is invariant with respect to $S^* \oplus Z^*$, so $\mathbf{T}_\Theta^* = (S^* \oplus Z^*)|\mathbf{K}_\Theta$.

We will also have the opportunity to use

$$\mathbf{H}_\Theta = \big(H^2(\mathcal{E}_*) \oplus \operatorname{clos}(\Delta H^2(\mathcal{E}))\big) \ominus \{\Theta h \oplus \Delta h : h \in H^2(\mathcal{E})\} \subset \mathbf{K}_\Theta. \tag{2.4}$$

$\mathbf{H}_\Theta$ is also invariant with respect to $S^* \oplus Z^*$. Obviously $\mathbf{H}_\Theta = \mathbf{K}_\Theta$ if and only if $\operatorname{clos}(\Delta H^2(\mathcal{E})) = \operatorname{clos}(\Delta L^2(\mathcal{E}))$. If this happens we will say that Θ is *distinguished.* In the scalar case $\dim \mathcal{E} = \dim \mathcal{E}_* = 1$ this is equivalent to Θ being an extreme point of the unit ball of $H^\infty(\mathcal{E}, \mathcal{E}_*)$, but the equivalence is not true in general (see [**7**, Section 9]).

2.3. The de Branges–Rovnyak space. Suppose P_1 is the orthogonal projection onto the first component of $H^2(\mathcal{E}_*) \oplus \operatorname{clos}(\Delta L^2(\mathcal{E}))$. The *de Branges–Rovnyak space* $\mathcal{H}(\Theta)$ associated to Θ is defined as a linear space as

$$\mathcal{H}(\Theta) = P_1(\mathbf{H}_\Theta) = P_1(\mathbf{K}_\Theta)$$

(the second equality being a consequence of the inclusion $\mathbf{K}_\Theta \ominus \mathbf{H}_\Theta \subset \{0\} \oplus \operatorname{clos}(\Delta L^2(\mathcal{E}))$). We have to endow it with a norm; this is done by showing that $P_1|\mathbf{H}_\Theta$ is one-to-one, and defining the norm on $\mathcal{H}(\Theta)$ precisely as to make it a unitary operator. This unitary operator will be denoted by π_Θ.

$\mathcal{H}(\Theta)$ is invariant with respect to S^*, and the restriction $X_\Theta := S^*|\mathcal{H}(\Theta)$ is unitarily equivalent (through π_Θ) to $(S^* \oplus Z^*)|\mathbf{H}_\Theta$. In case Θ is distinguished this means that X_Θ is unitarily equivalent to $\mathbf{T}_\Theta^*$; this is precisely the explanation of the equivalence of the Nagy–Foias and the de Branges–Rovnyak functional models. If Θ is not distinguished, then $\mathcal{H}(\Theta)$ is not a "good" model space, and one has to enlarge it to a more complicated space, usually denoted by $\mathcal{D}(\Theta)$, called the *complete de Branges–Rovnyak space* in [**4**]. We will discuss it shortly in Subsection 4.4; for more details, see [**1**, **7**].

The intrinsic study of the space $\mathcal{H}(\Theta)$ has attracted considerable interest mostly in the scalar case $\dim \mathcal{E} = \dim \mathcal{E}_* = 1$. A good reference is [**8**]; the forthcoming book [**3**] will include more recent results.

2.4. Reproducing kernels. As a space of analytic functions, $H^2(\mathcal{E})$ is a reproducing kernel space on the unit disc. Since the functions are vector valued, the reproducing kernels will be operator valued. More precisely, for any $x \in \mathcal{E}$, $f \in H^2(\mathcal{E})$, and $\lambda \in \mathbb{D}$ we have

$$\langle f(\lambda), x\rangle_{\mathcal{E}} = \langle f, k_\lambda x\rangle_{H^2(\mathcal{E})}, \quad \text{with } k_\lambda(z) = \frac{1}{1-\bar{\lambda} z} I_{\mathcal{E}}.$$

The de Branges–Rovnyak space $\mathcal{H}(\Theta)$ is also a reproducing kernel space on the unit disc—actually, the original approach of de Branges and Rovnyak makes extensive use of reproducing kernels. If $x \in \mathcal{E}_*$, then for $f \in \mathcal{H}(\Theta)$ and $\lambda \in \mathbb{D}$ we have

$$\langle f(\lambda), x\rangle_{\mathcal{E}_*} = \langle f, k_\lambda^\Theta x\rangle_{\mathcal{H}(\Theta)}, \text{ where } k_\lambda^\Theta(z) = \frac{1}{1-\bar{\lambda} z}(I - \Theta(z)\Theta(\lambda)^*). \tag{2.5}$$

In particular, $k_\lambda^\Theta x \in \mathcal{H}(\Theta)$ for all $x \in \mathcal{E}_*, \lambda \in \mathbb{D}$, and the linear span of all $k_\lambda^\Theta x$ is dense in $\mathcal{H}(\Theta)$.

Remember that $\pi_{\Theta}^{*} : \mathcal{H}(\Theta) \to \mathbf{H}_{\Theta}$ (the inverse of the projection P_1) is unitary. We want to compute its action on the reproducing kernels.

LEMMA 2.1. *If $x \in \mathcal{E}_*, \lambda \in \mathbb{D}$, then $\pi_{\Theta}^{*}(k_{\lambda}^{\Theta} x) = F_{\lambda}(x \oplus x)$, where*

$$F_{\lambda}(\zeta) = k_{\lambda}^{\Theta}(\zeta) \oplus -\frac{1}{1-\bar{\lambda}\zeta}\Delta(\zeta)\Theta(\lambda)^*.$$

PROOF. It is obvious that $F_{\lambda}(x\oplus x) \in H^2 \oplus \mathrm{clos}(\Delta H^2)$ and $k_{\lambda}^{\Theta} x = P_1 F_{\lambda}(x \oplus x)$. In order to ascertain that $F_{\lambda}(x \oplus x) \in \mathbf{H}_{\Theta}$, it remains to check that $F_{\lambda}(x \oplus x) \perp \{\Theta h \oplus \Delta h : h \in H^2(\mathcal{E})\}$. This is equivalent to

$$\Theta(\zeta)^* k_{\lambda}^{\Theta}(\zeta) x - \frac{1}{1-\bar{\lambda}\zeta}\Delta^2(\zeta)\Theta(\lambda)^* x \perp H^2(\mathcal{E}_*).$$

But

$$\Theta(\zeta)^* k_{\lambda}^{\Theta}(\zeta) x - \frac{1}{1-\bar{\lambda}\zeta}\Delta^2(\zeta)\Theta(\lambda)^* x = \frac{\bar{\zeta}}{\bar{\zeta}-\bar{\lambda}}(\Theta(\zeta)^* - \Theta(\lambda)^*)x.$$

For any $f \in H^2(\mathcal{E}_*)$ we have

$$\langle \frac{\bar{\zeta}}{\bar{\zeta}-\bar{\lambda}}(\Theta(\zeta)^* - \Theta(\lambda)^*)x, f\rangle = \langle x, \frac{\zeta}{\zeta-\lambda}(\Theta(\zeta) - \Theta(\lambda))f\rangle = \langle x, \zeta F f\rangle,$$

with $F(\zeta) = \frac{1}{\zeta-\lambda}(\Theta(\zeta) - \Theta(\lambda))$ an analytic bounded function. The presence of ζ in the right hand side term in the scalar product implies that the scalar product with the constant function x is 0, and thus the lemma is proved. □

3. A unitary operator

Suppose Θ is, as above, a pure function in $H^\infty(\mathcal{E}, \mathcal{E}_*)$ with $\|\Theta\|_\infty \le 1$. Then $\widetilde{\Theta}$ has the same properties; Θ and $\widetilde{\Theta}$ generate contractive multiplication operators $M_\Theta, M_{\widetilde{\Theta}}$. As a consequence of the analyticity of Θ we have $M_\Theta(H^2(\mathcal{E})) \subset H^2(\mathcal{E}_*)$, $M_{\widetilde{\Theta}}(H^2(\mathcal{E}_*)) \subset H^2(\mathcal{E})$.

The Julia operator $J(M_{\widetilde{\Theta}})$ is unitary from $L^2(\mathcal{E}_*) \oplus \mathrm{clos}(\Delta_{\widetilde{\Theta}^*} L^2(\mathcal{E}))$ to $L^2(\mathcal{E}) \oplus \mathrm{clos}(\Delta_{\widetilde{\Theta}} L^2(\mathcal{E}_*))$. On the other hand, the map $f \mapsto \hat{f}$ leaves $L^2(\mathcal{E}_*)$ invariant and maps $\mathrm{clos}(\Delta_\Theta L^2(\mathcal{E}))$ onto $\mathrm{clos}(\Delta_{\widetilde{\Theta}^*} L^2(\mathcal{E}))$. Let us denote by $\Omega : L^2(\mathcal{E}_*) \oplus \mathrm{clos}(\Delta_\Theta L^2(\mathcal{E})) \to L^2(\mathcal{E}) \oplus \mathrm{clos}(\Delta_{\widetilde{\Theta}} L^2(\mathcal{E}_*))$ the composition of these two operators; that is,

$$\Omega(f \oplus g) = J(M_{\widetilde{\Theta}})(\hat{f} \oplus \hat{g}) = \left[\widetilde{\Theta}\hat{f} + \Delta_{\widetilde{\Theta}^*}\hat{g}\right] \oplus \left[\Delta_{\widetilde{\Theta}}\hat{f} - \widetilde{\Theta}^*\hat{g}\right]. \tag{3.1}$$

THEOREM 3.1. *The unitary operator Ω maps $\mathbf{K}_\Theta$ onto $\mathbf{K}_{\widetilde{\Theta}}$, and*

$$\Omega \mathbf{T}_\Theta = \mathbf{T}_{\widetilde{\Theta}}^* \Omega.$$

PROOF. We have

$$\mathbf{K}_\Theta = \left[L^2(\mathcal{E}_*) \oplus \mathrm{clos}(\Delta_\Theta L^2(\mathcal{E}))\right] \ominus \left[(H^2_-(\mathcal{E}_*) \oplus \{0\}) \oplus \{\Theta h \oplus \Delta h : h \in H^2(\mathcal{E})\}\right],$$

$$\mathbf{K}_{\widetilde{\Theta}} = \left[L^2(\mathcal{E}) \oplus \mathrm{clos}(\Delta_{\widetilde{\Theta}} L^2(\mathcal{E}_*))\right] \ominus \left[(H^2_-(\mathcal{E}) \oplus \{0\}) \oplus \{\widetilde{\Theta} h \oplus \Delta_{\widetilde{\Theta}} h : h \in H^2(\mathcal{E}_*)\}\right].$$

Since the map $f \mapsto \hat{f}$ is unitary from $H^2_-(\mathcal{E}_*)$ to $H^2(\mathcal{E}_*)$, we have $\Omega(H^2_-(\mathcal{E}_*) \oplus \{0\}) = \{\widetilde{\Theta} h \oplus \Delta_{\widetilde{\Theta}} h : h \in H^2(\mathcal{E}_*)\}$. A more careful computation, using the equality $J(M_{\widetilde{\Theta}})^* = J(M_{\widetilde{\Theta}^*})$, leads to $\Omega^*(H^2_-(\mathcal{E}) \oplus \{0\}) = \{\Theta h \oplus \Delta h : h \in H^2(\mathcal{E})\}$, or $\Omega(\{\Theta h \oplus \Delta h : h \in H^2(\mathcal{E})\}) = H^2_-(\mathcal{E}) \oplus \{0\}$. Passing to orthogonal subspaces, it follows that $\Omega(\mathbf{K}_\Theta) = \mathbf{K}_{\widetilde{\Theta}}$.

Further on, using (3.1) one checks that

$$\Omega(Z \oplus Z) = (Z^* \oplus Z^*)\Omega,$$

(note that the different Zs denote multiplication with the variable on the corresponding various spaces). Since $\mathbf{T}_\Theta = P_{\mathbf{K}_\Theta} Z|\mathbf{K}_\Theta$, $\mathbf{T}_{\widetilde{\Theta}} = P_{\mathbf{K}_{\widetilde{\Theta}}} Z|\mathbf{K}_{\widetilde{\Theta}}$, and $\Omega(\mathbf{K}_{\widetilde{\Theta}}) = \mathbf{K}_\Theta$, it follows that $\Omega \mathbf{T}_{\widetilde{\Theta}} = \mathbf{T}_\Theta^* \Omega$. □

The operator Ω is the central character of this note. As shown by Theorem 3.1, it identifies the adjoint of a model operator with another model operator. The next section will further show its use in different instances related to model spaces.

4. Various roles of Ω

4.1. A contraction from $\mathcal{H}(\Theta)$ to $\mathcal{H}(\widetilde{\Theta})$. The next theorem is the main part of [**1**, Appendix, Th.5]. We provide a short proof.

THEOREM 4.1. *Suppose Θ is a contractive analytic function with values in $\mathcal{L}(\mathcal{E}, \mathcal{E}_*)$.*

(i) *For any $w \in \mathbb{D}$, if we define the analytic function $\kappa_w^\Theta : \mathbb{D} \to \mathcal{L}(\mathcal{E}, \mathcal{E}_*)$ by the formula*

$$\kappa_w^\Theta(z)x = \frac{1}{z-w}(\Theta(z) - \Theta(w))x,$$

then for any $x \in \mathcal{E}$ we have $\kappa_w^\Theta(z)x \in \mathcal{H}(\Theta)$.

(ii) *For any $w \in \mathbb{D}$ and $f \in \mathcal{H}(\Theta)$, the formula*

$$\langle (Wf)(w), y\rangle = \langle f, \kappa_{\bar{w}}^\Theta y\rangle_{\mathcal{H}(\Theta)} \tag{4.1}$$

defines an analytic function $Wf \in \mathcal{H}(\widetilde{\Theta})$, and the map $f \mapsto Wf$ is a contraction from $\mathcal{H}(\Theta)$ to $\mathcal{H}(\widetilde{\Theta})$.

PROOF. Applying (3.1) to $f \oplus g = F_\lambda(x \oplus x)$, computations lead to

$$\Omega(F_\lambda(x \oplus x)) = \left(\frac{1}{\zeta - \bar\lambda}(\widetilde{\Theta}(\zeta) - \widetilde{\Theta}(\bar\lambda))x\right) \oplus \left(\frac{1}{\zeta - \bar\lambda}\Delta_{\widetilde{\Theta}} x\right). \tag{4.2}$$

In particular, it follows that the first component is in $\mathcal{H}(\widetilde{\Theta})$. But the map $\Theta \mapsto \widetilde{\Theta}$ is an involution, and therefore applying this fact to $\widetilde{\Theta}$ it also follows that (i) is proved.

To prove (ii), we claim that W is actually $\pi_{\widetilde{\Theta}} \Omega \pi_\Theta^*$. Let us check first the formula fo reproducing kernels $k_\lambda^\Theta x$, $x \in \mathcal{E}_*$, $\lambda \in \mathbb{D}$. By Lemma 2.1, $\pi_\Theta^* k_\lambda^\Theta x = F_\lambda(x \oplus x)$, and so from (4.2) it follows that

$$\langle \pi_{\widetilde{\Theta}} \Omega \pi_\Theta^* k_\lambda^\Theta x, y\rangle = \langle \frac{1}{\zeta - \bar\lambda}(\widetilde{\Theta}(\zeta) - \widetilde{\Theta}(\bar\lambda))x, y\rangle. \tag{4.3}$$

On the other hand, if we apply (4.1) to $f = k_\lambda^\Theta x$, we obtain

$$\begin{aligned} \langle (W(k_\lambda^\Theta x))(w), y\rangle = \langle k_\lambda^\Theta x, \kappa_{\bar w}^\Theta y\rangle_{\mathcal{H}(\Theta)} &= \overline{\langle \frac{1}{\lambda - \bar w}(\Theta(\lambda) - \Theta(\bar w))y, x\rangle} \\ &= \langle \frac{1}{w - \bar\lambda}(\widetilde{\Theta}(w) - \widetilde{\Theta}(\bar\lambda))x, y\rangle. \end{aligned} \tag{4.4}$$

Comparing (4.3) and (4.4) shows that $W = \pi_{\widetilde{\Theta}} \Omega \pi_\Theta^*$ on the linear span of reproducing kernels. If $f \in \mathcal{H}(\Theta)$ is arbitrary, then we may take a sequence $f_n \to f$, f_n belonging to the linear span of reproducing kernels. Then obviously $\pi_{\widetilde{\Theta}} \Omega \pi_\Theta^* f_n \to \pi_{\widetilde{\Theta}} \Omega \pi_\Theta^* f$, while, on the other hand, $\langle (Wf_n)(w), y\rangle \to \langle (Wf)(w), y\rangle$ for all $w \in \mathbb{D}$

(convergence in norm implies convergence in points of $\mathbb{D}$). It follows that $Wf = \pi_{\widetilde{\Theta}}\Omega\pi_{\Theta}^* f$ for all $f \in \mathcal{H}(\Theta)$. Since Ω is unitary and $\pi_\Theta, \pi_{\widetilde{\Theta}}$ are contractions, (ii) is proved. □

Actually, one can say more.

THEOREM 4.2. *With the above notations,*

(i) *W is isometric if and only if $\widetilde{\Theta}$ is distinguished.*
(ii) *W is unitary if and only if Θ and $\widetilde{\Theta}$ are distinguished.*

PROOF. If we interchange Θ and $\widetilde{\Theta}$, then W is transformed in W^*. Therefore (i) implies (ii).

To prove (i), suppose first that $\widetilde{\Theta}$ is distinguished. then $\mathbf{H}_{\widetilde{\Theta}} = \mathbf{K}_{\widetilde{\Theta}}$, whence $\pi_{\widetilde{\Theta}}$ is unitary. Since Ω is unitary and π_{Θ}^* is isometric, it follows that W is isometric.

Conversely, suppose that W is isometric. This implies that $\Omega(\mathbf{H}_\Theta) \subset \mathbf{H}_{\widetilde{\Theta}}$. In particular, $\Omega F_\lambda(x \oplus x) \in \mathbf{H}_{\widetilde{\Theta}}$ for all $\lambda \in \mathbb{D}$, $x \in \mathcal{E}_*$, whence its second term in the direct sum decomposition is in $\operatorname{clos}(\Delta_{\widetilde{\Theta}} H^2(\mathcal{E}_*))$. By (4.2) this means that

$$\Delta_{\widetilde{\Theta}} \frac{\bar{\zeta}}{1 - \bar{\lambda}\bar{\zeta}} x \in \operatorname{clos}(\Delta_{\widetilde{\Theta}} H^2(\mathcal{E}_*))$$

for all $\lambda \in \mathbb{D}$. Since the functions $\frac{\bar{\zeta}}{1-\bar{\lambda}\bar{\zeta}} x$ span $L^2(\mathcal{E}_*) \ominus H^2(\mathcal{E}_*)$, we have $\Delta_{\widetilde{\Theta}}(L^2(\mathcal{E}_*) \ominus H^2(\mathcal{E}_*)) \subset \operatorname{clos}(\Delta_{\widetilde{\Theta}} H^2(\mathcal{E}_*))$. Therefore $\operatorname{clos}(\Delta_{\widetilde{\Theta}} H^2(\mathcal{E}_*)) = \operatorname{clos}(\Delta_{\widetilde{\Theta}} L^2(\mathcal{E}_*))$, which means that $\widetilde{\Theta}$ is distinguished. □

4.2. The case of scalar Θ. If $\dim \mathcal{E} = \dim \mathcal{E}_* = 1$, then Θ is actually a scalar function in the unit ball of H^∞, that we will denote by θ. As noted in Subsection 2.2, in this case θ is distinguished if and only if it is an extreme point of the unit ball of H^∞. Also, θ is an extreme point if and only if $\tilde{\theta}$ is an extreme point (this again is not true in general, as the simple example $\Theta(z) = (z \quad 0)$ shows). So the operator W in Subsection 4.1 is isometric if and only if it is unitary.

Suppose now that θ is inner. Then $\tilde{\theta}$ is also inner, $\Delta_\theta = \Delta_{\tilde{\theta}} = 0$, and $\mathbf{K}_\theta$ becomes the "classical" model space K_θ. Formula (3.1) defines a unitary operator $\omega : L^2 \to L^2$, $\omega(f) = \tilde{\theta}\hat{f}$, which maps K_θ onto $K_{\tilde{\theta}}$. Now, in this case we can consider also for functions in L^2 the map $f \mapsto \tilde{f}$, defined by the similar formula $\tilde{f}(\zeta) = \overline{f(\bar{\zeta})}$, which is a *conjugation*, that is, an isometric surjective antilinear operator. It interchanges K_θ and $K_{\tilde{\theta}}$; therefore, if we apply it after $\Omega : K_\theta \to K_{\tilde{\theta}}$, we obtain a conjugation C on K_θ. This turns out to be precisely the well known conjugation on model spaces given by $(Cf)(\zeta) = \theta(\zeta)\overline{\zeta f(\zeta)}$.

If θ is not inner, since $f \mapsto \tilde{f}$ interchanges also $\operatorname{clos} \Delta_\theta L^2$ and $\operatorname{clos} \Delta_{\tilde{\theta}} L^2$, it maps again $\mathbf{K}_\theta$ onto $\mathbf{K}_{\tilde{\theta}}$, and applying it after Ω yields a conjugation C on $\mathbf{K}_\theta$, given by the formula

$$(C(f \oplus g))(\zeta) = \left[\theta(\zeta)\overline{\zeta f(\zeta)} + \Delta(\zeta)\overline{\zeta g(\zeta)}\right] \oplus \left[\Delta(\zeta)\overline{\zeta g(\zeta)} - \overline{\theta(\zeta)}\zeta\overline{f(\zeta)}\right].$$

One checks that this conjugation satisfies the relation

$$\mathbf{T}_\theta^* = C\mathbf{T}_\theta C,$$

which shows that $\mathbf{T}_\theta$ is *complex symmetric* in the sense of [**5**]. The complex symmetry of the model operator in the scalar case is proved for inner θ in [**5**] (see also [**4**]) and for arbitrary θ in the unit ball of H^∞ in [**2**].

4.3. Conjugation on the model space. In the general case a model operator is not necessarily complex symmetric. A necessary and sufficient condition is given in [**2**]. First, we must have $\dim \mathcal{E} = \dim \mathcal{E}_*$, and thus we may assume that actually $\mathcal{E} = \mathcal{E}^*$. Then there exists a conjugation $\mathbf{C}$ on $\mathbf{K}_\Theta$ such that $\mathbf{T}_\Theta^* = \mathbf{C}\mathbf{T}_\Theta\mathbf{C}$ (that is, $\mathbf{T}_\Theta$ is complex symmetric) if and only if there is a conjugation J on $\mathcal{E}$ such that $\Theta(\lambda)^* = J\Theta(\lambda)J$ for all $\lambda \in \mathbb{D}$.

Based on the construction in Subsection 4.2, the alert reader should already guess how one obtains $\mathbf{C}$ from J. One starts by obtaining from J a conjugation $\mathbf{J}$ on $L^2(\mathcal{E})$ by the formula $\mathbf{J}(f)(\zeta) = J(f(\bar{\zeta}))$. This in turn leads to a conjugation $\mathbf{I}$ that interchanges $\mathbf{K}_\Theta$ and $\mathbf{K}_{\tilde{\Theta}}$, and finally one defines $\mathbf{C} = \mathbf{I} \circ \Omega$.

4.4. The space $\mathcal{D}(\Theta)$. As stated in Subsection 2.3, the general form of the complete de Branges–Rovnyak model space $\mathcal{D}(\Theta)$ is more complicated than just $\mathcal{H}(\Theta)$. Although the original definition is different, it is worth to note that one can give an equivalent description based on our operator Ω. Namely, $\mathcal{D}(\Theta)$ is a linear subspace of $H^2(\mathcal{E}_*) \oplus H^2(\mathcal{E})$, which as a set is precisely the image of the map

$$\xi \mapsto P_1\xi \oplus P_1(\Omega\xi), \quad \xi \in \mathbf{K}_\Theta.$$

This map is one-to-one on $\mathbf{K}_\Theta$, and the norm on $\mathcal{D}(\Theta)$ is defined by requiring that it becomes a unitary operator. This characterization of $\mathcal{D}(\Theta)$ is proved in [**7**, Theorem 12.2]. It leads to the equivalence of the Nagy–Foias and the de Branges–Rovnyak constructions in the general case.

Acknowledgements

Thanks are due to Emmanuel Fricain for a question that lead to the writing of this note.

References

[1] Louis de Branges and James Rovnyak, *Canonical models in quantum scattering theory*, Perturbation Theory and its Applications in Quantum Mechanics (Proc. Adv. Sem. Math. Res. Center, U.S. Army, Theoret. Chem. Inst., Univ. of Wisconsin, Madison, Wis., 1965), Wiley, New York, 1966, pp. 295–392. MR0244795 (39 #6109)

[2] Nicolas Chevrot, Emmanuel Fricain, and Dan Timotin, *The characteristic function of a complex symmetric contraction*, Proc. Amer. Math. Soc. **135** (2007), no. 9, 2877–2886 (electronic), DOI 10.1090/S0002-9939-07-08803-X. MR2317964 (2008c:47025)

E. Fricain, J. Mashreghi, *Theory of $\mathcal{H}(b)$ Spaces, vol. I and II*, New Monographs in Mathematics, Cambridge University Press, to appear.

[3] Javad Mashreghi and Emmanuel Fricain, *Exceptional sets for the derivatives of Blaschke products*, Proceedings of the St. Petersburg Mathematical Society. Vol. XIII, Amer. Math. Soc. Transl. Ser. 2, vol. 222, Amer. Math. Soc., Providence, RI, 2008, pp. 163–170. MR2433525 (2009m:30062)

[4] Stephan Ramon Garcia, *Conjugation and Clark operators*, Recent advances in operator-related function theory, Contemp. Math., vol. 393, Amer. Math. Soc., Providence, RI, 2006, pp. 67–111, DOI 10.1090/conm/393/07372. MR2198373 (2007b:47073)

[5] Stephan Ramon Garcia and Mihai Putinar, *Complex symmetric operators and applications*, Trans. Amer. Math. Soc. **358** (2006), no. 3, 1285–1315 (electronic), DOI 10.1090/S0002-9947-05-03742-6. MR2187654 (2006j:47036)

[6] Nikolai K. Nikolski, *Operators, functions, and systems: an easy reading. Vol. 2*, Mathematical Surveys and Monographs, vol. 93, American Mathematical Society, Providence, RI, 2002. Model operators and systems; Translated from the French by Andreas Hartmann and revised by the author. MR1892647 (2003i:47001b)

[7] N. K. Nikol′skiĭ and V. I. Vasyunin, *Notes on two function models*, The Bieberbach conjecture (West Lafayette, Ind., 1985), Math. Surveys Monogr., vol. 21, Amer. Math. Soc., Providence, RI, 1986, pp. 113–141, DOI 10.1090/surv/021/11. MR875237 (88f:47008)

[8] Donald Sarason, *Sub-Hardy Hilbert spaces in the unit disk*, University of Arkansas Lecture Notes in the Mathematical Sciences, 10, John Wiley & Sons, Inc., New York, 1994. A Wiley-Interscience Publication. MR1289670 (96k:46039)

[9] Béla Sz.-Nagy, Ciprian Foias, Hari Bercovici, and László Kérchy, *Harmonic analysis of operators on Hilbert space*, Revised and enlarged edition, Universitext, Springer, New York, 2010. MR2760647 (2012b:47001)

Institute of Mathematics of the Romanian Academy, P.O. Box 1-764, Bucharest 014700, Romania

E-mail address: Dan.Timotin@imar.ro

Contemporary Mathematics
Volume **638**, 2015
http://dx.doi.org/10.1090/conm/638/12813

Selected problems in classical function theory

Catherine Bénéteau and Dmitry Khavinson

Abstract. We discuss several problems in classical complex analysis that might appeal to graduate students and young researchers. Among them are possible extensions to multiply connected domains of the Neuwirth-Newman theorem regarding analytic functions with positive boundary values, characterizing domains by properties of best approximations of $\overline{z}$ by analytic functions in various metrics, and sharpening the celebrated Putnam inequality in the context of Toeplitz operators on Bergman spaces and the related isoperimetric inequalities, aka " isoperimetric sandwiches".

1. Introduction: Spaces of Analytic Functions

This paper is a selective survey of a few problems that are at the interface of complex analysis and geometry. We will be dealing with various classes of analytic functions in arbitrary domains, such as Hardy, Bergman, and Smirnov spaces. Let us begin by defining these spaces (see [**8, 9, 16**]).

Definition 1.1. *For $0 < p < \infty$, define the Bergman space of the disk to be*

$$A^p(\mathbb{D}) = \left\{ f \text{ analytic in } \mathbb{D} : \left(\int_{\mathbb{D}} |f(z)|^p dA(z) \right)^{\frac{1}{p}} =: \|f\|_p < \infty \right\},$$

where $dA = \frac{1}{\pi} dx dy$ denotes normalized area measure in the unit disk $\mathbb{D}$. The Bergman spaces $A^p(G)$ for an arbitrary domain G are defined in a similar way.

If instead of area measure, we consider line integrals on concentric circles, we get the Hardy spaces.

Definition 1.2. *For $0 < p < \infty$, define the Hardy space of the disk as*

$$H^p(\mathbb{D}) := \left\{ f \text{ analytic in } \mathbb{D} : \sup_{0<r<1} \frac{1}{2\pi} \int_0^{2\pi} |f(re^{it})|^p dt =: \|f\|_{H^p}^p < \infty \right\}.$$

When $p = \infty$, we define

$$H^\infty(\mathbb{D}) = \{ f \text{ analytic in } \mathbb{D} : \sup \{ |f(z)|, z \in \mathbb{D} \} =: \|f\|_\infty < \infty \}.$$

For arbitrary domains G, we define the Hardy spaces as follows (see [**11**]).

2010 *Mathematics Subject Classification.* Primary 30H05; Secondary 30E05.

The second author gratefully acknowledges partial support from the National Science Foundation under the grant DMS-0855597.

DEFINITION 1.3. *An analytic function $f(z)$ in G belongs to the Hardy class $H^p(G)$ for $0 < p < \infty$ if the subharmonic function $|f|^p$ has a harmonic majorant in G.*

Note that Hardy classes are conformally invariant, i.e., if $\varphi : K \to G$ is the conformal mapping of an n-connected circular domain K onto G, then $f \in H^p(G)$ if and only if $f \circ \varphi \in H^p(K)$. On the other hand, if one defines a space in a way analogous to that of Definition 1.2 for an arbitrary domain G, one gets a potentially different class of functions called *Smirnov* classes ([**8**, **14**]). More specifically, let G be an n-connected domain in the complex plane bounded by Jordan rectifiable curves $\gamma_1, \dots, \gamma_n$ and let $\Gamma = \bigcup_{i=1}^n \gamma_i$.

DEFINITION 1.4. *Let $0 < p < \infty$. An analytic function $f(z)$ in G is said to belong to the Smirnov class $E^p(G)$ if there exists a sequence of rectifiable curves $\{\Gamma_i\}$ in G converging to Γ such that*

$$\limsup_{i\to\infty} \int_{\Gamma_i} |f(z)|^p |dz| =: \|f\|_{E^p}^p < \infty.$$

Notice that the critical difference between Smirnov classes and Hardy classes is that the former are not conformally invariant. In addition, for $p \geq 1$, Hardy functions are represented by Poisson integrals of their boundary values, while Smirnov functions are represented by Cauchy integrals, i.e., in the former case the kernel is positive, in the latter, complex. This difference will allow for some interesting phenomena related to boundary values, which we discuss in the next section. In Section 3, we examine the concept of *analytic content*, which is the best approximation of the simplest anti-analytic function, namely $\bar{z}$, in an appropriate context. We will see that analytic content is tightly connected to the geometry of a domain. In Section 4, we discuss Putnam's inequality for Toeplitz operators on Bergman spaces. In each section, we state conjectures and mention some open problems.

2. Analytic Functions with Positive Boundary Values

For $p \geq 1$, it is well-known that functions in $H^p(\mathbb{D})$ cannot have real boundary values on the circle unless they are constants. This is because Hardy space functions ($p \geq 1$) can be represented as Poisson integrals of their boundary values, and therefore if they are real on the boundary, they are real in the whole domain, and thus must be constant. However, for $0 < p < 1$, there are many such functions, for example, $f(z) = i\frac{1+z}{1-z}$. A beautiful theorem of J. Neuwirth and D. J. Newman ([**24**]) shows that if we require the boundary values to be positive, we can go a little further:

THEOREM 2.1. *([**24**]) Let $f \in H^{1/2}(\mathbb{D})$ be such that $f(\zeta) \geq 0$ for almost every ζ on the unit circle $\mathbb{T}$. Then f is constant.*

Neuwirth and Newman pointed out that the value $1/2$ in the theorem is sharp, since the function $\frac{z}{(1+z)^2}$ is in $H^p(\mathbb{D})$ for all $0 < p < 1/2$ and has positive boundary values. If we consider arbitrary domains G that might have "bad boundaries", we can replace non-tangential boundary values with asymptotic boundary values (see [**19**]), and interpret "almost everywhere" as being understood with respect to harmonic measure, and then the theorem extends to $H^p(G)$.

An interesting question is, what happens for the Smirnov classes $E^p(G)$ when $p \geq 1$? The answer depends on the geometric character of the boundary. Recall that a finitely connected domain G is called Smirnov if the derivative of the conformal map from a circular domain onto G is an outer function (see [**8**]). It turns out that if the domain G is non-Smirnov, then the theorem is false, and indeed in such a domain, for every $0 < p < \infty$, there exist non-constant $f \in E^p(G)$ such that $0 \leq f(\zeta) \leq 1$ for almost every ζ on the boundary of the domain (see [**18**]). In the other extreme, if the domain G has smooth boundary Γ, then the classes $H^p(G)$ and $E^p(G)$ are equal (as sets), and therefore the situation in $E^p(G)$ is exactly the same as that of $H^p(G)$. However, if the domain is Smirnov and has singularities, these singularities allow for the construction of functions in E^p with real boundary values, for certain values of p. In fact, the values of p that allow for the construction of such functions are tightly connected in general to the geometric characteristic of the singularity. For more detail on the construction of such functions with real boundary values, see [**6, 7**] and the references therein.

In the case that there do exist non-trivial functions in E^p with real boundary values, if G is a simply connected Smirnov domain, L. DeCastro and D. Khavinson noted that the analogue of the Neuwirth-Newman theorem holds:

THEOREM 2.2. *([**7**]) Let G be a simply connected Smirnov domain with rectifiable boundary Γ. Let $p_0 \geq 1$ be defined as the smallest $p \geq 1$ such that $f \in E^p(G)$ and f has real boundary values a.e. on Γ imply that f is a constant. Then all $f \in E^{p_0/2}$ such that $f \geq 0$ a.e. on Γ are constants.*

The proof of this theorem is along the same lines as that of the original Neuwirth-Newman result, and is sketched here.

PROOF. Write $f(z) = B(z)S(z)F^2(z)$, where $B(z)$ is a generalized Blaschke product, $S(z)$ is a bounded singular inner function, and $F(z) \in E^{p_0}$ is an outer function. On Γ, since $f \geq 0$, we have that $B(z)S(z)F^2(z) = |f(z)|$. On the other hand, $|f(z)| = |F(z)|^2 = F(z)\overline{F(z)}$ a.e., and therefore $\overline{F(z)} = B(z)S(z)F(z) \in E^{p_0}(G)$. This implies that $F(z)+\overline{F(z)} \in E^{p_0}(G)$ and is real-valued, hence a constant. Thus, $f(z) = const \cdot B(z)S(z)$ is a bounded function with non-negative boundary values, hence a constant as well. □

In multiply connected domains, it is not clear whether the Neuwirth-Newman theorem holds. We can still write $f = QBSF^2$, where B is the generalized Blaschke product, S is a singular inner function, F^2 is an outer factor, $F \in E^{p_0}$, Q is an invertible bounded analytic function, and $|Q|$ is a local constant on Γ (see [**19**]). The problem is that $|B|$ and $|S|$ are local constants a.e. on the boundary of G. Hence, $f \geq 0$ a.e. on Γ only yields on Γ that $f = QBSF^2 = |QBS|F\overline{F}$, i.e., $\overline{F}$ coincides with *different* analytic functions on different boundary components. However, L. DeCastro and D. Khavinson showed (see [**7**]) that in an n - connected domain G of "cardioid type" (as pictured below) with m interior cusps on ∂G, all E^1-functions with positive boundary values are constants. (For such domains, $p_0 = 2$.)

It seems likely that the proof of this result will hold for finitely connected domains with finitely many corners, and thus, the authors of that paper conjectured that the Neuwirth-Newman theorem holds in multiply connected domains:

CONJECTURE 2.1. *([**7**]) Let G be a finitely connected Smirnov domain. If $p_0 \geq 1$ is the smallest index for which all functions in E^{p_0} with real boundary*

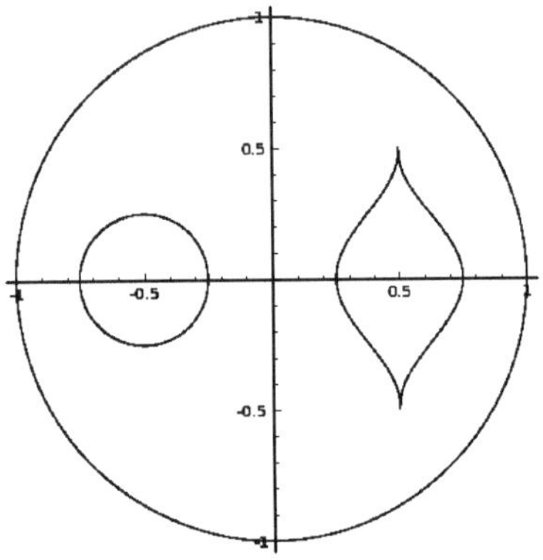

values are constants, then all $E^{p_0/2}$ functions with positive boundary values are constants.

3. Analytic Content

Let us now turn to a discussion of the approximation of $\bar{z}$ by analytic functions from different classes in domains with analytic boundaries. The main focus will be the concept of *analytic content.*

3.1. Bounded Functions. Let G be a finitely connected region in $\mathbb{C}$ with boundary Γ consisting of n simple closed analytic curves γ_j, $j = 1, \dots, n$.

DEFINITION 3.1. *The analytic content of a domain G is*

$$\lambda(G) := \inf_{\phi \in H^\infty(G)} \| \bar{z} - \phi \|_{L^\infty(\Gamma)} \, .$$

We often call ϕ the *best approximation* of $\bar{z}$ in $H^\infty(G)$. The analytic content turns out to have an interesting relationship with certain geometric features of the domain, as can be seen by the following theorem.

THEOREM 3.1. *(**[2, 13, 21]**) Let G be a finitely connected region whose boundary is analytic, and let A and P be the area and perimeter of G, respectively. Then*

$$\frac{2A}{P} \le \lambda(G) \le \sqrt{\frac{A}{\pi}}.$$

Moreover, $\lambda(G) = \sqrt{\frac{A}{\pi}}$ if and only if G is a disk.

Note that by ignoring the analytic content in the inequality stated above, one recovers the classical isoperimetric inequality, namely that $P^2 \ge 4\pi A$. In addition, the theorem states that the upper bound is achieved if and only if G is a disk, and therefore a natural question is, for which G does the equality $\frac{2A}{P} = \lambda(G)$ hold? The following theorem gives some equivalent forms for the achievement of the lower bound.

THEOREM 3.2. *(**[21]**) Let G be a finitely connected region whose boundary is analytic, and let A and P be the area and perimeter of G. The following are equivalent:*

(i) $\lambda = \frac{2A}{P}$;

(ii) There is $\phi \in H^\infty(G)$ *such that* $\bar{z}(s) - i\lambda \frac{d\bar{z}}{ds} = \phi(z(s))$ *on* Γ*, where* s *is the arc-length parameter;*

(iii) $\frac{1}{A}\int_G f dA = \frac{1}{P}\int_\Gamma f ds$ *for all* $f \in H^\infty(G)$*.*

Notice that *(iii)* holds for annuli $G = \{r < |z| < R\}$! Therefore, the lower bound does indeed hold for regions other than disks. However, D. Khavinson proved in [**21**] that if the domain is simply connected, then it is a disk. On the other hand, if the domain is finitely connected, then *(ii)* implies that if Γ contains a circular arc, G is a disk or an annulus. In [**21**], the author asked whether these are the only two possibilities. It turns out that the answer is yes! This has been recently proved in [**1**].

3.2. Smirnov Classes. If we consider the Smirnov classes $E^p(G)$ instead of bounded analytic functions in G, we can generalize the notion of analytic content. In what follows, $L^p(\Gamma) := L^p(\Gamma, ds)$, where s is the arc length parameter.

DEFINITION 3.2. *For* $p \geq 1$*, the Smirnov analytic content of a domain* G *is*

$$\lambda_{E^p}(G) := \inf_{\phi \in E^p(G)} \| \bar{z} - \phi \|_{L^p(\Gamma)} .$$

This extremal quantity turns out to be equal to another, often referred to as the "dual extremal problem":

$$\sup_{f \in E_1^q(G)} \left| \int_\Gamma \bar{z} f(z) dz \right| .$$

Here, $E_1^q(G)$ refers to the unit ball in $E^q(G)$, where $1/p + 1/q = 1$. One might ask if there are upper and lower bounds for the analytic content for the Smirnov classes similar to those in Theorem 3.1, and indeed there are.

THEOREM 3.3. *([**15**]) Let* A, P *denote the area and perimeter of a finitely connected domain* G*. For* $p \geq 1$, $q = \frac{p}{p-1}$*, we have*

$$\frac{2A}{\sqrt[q]{P}} \leq \lambda_{E^p} \leq \sqrt{\frac{A}{\pi}} P^{\frac{1}{p}} .$$

Again, one might ask, are disks and annuli the only extremal domains for all λ_{E^p}, $p \geq 1$? In addition, do the extremal functions for λ_{E^p} characterize the domain G? The following theorem gives some insight into the second question.

THEOREM 3.4. *([**15**]) Let* $\Gamma := \partial G$ *be real analytic and* $p \geq 1$*. If the best approximation to* $\bar{z}$ *in* E^p *is a constant, then* G *is a disk.*

Sketch of the proof: Without loss of generality, let's assume that the best approximation is zero. Then one can show that $0 \in G$ and that the extremal function f^* for the dual problem satisfies $|f^*| \leq 1$ in G and $|f^*| = 1$ on the boundary, and the duality relationship

$$f^* \bar{z} dz = const |z|^p ds \quad \text{on}\, \Gamma$$

holds, where we can take the constant to be positive. Dividing by z yields

$$\frac{f^*(z)}{z} dz = const |z|^{p-2} ds. \tag{3.1}$$

For $p = 1$, using regularity results for extremals (see [**23**]) in order to apply the argument principle, if f^* is not constant, one can show that the left hand side of (3.1) has a non-trivial increment of its argument, while the right hand side doesn't (because it's positive), which is a contradiction. Therefore, we conclude that f^* is a unimodular constant. Now again using regularity of the boundary and parametrizing $z = r(\theta)e^{i\theta}$, and using the duality relationship (3.1) gives, after some simple calculus, that $dr/d\theta = 0$, and hence Γ consists of circles centered at the origin. Using the duality equation one last time shows that since dz/ds must have the same sign on both circles, then there can only be one circle, and hence, G is a disk. The case $p > 1$ is more complicated, and in particular, the case that $p \in \mathbb{N}$ has to be treated separately. For details, see [**15**].

Note that this theorem *proves* that the domain is simply connected. If we assume G to be simply connected to begin with, the regularity hypothesis (that is, the analyticity of the boundary) can be relaxed significantly to assume merely that G is a Smirnov domain, by appealing to the following theorem.

THEOREM 3.5. *([**10**]) Let G be a Jordan domain in $\mathbb{C}$ containing 0 and with the rectifiable boundary Γ satisfying the Smirnov condition. Suppose the harmonic measure on Γ with respect to the origin equals $c|z|^\alpha ds$ for $z \in \Gamma$, where ds denotes arclength measure on Γ, $\alpha \in \mathbb{R}$ and c is a positive constant. Then*

(i) For $\alpha = -2$, the solutions are precisely all disks G containing 0.

(ii) For $\alpha = -3, -4, -5, \ldots$ there are solutions G which are not disks.

(iii) For all other values of α, the only solutions are disks centered at 0.

The conclusion of Theorem 3.4 then follows, because the left hand side of (3.1) is a constant multiple of harmonic measure at the origin, and since in our case, $\alpha = p - 2$ with $p > 1$ so $\alpha > -1$, part (iii) of Theorem 3.5 applies, giving that G is a disk.

What happens in the finitely connected case is not known, and thus leads to the following problem.

PROBLEM 3.1. *Extend Theorem 3.4 to finitely connected Smirnov domains. In particular, do the hypotheses of that theorem imply that G is simply connected?*

Notice that in the case $0 < p < 1$, we can still define analytic content, but we lose duality (since in that case E^p is not a Banach space), and so it is not clear whether you might get a type of "duality equation" on the boundary holding for the extremal. Thus one might consider the following.

PROBLEM 3.2. *What can be said about analytic content in E^p spaces for $0 < p < 1$? Are there estimates similar to those in Theorem 3.3?*

In a similar manner as before, one might ask whether best approximations of $\bar{z}$ characterize annuli. The following theorem gives the answer when $p = 1$.

THEOREM 3.6. *([**15**]) Let $\Gamma := \partial G$ be real analytic and $p = 1$. If the best approximation to $\bar{z}$ in E^1 is a rational function $g(z) = \frac{c}{z-a}$, then G is an annulus centered at a.*

CONJECTURE 3.1. *Theorem 3.6 holds for all $p > 1$ and all finitely connected Smirnov domains.*

The following problem is completely unknown territory, and it is easy to see that the study of such domains leads to a larger class than the well-known quadrature domains.

PROBLEM 3.3. *Study domains where best approximations of $\bar{z}$ in E^p are, say, rational functions.*

3.3. Bergman Spaces. Let us now discuss analytic content in the context of Bergman spaces.

DEFINITION 3.3. *For $p \geq 1$, the Bergman space analytic content of a domain G is*

$$\lambda_{A^p}(G) := \inf_{\phi \in A^p(G)} \| \bar{z} - \phi \|_{A^p(G)} .$$

THEOREM 3.7. *([**15**]) Let G be a Smirnov domain and let $p \geq 1$. Then*
(i) If the best approximation of $\bar{z}$ in A^p is a constant, then G is a disk.
(ii) If the best approximation of $\bar{z}$ in A^p is $g(z) = \frac{c}{z-a}$, then G is an annulus centered at a.

Sketch of proof. For (i), assume for the sake of brevity that $p > 1$ and that the best approximation of $\bar{z}$ is 0. Recall that Khavin's lemma (see [**27**]) states that the annihilator $(A^p)^\perp$ of A^p inside $L^q(G, dA)$ is given by the $\bar{z}$ derivatives of functions in the standard Sobolev space $W_0^{1,q}$, where p and q are conjugate indices. Therefore, the dual problem in this context states that

$$\inf_{\phi \in A^p(G)} \| \bar{z} - \phi \|_{A^p(G)} = \sup_{u \in W_0^{1,q}, \|u_{\bar{z}}\|_q \leq 1} \left| \int_G \bar{z} \frac{\partial u}{\partial \bar{z}} \, dA(z) \right| .$$

The duality relationship then yields that for $z \in G$,

$$\frac{\partial u}{\partial \bar{z}} = const \frac{|z|^p}{\bar{z}}, \, u \in W_0^{1,q}$$

where u is a solution of the dual problem. Integrating with respect to $\bar{z}$ gives (in G):

$$u = const|z|^p + h, \text{ where } h \in H^\infty(G).$$

Since $u = 0$ on Γ, we get that $h|_\Gamma$ is real-valued and therefore constant on Γ, and therefore $|z|$ is constant on Γ, hence Γ is a disk. For the proof of (ii), see [**15**].

Note that this proof is easier in the context of Bergman spaces, because the duality relationship holds in the whole domain G.

PROBLEM 3.4. *What are the isoperimetric "sandwich" estimates for λ_{A^p}?*

Nothing is known about the following problem.

PROBLEM 3.5. *What can be said about domains with other rational best approximations of $\bar{z}$ in A^p? For example, if the best approximation is a rational function of degree 2, what is the corresponding domain?*

4. Putnam's Inequality for Toeplitz Operators in Bergman Spaces

Let us now turn to a discussion of isoperimetric inequalities in the context of operator theory. Recall that if T is a bounded linear operator on a Hilbert space, then T is called hyponormal if $[T^*, T] := T^*T - TT^* \geq 0$. Putnam's inequality (see [**26**]) applied to T then states that

$$\|[T^*, T]\| \leq \frac{Area(sp(T))}{\pi}$$

where $sp(T)$ denotes the spectrum of T. In particular, if ϕ is analytic in a neighborhood of the finitely connected domain G and $T := T_\phi : E^2 \to E^2$ is defined by

$Tf = \phi f$, (T is called an analytic Toeplitz operator), then Putnam's inequality in this context states that

$$\|[T^*, T]\| \le \frac{Area(\phi(G))}{\pi}.$$

In [**20**], the author gave a lower bound:

$$\frac{4(Area(\phi(G)))^2}{\|\phi'\|^2_{E^2(G)} \cdot P} \le \|[T^*, T]\|,$$

where $\partial(\phi(G)) =: \Gamma$, and $P := P(\Gamma) =$ perimeter of $\phi(G)$. Putting the above two inequalities together and taking $\phi(z) = z$ gives $P^2 \ge 4\pi A$, the classical isoperimetric inequality. If $\phi(z) = z$ and $G = \phi(G) = \mathbb{D}$, then equality is achieved, and thus, Putnam's inequality in the E^2 context is sharp in the sense that there exists an operator on E^2 for which Putnam's inequality becomes equality.

One might then ask what happens in spaces other than E^2. The authors in [**12**] explore this question for Bergman spaces, following the paper [**3**]. Without loss of generality, we can assume there exists a measure μ in $\mathbb{C}$ such that T is unitarily equivalent to the operator T_z of multiplication by z on $L^2_a(\mu)$, the closure (in $L^2(\mu)$) of functions analytic in a neighborhood of the support of μ (see [**3**]). Letting K be the support of the measure μ and G the polynomial hull of K, a standard Hilbert space calculation then shows that

$$\|[T_z^*, T_z]\| = \sup_{\|g\|_2=1} \left\{ \inf_{f\in H^\infty(G)} \|\bar{z}g - f\|_2^2 \right\}.$$

Taking $f = gh$ then gives this right hand side less than or equal to

$$\inf_{h\in H^\infty(G)} \|\bar{z} - h\|^2_\infty =: \lambda^2(G).$$

Here $\lambda := \lambda(G)$ is the analytic content for bounded functions from Definition 3.1. Thus, Theorem 3.1 and Putnam's inequality give that $\|[T_z^*, T_z]\| \le \lambda^2 \le A(G)/\pi$, and therefore equality is attained in Putnam's inequality only if the spectrum $sp(T)$ is a disk and the spectral measure "sits" on the circumference. Thus it is clear that in the context of Bergman spaces, for example, equality can never be attained in Putnam's inequality. A calculation (straightforward but tedious!) reveals that in $A^2(\mathbb{D})$, $\|[T_z^*, T_z]\| = 1/2$, that is, the upper bound is two times smaller than the one in the general Putnam inequality. One might ask, then, should Putnam's inequality in this context be corrected by a factor $1/2$?

In exploring this question, the authors of [**4**] considered the torsional rigidity ρ of a domain G, which measures the resilience of the beam of cross section G to twisting. In terms of a "Rayleigh type" quotient,

$$\rho := \sup_{\psi\in C_0^\infty} \left(\frac{2\|\psi\|_1}{\|\nabla\psi\|_2}\right)^2.$$

They then proved the following.

THEOREM 4.1. *([**4**])*

$$\frac{\rho}{Area(\phi(G))} \le \|[T_\phi^*, T_\phi]\|.$$

Hence, taking $\phi = z$ and using the upper bound given by Putnam's inequality gives the "isoperimetric sandwich"

$$\rho \leq \frac{(Area(G))^2}{\pi}.$$

The estimate in the above theorem $\rho \leq \frac{(\mathrm{Area}(G))^2}{\pi}$ was missing by a factor of 2 the celebrated Saint-Venant inequality conjectured in 1856, which was first proved by G. Polya in 1948. This prompted the following conjecture.

CONJECTURE 4.1. *([**4**]) For the Bergman space, $\|[T_z^*, T_z]\| \leq \frac{Area(G)}{2\pi}$.*

For simply connected domains G, this conjecture is now a theorem! (See [**25**].) Hence, this leads to a new proof of Saint-Venant's Inequality that $\rho \leq \frac{(\mathrm{Area}(G))^2}{2\pi}$. The proof in [**25**] is tour de force calculation with power series. This is why the statement is restricted to simply connected domains. The authors of [**12**] noted that refining Olsen and Reguera's proof implies that the equality for the self-commutator upper bound in simply connected domains holds only for disks. This yields an alternative proof that Saint-Venant's inequality becomes equality only for disks. We are thus left with a host of interesting problems to investigate.

PROBLEM 4.1. *Find the "book" proof of the Olsen - Reguera theorem in* [**25**], *freeing it from the power series calculation and extending the result to arbitrary domains.*

PROBLEM 4.2. *Is the sharp upper bound for the A^2-content equal to $\frac{1}{2}\sqrt{\frac{Area(G)}{\pi}}$?*

PROBLEM 4.3. *What is the sharp lower bound for the A^2-content expressed in terms of geometric characteristics (e.g., area, perimeter, principal frequency) of the domain?*

PROBLEM 4.4. *Refine the "isoperimetric sandwich" inequalities for $\|[T^*, T]\|$ to include the connectivity of the domain.*

This last problem is virtually unexplored territory. In his thesis in the 70s ([**17**]), S. Jacobs refined Carleman's celebrated inequality ([**5**]) bounding the A^2 norm of g in terms of the E^1 norm of g for multiply connected domains. In [**22**], there is a result connecting geometric characteristics of the domain G (area, perimeter, connectivity, and analytic content) with the mapping properties of φ, the best approximation of $\bar{z}$, and the mapping properties of the extremal function in the dual problem.

References

[1] A. Abanov, C. Bénéteau, D. Khavinson, and R. Teodorescu, Extremal Domains for Analytic Content, in preparation.

[2] H. Alexander, *Projections of polynomial hulls*, J. Functional Analysis **13** (1973), 13–19. MR0338444 (49 #3209)

[3] S. Axler and J. H. Shapiro, *Putnam's theorem, Alexander's spectral area estimate, and VMO*, Math. Ann. **271** (1985), no. 2, 161–183, DOI 10.1007/BF01455985. MR783550 (87b:30053)

[4] S. Bell, T. Ferguson, E. Lundberg, Self-commutators of Toeplitz operators and isoperimetric inequalities, Proc. Royal Irish Acad., to appear.

[5] T. Carleman, *Zur Theorie der Minimalflächen* (German), Math. Z. **9** (1921), no. 1-2, 154–160, DOI 10.1007/BF01378342. MR1544458

[6] L. De Castro and D. Khavinson, *Analytic functions in Smirnov classes E^p with real boundary values*, Complex Anal. Oper. Theory **7** (2013), no. 1, 101–106, DOI 10.1007/s11785-011-0174-x. MR3010790

[7] L. De Castro and D. Khavinson, *Analytic functions in Smirnov classes E^p with real boundary values II*, Anal. Math. Phys. **3** (2013), no. 1, 21–35, DOI 10.1007/s13324-012-0036-3. MR3015628

[8] P. L. Duren, *Theory of H^p spaces*, Pure and Applied Mathematics, Vol. 38, Academic Press, New York-London, 1970. MR0268655 (42 #3552)

[9] P. Duren and A. Schuster, *Bergman spaces*, Mathematical Surveys and Monographs, vol. 100, American Mathematical Society, Providence, RI, 2004. MR2033762 (2005c:30053)

[10] P. Ebenfelt, D. Khavinson, and H. S. Shapiro, *A free boundary problem related to single-layer potentials*, Ann. Acad. Sci. Fenn. Math. **27** (2002), no. 1, 21–46. MR1884347 (2002m:35227)

[11] S. D. Fisher, *Function theory on planar domains*, Pure and Applied Mathematics (New York), John Wiley & Sons, Inc., New York, 1983. A second course in complex analysis; A Wiley-Interscience Publication. MR694693 (85d:30001)

[12] M. Fleeman and D. Khavinson, Extremal domains for self-commutators in the Bergman space, preprint, arXiv:1311.1842, 2013.

[13] T. W. Gamelin and D. Khavinson, *The isoperimetric inequality and rational approximation*, Amer. Math. Monthly **96** (1989), no. 1, 18–30, DOI 10.2307/2323251. MR979592 (90d:30108)

[14] G. M. Goluzin, *Geometric theory of functions of a complex variable*, Translations of Mathematical Monographs, Vol. 26, American Mathematical Society, Providence, R.I., 1969. MR0247039 (40 #308)

[15] Z. Guadarrama and D. Khavinson, *Approximating $\bar{z}$ in Hardy and Bergman norms*, Banach spaces of analytic functions, Contemp. Math., vol. 454, Amer. Math. Soc., Providence, RI, 2008, pp. 43–61, DOI 10.1090/conm/454/08826. MR2408234 (2009e:30085)

[16] H. Hedenmalm, B. Korenblum, and K. Zhu, *Theory of Bergman spaces*, Graduate Texts in Mathematics, vol. 199, Springer-Verlag, New York, 2000. MR1758653 (2001c:46043)

[17] S. Jacobs, An isoperimetric inequality for functions analytic in multiply connected domains, Institut Mittag-Leffler, Report, 1972.

[18] D. Khavinson, *Remarks concerning boundary properties of analytic functions of E_p-classes*, Indiana Univ. Math. J. **31** (1982), no. 6, 779–787, DOI 10.1512/iumj.1982.31.31054. MR674867 (84b:30037)

[19] D. Khavinson, *Factorization theorems for different classes of analytic functions in multiply connected domains*, Pacific J. Math. **108** (1983), no. 2, 295–318. MR713739 (85j:30009)

[20] D. Khavinson, *A note on Toeplitz operators*, Banach spaces (Columbia, Mo., 1984), Lecture Notes in Math., vol. 1166, Springer, Berlin, 1985, pp. 89–94, DOI 10.1007/BFb0074697. MR827763

[21] D. Khavinson, *Symmetry and uniform approximation by analytic functions*, Proc. Amer. Math. Soc. **101** (1987), no. 3, 475–483, DOI 10.2307/2046392. MR908652 (88j:30083)

[22] D. Khavinson and D. Luecking, *On an extremal problem in the theory of rational approximation*, J. Approx. Theory **50** (1987), no. 2, 127–132, DOI 10.1016/0021-9045(87)90003-7. MR888294 (88g:41013)

[23] S. Ya. Khavinson, *Foundations of the Theory of Extremal Problems for Bounded Analytic Functions and various generalizations of them*, Amer. Math. Soc. Transl. (2), 129 (1986), 1–56.

[24] J. Neuwirth and D. J. Newman, *Positive $H^{1/2}$ functions are constants*, Proc. Amer. Math. Soc. **18** (1967), 958. MR0213576 (35 #4436)

[25] J-F. Olsen and M. Reguera, On a sharp estimate for Hankel operators and Putnam's inequality, arXiv:1305.5193v1, 2013.

[26] C. R. Putnam, *An inequality for the area of hyponormal spectra*, Math. Z. **116** (1970), 323–330. MR0270193 (42 #5085)

[27] H. S. Shapiro, *The Schwarz function and its generalization to higher dimensions*, University of Arkansas Lecture Notes in the Mathematical Sciences, 9, John Wiley & Sons, Inc., New York, 1992. A Wiley-Interscience Publication. MR1160990 (93g:30059)

Department of Mathematics and Statistics, University of South Florida, 4202 E. Fowler Avenue, CMC342, Tampa Florida 33620
E-mail address: cbenetea@usf.edu

Department of Mathematics and Statistics, University of South Florida, 4202 E. Fowler Avenue, CMC342, Tampa Florida 33620
E-mail address: dkhavins@usf.edu

Contemporary Mathematics
Volume **638**, 2015
http://dx.doi.org/10.1090/conm/638/12814

The linear bound for Haar multiplier paraproducts

Kelly Bickel, Eric T. Sawyer, and Brett D. Wick

ABSTRACT. We study the natural resolution of the conjugated Haar multiplier T_σ:

$$M_{w^{\frac{1}{2}}} T_\sigma M_{w^{-\frac{1}{2}}} = \left(\mathsf{P}^{(0,1)}_{\widehat{w^{\frac{1}{2}}}} + \mathsf{P}^{(1,0)}_{\widehat{w^{\frac{1}{2}}}} + \mathsf{P}^{(0,0)}_{\langle w^{\frac{1}{2}}\rangle}\right) T_\sigma \left(\mathsf{P}^{(0,1)}_{\widehat{w^{-\frac{1}{2}}}} + \mathsf{P}^{(1,0)}_{\widehat{w^{-\frac{1}{2}}}} + \mathsf{P}^{(0,0)}_{\langle w^{-\frac{1}{2}}\rangle}\right),$$

where each $M_{w^{\pm\frac{1}{2}}}$ is decomposed into its canonical paraproduct decomposition. We prove that each constituent operator obtained from this resolution has a linear bound on $L^2(\mathbb{R}^d; w)$ in terms of the A_2 characteristic of w. The main tools used are a "product formula" for Haar coefficients, the Carleson Embedding Theorem, the linear bound for the square function, and the well-known linear bound of T_σ on $L^2(w)$.

Contents

1. Introduction and Statement of Main Results

Let $L^2 \equiv L^2(\mathbb{R}^d)$ denote the space of square integrable functions over $\mathbb{R}^d$. For a weight w, i.e., a positive locally integrable function on $\mathbb{R}^d$, we set $L^2(w) \equiv L^2(\mathbb{R}^d; w)$. In particular, we will be interested in A_2 weights, which are defined by:

$$[w]_{A_2} \equiv \sup_I \langle w \rangle_I \langle w^{-1} \rangle_I,$$

where $\langle w \rangle_I$ denotes the average of w over a cube I.

An operator T is bounded on $L^2(w)$ if and only if $M_{w^{\frac{1}{2}}} T M_{w^{-\frac{1}{2}}}$ - the conjugation of T by the multiplication operators $M_{w^{\pm\frac{1}{2}}}$ - is bounded on L^2. Moreover, the operator norms are equal:

$$\|T\|_{L^2(w)\to L^2(w)} = \left\| M_{w^{\frac{1}{2}}} T M_{w^{-\frac{1}{2}}} \right\|_{L^2\to L^2}.$$

2010 *Mathematics Subject Classification.* Primary 42-XX, 42B99.

Research supported in part by a NSERC Grant.
The first and third authors' research was supported in part by National Science Foundation DMS grant # 0955432.

In the case that T is a dyadic operator adapted to a dyadic grid $\mathcal{D}$, it is natural to study weighted norm properties of T by decomposing the multiplication operators $M_{w^{\pm\frac{1}{2}}}$ into their canonical paraproduct decompositions relative to the grid $\mathcal{D}$, i.e.

$$M_{w^{\pm\frac{1}{2}}} f = \mathsf{P}^{(0,1)}_{\widehat{w^{\pm\frac{1}{2}}}} f + \mathsf{P}^{(1,0)}_{\widehat{w^{\pm\frac{1}{2}}}} f + \mathsf{P}^{(0,0)}_{\langle w^{\pm\frac{1}{2}}\rangle} f$$

(the paraproduct operators are defined in the next section) and then decomposing $M_{w^{\frac{1}{2}}} T M_{w^{-\frac{1}{2}}}$ into the nine canonical individual paraproduct composition operators:

$$\begin{aligned} (1.1)\ M_{w^{\frac{1}{2}}} T M_{w^{-\frac{1}{2}}} &= \left(\mathsf{P}^{(0,1)}_{\widehat{w^{\frac{1}{2}}}} + \mathsf{P}^{(1,0)}_{\widehat{w^{\frac{1}{2}}}} + \mathsf{P}^{(0,0)}_{\langle w^{\frac{1}{2}}\rangle}\right) T \left(\mathsf{P}^{(0,1)}_{\widehat{w^{-\frac{1}{2}}}} + \mathsf{P}^{(1,0)}_{\widehat{w^{-\frac{1}{2}}}} + \mathsf{P}^{(0,0)}_{\langle w^{-\frac{1}{2}}\rangle}\right) \\ &\equiv Q^{(0,1),(0,1)}_{T,w} + Q^{(0,1),(1,0)}_{T,w} + Q^{(0,1),(0,0)}_{T,w} \\ &\quad + Q^{(1,0),(0,1)}_{T,w} + Q^{(1,0),(1,0)}_{T,w} + Q^{(1,0),(0,0)}_{T,w} \\ &\quad + Q^{(0,0),(0,1)}_{T,w} + Q^{(0,0),(1,0)}_{T,w} + Q^{(0,0),(0,0)}_{T,w}, \end{aligned}$$

where $Q^{(\varepsilon_1,\varepsilon_2),(\varepsilon_3,\varepsilon_4)}_{T,w}$ is defined in the obvious way from the expression above. If one could show that the operator norms of the $Q^{(\varepsilon_1,\varepsilon_2),(\varepsilon_3,\varepsilon_4)}_{T,w}$ are linear in the A_2 characteristic,

$$\left\| Q^{(\varepsilon_1,\varepsilon_2),(\varepsilon_3,\varepsilon_4)}_{T,w} \right\|_{L^2 \to L^2} \lesssim [w]_{A_2}$$

it then becomes reasonable to expect that the canonical decomposition of a dyadic operator T into its paraproduct compositions $Q^{(\varepsilon_1,\varepsilon_2),(\varepsilon_3,\varepsilon_4)}_{T,w}$ will inherit the salient properties of T without losing anything of importance. Of course, these dyadic paraproduct compositions can be expected to yield to structured dyadic proof strategies.

This idea has been successfully used to study decompositions of the Hilbert transform in [**6**]. We now extend this idea to the martingale transforms. Specifically, let $\{\sigma_I\}_{I\in\mathcal{D}}$ denote a sequence of $2^d - 1 \times 2^d - 1$ diagonal matrices indexed by the dyadic cubes with diagonal entries denoted by $(\sigma_I)_{\alpha\alpha} \equiv \sigma_{I,\alpha}$ for $\alpha = 1, \ldots, 2^d - 1$. Define

$$T_\sigma f \equiv \sum_{I\in\mathcal{D}} \left(\sigma_I \widehat{f}(I)\right) \cdot h_I \qquad \forall f \in L^2,$$

where h_I is the vector of Haar functions adapted to the cube I and $\widehat{f}(I)$ is the vector of Haar coefficients associated to the function f. For precise definitions of these Haar objects, see Section 2 and for a precise definition of T_σ, see the beginning of Section 3. It is well known and simple to see that

$$\|T_\sigma\|_{L^2 \to L^2} \le \|\sigma\|_\infty,$$

where

$$\|\sigma\|_\infty \equiv \sup_{I\in\mathcal{D}} \sup_{1\le\alpha\le 2^d-1} |\sigma_{I,\alpha}|.$$

A similar norm bound holds for $L^2(w)$. Specifically, in [**10**], J. Wittwer established the following result in one-dimension and using related arguments, D. Chung obtained the d-dimensional analogue in [**1**].

THEOREM 1.1 (Linear Bound for Martingale Transforms).

$$\|T_\sigma\|_{L^2(w)\to L^2(w)} \lesssim [w]_{A_2} \|\sigma\|_\infty.$$

Wittwer only established the result for the case where each $\sigma_I \in \{\pm 1\}$, but the general case follows using the same arguments. F. Nazarov, S. Treil, and A. Volberg obtained more general results in [**3**, **4**], where they showed that certain testing conditions are sufficient to determine when Haar multipliers and related operators are bounded from $L^2(w)$ to $L^2(v)$.

Here, we study the paraproduct decomposition of T_σ and establish the following result:

THEOREM 1.2. *let $\{\sigma_I\}_{I\in\mathcal{D}}$ denote a sequence of $2^d-1\times 2^d-1$ diagonal matrices indexed by the dyadic cubes and let w be an A_2 weight. Then, each paraproduct composition in the canonical resolution of T_σ can be controlled by a linear power of $[w]_{A_2}$, i.e.*

$$\left\| Q_{T_\sigma,w}^{(\varepsilon_1,\varepsilon_2),(\varepsilon_3,\varepsilon_4)} \right\|_{L^2\to L^2} \lesssim [w]_{A_2} \|\sigma\|_\infty ,$$

for each $Q_{T_\sigma,w}^{(\varepsilon_1,\varepsilon_2),(\varepsilon_3,\varepsilon_4)}$ in (1.1).

The proof relies heavily on arguments appearing in [**6**], especially the use of a "product formula" for Haar coefficients, the Carleson Embedding Theorem, and the linear bound for the square function. Crucial to the approach taken in this paper is the introduction of Wilson's Haar basis in $\mathbb{R}^d$, [**9**], and certain modifications of Chung obtained in [**1**]. The approach carried out in [**6**] is dependent upon the Haar system in $\mathbb{R}$ and the modifications in [**1**, **9**] are crucial for our analysis in this note. To handle the final resolvent paraproduct, we must rely on Theorem 1.1. Obtaining the bound independent of Theorem 1.1 is currently an open question.

2. Notation and Useful Facts

Before proving our main result, we collect necessary notation and estimates. Throughout this paper $A \equiv B$ means that the expressions are equal by definition, and $A \lesssim B$ means that there exists a constant c_d, which may depend on the dimension d, such that $A \le c_d B$.

Let $\mathcal{D}$ denote the usual dyadic grid of cubes in $\mathbb{R}^d$. For $I \in \mathcal{D}$, let $\mathcal{C}_1(I)$ denote the 2^d children of I. Note that each child $J \in \mathcal{C}_1(I)$ satisfies $|J| = 2^{-d}|I|$. Further, for $d \in \mathbb{N}$, set

$$\Gamma_d \equiv \{0,1\}^d \setminus \{(1,\ldots,1)\},$$

and fix an enumeration of this set for the rest of the paper. Elements of this set will be denoted by lowercase greek letters.

2.1. Wilson's Haar System.

While we would like to use the standard Haar system in the analysis below, it is more convenient to use an orthonormal system developed by M. Wilson in [**9**]. To construct it, we need the following lemma. It is worth mentioning that property (i) did not appear in Wilson's original lemma but was added by D. Chung in [**1**].

LEMMA 2.1 (Wilson, [**9**, Lemma 2]). *Let $I \in \mathcal{D}$. Then there are $2^d - 1$ pairs of sets $\left\{(E^1_{\alpha,I}, E^2_{\alpha,I})\right\}_{\alpha\in\Gamma_d}$ such that*

(i) *For each $\alpha \in \Gamma_d$, $\left|E^1_{\alpha,I}\right| = \left|E^2_{\alpha,I}\right|$;*
(ii) *For each α and $s = 1, 2$, $E^s_{\alpha,I}$ is a non-empty union of cubes from $\mathcal{C}_1(I)$;*
(iii) *For each α, $E^1_{\alpha,I} \cap E^2_{\alpha,I} = \emptyset$;*
(iv) *For every $\alpha \neq \beta$ one of the following must hold:*

(a) $E^1_{\alpha,I} \cup E^2_{\alpha,I}$ *is entirely contained in either* $E^1_{\beta,I}$ *or* $E^2_{\beta,I}$;
(b) $E^1_{\beta,I} \cup E^2_{\beta,I}$ *is entirely contained in either* $E^1_{\alpha,I}$ *or* $E^2_{\alpha,I}$;
(c) $\left(E^1_{\beta,I} \cup E^2_{\beta,I}\right) \cap \left(E^1_{\alpha,I} \cup E^2_{\alpha,I}\right) = \emptyset$.

Set $E_{\alpha,I} \equiv E^1_{\alpha,I} \cup E^2_{\alpha,I}$. It is important to observe that $|E_{\alpha,I}| \approx |I|$ for some dimensional constants. Further, given any $E_{\alpha,I}$ and $E_{\beta,J}$, it follows from the properties of $\mathcal{D}$ and (iv) that one of the following must hold: $E_{\alpha,I} \subsetneq E_{\beta,J}$, $E_{\beta,J} \subsetneq E_{\alpha,I}$, or $E_{\alpha,I} = E_{\beta,J}$. Given this collection of sets, we can now introduce Wilson's Haar system for $L^2(w)$. Fix $\alpha \in \Gamma_d$, $I \in \mathcal{D}$ and define:

$$h_I^{w,\alpha} \equiv \frac{1}{\sqrt{w(E_{\alpha,I})}} \left[\frac{\sqrt{w(E^1_{\alpha,I})}}{\sqrt{w(E^2_{\alpha,I})}} 1_{E^2_{\alpha,I}} - \frac{\sqrt{w(E^2_{\alpha,I})}}{\sqrt{w(E^1_{\alpha,I})}} 1_{E^1_{\alpha,I}} \right]. \tag{2.1}$$

It is easy to show that $\{h_I^{w,\alpha}\}_{\alpha\in\Gamma_d, I\in\mathcal{D}}$ is an orthonormal system in $L^2(w)$. When the weight $w \equiv 1$, we denote this collection of functions by $\{h_I^\alpha\}_{\alpha\in\Gamma_d, I\in\mathcal{D}}$. Note that each function h_I^α has a fixed sign on each child cube $I' \in \mathcal{C}_1(I)$. Now, for a fixed dyadic cube J, set

$$\begin{aligned} \widehat{f}(J,\alpha) &\equiv \langle f, h_J^\alpha\rangle_{L^2} \quad \forall \alpha \in \Gamma_d, \\ \widehat{f}(J) &\equiv \left(\widehat{f}(J,\alpha)\right)_{\alpha\in\Gamma_d}, \\ h_J &\equiv (h_J^\alpha)_{\alpha\in\Gamma_d}, \\ \triangle_J f &\equiv \widehat{f}(J)\cdot h_J = \sum_{\alpha\in\Gamma_d} \widehat{f}(J,\alpha)\, h_J^\alpha . \end{aligned}$$

This means that $\widehat{f}(J)$ is the vector of Haar coefficients of the function f, and h_J is the vector of Haar functions. It is easy to see that the set $\{h_I^\alpha\}_{I\in\mathcal{D},\alpha\in\Gamma_d}$ is an orthonormal basis for L^2 and so, $f = \sum_{I\in\mathcal{D}} \triangle_I f = \sum_{I\in\mathcal{D}} \widehat{f}(I)\cdot h_I$. This implies

$$\|f\|^2_{L^2} = \sum_{I\in\mathcal{D}} \left|\widehat{f}(I)\right|^2 = \sum_{I\in\mathcal{D}} \sum_{\alpha\in\Gamma_d} \left|\widehat{f}(I,\alpha)\right|^2 = \sum_{\alpha\in\Gamma_d} \sum_{I\in\mathcal{D}} \left|\widehat{f}(I,\alpha)\right|^2 .$$

Now given a set E, define $h_E^1 \equiv \frac{1_E}{|E|}$, so that the function is L^1 normalized. We also set $\langle f\rangle_E \equiv \langle f, h_E^1\rangle_{L^2}$. The Wilson Haar system has the standard martingale property that the average of f over each $E_{\alpha,I}$ satisfies

$$\langle f\rangle_{E_{\alpha,I}} = \sum_{J\in\mathcal{D}} \sum_{\beta\in\Gamma_d} \widehat{f}(J,\beta) \left\langle h_J^\beta, h^1_{E_{\alpha,I}} \right\rangle_{L^2} = \sum_{J:J\supseteq I} \sum_{\beta: E_{J,\beta}\supsetneq E_{\alpha,I}} \widehat{f}(J,\beta) \left\langle h_J^\beta, h^1_{E_{\alpha,I}} \right\rangle_{L^2} .$$

A fundamental tool in our study will be the product formula for Haar expansions in L^2. A version of this previously appeared in [**7**]. Specifically, given two functions f and g in L^2, we can expand them with respect to this Haar basis and formally obtain

$$\begin{aligned} fg &= \left(\sum_{I\in\mathcal{D}}\sum_{\alpha\in\Gamma_d}\widehat{f}(I,\alpha)h_I^\alpha\right)\times\left(\sum_{J\in\mathcal{D}}\sum_{\beta\in\Gamma_d}\widehat{g}(J,\beta)h_J^\beta\right) \\ &= \sum_{\alpha\in\Gamma_d}\sum_{I\in\mathcal{D}}\langle g\rangle_{E_{\alpha,I}}\widehat{f}(I,\alpha)h_I^\alpha+\sum_{\alpha\in\Gamma_d}\sum_{I\in\mathcal{D}}\langle f\rangle_{E_{\alpha,I}}\widehat{g}(I,\alpha)h_I^\alpha \\ &\qquad +\sum_{\alpha\in\Gamma_d}\sum_{I\in\mathcal{D}}\widehat{g}(I,\alpha)\widehat{f}(I,\alpha)h^1_{E_{\alpha,I}}, \end{aligned}$$

where we use the fact that

$$h_I^\alpha h_J^\beta=\left\langle h_J^\beta,h^1_{E_{\alpha,I}}\right\rangle_{L^2}h_I^\alpha \text{ whenever } E_{\alpha,I}\subsetneq E_{\beta,J}.$$

Although the product formula above does not necessarily make sense for arbitrary $f,g\in L^2$, it is well-defined if f,g are finite linear combinations of Haar functions. Moreover, for $J\in\mathcal{D}$ and $\beta\in\Gamma_d$, we have

(2.2)
$$\widehat{fg}\,(J,\beta)=\sum_{\alpha\in\Gamma_d}\sum_{I\in\mathcal{D}}\widehat{f}(I,\alpha)\widehat{g}(I,\alpha)\left\langle h^1_{E_{\alpha,I}},h_J^\beta\right\rangle_{L^2}+\widehat{f}\,(J,\beta)\,\langle g\rangle_{E_{\beta,J}}+\widehat{g}\,(J,\beta)\,\langle f\rangle_{E_{\beta,J}}.$$

For finite linear combinations of Haar functions, this is obtained by simply calculating the Haar coefficient corresponding to $\beta\in\Gamma_d$ and $J\in\mathcal{D}$ in formula for the product fg. If f,g are locally in L^2, we can approximate them on $E_{\beta,J}$ by finite linear combinations of Haar functions and still deduce (2.2). Primarily, we will use version (2.2) of the product formula. However, it should be noted that support conditions in the first term actually imply that

$$\begin{aligned} \widehat{fg}\,(J,\beta) = \sum_{I:I\subseteq J}\ \sum_{\alpha:E_{\alpha,I}\subsetneq E_{\beta,J}}\widehat{f}(I,\alpha)\widehat{g}(I,\alpha)\left\langle h^1_{E_{\alpha,I}},h_J^\beta\right\rangle_{L^2}+\widehat{f}\,(J,\beta)\,\langle g\rangle_{E_{\beta,J}} \\ +\widehat{g}\,(J,\beta)\,\langle f\rangle_{E_{\beta,J}}. \end{aligned}$$

Motivated by these product decompositions, we consider the following dyadic operators. They will be of fundamental importance in this paper. Give a sequence of numbers $a=\{a_{I,\alpha}\}_{I\in\mathcal{D},\alpha\in\Gamma_d}$ indexed by $I\in\mathcal{D}$ and $\alpha\in\Gamma_d$, we define the following paraproduct type operators:

$$\begin{aligned} \mathsf{P}_a^{(0,0)}f &\equiv \sum_{\alpha\in\Gamma_d}\sum_{I\in\mathcal{D}}a_{I,\alpha}\,\widehat{f}(I,\alpha)h_I^\alpha \\ \mathsf{P}_a^{(0,1)}f &\equiv \sum_{\alpha\in\Gamma_d}\sum_{I\in\mathcal{D}}a_{I,\alpha}\,\langle f\rangle_{E_{\alpha,I}}\,h_I^\alpha \\ \mathsf{P}_a^{(1,0)}f &\equiv \sum_{\alpha\in\Gamma_d}\sum_{I\in\mathcal{D}}a_{I,\alpha}\widehat{f}(I,\alpha)h^1_{E_{\alpha,I}}. \end{aligned}$$

It is easy to see that the operator M_g of multiplication by g can formally be written as

$$M_gf=\mathsf{P}_{\langle g\rangle}^{(0,0)}f+\mathsf{P}_{\widehat{g}}^{(0,1)}f+\mathsf{P}_{\widehat{g}}^{(1,0)}f \tag{2.3}$$

where $\langle g\rangle\equiv\{\langle g\rangle_{E_{\alpha,I}}\}_{I\in\mathcal{D},\alpha\in\Gamma_d}$ and $\widehat{g}\equiv\{\widehat{g}(I,\alpha)\}_{I\in\mathcal{D},\alpha\in\Gamma_d}$. We will use (2.3) to decompose the operators $M_{w^{\pm\frac{1}{2}}}$.

2.1.1. *Disbalanced Haars.* At points in our later arguments, we will use disbalanced Haar functions. To do so, we require some additional notation. Fixing a dyadic cube J, a weight w on $\mathbb{R}^d$, and $\beta \in \Gamma_d$, we set

$$C_J(w,\beta) \equiv \sqrt{\frac{\langle w\rangle_{E^1_{\beta,J}} \langle w\rangle_{E^2_{\beta,J}}}{\langle w\rangle_{E_{\beta,J}}}} \quad \text{and} \quad D_J(w,\beta) \equiv \frac{\widehat{w}(J,\beta)}{\langle w\rangle_{E_{\beta,J}}}. \tag{2.4}$$

Then we have

$$h^\beta_J = C_J(w,\beta) h^{w,\beta}_J + D_J(w,\beta) h^1_{E_{\beta,J}}$$

where $\{h^{w,\alpha}_I\}_{I\in\mathcal{D},\alpha\in\Gamma_d}$ is the $L^2(w)$ orthonormal system defined in (2.1). To see this, we use the two equations

$$\int_J h^{w,\beta}_J w = 0 \quad \text{and} \quad \int_J \left(h^{w,\beta}_J\right)^2 w = 1$$

to solve for $C_J(w,\beta)$ and $D_J(w,\beta)$. The claimed formula for $D_J(w,\beta)$ follows immediately from the condition that $h^{w,\beta}_J$ have integral zero. Using the second condition and the formula for $D_J(w,\beta)$, one can easily prove that

$$C_J(w,\beta) = \sqrt{\frac{\langle w\rangle^2_{E_{\beta,J}} - |E_{\beta,J}|^{-1}\,\widehat{w}(J,\beta)^2}{\langle w\rangle_{E_{\beta,J}}}}.$$

Using basic manipulations, and the fact that $|E_{\beta,J}| = 2|E^j_{\beta,J}|$ for $j = 1,2$, we have

$$\begin{aligned}
&\langle w\rangle^2_{E_{\beta,J}} - |E_{\beta,J}|^{-1}\,\widehat{w}(J,\beta)^2 \\
&\quad = \frac{1}{|E_{\beta,J}|^2}\left(\left(w(E^1_{\beta,J}) + w(E^2_{\beta,J})\right)^2 - \left(w(E^2_{\beta,J}) - w(E^1_{\beta,J})\right)^2\right) \\
&\quad = \frac{4}{|E_{\beta,J}|^2} w(E^1_{\beta,J}) w(E^2_{\beta,J}) = \langle w\rangle_{E^1_{\beta,J}} \langle w\rangle_{E^2_{\beta,J}}.
\end{aligned}$$

This gives the desired formula for $C_J(w,\beta)$. A useful observation is

$$C_J(w,\beta)^2 \le 4\langle w\rangle_{E_{\beta,J}} \le 2^{d+1}\langle w\rangle_J, \tag{2.5}$$

which follows since each $\langle w\rangle_{E^j_{\beta,J}} \le 2\langle w\rangle_{E_{\beta,J}}$ and as $E_{\beta,J}$ contains at least two children of J, $\langle w\rangle_{E_{\beta,J}} \le 2^{d-1}\langle w\rangle_J$.

2.1.2. *Carleson Embedding Theorem.* A major tool in this paper is the following modification of the standard Carleson Embedding Theorem to the sets $\{E_{\alpha,I}\}_{I\in\mathcal{D},\alpha\in\Gamma_d}$. It appears in [**1**, Theorem 4.3]:

THEOREM 2.2 (Modified Carleson Embedding Theorem). *Let w be a weight on $\mathbb{R}^d$ and let $\{a_{\alpha,I}\}_{I\in\mathcal{D},\alpha\in\Gamma_d}$ be a sequence of nonnegative numbers. Then, there is a constant $A > 0$ such that*

$$\frac{1}{|E_{\alpha,I}|}\sum_{J\subset I}\sum_{\beta: E_{\beta,J}\subset E_{\alpha,I}} a_{\beta,J}\langle w\rangle^2_{E_{\beta,J}} \le A\langle w\rangle_{E_{\alpha,I}} \qquad \forall I\in\mathcal{D},\ \alpha\in\Gamma_d,$$

if and only if

$$\sum_{I\in\mathcal{D}}\sum_{\alpha\in\Gamma_d} a_{\alpha,I}\left\langle w^{\frac{1}{2}} f\right\rangle^2_{E_{\alpha,I}} \lesssim A\|f\|^2_{L^2} \quad \forall f\in L^2.$$

The proof strategy employed in [**1**] to deduce this is based on the Bellman technique; however there are other methods by which the interested reader can arrive at the Theorem above.

2.2. Square Function Estimates.

2.2.1. *Square Function Bound.* Define the dyadic square function S on L^2 by

$$Sf(x)^2 = \sum_{I \in \mathcal{D}} |\widehat{f}(I)|^2 h_I^1(x) = \sum_{I \in \mathcal{D}} \sum_{\alpha \in \Gamma_d} |\widehat{f}(I,\alpha)|^2 h_I^1(x).$$

It is clear from the definition that $\|Sf\|_{L^2} = \|f\|_{L^2}$. Versions of the square function have been studied in the weighted setting $L^2(w)$ and it has been shown that a linear bound in terms of the A_2 characteristic holds. We point the interested readers to [**2**]. In the one dimensional case we refer the readers to [**8, 10**].

For our needs, we require a slightly different formulation and so provide an alternate proof of this fact. Using the arguments from Petermichl and Pott [**5**], we prove

THEOREM 2.3. *Let w be an A_2 weight in $\mathbb{R}^d$. Then*

$$\|Sf\|_{L^2(w)} \lesssim [w]_{A_2} \|f\|_{L^2(w)} \quad \forall f \in L^2(w).$$

PROOF. As in [**5**], without loss of generality, we can assume w and w^{-1} are bounded so long as the bounds do not appear in our final estimates. We first prove the lower bound:

$$\|f\|^2_{L^2(w)} \lesssim [w]_{A_2} \|Sf\|^2_{L^2(w)} \qquad \forall f \in L^2(w). \tag{2.6}$$

To this end, define the discrete multiplication operator $D_w : L^2 \to L^2$ by

$$D_w : h_I^\alpha \mapsto \langle w \rangle_I h_I^\alpha \quad \forall I \in \mathcal{D}, \alpha \in \Gamma_d$$

and let M_w denote multiplication by w. Then, we can rewrite (2.6) as:

$$\langle M_w f, f \rangle_{L^2} \lesssim [w]_{A_2} \langle D_w f, f \rangle_{L^2}, \qquad \forall f \in L^2. \tag{2.7}$$

First, since w and w^{-1} are bounded, D_w and M_w are bounded and invertible with $M_w^{-1} = M_{w^{-1}}$ and D_w^{-1} defined by

$$D_w^{-1} : h_I^\alpha \mapsto \langle w \rangle_I^{-1} h_I^\alpha \quad \forall I \in \mathcal{D}, \alpha \in \Gamma_d.$$

As in [**5**], one can convert (2.7) to the equivalent inverse inequality:

$$\langle D_w^{-1} f, f \rangle_{L^2} \lesssim [w]_{A_2} \langle M_w^{-1} f, f \rangle_{L^2}, \qquad \forall f \in L^2. \tag{2.8}$$

So, we need to establish:

$$\sum_{I \in \mathcal{D}} \langle w \rangle_I^{-1} \left| \widehat{f}(I) \right|^2 \lesssim [w]_{A_2} \|f\|^2_{L^2(w^{-1})} \quad \forall f \in L^2.$$

As in [**5**], our first step is to rewrite the sums using disbalanced Haar functions adapted to w using (2.4). To do so, fix a cube J and $\alpha \in \Gamma_d$ and recall that

$$C_J(w,\alpha) \equiv \sqrt{\frac{\langle w \rangle_{E^1_{\alpha,J}} \langle w \rangle_{E^2_{\alpha,J}}}{\langle w \rangle_{E_{\alpha,J}}}} \quad \text{and} \quad D_J(w,\alpha) \equiv \frac{\widehat{w}(J,\alpha)}{\langle w \rangle_{E_{\alpha,J}}}.$$

Then we have

$$h_J^\alpha = C_J(w,\alpha) h_J^{w,\alpha} + D_J(w,\alpha) h^1_{E_{\alpha,J}},$$

where $\{h_J^{w,\alpha}\}_{J\in\mathcal{D},\alpha\in\Gamma_d}$ is the previously-defined $L^2(w)$ orthonormal system. Returning to the sum in question, we use the disbalanced Haar functions to write:

$$\begin{aligned}\sum_{I\in\mathcal{D}}\langle w\rangle_I^{-1}\left|\widehat{f}(I)\right|^2 &= \sum_{I\in\mathcal{D}}\sum_{\alpha\in\Gamma_d}\langle w\rangle_I^{-1}\left\langle f,h_I^{\alpha}\right\rangle_{L^2}^2\\ &= \sum_{I\in\mathcal{D}}\sum_{\alpha\in\Gamma_d} C_I(w,\alpha)^2\langle w\rangle_I^{-1}\left\langle f,h_I^{w,\alpha}\right\rangle_{L^2}^2\\ &\quad+2\sum_{I\in\mathcal{D}}\sum_{\alpha\in\Gamma_d} C_I(w,\alpha)D_I(w,\alpha)\langle w\rangle_I^{-1}\left\langle f,h_I^{w,\alpha}\right\rangle_{L^2}\langle f\rangle_{E_{\alpha,I}}\\ &\quad+\sum_{I\in\mathcal{D}}\sum_{\alpha\in\Gamma_d} D_I(w,\alpha)^2\langle w\rangle_I^{-1}\langle f\rangle_{E_{\alpha,I}}^2\\ &= S_1+S_2+S_3.\end{aligned}$$

Since by (2.5), each $C_I(w,\alpha)^2\lesssim\langle w\rangle_I$, we can conclude that each $C_I(w,\alpha)^2\langle w\rangle_I^{-1}\lesssim 1$. This means

$$S_1\lesssim\sum_{I\in\mathcal{D}}\sum_{\alpha\in\Gamma_d}|\langle f,h_I^{w,\alpha}\rangle_{L^2}|^2=\sum_{I\in\mathcal{D}}\sum_{\alpha\in\Gamma_d}\left|\langle w^{-1}f,h_I^{w,\alpha}\rangle_{L^2(w)}\right|^2\le\|f\|_{L^2(w^{-1})}^2.$$

Observe that

$$\begin{aligned}&S_2\\ &\lesssim\left(\sum_{I\in\mathcal{D}}\sum_{\alpha\in\Gamma_d} C_I(w,\alpha)^2\langle w\rangle_I^{-1}\left|\langle f,h_I^{w,\alpha}\rangle_{L^2}\right|^2\right)^{\frac12}\left(\sum_{I\in\mathcal{D}}\sum_{\alpha\in\Gamma_d} D_I(w,\alpha)^2\langle w\rangle_I^{-1}\langle f\rangle_{E_{\alpha,I}}^2\right)^{\frac12}.\end{aligned}$$

The first part of the product is the square root of S_1 and the second part is the square root of S_3. Thus, the proof is reduced to controlling S_3. We use the modified Carleson Embedding Theorem. To apply it, we need

$$\begin{aligned}&\frac{1}{|E_{\alpha,I}|}\sum_{J\subset I}\sum_{\beta:E_{\beta,J}\subset E_{\alpha,I}} D_J(w,\beta)^2\langle w\rangle_J^{-1}\langle w\rangle_{E_{\beta,J}}^2\\ &\quad=\frac{1}{|E_{\alpha,I}|}\sum_{J\subset I}\sum_{\beta:E_{\beta,J}\subset E_{\alpha,I}}\frac{\widehat{w}(J,\beta)^2}{\langle w\rangle_{E_{\beta,J}}^2}\langle w\rangle_J^{-1}\langle w\rangle_{E_{\beta,J}}^2\\ &\quad\lesssim\frac{1}{|E_{\alpha,I}|}\sum_{J\subset I}\sum_{\beta:E_{\beta,J}\subset E_{\alpha,I}}\widehat{w}(J,\beta)^2\langle w\rangle_{E_{\beta,J}}^{-1}\\ &\quad\lesssim[w]_{A_2}\langle w\rangle_{E_{\alpha,I}},\end{aligned}$$

where the last inequality appears in [**1**, Proposition 4.9, (4.17)]. Then, the modified Carleson Embedding Theorem implies

$$S_3=\sum_{I\in\mathcal{D}}\sum_{\alpha\in\Gamma_d} D_I(w,\alpha)^2\langle w\rangle_I^{-1}\left\langle fw^{\frac12}w^{-\frac12}\right\rangle_{E_{\alpha,I}}^2\lesssim[w]_{A_2}\|fw^{-\frac12}\|_{L^2}^2=[w]_{A_2}\|f\|_{L^2(w^{-1})}^2,$$

which proves the lower square function bound. Given (2.6) for every A_2 weight, the upper square function bound follows almost immediately. Now, for w with w,w^{-1} bounded, the desired inequality is equivalent to

$$\langle D_w f,f\rangle_{L^2}\lesssim[w]_{A_2}^2\langle M_w f,f\rangle_{L^2},\quad\forall f\in L^2.$$

To prove that, we require the following operator inequality

$$D_w \le [w]_{A_2} \left(D_{w^{-1}}\right)^{-1}.$$

This is immediate since the trivial inequality $\langle w\rangle_I \le [w]_{A_2} \langle w^{-1}\rangle_I^{-1}$ implies

$$\langle D_w f, f\rangle_{L^2} = \sum_{I\in\mathcal{D}} \langle w\rangle_I \left|\widehat{f}(I)\right|^2 \le [w]_{A_2} \sum_{I\in\mathcal{D}} \langle w^{-1}\rangle_I^{-1} \left|\widehat{f}(I)\right|^2 = [w]_{A_2} \langle (D_{w^{-1}})^{-1} f, f\rangle_{L^2}.$$

Combining that estimate with (2.8) applied to w^{-1} gives:

$$\begin{aligned}\langle D_w f, f\rangle_{L^2} &\le [w]_{A_2} \langle (D_{w^{-1}})^{-1} f, f\rangle_{L^2} \lesssim [w]_{A_2}^2 \langle (M_{w^{-1}})^{-1} f, f\rangle_{L^2} \\ &= [w]_{A_2}^2 \langle M_w f, f\rangle_{L^2}, \qquad \forall f \in L^2,\end{aligned}$$

which completes the proof. □

2.2.2. *Key Estimates Deduced from the Square Function.* As we have shown above, for $w \in A_2$,

$$\|Sf\|_{L^2(w)}^2 \lesssim [w]_{A_2}^2 \|f\|_{L^2(w)}^2 .$$

Applying this inequality to $f = w^{-\frac{1}{2}} \mathbf{1}_{E_{\alpha,I}}$ for $I \in \mathcal{D}, \alpha \in \Gamma_d$ and using some trivial estimates yields the following:

$$\sum_{J\subseteq I} \sum_{\beta: E_{\beta,J}\subseteq E_{\alpha,I}} \left|\widehat{w^{-\frac{1}{2}}}(J,\beta)\right|^2 \langle w\rangle_J \lesssim [w]_{A_2}^2 |E_{\alpha,I}| \quad \forall I \in \mathcal{D}, \alpha \in \Gamma_d. \tag{2.9}$$

A trivial consequence of (2.9) is the following:

$$\sum_{J\subseteq I} \sum_{\beta: E_{\beta,J}\subseteq E_{\alpha,I}} \left|\widehat{w^{-\frac{1}{2}}}(J,\beta)\right|^2 \left\langle w^{\frac{1}{2}}\right\rangle_{E_{\beta,J}}^2 \lesssim [w]_{A_2}^2 |E_{\alpha,I}| \quad \forall I \in \mathcal{D}, \alpha \in \Gamma_d, \tag{2.10}$$

since $\left\langle w^{\frac{1}{2}}\right\rangle_{E_{\beta,J}}^2 \lesssim \left\langle w^{\frac{1}{2}}\right\rangle_J^2 \le \langle w\rangle_J$. Applying the linear bound of the square function to $w^{-1}\mathbf{1}_{E_{\alpha,I}}$, again using trivial estimates, yields

$$\sum_{J\subseteq I} \sum_{\beta: E_{\beta,J}\subseteq E_{\alpha,I}} \left|\widehat{w^{-1}}(J,\beta)\right|^2 \langle w\rangle_{E_{\beta,J}} \lesssim [w]_{A_2}^2 w^{-1}(E_{\alpha,I}) \quad \forall I \in \mathcal{D}, \alpha \in \Gamma_d. \tag{2.11}$$

Because of the symmetry of the A_2 condition, we additionally have these estimates with the roles of w and w^{-1} interchanged. These estimates will all play a fundamental role when applying the modified Carleson Embedding Theorem.

3. Linear Bound for Haar Multipliers

We now turn to proving Thereom 1.2. Given a sequence $\sigma = \{\sigma_{I,\alpha}\}_{I\in\mathcal{D},\alpha\in\Gamma_d}$ we define the Haar multiplier by

$$T_\sigma f \equiv \sum_{\alpha\in\Gamma_d} \sum_{I\in\mathcal{D}} \sigma_{I,\alpha} \widehat{f}(I,\alpha) h_I^\alpha \qquad \forall f \in L^2.$$

We must show that

$$\left\| Q_{T_\sigma,w}^{(\varepsilon_1,\varepsilon_2),(\varepsilon_3,\varepsilon_4)} \right\|_{L^2\to L^2} \lesssim [w]_{A_2} \|\sigma\|_\infty$$

where the operators $Q_{T_\sigma,w}^{(\varepsilon_1,\varepsilon_2),(\varepsilon_3,\varepsilon_4)}$ are defined via (1.1) and the canonical decomposition of $M_{w^{\pm\frac{1}{2}}}$ into paraproducts is given in (2.3).

3.1. Estimating the Easy Terms. There are four easy terms that arise from (1.1). They are easy because the composition of the paraproducts reduce to classical paraproduct type operators. The terms are:

$$\mathsf{P}^{(1,0)}_{\widehat{w^{\frac{1}{2}}}} T_\sigma \mathsf{P}^{(0,1)}_{\widehat{w^{-\frac{1}{2}}}}; \tag{3.1}$$

$$\mathsf{P}^{(1,0)}_{\widehat{w^{\frac{1}{2}}}} T_\sigma \mathsf{P}^{(0,0)}_{\langle w^{-\frac{1}{2}}\rangle}; \tag{3.2}$$

$$\mathsf{P}^{(0,0)}_{\langle w^{\frac{1}{2}}\rangle} T_\sigma \mathsf{P}^{(0,1)}_{\widehat{w^{-\frac{1}{2}}}}; \tag{3.3}$$

$$\mathsf{P}^{(0,0)}_{\langle w^{\frac{1}{2}}\rangle} T_\sigma \mathsf{P}^{(0,0)}_{\langle w^{-\frac{1}{2}}\rangle}. \tag{3.4}$$

For these terms, we proceed by computing the norm of the operators in question by using duality. Key to this will be the application of the modified Carleson Embedding Theorem.

3.1.1. *Estimating* $P^{(1,0)}_{\widehat{w^{\frac{1}{2}}}} T_\sigma P^{(0,1)}_{\widehat{w^{-\frac{1}{2}}}}$. Fix $\phi, \psi \in L^2$ and observe that

$$\begin{aligned} P^{(1,0)}_{\widehat{w^{\frac{1}{2}}}} T_\sigma P^{(0,1)}_{\widehat{w^{-\frac{1}{2}}}} \phi &= P^{(1,0)}_{\widehat{w^{\frac{1}{2}}}} \sum_{\alpha\in\Gamma_d} \sum_{I\in\mathcal{D}} \sigma_{I,\alpha} \langle \phi \rangle_{E_{\alpha,I}} \widehat{w^{-\frac{1}{2}}}(I,\alpha) h_I^\alpha \\ &= \sum_{\alpha\in\Gamma_d} \sum_{I\in\mathcal{D}} \sigma_{I,\alpha} \widehat{w^{\frac{1}{2}}}(I,\alpha) \widehat{w^{-\frac{1}{2}}}(I,\alpha) \langle \phi \rangle_{E_{\alpha,I}} h^1_{E_{\alpha,I}}. \end{aligned}$$

Then, we can calculate:

$$\begin{aligned} \left| \left\langle P^{(1,0)}_{\widehat{w^{\frac{1}{2}}}} T_\sigma P^{(0,1)}_{\widehat{w^{-\frac{1}{2}}}} \phi, \psi \right\rangle_{L^2} \right| &= \left| \sum_{\alpha\in\Gamma_d} \sum_{I\in\mathcal{D}} \sigma_{I,\alpha} \widehat{w^{\frac{1}{2}}}(I,\alpha) \widehat{w^{-\frac{1}{2}}}(I,\alpha) \langle \phi \rangle_{E_{\alpha,I}} \langle \psi \rangle_{E_{\alpha,I}} \right| \\ &\leq \|\sigma\|_\infty \left(\sum_{\alpha\in\Gamma_d} \sum_{I\in\mathcal{D}} |\widehat{w^{\frac{1}{2}}}(I,\alpha) \widehat{w^{-\frac{1}{2}}}(I,\alpha)| \langle \phi \rangle^2_{E_{\alpha,I}} \right)^{\frac{1}{2}} \\ &\quad \times \left(\sum_{\alpha\in\Gamma_d} \sum_{I\in\mathcal{D}} |\widehat{w^{\frac{1}{2}}}(I,\alpha) \widehat{w^{-\frac{1}{2}}}(I,\alpha)| \langle \psi \rangle^2_{E_{\alpha,I}} \right)^{\frac{1}{2}} \\ &\lesssim \|\sigma\|_\infty [w]^{\frac{1}{2}}_{A_2} \|\phi\|_{L^2} \|\psi\|_{L^2}, \end{aligned}$$

where the last inequality follows from the Carleson Embedding Theorem. It applies here, since Cauchy-Schwarz gives:

$$\begin{aligned} \frac{1}{|E_{\alpha,I}|} \sum_{J\subset I} \sum_{\beta: E_{\beta,J}\subset E_{\alpha,I}} \left| \widehat{w^{\frac{1}{2}}}(J,\beta) \widehat{w^{-\frac{1}{2}}}(J,\beta) \right| &\leq \frac{1}{|E_{\alpha,I}|} \left\| w^{\frac{1}{2}} \mathbf{1}_{E_{\alpha,I}} \right\|_{L^2} \left\| w^{-\frac{1}{2}} \mathbf{1}_{E_{\alpha,I}} \right\|_{L^2} \\ &= \left(\langle w \rangle_{E_{\alpha,I}} \langle w^{-1} \rangle_{E_{\alpha,I}} \right)^{\frac{1}{2}} \\ &\lesssim [w]^{\frac{1}{2}}_{A_2}. \end{aligned} \tag{3.5}$$

Taking the supremum over all $\phi, \psi \in L^2$ and using duality gives

$$\left\| P^{(1,0)}_{\widehat{w^{\frac{1}{2}}}} T_\sigma P^{(0,1)}_{\widehat{w^{-\frac{1}{2}}}} \right\|_{L^2\to L^2} \lesssim \|\sigma\|_\infty [w]^{\frac{1}{2}}_{A_2} \leq \|\sigma\|_\infty [w]_{A_2}.$$

3.1.2. *Estimating* $P^{(1,0)}_{\widehat{w^{\frac{1}{2}}}} T_\sigma P^{(0,0)}_{\langle w^{-\frac{1}{2}}\rangle}$ *and* $P^{(0,0)}_{\langle w^{\frac{1}{2}}\rangle} T_\sigma P^{(0,1)}_{\widehat{w^{-\frac{1}{2}}}}$. As these two operators are symmetric, very similar arguments can be used to control both of them. Thus, we only provide details for the first operator. Observe that

$$\begin{aligned} P^{(1,0)}_{\widehat{w^{\frac{1}{2}}}} T_\sigma P^{(0,0)}_{\langle w^{-\frac{1}{2}}\rangle}\phi &= P^{(1,0)}_{\widehat{w^{\frac{1}{2}}}} \sum_{\alpha\in\Gamma_d}\sum_{I\in\mathcal{D}} \sigma_{I,\alpha}\langle w^{-\frac{1}{2}}\rangle_{E_{\alpha,I}}\widehat{\phi}(I,\alpha)h_I^\alpha \\ &= \sum_{\alpha\in\Gamma_d}\sum_{I\in\mathcal{D}} \sigma_{I,\alpha}\langle w^{-\frac{1}{2}}\rangle_{E_{\alpha,I}}\widehat{w^{\frac{1}{2}}}(I,\alpha)\widehat{\phi}(I,\alpha)\, h^1_{E_{\alpha,I}}. \end{aligned}$$

Fixing $\phi,\psi\in L^2$, we can calculate

$$\begin{aligned} &\left|\left\langle P^{(1,0)}_{\widehat{w^{\frac{1}{2}}}} T_\sigma P^{(0,0)}_{\langle w^{-\frac{1}{2}}\rangle}\phi,\psi\right\rangle_{L^2}\right| \\ &\quad= \left|\sum_{\alpha\in\Gamma_d}\sum_{I\in\mathcal{D}} \sigma_{I,\alpha}\langle w^{-\frac{1}{2}}\rangle_{E_{\alpha,I}}\widehat{w^{\frac{1}{2}}}(I,\alpha)\widehat{\phi}(I,\alpha)\langle\psi\rangle_{E_{\alpha,I}}\right| \\ &\quad\le \|\sigma\|_\infty\|\phi\|_{L^2}\left(\sum_{\alpha\in\Gamma_d}\sum_{I\in\mathcal{D}}\left|\widehat{w^{\frac{1}{2}}}(I,\alpha)\right|^2\langle w^{-\frac{1}{2}}\rangle^2_{E_{\alpha,I}}\langle\psi\rangle^2_{E_{\alpha,I}}\right)^{\frac{1}{2}} \\ &\quad\lesssim \|\sigma\|_\infty[w]_{A_2}\|\phi\|_{L^2}\|\psi\|_{L^2}, \end{aligned}$$

where the last inequality follows from the Carleson Embedding Theorem and estimate (2.10). Again, taking the supremum over all $\phi,\psi\in L^2$ and using duality gives

$$\left\|P^{(1,0)}_{\widehat{w^{\frac{1}{2}}}} T_\sigma P^{(0,0)}_{\langle w^{-\frac{1}{2}}\rangle}\right\|_{L^2\to L^2} \lesssim \|\sigma\|_\infty[w]_{A_2}.$$

3.1.3. *Estimating* $P^{(0,0)}_{\langle w^{\frac{1}{2}}\rangle} T_\sigma P^{(0,0)}_{\langle w^{-\frac{1}{2}}\rangle}$. Fixing $\phi,\psi\in L^2$, observe that

$$\begin{aligned} P^{(0,0)}_{\langle w^{\frac{1}{2}}\rangle} T_\sigma P^{(0,0)}_{\langle w^{-\frac{1}{2}}\rangle}\phi &= P^{(0,0)}_{\langle w^{\frac{1}{2}}\rangle}\sum_{\alpha\in\Gamma_d}\sum_{I\in\mathcal{D}}\sigma_{I,\alpha}\left\langle w^{-\frac{1}{2}}\right\rangle_{E_{\alpha,I}}\widehat{\phi}(I,\alpha)h_I^\alpha \\ &= \sum_{\alpha\in\Gamma_d}\sum_{I\in\mathcal{D}}\sigma_{I,\alpha}\left\langle w^{\frac{1}{2}}\right\rangle_{E_{\alpha,I}}\left\langle w^{-\frac{1}{2}}\right\rangle_{E_{\alpha,I}}\widehat{\phi}(I,\alpha)h_I^\alpha. \end{aligned}$$

This means we can calculate

$$\begin{aligned} \left|\left\langle P^{(0,0)}_{\langle w^{\frac{1}{2}}\rangle} T_\sigma P^{(0,0)}_{\langle w^{-\frac{1}{2}}\rangle}\phi,\psi\right\rangle_{L^2}\right| &= \left|\sum_{\alpha\in\Gamma_d}\sum_{I\in\mathcal{D}}\sigma_{I,\alpha}\left\langle w^{\frac{1}{2}}\right\rangle_{E_{\alpha,I}}\left\langle w^{-\frac{1}{2}}\right\rangle_{E_{\alpha,I}}\widehat{\phi}(I,\alpha)\widehat{\psi}(I,\alpha)\right| \\ &\lesssim \|\sigma\|_\infty\sup_{I\in\mathcal{D}}\left(\left\langle w^{\frac{1}{2}}\right\rangle_I\left\langle w^{-\frac{1}{2}}\right\rangle_I\right)\|\phi\|_{L^2}\|\psi\|_{L^2} \\ &\le \|\sigma\|_\infty[w]^{\frac{1}{2}}_{A_2}\|\phi\|_{L^2}\|\psi\|_{L^2}. \end{aligned}$$

Taking the supremum over all $\phi,\psi\in L^2$ and using duality gives the desired linear norm bound. This concludes the proof for the easy terms.

3.2. Estimating the Hard Terms. There are five remaining terms to be controlled. These include the four difficult terms:

$$\mathsf{P}^{(0,1)}_{\widehat{w^{\frac{1}{2}}}} T_\sigma \mathsf{P}^{(0,1)}_{\widehat{w^{-\frac{1}{2}}}}; \tag{3.6}$$

$$\mathsf{P}^{(0,1)}_{\widehat{w^{\frac{1}{2}}}} T_\sigma \mathsf{P}^{(0,0)}_{\langle w^{-\frac{1}{2}}\rangle}; \tag{3.7}$$

$$\mathsf{P}^{(1,0)}_{\widehat{w^{\frac{1}{2}}}} T_\sigma \mathsf{P}^{(1,0)}_{\widehat{w^{-\frac{1}{2}}}}; \tag{3.8}$$

$$\mathsf{P}^{(0,0)}_{\langle w^{\frac{1}{2}}\rangle} T_\sigma \mathsf{P}^{(1,0)}_{\widehat{w^{-\frac{1}{2}}}}. \tag{3.9}$$

To estimate terms (3.6) and (3.8) we will rely on disbalanced Haar functions adapted to the weights w and w^{-1}. For these terms, we also compute the norms using duality and frequent application of the modified Carleson Embedding Theorem. The proof of the estimates for these terms is carried out in subsection 3.2.1. Terms (3.7) and (3.9) will be handled via a similar method; their analysis appears in subsection 3.2.2.

The remaining term is the one for which T_σ can not be absorbed into one of the paraproducts. Namely, we need to control the following expression:

$$\mathsf{P}^{(0,1)}_{\widehat{w^{\frac{1}{2}}}} T_\sigma \mathsf{P}^{(1,0)}_{\widehat{w^{-\frac{1}{2}}}}. \tag{3.10}$$

To handle this term, we must rely on Therem 1.1 and the computed linear bounds for the other eight paraproduct compositions. This leaves the open the question of whether there is an independent proof of the linear bound for (3.10). This is discussed further in subsection 3.2.3.

3.2.1. *Estimating* $P^{(0,1)}_{\widehat{w^{\frac{1}{2}}}} T_\sigma P^{(0,1)}_{\widehat{w^{-\frac{1}{2}}}}$ *and* $P^{(1,0)}_{\widehat{w^{\frac{1}{2}}}} T_\sigma P^{(1,0)}_{\widehat{w^{-\frac{1}{2}}}}$. Similar arguments handle both terms and so, we restrict attention to the first one. Fix $\phi, \psi \in L^2$. Observe that basic manipulations and the product formula (2.2) for Haar coefficients give

$$\begin{aligned}
&\left\langle P^{(0,1)}_{\widehat{w^{\frac{1}{2}}}} T_\sigma P^{(0,1)}_{\widehat{w^{-\frac{1}{2}}}} \phi, \psi \right\rangle_{L^2} \\
&\quad = \sum_{\alpha \in \Gamma_d} \sum_{I \in \mathcal{D}} \sigma_{I,\alpha} \langle \phi \rangle_{E_{\alpha,I}} \widehat{w^{-\frac{1}{2}}}(I,\alpha) \sum_{\beta \in \Gamma_d} \sum_{J \in \mathcal{D}} \widehat{w^{\frac{1}{2}}}(J,\beta) \widehat{\psi}(J,\beta) \left\langle h_I^\alpha, h^1_{E_{\beta,J}} \right\rangle_{L^2} \\
&\quad = \sum_{\alpha \in \Gamma_d} \sum_{I \in \mathcal{D}} \sigma_{I,\alpha} \langle \phi \rangle_{E_{\alpha,I}} \widehat{w^{-\frac{1}{2}}}(I,\alpha) \\
&\qquad \times \left(\widehat{\psi w^{\frac{1}{2}}}(I,\alpha) - \widehat{\psi}(I,\alpha) \left\langle w^{\frac{1}{2}} \right\rangle_{E_{\alpha,I}} - \widehat{w^{\frac{1}{2}}}(I,\alpha) \langle \psi \rangle_{E_{\alpha,I}} \right) \\
&\quad \equiv T_1 + T_2 + T_3.
\end{aligned}$$

We will show that each $|T_j| \lesssim \|\sigma\|_\infty [w]_{A_2} \|\phi\|_{L^2} \|\psi\|_{L^2}$. The bounds for T_2 and T_3 follow easily. First, observe that

$$\begin{aligned}
|T_2| &\le \|\sigma\|_\infty \sum_{\alpha \in \Gamma_d} \sum_{I \in \mathcal{D}} \left| \langle \phi \rangle_{E_{\alpha,I}} \widehat{w^{-\frac{1}{2}}}(I,\alpha) \widehat{\psi}(I,\alpha) \left\langle w^{\frac{1}{2}} \right\rangle_{E_{\alpha,I}} \right| \\
&\le \|\sigma\|_\infty \|\psi\|_{L^2} \left(\sum_{\alpha \in \Gamma_d} \sum_{I \in \mathcal{D}} \left| \widehat{w^{-\frac{1}{2}}}(I,\alpha) \right|^2 \left\langle w^{\frac{1}{2}} \right\rangle^2_{E_{\alpha,I}} \langle \phi \rangle^2_{E_{\alpha,I}} \right)^{\frac{1}{2}} \\
&\lesssim \|\sigma\|_\infty [w]_{A_2} \|\psi\|_{L^2} \|\phi\|_{L^2}.
\end{aligned}$$

The last inequality follows via the Carleson Embedding Theorem and the square function estimate (2.10). For T_3, the computations are similarly straightforward:

$$\begin{aligned}
|T_3| &\leq \|\sigma\|_\infty \sum_{\alpha\in\Gamma_d}\sum_{I\in\mathcal{D}} \left|\widehat{w^{-\frac{1}{2}}}(I,\alpha)\widehat{w^{\frac{1}{2}}}(I,\alpha)\langle\phi\rangle_{E_{\alpha,I}}\langle\psi\rangle_{E_{\alpha,I}}\right| \\
&\leq \|\sigma\|_\infty \left(\sum_{\alpha\in\Gamma_d}\sum_{I\in\mathcal{D}} |\widehat{w^{-\frac{1}{2}}}(I,\alpha)\widehat{w^{\frac{1}{2}}}(I,\alpha)|\langle\phi\rangle^2_{E_{\alpha,I}}\right)^{\frac{1}{2}} \\
&\quad\times \left(\sum_{\alpha\in\Gamma_d}\sum_{I\in\mathcal{D}} |\widehat{w^{-\frac{1}{2}}}(I,\alpha)\widehat{w^{\frac{1}{2}}}(I,\alpha)|\langle\psi\rangle^2_{E_{\alpha,I}}\right)^{\frac{1}{2}} \\
&\lesssim \|\sigma\|_\infty [w]_{A_2}^{\frac{1}{2}}\|\phi\|_{L^2}\|\psi\|_{L^2}.
\end{aligned}$$

Here, the last inequality follows from two applications of the Carleson Embedding Theorem using the estimate given in (3.5). Estimating T_1 requires the use of disbalanced Haar functions. We expand the Haar functions in the sum using two disbalanced systems, one associated to w and one associated to w^{-1}, as follows:

$$\begin{aligned}
T_1 &= \sum_{\alpha\in\Gamma_d}\sum_{I\in\mathcal{D}} \sigma_{I,\alpha}\langle\phi\rangle_{E_{\alpha,I}}\widehat{w^{-\frac{1}{2}}}(I,\alpha)\widehat{\psi w^{\frac{1}{2}}}(I,\alpha) \\
&= \sum_{\alpha\in\Gamma_d}\sum_{I\in\mathcal{D}} \sigma_{I,\alpha}\langle\phi\rangle_{E_{\alpha,I}}\left\langle w^{-\frac{1}{2}}, C_I(w^{-1},\alpha)h_I^{w^{-1},\alpha} + D_I(w^{-1},\alpha)h^1_{E_{\alpha,I}}\right\rangle_{L^2} \\
&\qquad\times\left\langle \psi w^{\frac{1}{2}}, C_I(w,\alpha)h_I^{w,\alpha} + D_I(w,\alpha)h^1_{E_{\alpha,I}}\right\rangle_{L^2} \\
&= \sum_{\alpha\in\Gamma_d}\sum_{I\in\mathcal{D}} \sigma_{I,\alpha}\langle\phi\rangle_{E_{\alpha,I}}C_I(w^{-1},\alpha)C_I(w,\alpha)\left\langle w^{-\frac{1}{2}}, h_I^{w^{-1},\alpha}\right\rangle_{L^2}\left\langle \psi w^{\frac{1}{2}}, h_I^{w,\alpha}\right\rangle_{L^2} \\
&\quad+ \sum_{\alpha\in\Gamma_d}\sum_{I\in\mathcal{D}} \sigma_{I,\alpha}\langle\phi\rangle_{E_{\alpha,I}}C_I(w^{-1},\alpha)D_I(w,\alpha)\left\langle w^{-\frac{1}{2}}, h_I^{w^{-1},\alpha}\right\rangle_{L^2}\left\langle \psi w^{\frac{1}{2}}\right\rangle_{E_{\alpha,I}} \\
&\quad+ \sum_{\alpha\in\Gamma_d}\sum_{I\in\mathcal{D}} \sigma_{I,\alpha}\langle\phi\rangle_{E_{\alpha,I}}D_I(w^{-1},\alpha)C_I(w,\alpha)\left\langle w^{-\frac{1}{2}}\right\rangle_{E_{\alpha,I}}\left\langle \psi w^{\frac{1}{2}}, h_I^{w,\alpha}\right\rangle_{L^2} \\
&\quad+ \sum_{\alpha\in\Gamma_d}\sum_{I\in\mathcal{D}} \sigma_{I,\alpha}\langle\phi\rangle_{E_{\alpha,I}}D_I(w^{-1},\alpha)D_I(w,\alpha)\left\langle w^{-\frac{1}{2}}\right\rangle_{E_{\alpha,I}}\left\langle \psi w^{\frac{1}{2}}\right\rangle_{E_{\alpha,I}} \\
&\equiv S_1 + S_2 + S_3 + S_4.
\end{aligned}$$

Now, we show each $|S_j| \lesssim \|\sigma\|_\infty [w]_{A_2} \|\phi\|_{L^2} \|\psi\|_{L^2}$, which gives the bound for T_1. Observe that by (2.5),

$$\begin{aligned}
&|S_1| \\
&\leq \|\sigma\|_\infty \sum_{\alpha\in\Gamma_d}\sum_{I\in\mathcal{D}} \left| \langle\phi\rangle_{E_{\alpha,I}} C_I(w^{-1},\alpha) C_I(w,\alpha) \left\langle w^{-\frac{1}{2}}, h_I^{w^{-1},\alpha}\right\rangle_{L^2} \left\langle \psi w^{-\frac{1}{2}}, h_I^{w,\alpha}\right\rangle_{L^2(w)} \right| \\
&\lesssim \|\sigma\|_\infty \left\| \psi w^{-\frac{1}{2}} \right\|_{L^2(w)} \left(\sum_{\alpha\in\Gamma_d}\sum_{I\in\mathcal{D}} \langle w^{-1}\rangle_I \langle w\rangle_I \left\langle w^{-\frac{1}{2}}, h_I^{w^{-1},\alpha}\right\rangle_{L^2}^2 \langle\phi\rangle_{E_{\alpha,I}}^2 \right)^{\frac{1}{2}} \\
&\lesssim \|\sigma\|_\infty [w]_{A_2}^{\frac{1}{2}} \|\psi\|_{L^2} \left(\sum_{\alpha\in\Gamma_d}\sum_{I\in\mathcal{D}} \left\langle w^{-\frac{1}{2}}, h_I^{w^{-1},\alpha}\right\rangle_{L^2}^2 \langle\phi\rangle_{E_{\alpha,I}}^2 \right)^{\frac{1}{2}} \\
&\lesssim \|\sigma\|_\infty [w]_{A_2}^{\frac{1}{2}} \|\psi\|_{L^2} \|\phi\|_{L^2},
\end{aligned}$$

where the last inequality followed via the Carleson Embedding Theorem using the estimate

$$\begin{aligned}
\sum_{J\subset I}\sum_{\beta: E_{\beta,J}\subset E_{\alpha,I}} \left\langle w^{-\frac{1}{2}}, h_J^{w^{-1},\beta}\right\rangle_{L^2}^2 &= \sum_{J\subset I}\sum_{\beta: E_{\beta,J}\subset E_{\alpha,I}} \left\langle w^{\frac{1}{2}}, h_J^{w^{-1},\beta}\right\rangle_{L^2(w^{-1})}^2 \\
&\leq \left\| w^{\frac{1}{2}} \mathbf{1}_{E_{\alpha,I}} \right\|_{L^2(w^{-1})}^2 = |E_{\alpha,I}|.
\end{aligned}$$

The calculation for S_2 is also straightforward:

$$\begin{aligned}
|S_2| &\leq \|\sigma\|_\infty \sum_{\alpha\in\Gamma_d}\sum_{I\in\mathcal{D}} \left| \langle\phi\rangle_{E_{\alpha,I}} C_I(w^{-1},\alpha) D_I(w,\alpha) \left\langle w^{-\frac{1}{2}}, h_I^{w^{-1},\alpha}\right\rangle_{L^2} \left\langle \psi w^{\frac{1}{2}}\right\rangle_{E_{\alpha,I}} \right| \\
&\lesssim \|\sigma\|_\infty \left(\sum_{\alpha\in\Gamma_d}\sum_{I\in\mathcal{D}} \left\langle w^{-\frac{1}{2}}, h_I^{w^{-1},\alpha}\right\rangle_{L^2}^2 \langle\phi\rangle_{E_{\alpha,I}}^2 \right)^{\frac{1}{2}} \\
&\quad \times \left(\sum_{\alpha\in\Gamma_d}\sum_{I\in\mathcal{D}} \langle w^{-1}\rangle_{E_{\alpha,I}} \frac{|\widehat{w}(I,\alpha)|^2}{\langle w\rangle_{E_{\alpha,I}}^2} \left\langle \psi w^{\frac{1}{2}}\right\rangle_{E_{\alpha,I}}^2 \right)^{\frac{1}{2}} \\
&\lesssim \|\sigma\|_\infty [w]_{A_2} \|\phi\|_{L^2} \|\psi\|_{L^2},
\end{aligned}$$

where we use the Carleson Embedding Theorem twice. The application for the ϕ term follows as in the estimate for S_1, while the application for ψ follows from the square function estimate (2.11). Similarly, for S_3, we can calculate

$$\begin{aligned}
|S_3| &\leq \|\sigma\|_\infty \sum_{\alpha\in\Gamma_d}\sum_{I\in\mathcal{D}} \left| \langle\phi\rangle_{E_{\alpha,I}} D_I(w^{-1},\alpha) C_I(w,\alpha) \left\langle w^{-\frac{1}{2}} \right\rangle_{E_{\alpha,I}} \left\langle \psi w^{\frac{1}{2}}, h_I^{w,\alpha} \right\rangle_{L^2} \right| \\
&\lesssim \|\sigma\|_\infty \left(\sum_{\alpha\in\Gamma_d}\sum_{I\in\mathcal{D}} \frac{\left|\widehat{w^{-1}}(I,\alpha)\right|^2}{\langle w^{-1}\rangle_{E_{\alpha,I}}^2} \langle\phi\rangle_{E_{\alpha,I}}^2 \right)^{\frac{1}{2}} \\
&\quad \times \left(\sum_{\alpha\in\Gamma_d}\sum_{I\in\mathcal{D}} \langle w\rangle_I \left\langle w^{-\frac{1}{2}} \right\rangle_I^2 \left\langle \psi w^{-\frac{1}{2}}, h_I^{w,\alpha} \right\rangle_{L^2(w)}^2 \right)^{\frac{1}{2}} \\
&\lesssim \|\sigma\|_\infty [w]_{A_2}^{\frac{1}{2}} \left\| \psi w^{-\frac{1}{2}} \right\|_{L^2(w)} \left(\sum_{\alpha\in\Gamma_d}\sum_{I\in\mathcal{D}} \frac{\left|\widehat{w^{-1}}(I,\alpha)\right|^2}{\langle w^{-1}\rangle_{E_{\alpha,I}}^2} \langle\phi\rangle_{E_{\alpha,I}}^2 \right)^{\frac{1}{2}} \\
&\lesssim \|\sigma\|_\infty [w]_{A_2} \|\phi\|_{L^2} \|\psi\|_{L^2},
\end{aligned}$$

where the last inequality follows from the Carleson Embedding Theorem. It applies here since:

$$\sum_{J\subset I}\sum_{\beta: E_{\beta,J}\subset E_{\alpha,I}} \frac{\left|\widehat{w^{-1}}(J,\beta)\right|^2}{\langle w^{-1}\rangle_{E_{\beta,J}}^2} \lesssim [w]_{A_2} |E_{\alpha,I}| \quad \forall\alpha\in\Gamma_d \quad \forall I\in\mathcal{D}. \tag{3.11}$$

To see that (3.11) holds, it is then a simple application of Cauchy-Schwarz and the following estimates:

$$\sum_{J\subset I}\sum_{\beta: E_{\beta,J}\subset E_{\alpha,I}} \frac{\left|\widehat{w^{-1}}(J,\beta)\right|^2}{\langle w^{-1}\rangle_{E_{\beta,J}}} \lesssim [w]_{A_2} w^{-1}(E_{\alpha,I}) \quad \forall\alpha\in\Gamma_d \quad \forall I\in\mathcal{D}; \tag{3.12}$$

$$\sum_{J\subset I}\sum_{\beta: E_{\beta,J}\subset E_{\alpha,I}} \frac{\left|\widehat{w^{-1}}(J,\beta)\right|^2}{\langle w^{-1}\rangle_{E_{\beta,J}}^3} \lesssim w(E_{\alpha,I}) \quad \forall\alpha\in\Gamma_d \quad \forall I\in\mathcal{D}. \tag{3.13}$$

The proofs of (3.12) and (3.13) can be found in [**1**, Proposition 4.9, Equation (4.17)] and [**1**, Proposition 4.7, Equation (4.11)] respectively.

Lastly, the estimate for S_4 is computed as follows:

$$\begin{aligned}
|S_4| &\leq \|\sigma\|_\infty \sum_{\alpha\in\Gamma_d}\sum_{I\in\mathcal{D}} \left| \langle\phi\rangle_{E_{\alpha,I}} D_I(w^{-1},\alpha) D_I(w,\alpha) \left\langle w^{-\frac{1}{2}} \right\rangle_{E_{\alpha,I}} \left\langle \psi w^{\frac{1}{2}} \right\rangle_{E_{\alpha,I}} \right| \\
&\leq \|\sigma\|_\infty \left(\sum_{\alpha\in\Gamma_d}\sum_{I\in\mathcal{D}} \left| \widehat{w}(I,\alpha)\widehat{w^{-1}}(I,\alpha) \right| \langle\phi\rangle_{E_{\alpha,I}}^2 \right)^{\frac{1}{2}} \\
&\quad \times \left(\sum_{\alpha\in\Gamma_d}\sum_{I\in\mathcal{D}} \frac{\left| \widehat{w}(I,\alpha)\widehat{w^{-1}}(I,\alpha) \right|}{\langle w\rangle_{E_{\alpha,I}}^2 \langle w^{-1}\rangle_{E_{\alpha,I}}} \left\langle \psi w^{\frac{1}{2}} \right\rangle_{E_{\alpha,I}}^2 \right)^{\frac{1}{2}} \\
&= \|\sigma\|_\infty [w]_{A_2} \|\phi\|_{L_2} \|\psi\|_{L^2},
\end{aligned}$$

where the Carleson Embedding Theorem is used twice. The application for the ϕ term uses

$$\sum_{J\subset I}\sum_{\beta:E_{\beta,J}\subset E_{\alpha,I}}\left|\widehat{w}(J,\beta)\,\widehat{w^{-1}}(J,\beta)\right|\lesssim [w]_{A_2}\left|E_{\alpha,I}\right|\quad\forall\alpha\in\Gamma_d\quad\forall I\in\mathcal{D}.$$

As stated, the proof of this is found in [**1**, Equation (6.3)]. The one-dimensional version is established in [**10**, Lemma 4.7]. The application for the ψ term uses

$$\sum_{J\subset I}\sum_{\beta:E_{\beta,J}\subset E_{\alpha,I}}\frac{\left|\widehat{w}(J,\beta)\,\widehat{w^{-1}}(J,\beta)\right|}{\left\langle w^{-1}\right\rangle_{E_{\beta,J}}}\lesssim [w]_{A_2}\,w\left(E_{\alpha,I}\right)\quad\forall\alpha\in\Gamma_d\quad\forall I\in\mathcal{D}.$$

As stated, the proof of this is found in [**1**, Equation (6.4)].

This concludes the proof of the estimates for T_1,T_2,T_3. By taking the supremum over $\phi,\psi\in L^2$ and using duality, we conclude that

$$\left\|P^{(0,1)}_{\widehat{w^{\frac12}}}T_\sigma P^{(0,1)}_{\widehat{w^{-\frac12}}}\right\|_{L^2\to L^2}\lesssim\|\sigma\|_\infty[w]_{A_2}.$$

3.2.2. *Estimating* $P^{(0,1)}_{\widehat{w^{\frac12}}}T_\sigma P^{(0,0)}_{\langle w^{-\frac12}\rangle}$ *and* $P^{(0,0)}_{\langle w^{\frac12}\rangle}T_\sigma P^{(1,0)}_{\widehat{w^{-\frac12}}}$. We only discuss the first operator, as the estimates for the second one follow via similar arguments. Fix $\phi,\psi\in L^2$. We first simplify using basic manipulations and the product formula for Haar coefficients, (2.2), as follows:

$$\begin{aligned}\left\langle P^{(0,1)}_{\widehat{w^{\frac12}}}T_\sigma P^{(0,0)}_{\langle w^{-\frac12}\rangle}\phi,\psi\right\rangle_{L^2}&=\sum_{\alpha\in\Gamma_d}\sum_{I\in\mathcal{D}}\sigma_{I,\alpha}\left\langle w^{-\frac12}\right\rangle_{E_{\alpha,I}}\widehat{\phi}(I,\alpha)\\&\quad\times\sum_{\beta\in\Gamma_d}\sum_{J\in\mathcal{D}}\widehat{w^{\frac12}}(J,\beta)\widehat{\psi}(J,\beta)\left\langle h_I^\alpha,h^1_{E_{\beta,J}}\right\rangle_{L^2}\\&=\sum_{\alpha\in\Gamma_d}\sum_{I\in\mathcal{D}}\sigma_{I,\alpha}\left\langle w^{-\frac12}\right\rangle_{E_{\alpha,I}}\widehat{\phi}(I,\alpha)\\&\quad\times\left(\widehat{\psi w^{\frac12}}(I,\alpha)-\widehat{\psi}(I,\alpha)\left\langle w^{\frac12}\right\rangle_{E_{\alpha,I}}-\widehat{w^{\frac12}}(I,\alpha)\langle\psi\rangle_{E_{\alpha,I}}\right)\\&\equiv T_1+T_2+T_3.\end{aligned}$$

As in the previous case, we show that each $|T_j|\lesssim\|\sigma\|_\infty[w]_{A_2}\|\phi\|_{L^2}\|\psi\|_{L^2}$. The estimates for T_2 and T_3 follow easily. Observe that

$$\begin{aligned}|T_2|&\le\|\sigma\|_\infty\sum_{\alpha\in\Gamma_d}\sum_{I\in\mathcal{D}}\left|\left\langle w^{-\frac12}\right\rangle_{E_{\alpha,I}}\left\langle w^{\frac12}\right\rangle_{E_{\alpha,I}}\widehat{\phi}(I,\alpha)\widehat{\psi}(I,\alpha)\right|\\&\lesssim\|\sigma\|_\infty[w]^{\frac12}_{A_2}\sum_{\alpha\in\Gamma_d}\sum_{I\in\mathcal{D}}\left|\widehat{\phi}(I,\alpha)\widehat{\psi}(I,\alpha)\right|\\&\le\|\sigma\|_\infty[w]^{\frac12}_{A_2}\|\phi\|_{L^2}\|\psi\|_{L^2},\end{aligned}$$

and similarly,

$$
\begin{aligned}
|T_3| &\leq \|\sigma\|_\infty \sum_{\alpha\in\Gamma_d}\sum_{I\in\mathcal{D}} \left|\left\langle w^{-\frac{1}{2}}\right\rangle_{E_{\alpha,I}} \langle\psi\rangle_{E_{\alpha,I}}\widehat{\phi}(I,\alpha)\widehat{w^{\frac{1}{2}}}(I,\alpha)\right| \\
&\leq \|\sigma\|_\infty\|\phi\|_{L^2}\left(\sum_{\alpha\in\Gamma_d}\sum_{I\in\mathcal{D}}\left\langle w^{-\frac{1}{2}}\right\rangle_{E_{\alpha,I}}^2 \left|\widehat{w^{\frac{1}{2}}}(I,\alpha)\right|^2 \langle\psi\rangle_{E_{\alpha,I}}^2\right)^{\frac{1}{2}} \\
&\lesssim \|\sigma\|_\infty[w]_{A_2}\|\phi\|_{L^2}\|\psi\|_{L^2},
\end{aligned}
$$

where the last inequality followed via an application of the Carleson Embedding Theorem using (2.10).

To estimate T_1, we rewrite the term $\left\langle \psi w^{\frac{1}{2}}, h_I^\alpha\right\rangle_{L^2}$ using disbalanced Haar functions adapted to w as follows:

(3.14)

$$
\begin{aligned}
T_1 &= \sum_{\alpha\in\Gamma_d}\sum_{I\in\mathcal{D}}\sigma_{I,\alpha}\left\langle w^{-\frac{1}{2}}\right\rangle_{E_{\alpha,I}}\widehat{\phi}(I,\alpha)\widehat{\psi w^{\frac{1}{2}}}(I,\alpha) \\
&= \sum_{\alpha\in\Gamma_d}\sum_{I\in\mathcal{D}}\sigma_{I,\alpha}\left\langle w^{-\frac{1}{2}}\right\rangle_{E_{\alpha,I}}\widehat{\phi}(I,\alpha)\left\langle \psi w^{\frac{1}{2}}, C_I(w,\alpha)h_I^{w,\alpha}+D_I(w,\alpha)h_{E_{\alpha,I}}^1\right\rangle_{L^2} \\
&= \sum_{\alpha\in\Gamma_d}\sum_{I\in\mathcal{D}}\sigma_{I,\alpha}C_I(w,\alpha)\left\langle w^{-\frac{1}{2}}\right\rangle_{E_{\alpha,I}}\widehat{\phi}(I,\alpha)\left\langle \psi w^{\frac{1}{2}}, h_I^{w,\alpha}\right\rangle_{L^2} \\
&\quad+\sum_{\alpha\in\Gamma_d}\sum_{I\in\mathcal{D}}\sigma_{I,\alpha}D_I(w,\alpha)\left\langle w^{-\frac{1}{2}}\right\rangle_{E_{\alpha,I}}\widehat{\phi}(I,\alpha)\left\langle \psi w^{\frac{1}{2}}\right\rangle_{E_{\alpha,I}} \\
&= S_1+S_2.
\end{aligned}
$$

Now, we show each $|S_j|\lesssim \|\sigma\|_\infty[w]_{A_2}\|\phi\|_{L^2}\|\psi\|_{L^2}$, which will give the estimate for T_1. First, consider S_1:

$$
\begin{aligned}
|S_1| &= \left|\sum_{\alpha\in\Gamma_d}\sum_{I\in\mathcal{D}}\sigma_{I,\alpha}C_I(w,\alpha)\left\langle w^{-\frac{1}{2}}\right\rangle_{E_{\alpha,I}}\widehat{\phi}(I,\alpha)\left\langle \psi w^{\frac{1}{2}}, h_I^{w,\alpha}\right\rangle_{L^2}\right| \\
&\leq \|\sigma\|_\infty\sum_{\alpha\in\Gamma_d}\sum_{I\in\mathcal{D}}\left|C_I(w,\alpha)\left\langle w^{-\frac{1}{2}}\right\rangle_{E_{\alpha,I}}\widehat{\phi}(I,\alpha)\left\langle \psi w^{\frac{1}{2}}, h_I^{w,\alpha}\right\rangle_{L^2}\right| \\
&\lesssim \|\sigma\|_\infty\|\phi\|_{L^2}\left(\sum_{\alpha\in\Gamma_d}\sum_{I\in\mathcal{D}}\langle w\rangle_I\left\langle w^{-\frac{1}{2}}\right\rangle_{E_{\alpha,I}}^2\left\langle \psi w^{-\frac{1}{2}}, h_I^{w,\alpha}\right\rangle_{L^2(w)}^2\right)^{\frac{1}{2}} \\
&\lesssim \|\sigma\|_\infty\,[w]_{A_2}^{\frac{1}{2}}\,\|\phi\|_{L^2}\left(\sum_{\alpha\in\Gamma_d}\sum_{I\in\mathcal{D}}\left\langle \psi w^{-\frac{1}{2}}, h_I^{w,\alpha}\right\rangle_{L^2(w)}^2\right)^{\frac{1}{2}} \\
&\leq \|\sigma\|_\infty[w]_{A_2}^{\frac{1}{2}}\|\phi\|_{L^2}\left\|\psi w^{-\frac{1}{2}}\right\|_{L^2(w)} \\
&= \|\sigma\|_\infty[w]_{A_2}^{\frac{1}{2}}\|\phi\|_{L^2}\|\psi\|_{L^2}.
\end{aligned}
$$

Above we used (2.5) coupled with the A_2 condition. Lastly, we estimate S_2 as follows:

$$\begin{aligned}|S_2| &\le \left|\sum_{\alpha\in\Gamma_d}\sum_{I\in\mathcal{D}} \sigma_{I,\alpha} D_I(w,\alpha)\left\langle w^{-\frac{1}{2}}\right\rangle_{E_{\alpha,I}} \widehat{\phi}(I,\alpha)\left\langle \psi w^{\frac{1}{2}}\right\rangle_{E_{\alpha,I}}\right| \\ &\le \|\sigma\|_\infty \sum_{\alpha\in\Gamma_d}\sum_{I\in\mathcal{D}} \left|D_I(w,\alpha)\left\langle w^{-\frac{1}{2}}\right\rangle_{E_{\alpha,I}} \widehat{\phi}(I,\alpha)\left\langle \psi w^{\frac{1}{2}}\right\rangle_{E_{\alpha,I}}\right| \\ &\le \|\sigma\|_\infty \|\phi\|_{L^2}\left(\sum_{\alpha\in\Gamma_d}\sum_{I\in\mathcal{D}} \frac{|\widehat{w}(I,\alpha)|^2}{\langle w\rangle^2_{E_{\alpha,I}}}\left\langle w^{-\frac{1}{2}}\right\rangle^2_{E_{\alpha,I}}\left\langle \psi w^{\frac{1}{2}}\right\rangle^2_{E_{\alpha,I}}\right)^{\frac{1}{2}} \\ &\lesssim \|\sigma\|_\infty [w]_{A_2}\|\phi\|_{L^2}\,\|\psi\|_{L^2},\end{aligned}$$

where the third inequality follows via an application of the Carleson Embedding Theorem using estimate (2.11). This establishes that each $|T_j| \lesssim \|\sigma\|_\infty [w]_{A_2}\|\phi\|_{L^2}\|\psi\|_{L^2}$. Since $\phi,\psi \in L^2$ were arbitrary, we can use duality to conclude

$$\left\| P^{(0,1)}_{\widehat{w^{\frac{1}{2}}}} T_\sigma P^{(0,0)}_{\langle w^{-\frac{1}{2}}\rangle}\right\|_{L^2\to L^2} \lesssim \|\sigma\|_\infty [w]_{A_2},$$

which finishes the proof of these terms.

3.2.3. *Estimating* $P^{(0,1)}_{\widehat{w^{\frac{1}{2}}}} T_\sigma P^{(1,0)}_{\widehat{w^{-\frac{1}{2}}}}$. To obtain estimates on the final term $P^{(0,1)}_{\widehat{w^{\frac{1}{2}}}} T_\sigma P^{(1,0)}_{\widehat{w^{-\frac{1}{2}}}}$ we simply use (1.1) to observe that

$$\begin{aligned}\left\| P^{(0,1)}_{\widehat{w^{\frac{1}{2}}}} T_\sigma P^{(1,0)}_{\widehat{w^{-\frac{1}{2}}}}\phi\right\|_{L^2} &\le \left\| M_{w^{\frac{1}{2}}} T_\sigma M_{w^{-\frac{1}{2}}}\phi\right\|_{L^2} \\ &\quad + \left\| \mathsf{P}^{(1,0)}_{\widehat{w^{\frac{1}{2}}}} T_\sigma \mathsf{P}^{(0,1)}_{\widehat{w^{-\frac{1}{2}}}}\phi\right\|_{L^2} + \left\| \mathsf{P}^{(1,0)}_{\widehat{w^{\frac{1}{2}}}} T_\sigma \mathsf{P}^{(0,0)}_{\langle w^{-\frac{1}{2}}\rangle}\phi\right\|_{L^2} \\ &\quad + \left\| \mathsf{P}^{(0,0)}_{\langle w^{\frac{1}{2}}\rangle} T_\sigma \mathsf{P}^{(0,1)}_{\widehat{w^{-\frac{1}{2}}}}\phi\right\|_{L^2} + \left\| \mathsf{P}^{(0,0)}_{\langle w^{\frac{1}{2}}\rangle} T_\sigma \mathsf{P}^{(0,0)}_{\langle w^{-\frac{1}{2}}\rangle}\phi\right\|_{L^2} \\ &\quad + \left\| \mathsf{P}^{(0,0)}_{\langle w^{\frac{1}{2}}\rangle} T_\sigma \mathsf{P}^{(0,1)}_{\widehat{w^{-\frac{1}{2}}}}\phi\right\|_{L^2} + \left\| \mathsf{P}^{(0,1)}_{\widehat{w^{\frac{1}{2}}}} T_\sigma \mathsf{P}^{(0,1)}_{\widehat{w^{-\frac{1}{2}}}}\phi\right\|_{L^2} \\ &\quad + \left\| \mathsf{P}^{(0,1)}_{\widehat{w^{\frac{1}{2}}}} T_\sigma \mathsf{P}^{(0,0)}_{\langle w^{-\frac{1}{2}}\rangle}\phi\right\|_{L^2} + \left\| \mathsf{P}^{(1,0)}_{\widehat{w^{\frac{1}{2}}}} T_\sigma \mathsf{P}^{(1,0)}_{\widehat{w^{-\frac{1}{2}}}}\phi\right\|_{L^2} \\ &\quad + \left\| \mathsf{P}^{(0,1)}_{\widehat{w^{\frac{1}{2}}}} T_\sigma \mathsf{P}^{(0,0)}_{\langle w^{-\frac{1}{2}}\rangle}\phi\right\|_{L^2} + \left\| \mathsf{P}^{(0,0)}_{\langle w^{\frac{1}{2}}\rangle} T_\sigma \mathsf{P}^{(1,0)}_{\widehat{w^{-\frac{1}{2}}}}\phi\right\|_{L^2} \\ &\lesssim [w]_{A_2}\|\phi\|_{L^2},\end{aligned}$$

using Theorem 1.1 and our previous computations. This proof strategy motivates the open question:

QUESTION 3.1. Is there a proof of the linear bound for the final paraproduct composition $P^{(0,1)}_{\widehat{w^{\frac{1}{2}}}} T_\sigma P^{(1,0)}_{\widehat{w^{-\frac{1}{2}}}}$ that does not rely on the linear bound of T_σ on $L^2(w)$?

To be precise, fix $\phi, \psi \in L^2$. Then, the term of interest is

$$
\begin{aligned}
&\left\langle P^{(0,1)}_{\widehat{w^{\frac{1}{2}}}} T_\sigma P^{(1,0)}_{\widehat{w^{-\frac{1}{2}}}} \phi, \psi \right\rangle_{L^2} \\
&= \left\langle P^{(0,1)}_{\widehat{w^{\frac{1}{2}}}} T_\sigma \left(\sum_{\alpha\in\Gamma_d} \sum_{I\in\mathcal{D}} \widehat{w^{-\frac{1}{2}}}(I,\alpha)\, \widehat{\phi}(I,\alpha) h^1_{E_{\alpha,I}} \right), \psi \right\rangle_{L^2} \\
&= \sum_{\alpha,\beta\in\Gamma_d} \sum_{I,J\in\mathcal{D}} \widehat{w^{-\frac{1}{2}}}(I,\alpha)\, \widehat{\phi}(I,\alpha)\, \widehat{w^{\frac{1}{2}}}(J,\beta)\, \widehat{\psi}(J,\beta) \left\langle T_\sigma h^1_{E_{\alpha,I}}, h^1_{E_{\beta,J}} \right\rangle_{L^2} \\
&= \sum_{\alpha,\beta,\gamma\in\Gamma_d} \sum_{I,J,K\in\mathcal{D}} \widehat{w^{-\frac{1}{2}}}(I,\alpha)\, \widehat{\phi}(I,\alpha)\, \widehat{w^{\frac{1}{2}}}(J,\beta)\, \widehat{\psi}(J,\beta) \sigma_{K,\gamma}\, \widehat{h^1_{E_{\alpha,I}}}(K,\gamma) \widehat{h^1_{E_{\beta,J}}}(K,\gamma) \\
&= \sum_{\gamma\in\Gamma_d} \sum_{K\in\mathcal{D}} \sigma_{K,\gamma} \left(\sum_{\alpha\in\Gamma_d} \sum_{I\in\mathcal{D}} \widehat{w^{-\frac{1}{2}}}(I,\alpha)\, \widehat{\phi}(I,\alpha) \widehat{h^1_{E_{\alpha,I}}}(K,\gamma) \right) \\
&\qquad \times \left(\sum_{\beta\in\Gamma_d} \sum_{J\in\mathcal{D}} \widehat{w^{\frac{1}{2}}}(J,\beta)\, \widehat{\psi}(J,\beta) \widehat{h^1_{E_{\beta,J}}}(K,\gamma) \right).
\end{aligned}
$$

Currently, our tools seem unequal to the task of bounding this term without recourse to the bound for T_σ. However, given such arguments, one would also obtain a new proof of the linear bound for T_σ on $L^2(w)$.

References

[1] Daewon Chung, *Weighted inequalities for multivariable dyadic paraproducts*, Publ. Mat. **55** (2011), no. 2, 475–499, DOI 10.5565/PUBLMAT_55211_10. MR2839452 (2012m:42015) ↑268, 269, 272, 273, 274, 281, 282

[2] David Cruz-Uribe, José María Martell, and Carlos Pérez, *Sharp weighted estimates for classical operators*, Adv. Math. **229** (2012), no. 1, 408–441, DOI 10.1016/j.aim.2011.08.013. MR2854179 (2012k:42020) ↑273

[3] F. Nazarov, S. Treil, and A. Volberg, *Two weight inequalities for individual Haar multipliers and other well localized operators*, Math. Res. Lett. **15** (2008), no. 3, 583–597, DOI 10.4310/MRL.2008.v15.n3.a16. MR2407233 (2009e:42031) ↑269

[4] F. Nazarov, S. Treil, and A. Volberg, *The Bellman functions and two-weight inequalities for Haar multipliers*, J. Amer. Math. Soc. **12** (1999), no. 4, 909–928, DOI 10.1090/S0894-0347-99-00310-0. MR1685781 (2000k:42009) ↑269

[5] S. Petermichl and S. Pott, *An estimate for weighted Hilbert transform via square functions*, Trans. Amer. Math. Soc. **354** (2002), no. 4, 1699–1703 (electronic), DOI 10.1090/S0002-9947-01-02938-5. MR1873024 (2002j:42010) ↑273

[6] S. Pott, M. C. Reguera, E. T. Sawyer, and B. D. Wick, *The Linear Bound for the Natural Weighted Resolution of the Haar Shift* (2013), available at `http://arxiv.org/abs/1308.5349`. ↑268, 269

[7] E. Sawyer, C.-Y. Shen, and I. Uriarte-Tuero, *The two weight theorem for the vector of Riesz transforms: An Expanded version*, available at `http://arxiv.org/abs/1302.5093v3`. ↑270

[8] S. Hukovic, S. Treil, and A. Volberg, *The Bellman functions and sharp weighted inequalities for square functions*, Complex analysis, operators, and related topics, Oper. Theory Adv. Appl., vol. 113, Birkhäuser, Basel, 2000, pp. 97–113. MR1771755 (2001j:42012) ↑273

[9] J. Michael Wilson, *Paraproducts and the exponential-square class*, J. Math. Anal. Appl. **271** (2002), no. 2, 374–382, DOI 10.1016/S0022-247X(02)00121-X. MR1923641 (2003m:42033) ↑269

[10] Janine Wittwer, *A sharp estimate on the norm of the martingale transform*, Math. Res. Lett. **7** (2000), no. 1, 1–12, DOI 10.4310/MRL.2000.v7.n1.a1. MR1748283 (2001e:42022) ↑268, 273, 282

Department of Mathematics, Bucknell University, 1 Dent Drive, Lewisburg, PA 17837
E-mail address: kelly.bickel@bucknell.edu

Department of Mathematics, McMaster University, Hamilton, Canada
E-mail address: sawyer@mcmaster.ca

School of Mathematics, Georgia Institute of Technology, 686 Cherry Street, Atlanta, Georgia 30332-0160
E-mail address: wick@math.gatech.edu
URL: www.math.gatech.edu/~wick

Contemporary Mathematics
Volume **638**, 2015
http://dx.doi.org/10.1090/conm/638/12815

Transitivity and bundle shifts

Ronald G. Douglas and Anjian Xu

ABSTRACT. A subalgebra A of the algebra $B(\mathcal{H})$ of bounded linear operators on a separable Hilbert space $\mathcal{H}$ is said to be catalytic if every transitive subalgebra $\mathcal{T} \subset B(\mathcal{H})$ containing it is strongly dense. We show that for a hypo-Dirichlet or logmodular algebra, $A = H^\infty(m)$ acting on a generalized Hardy space $H^2(m)$ for a representing measure m that defines a reproducing kernel Hilbert space is catalytic.

For the case of a nice finitely-connected domain, we show that the "holomorphic functions" of a bundle shift yields a catalytic algebra, thus generalizing a result of Bercovici, Foias, Pearcy and the first author (2010).

1. Introduction

In this paper, let $\mathcal{H}$ denote a complex separable Hilbert space, and $B(\mathcal{H})$ the algebra of all bounded linear operators on $\mathcal{H}$. A unital subalgebra $\mathcal{A}$ of $B(\mathcal{H})$ is called *transitive* if it has only trivial invariant subspaces. The transitive algebra problem asks if every transitive algebra $\mathcal{A} \subset B(\mathcal{H})$ is strongly dense in $B(\mathcal{H})$. An operator, or a set of operators, will be called *catalytic* if any transitive operator algebra containing it is strong dense in $B(\mathcal{H})$. It is Arveson [**6**] who first stated explicitly the problem, and developed the main tool, an inductive strategy, for studying transitive algebras. In the same paper, he proved that both the unilateral shift with multiplicity one and a non scalar Hermitian operator of multiplicity one are catalytic. Later, Richter [**15**] proved that the Dirichlet shift is catalytic, and Nordgren [**14**] generalized Arveson's result to unilateral shifts with finite multiplicities. Cheng, Guo and Wang [**8**] proved that the coordinate multiplication operators on functional Hilbert spaces with complete Nevanlinna-Pick kernels are catalytic. The invariant subspace problem asks if a singly-generated algebra acting on a separable Hilbert space $\mathcal{H}$ can be transitive. Catalytic operators must have nontrivial invariant subspaces.

Let R be a finitely-connected, bounded domain in the complex plane whose boundary ∂R consists of $n+1$ nonintersecting smooth Jordan curves. The Hardy space $H^2(R)$ over R is defined to be the space of analytic functions f on R such that the subharmonic function $|f|^2$ is majorized by a harmonic function u on R.

2010 *Mathematics Subject Classification.* Primary 47B35; Secondary 15A15.
Key words and phrases. Shift operators, multiply-connected domain, transitive algebras.
The second author was supported in part by NSFC Grant #11271388.

For each fixed point $t \in R$, there is a norm $\|\cdot\|_t$ on $H^2(R)$ defined by

$$\|f\|_t = \inf_u \{u(t)^{\frac{1}{2}} \mid u \text{ is a harmonic majorant of } |f|^2\}.$$

As in the case of R equal to the unit disc $\mathbb{D}$, one can define an isomorphic space of functions on ∂R using boundary values. Let m_t be the harmonic measure for the point $t \in R$; that is, $f(t) = \int_{\partial R} f dm_t$ for f bounded and harmonic on R. Let $L^2(\partial R, m_t)$ be the space of square-integrable complex-valued measurable functions on ∂R defined with respect to m_t. Then $H^2(\partial R, m_t)$ is defined to be the set of functions $f \in L^2(\partial R, m_t)$ such that $\int_{\partial R} f(z)g(z)dm_t = 0$ for every g that is analytic in a neighborhood of the closure of R. Although the norm on $H^2(\partial R, m_t)$ depends on the fixed point $t \in R$, the spaces of functions for both $H^2(\partial R, m_t)$ and $L^2(\partial R, m_t)$ are independent of t, and these norms are boundedly equivalent for all $t \in R$. We fix $t \in R$ and use $H^2(\partial R) = H^2(\partial R, m_t)$ omitting reference to it.

For any point $w \in R$, the point evaluation functional on $H^2(R)$, defined by $ev_w(f) = f(w)$, is a bounded linear functional. Thus $H^2(R)$ is a reproducing kernel Hilbert space for R. We use k_w to denote the reproducing kernel function for $w \in R$; that is, $ev_w(f) = \langle f, k_w \rangle$. Using the isomorphism of $H^2(R)$ and $H^2(\partial R)$, we see that $H^2(\partial R)$ is a reproducing kernel Hilbert space also. Let $\hat{k}_\lambda \in H^2(\partial R)$ denote the reproducing kernel function for $H^2(\partial R)$ at $\lambda \in R$; that is, $\langle f, \hat{k}_\lambda \rangle = f(\lambda)$ for $f \in H^2(\partial R)$. (Of course, for the identification of $H^2(R)$ and $H^2(\partial R)$ to be an isomorphism, one must use the same point $t \in R$ in defining both norms. One can also use the function $w \to \langle g, \hat{k}_w \rangle$ for $g \in H^2(\partial R)$ to identify $H^2(R)$ and $H^2(\partial R)$.)

We define an operator T_z on $H^2(R)$ by $(T_z f)(z) = zf(z)$ for every $f \in H^2(R)$ and $z \in R$. Furthermore, the isomorphism of $H^2(R)$ and $H^2(\partial R)$ induces a corresponding operator, also denoted T_z, on $H^2(\partial R)$. And we define N_z on $L^2(\partial R)$ by $(N_z f)(z) = zf(z)$ for $z \in \partial R$. It can be seen that T_z is a pure subnormal operator [**3**] and N_z is the minimal normal extension of T_z on $H^2(\partial R)$. Let T_φ be the operator on $H^2(R)$ defined by $(T_\varphi f)(z) = \varphi(z)f(z)$ for every $f \in H^2(R)$ and $\varphi \in H^\infty(R)$. We use T_φ to denote the corresponding operator on $H^2(\partial R)$.

Similarly, for a Hilbert space $\mathcal{H}$, we can define an $\mathcal{H}$-valued Hardy space, $H^2_{\mathcal{H}}(R)$, which is the space of $\mathcal{H}$-valued analytic functions $f : R \to \mathcal{H}$ such that the subharmonic function $\|f(z)\|^2_{\mathcal{H}}$ is majorized by a harmonic function u on R. We define two corresponding operators, $(T_{\mathcal{H}} f)(z) = zf(z)$ for $f \in H^2_{\mathcal{H}}(R)$ and $z \in R$, and $N_{\mathcal{H}}$ on $L_{\mathcal{H}}(\partial R)$, $(N_{\mathcal{H}} f)(z) = zf(z)$ for $f \in L^2_{\mathcal{H}}(\partial R)$ and $z \in \partial R$. Now $H^2_{\mathcal{H}}(R)$ is a reproducing kernel Hilbert space, and we use $k^{\mathcal{H}}_\lambda \in B(\mathcal{H})$ to represent the reproducing kernel at $\lambda \in R$; that is, $\langle f(\lambda), h \rangle_{\mathcal{H}} = \langle f, k^{\mathcal{H}}_\lambda h \rangle_{H^2_{\mathcal{H}}(R)}$ for $f \in H^2_{\mathcal{H}}(R)$ and $h \in \mathcal{H}$. For more information about function theory on finitely connected domains, one can see [**1, 16, 17**].

Let E be a Hermitian holomorphic vector bundle over R. A section of E is a holomorphic function f from R into E such that $p(f(z)) = z$ for all $z \in R$, where $p : E \to R$ is the projection map [**18**]. The set of all holomorphic sections of E is denoted by $\Gamma_a(E)$ where the subscript "a" represents "analytic". For E a rank n Hermitian vector bundle over R, a unitary coordinate cover for E is a covering $\{U_s, \varphi_s\}$ with $\varphi_s : U_s \times \mathbb{C}^n \to E|_{U_s}$ such that for each s and $z \in U_s$, the fiber map $\varphi^z_s : \mathbb{C}^n \to E_z$, is unitary. The unitary coordinate cover $\{U_s, \varphi_s\}$ is said to be *flat* if the transition functions, $\varphi_{st} = \varphi_s^{-1}\varphi_t$ on $U_s \cap U_t$ for all s and t, are constant. A flat unitary vector bundle is a vector bundle with a flat unitary coordinate covering.

If E is a flat unitary vector bundle over the finitely-connected domain R with fiber $\mathcal{E}$ and coordinate covering $\{U_s, \varphi_s\}$ and f is a holomorphic section of E, then for $z \in U_s \cap U_t$, the operator $(\varphi_t^z)^{-1}\varphi_s^z$ is unitary so that $\|(\varphi_t^z)^{-1}f(z)\| = \|(\varphi_s^z)^{-1}f(z)\|$. This means that there is a function on R defined by $h_f^E(z) = \|(\varphi_s^z)^{-1}f(z)\|_E$, where $z \in U_s$. One defines $H_E^2(R)$ to be the space of holomorphic sections f of E such that $(h_f^E(z))^2$ is majorized by a harmonic function, then $H_E^2(R)$ is a Hilbert space. $H_E^2(R)$ is invariant under multiplication by any bounded analytic function on R. The operator T_E on $H_E^2(R)$, defined by $(T_E f)(z) = zf(z)$ for $z \in R$, is called a *bundle shift* over R. These objects are studied by Abrahamse and the first author [**3**].

Finally, let $\mathcal{J}_E$ be the subalgebra of $B(H_E^2(R))$ of operators T_Φ, where Φ is a bundle map on E which extends to an open set containing the closure of R.

In this paper, first for a hypo-Dirichlet algebra (see section 2.1 for the definition), we prove that an operator algebra on $H^2(m)$ containing $H^\infty(m)$ is strongly dense in $B(H^2(m))$, which implies that any operator algebra containing $\mathcal{Q}(T_z)$ is strongly dense in $B(H^2(R))$, where $\mathcal{Q}$ is the function algebra consisting of all rational analytic functions with poles outside the closure of R. Thus this algebra is catalytic. We next prove a similar result for a logmodular algebra. Second, we show that an analogous catalytic result also holds for the case of bundle shifts, which generalizes the result of Bercovici, Foias, Pearcy and the first author [**7**], where they proved that $\mathcal{Q}(T_z)$ is catalytic by constructing a modulus automorphic factorization of a function in $H^2(R)$. We do this by proving that the algebra $\mathcal{A} \otimes M_n(\mathbb{C})$ is catalytic on $\mathcal{K} \otimes \mathbb{C}^n$ if $\mathcal{A}$ is catalytic on $\mathcal{K}$ for any positive integer n. This fact was observed by Nordgren [**14**].

We conclude this introduction by recalling that this approach doesn't work for the Bergman shift, and hence are limited in scope [**10**].

Acknowledgments. Research of the second author was carried out during a year-long visit to the Department of Mathematics at Texas A&M University, which was supported by the China scholarship council and the Department of Mathematics at Chongqing University of Technology.

We thank Carl M. Pearcy for his comments and corrections.

2. Function Algebras

Let X be a compact Hausdorff space and $C(X)$ be the algebra of all continuous complex-valued functions on X. A subalgebra A of $C(X)$ is called a *uniform algebra* on X if it contains the constant functions and separates the points of X. One can see Gamelin's book [**11**] for details on uniform algebras. The space of real parts of the functions in A is denoted by ReA, the set of invertible elements in A by A^{-1}, and $\log|A^{-1}|$ denotes the set, $\{\log|f|, f \in A^{-1}\}$.

Let φ be a multiplicative linear functional on A. Then there exist positive measures m on X such that

$$\varphi(f) = \int_X f dm, \quad \text{for } f \in A,$$

which are called *representing measures* for φ. Usually, the representing measure is not unique. An *Arens-Singer measure* for φ is a positive measure m on X such that

$$\log|\varphi(f)| = \int_X \log|f|dm, \quad \text{for } f \in A^{-1}.$$

An Arens-Singer measure for φ is also a representing measure for φ. In fact, we have

$$\begin{aligned} \textstyle\int_X f dm &= \int_X (u+iv)dm = \int_X \log e^u dm + i\int_X \log e^v dm \\ &= \log e^{\varphi(u)} + i\log e^{\varphi(v)} = \varphi(u) + i\varphi(v) = \varphi(f) \end{aligned}$$

for $f = u+iv$. An Arens-Singer measure always exists (cf. [**4**, **12**] and the references therein).

2.1. Hypo-Dirichlet Algebras. The uniform algebra A is called a *hypo-Dirichlet algebra* if there exists a finite set of elements $\{f_1, \cdots, f_s\}$ in A^{-1} such that the linear span of ReA and $\log|f_1|, \cdots, \log|f_s|$ is uniformly dense in $ReC(X)$, where s is the minimal number possible. Ahern and Sarason studied hypo-Dirichlet algebras in [**4**].

For a multiplicative functional φ on A, we use $\mathcal{M}(\varphi)$ to denote the set of representing measures for φ, and S to denote the real linear span of all differences between pairs of measures in $\mathcal{M}(\varphi)$. For a hypo-Dirichlet algebra, one can show that $\dim S = s$. Further, there is only one Aren-Singer measure on a hypo-Dirichlet algebra.

For a fixed measure m on X corresponding to the multiplicative linear functional φ on A, let A_0 denote the kernel of the functional φ, $H^2(m)$ denote the closure of A in $L^2(m)$, and let $H^\infty(m)$ be the weak-star closure of A in $L^\infty(m)$. Ahern and Sarason [**4**] showed that the orthogonal complement N of $A + \bar{A}$ in $L^2(m)$ is a subspace of L^∞. Moreover $L^2(m)$ has the following decomposition, $L^2(m) = H^2(m) \oplus \overline{H_0^2(m)} \oplus N$, where $\overline{H_0^2(m)}$ is the complex conjugates of the functions in $H_0^2(m)$, the functions which are annulled by m.

If the difference of two multiplicative functionals φ, ψ on A has norm less than 2, then we say that φ and ψ lie in the same *part* of the maximal ideal space $\mathcal{M}_A$ of A, which defines an equivalence relation on $\mathcal{M}_A$ and partitions $\mathcal{M}_A$ into the disjoint union of parts. In the case of a hypo-Dirichlet algebra, the Arens-Singer measures of two functionals in the same part are mutually, boundedly absolutely continuous [**11**].

An element x of a convex subset K of a vector space V is said to be a *core point* of K if whenever $y \in V$ is such that $x+y \in K$, then $x - \epsilon y \in K$ for $\epsilon > 0$ sufficiently small. A measure m is called a *core measure* for φ if m is a core point of $\mathcal{M}(\varphi)$. In particular, one can show that the harmonic measure on ∂R, corresponding to each $z \in R$, is a core measure for φ.

With these notions defined, we can state a basic lemma on representations of functions in a generalized Hardy space by combining Theorem V.4.2 and Theorem

V.4.3 in Gamelin[**11**]. The interested reader can consult Gamelin [**11**] for further information.

LEMMA 2.1. *Let A be a function algebra on a compact metric space X, and φ be a multiplicative linear functional on A. If the space of representing measures for φ is finite dimensional and if m is a core measure, then every function $f \in H^2(m)$ can be expressed as a quotient of two $H^\infty(m)$ functions.*

2.2. Logmodular Algebras. A uniform algebra A is a *Dirichlet algebra* on X if ReA is uniformly dense in $ReC(X)$; A is a *logmodular algebra* on X if $\log|A^{-1}|$ is uniformly dense in $ReC(X)$. A Dirichlet algebra is a logmodular algebra since $ReA \subseteq \log|A^{-1}|$. One can consult Hoffman[**12**] for further results about logmodular algebras.

For every multiplicative linear functional φ on a logmodular algebra A, there is a unique representing measure m on X. Let A_0 denote the kernel of the functional φ also, $H^2(m)$ denote the closure of A in $L^2(m)$, and $H^\infty(m)$ be the weak-star closure of A in $L^\infty(m)$. A function g in $H^2(m)$ is called an *outer function* if Ag is dense in $H^2(m)$. Equivalently, g is outer if $\int \log|g|dm = \log|\int gdm| > -\infty$.

Now we have the following lemma whose proof is similar to that of the classic Hardy space $H^2(\mathbb{D})$ on $\mathbb{D}$ [**13**], and Theorem V.4.3. in Gamelin [**11**].

LEMMA 2.2. *Let A be a logmodular algebra on X. For every multiplicative linear functional φ on A, let m be the unique representing measure for φ on X. For every function $f \in H^2(m)$, there exist two functions $f_1 \in H^\infty(m)$, $f_2 \in H^\infty(m)^{-1}$ such that $f = \frac{f_1}{f_2}$.*

PROOF. We have the orthogonal decomposition [**12**] of the space $L^2(m) = H^2(m) \oplus \overline{H_0^2(m)}$, where $H_0^2(m)$ is the closure of A_φ in $H^2(m)$. Note that for every function $u \in ReH^2(m)$, there exists a unique function $v \in H_0^2(m)$ such that $u + iv \in H^2(m)$. Therefore, for $f \in H^2(m)$, there exists a function v such that $\log^+|f| + iv \in H^2(m)$ where $\log^+|f| = \max\{\log|f|, 0\}$. If we define $g = e^{-(\log^+|f|+iv)}$, then g is invertible in $H^\infty(m)$ ([**11**]), hence fg is bounded also. So we have the representation lemma for functions in $H^2(m)$ by setting $f_1 = fg$ and $f_2 = g$. □

This Lemma generalizes the corresponding result for the classic Hardy space $H^2(\mathbb{D})$ on the unit disc.

3. Catalytic in Function algebras

Let $\mathcal{H}$ denote an infinite dimensional Hilbert space, and $B(\mathcal{H})$ be the algebra of bounded linear operators on $\mathcal{H}$. Let $\mathcal{A}$ be a subalgebra of $B(\mathcal{H})$, and let n be a positive integer. Then $\mathcal{A}$ is called *n-transitive* if for any choice of elements $x_1, \cdots, x_n, y_1, \cdots, y_n \in \mathcal{H}$ with $\{x_1, \cdots, x_n\}$ linearly independent, there exists a sequence $A_i \in \mathcal{A}$ such that

$$\lim_j A_j x_k = y_k, \quad 1 \le k \le n.$$

Further, a unital subalgebra $\mathcal{A}$ of $B(\mathcal{H})$ is said to be *transitive* if it is 1-transitive or, equivalently, it has only trivial invariant subspaces; that is, only $\{0\}$ and $\mathcal{H}$. It is easy to show that a subalgebra $\mathcal{A}$ is strongly dense in $B(\mathcal{H})$ if and only if $\mathcal{A}$ is n-transitive for every $n \ge 1$. This equivalence is due essentially to Arveson [**6**].

As mentioned in the introduction, Arveson introduced an inductive strategy to show that transitive algebras are strongly dense. The first step in this approach is the content of the following lemma.

LEMMA 3.1. *(Arveson's Corollary 2.5.*[**6**]*) Let $\mathcal{A}$ be a transitive subalgebra of $B(\mathcal{H})$, where $\mathcal{H}$ is a Hilbert space. Then $\mathcal{A}$ is not 2-transitive, if and only if, there exists a closed, densely defined, non-scalar linear transformation on $\mathcal{H}$ that commutes with $\mathcal{A}$.*

The further induction steps are all the same and are subsumed in the following lemma.

LEMMA 3.2. *(Arveson's Corollary 2.5.*[**6**]*) Let A be a subalgebra of $B(\mathcal{H})$ for a Hilbert space $\mathcal{H}$. For $N \geq 2$, suppose that $\mathcal{A}$ is N-transitive but not $(N+1)$-transitive. Then there exists N linear transformations $T_1, \cdots, T_N$, defined on a common linear domain $\mathcal{D}$ dense in $\mathcal{H}$, such that*

(i) each T_i commutes with $\mathcal{A}$;

(ii) no T_i is closable; (that is, the closure of the graph of T_i is not the graph of a transformation); and

(iii) $\{(x, T_1x, \cdots, T_Nx) | \, x \in \mathcal{D}\}$ is a proper closed subspace of $\mathcal{H}^{(N+1)}$.

For a general hypo-Dirichlet algebra A, a part of M_A is defined as a analytic part if it contains more than one point. Firstly we use Lemma 2.1 to prove the following proposition.

PROPOSITION 3.3. *Let A be a general hypo-Dirichlet algebra on X, φ be a multiplicative linear functional on A, and assume that the representing measure m corresponding to φ is a core measure, and $H^2(m)$ is a reproducing kernel Hilbert space. If $\mathcal{A}$ is an operator algebra containing*

$$\{T_h \in B(H^2(m)) | \; h \in H^\infty(m)\},$$

T is a map from a dense $\mathcal{A}$-invariant submanifold $\mathcal{D}$ of $H^2(m)$ to $H^2(m)$, and suppose $TBf = BTf, f \in \mathcal{D}$, $B \in \mathcal{A}$. Then there exists a function of the form $h = \frac{f_1}{f_2}$, with $f_1, f_2 \in H^\infty(m)$, such that $Tf = hf$ for all $f \in \mathcal{D}$. Moreover, T is linear and closable.

PROOF. Firstly, we prove that $\frac{Tf}{f}$ is equal to $\frac{Tg}{g}$ for every pair $f, g \in \mathcal{D}$, $f, g \neq 0$, the zero function. By the statement of Lemma 2.1, there exist functions $f_1, f_2, g_1, g_2 \in H^\infty(m)$, f_2 and g_2 invertible in $H^\infty(m)$ such that $f = \frac{f_1}{f_2}$ and $g = \frac{g_1}{g_2}$. We want to prove $\frac{Tf}{f} = \frac{Tg}{g}$, or $g_1 f_2 Tf = g_2 f_1 Tg$. But from the commutativity of T with $T_{g_1 f_2}$ and $T_{g_2 f_1}$, it follows that $g_1 f_2 Tf = T_{g_1 f_2} Tf = Tg_1 f_2 f = Tg_2 f_1 g = T_{g_2 f_1} Tg = g_2 f_1 Tg$, so we have $\frac{Tf}{f} = \frac{Tg}{g}$ by dividing the last equation by $f_1 g_2$. We have $Tf = hf$ by letting $h = \frac{Tf}{f}$. h is the quotient of two functions in $H^2(m)$, then since every function in $H^2(m)$ can be expressed as a quotient of two functions in $H^\infty(m)$. Hence h can be written as a quotient $\frac{h_1}{h_2}$ of two functions $h_1, h_2 \in H^\infty(m)$.

Secondly, we prove that the operator defined by $Tf = hf$ on $\mathcal{D} = \{f \in H^2(m) | \;\; hf \in H^2(m)\}$ is closable. It is clear that the domain $\mathcal{D}$ is invariant under linear operations in $H^2(m)$. Further for any sequence $\{f_n\} \subset \mathcal{D}$ such that $\lim_{n\to\infty} f_n = f$ and $\lim_{n\to\infty} hf_n = g$, $f, g \in H^2(m)$, we see that $g = hf$ as follows. We assume that the reproducing kernel for λ in analytic parts of M_A is denoted by k_λ^m,

$\lim_{n\to\infty} f_n(\lambda) = \lim_{n\to\infty}\langle f_n, k_\lambda^m\rangle = \langle f, k_\lambda^m\rangle = f(\lambda)$, and $\lim_{n\to\infty} f_n h(\lambda) = \lim_{n\to\infty}\langle hf_n, k_\lambda^m\rangle = \langle g, k_\lambda^m\rangle = g(\lambda)$. So $g(\lambda) = h(\lambda)f(\lambda)$. This means that $hf \in H^2(m)$ and $g = hf$; that is, T is closable. □

LEMMA 3.4. *Let T be the operator defined in Proposition 3.3, $k_\lambda^m(z)$ be the kernel function at the point λ in analytic parts of M_A. Then k_λ^m is in the domain of the operator T^*.*

PROOF. For any $f \in \mathcal{D}$,

$$\langle Tf, k_\lambda\rangle = \langle hf, k_\lambda^m\rangle = h(\lambda)f(\lambda) = h(\lambda)\langle f, k_\lambda^m\rangle = \langle f, \overline{h(\lambda)}\rangle = \langle f, T^*k_\lambda^m\rangle,$$

it is continuous in f. This proves that k_λ^m is in the domain of T^* and $T^*k_\lambda^m = \overline{h(\lambda)}k_\lambda^m$. □

THEOREM 3.5. *Let A be a general hypo-Dirichlet algebra, and φ be a multiplicative linear functional. If the representing measure m corresponding to φ is a core measure, and $H^2(m)$ is a reproducing kernel Hilbert space. Then $H^\infty(m)$ is catalytic on $H^2(m)$.*

PROOF. Suppose $\mathcal{A}$ is a transitive subalgebra of $B(H^2(m))$ containing $H^\infty(m)$ and T is a closed, densely defined operator commuting with $\mathcal{A}$. Then T^* commutes with $\mathcal{A}^*$. For λ in analytic parts of M_A, by Lemma 3.4, k_λ^m is in the domain of T^* and $T^*k_\lambda^m = \overline{h(\lambda)}k_\lambda^m$, so that for every $B \in \mathcal{A}$,

$$\overline{h(\lambda)}B^*k_\lambda^m = B^*\overline{h(\lambda)}k_\lambda^m = B^*T^*k_\lambda^m = T^*B^*k_\lambda^m.$$

This shows that the nonempty submanifold $\{f \in \mathcal{D}(T^*)|\, T^*f = \overline{h(\lambda)}f\}$ is invariant under $\mathcal{A}^*$. Since $\mathcal{A}^*$ is transitive, this manifold is dense. Thus $T^* = \overline{h(\lambda)}I$, and so T is a scalar. This shows that $\mathcal{A}$ is 2-transitive.

Every densely defined linear transformation that commutes with $\mathcal{A}$ is closable by Proposition 3.3, hence by Arveson's Lemma 3.2, $\mathcal{A}$ is N-transitive for every $N > 2$, which shows that $\mathcal{A}$ is strongly dense in $B(H^2(m))$. Therefore $H^\infty(m)$ is catalytic. □

For logmodular algebras, we can use Lemma 2.2 to prove in a similar way the following theorem by proving analogues of Proposition 3.3, Lemma 3.4 and finally Theorem 3.5.

THEOREM 3.6. *Let A be a general logmodular algebra, φ be a multiplicative linear functional, m be the representing measure corresponding to φ. If $H^2(m)$ is a reproducing kernel Hilbert space, then $H^\infty(m)$ is catalytic on $H^2(m)$.*

Note that the domain for $H^2(m)$ can be identified with a subset of the maximal ideal space for $H^\infty(m)$.

4. Finitely-connected domains

We now return to the case of the finitely-connected domain $R \subset \mathbb{C}$ whose boundary ∂R consists of $(n+1)$ nonintersecting smooth Jordan curves. Here we can be more precise. In particular, $H^\infty(R)$ can be shown to be a hypo-Dirichlet algebra on ∂R with $s = n$. For any point $z \in R$, the harmonic measure $m(= m_z)$, supported on ∂R, corresponding to evaluation at z is the unique Arens-Singer measure for the functional φ on $H^\infty(R)$ of evaluation at z. For any two points $z_1, z_2 \in R$, the corresponding functionals lie in the same part, and they are mutually boundedly absolutely continuous. Moreover, in this case, the measures in S can be described explicitly. In fact, a basis for the homology of R can be used to give a linear basis for S [**4**].

PROPOSITION 4.1. *Let R be a finitely-connected domain whose boundary, ∂R, consists of $n+1$ nonintersecting smooth Jordan curves. If $f \in H^2(R)$, then $f = \frac{g}{h}$, where g and h belong to $H^\infty(R)$, and h is invertible in $H^\infty(R)$.*

PROOF. Let A be the algebra of all function on ∂R that can be uniformly approximated by rational functions with poles off R. It is well known that A is a hypo-Dirichlet algebra [**4**]. For a point $z \in R$, the corresponding evaluation functional φ is a multiplicative linear functional. The space of representing measures M_φ has dimension n. So there exists a core measure denoted by m corresponding to φ [**11**, p106]. In this case, $H^\infty(R) = H^\infty(m)$, $H^2(m)$ and $H^2(R)$ are boundedly equivalent. Then the proposition follows from Proposition 3.3. □

For finitely-connected domains and the algebra A in Proposition 4.1, the analytic part of M_A is the open set R. So we have the following proposition which can be proved just as Proposition 3.3.

PROPOSITION 4.2. *For the finitely-connected domain R whose boundary ∂R consists of $(n+1)$ nonintersecting smooth Jordan curves, let $\mathcal{A}$ be an operator algebra containing*

$$\{T_h \in B(H^2(R))|\ h \in H^\infty(R)\}.$$

If T is a map of a dense $\mathcal{A}$-invariant submanifold $\mathcal{D}$ of $H^2(R)$ to $H^2(R)$, such that $TAf = ATf$, for $f \in \mathcal{D}$, $A \in \mathcal{A}$. Then there exists a meromorphic function of the form $h = \frac{f_1}{f_2}$, with $f_1, f_2 \in H^\infty(R)$, such that $Tf = hf$ for all $f \in \mathcal{D}$. Moreover, T is linear and closable.

The following lemma follows in the same way as Lemma 3.4.

LEMMA 4.3. *Let T be the closed operator defined in Proposition 4.2, and $k_\lambda(z)$ be the kernel function at the point $\lambda \in R$. Then k_λ is in the domain of the operator T^*.*

Now we are ready to obtain the catalytic result for finitely-connected domains by combining the method of proof in Theorem 3.5 and Proposition 4.1, Proposition 4.2, and Lemma 4.3.

THEOREM 4.4. *Let R be a finitely-connected domain whose boundary ∂R consists of $n+1$ nonintersecting smooth Jordan curves. Then the algebra $\mathcal{Q}(T_z)$ on $H^2(\partial R)$ is catalytic, where $\mathcal{Q}$ is the space of all rational analytic functions with poles outside the closure of R.*

In [**3**] Abrahamse and the first author proved that if E is a flat unitary vector bundle over a finitely-connected domain R, whose boundary ∂R consists of $n+1$ nonintersecting smooth Jordan curves, with fiber $\mathcal{H}$, then the bundle shift T_E is similar to $T_\mathcal{H}$. Thus to show that T_E is catalytic, it is sufficient to prove the vector-valued case or that $T_\mathcal{H}$ is catalytic.

Let $\mathcal{J}_E(R)$ be the subalgebra of $B(H^2_E(R))$ of operators T_Φ, where Φ is a bundle map on E which extends to an open set containing the closure of R. Now, in particular, we will prove $\mathcal{J}_E(R)$ is catalytic by first showing the following general result.

PROPOSITION 4.5. *Let $\mathcal{K}$ be a Hilbert space. If $\mathcal{A}$ is a catalytic algebra on $\mathcal{K}$, then $\mathcal{A} \otimes M_n(\mathbb{C})$ is a catalytic algebra on $\mathcal{K} \otimes \mathbb{C}^n$ for any positive integer n.*

PROOF. Suppose that the transitive algebra $\mathcal{T}$ on $\mathcal{K} \otimes \mathbb{C}^n$ contains $\mathcal{A} \otimes M_n(\mathbb{C})$. We use $\{e_k\}$ to denote the orthonormal basis of $\mathbb{C}^n$. Let E_{ij} be the matrix unit operator $e_i \otimes e_j$ on $\mathbb{C}^n$, and $\widehat{E}_{ij} = I_{\mathcal{K}} \otimes E_{ij}$.

Let T be a densely defined linear transformation on $\mathcal{D} \subset \mathcal{K} \otimes \mathbb{C}^n$ which commutes with $\mathcal{T}$. Note that $\mathcal{A} \otimes M_n(\mathbb{C})$ contains $\{\widehat{E}_{ij}\}$, then the operator $\widehat{E}_{jj} T \widehat{E}_{ii}$ can be seen as a densely defined linear transform from $\mathcal{K} \otimes \mathbb{C}e_i$ to $\mathcal{K} \otimes \mathbb{C}e_j$ which are both isomorphic to $\mathcal{K}$. Then it is easy to see that $\widehat{E}_{jj} T \widehat{E}_{ii}$ commutes with $\mathcal{A}$. Further, $\widehat{E}_{jj} \mathcal{T} \widehat{E}_{ii}$ is transitive on $\mathcal{K}$ since $\mathcal{T}$ is transitive and contains $\mathcal{A}$. Thus $\widehat{E}_{jj} \mathcal{T} \widehat{E}_{ii}$ is strong dense in $B(\mathcal{K})$ since $\mathcal{A}$ is catalytic, which shows that $\widehat{E}_{jj} T \widehat{E}_{ii}$ is a constant $a_{ij} \in \mathbb{C}$. So T has a form $I_{\mathcal{K}} \otimes A$, where $A = (a_{ij})$ is an operator on $\mathbb{C}^n$, it shows that T is closable. By Arveson's Lemma 3.2, $\mathcal{T}$ is n-transitive for $n \geq 2$ if $\mathcal{T}$ is 2-transitive.

Moreover, since $T = I_{\mathcal{K}} \otimes A$ commutes with $\mathcal{A} \otimes M_n(\mathbb{C})$, it means $A = \lambda I_{\mathbb{C}^n}$ because A commutes $M_n(\mathbb{C})$ for some $\lambda \in \mathbb{C}$. So, $T = \lambda I_{\mathcal{K} \otimes \mathbb{C}^n}$, that is, $\mathcal{T}$ is 2-transitive by Arveson's Lemma 3.1. Then $\mathcal{T}$ is strong dense in $B(\mathcal{K} \otimes M_n(\mathbb{C}))$, that is $\mathcal{A} \otimes M_n(\mathbb{C})$ is catalytic. □

Note that we can use the same proof to show that if $\mathcal{A}$ is a catalytic algebra on a Hilbert space $\mathcal{K}$, then $\mathcal{A} \otimes B(\mathcal{K}_1)$ is also a catalytic algebra for any separable Hilbert space $\mathcal{K}_1$. Actually, the proof shows that $A \otimes K(\mathcal{K}_1)$ is catalytic, where $K(\mathcal{K}_1)$ denotes the algebra of compact operators. A more interesting question is whether $A \otimes I_{\mathcal{K}_1}$ is catalytic.

PROPOSITION 4.6. *1) Suppose that $\mathcal{A}$ is a catalytic algebra acting on $\mathcal{K}$. Then every larger algebra acting on $\mathcal{K}$ is also catalytic.*

2) Suppose that $\mathcal{A}$ is a catalytic algebra containing identity on $\mathcal{K}$. Then the direct sum $\mathcal{A}^{(n)}$ of $\mathcal{A}$ with itself on $\mathcal{K}^{(n)}$ is also catalytic on $\mathcal{K}^{(n)}$.

PROOF. 1) is obvious. For 2), let $\mathcal{B}$ be a transitive algebra on $\mathcal{K}^{(n)}$ containing $\mathcal{A}^{(n)}$ and let $\widehat{E}_{11}$ be the projection in $\mathcal{A}^{(n)}$ onto the first copy $\mathcal{A}$. Then $\widehat{E}_{11} \mathcal{A}^{(n)} \widehat{E}_{11}$ is a subalgebra of $\mathcal{B}$ isomorphic to $\mathcal{A}$ acting on $\mathcal{K}$, and $\widehat{E}_{11} \mathcal{B} \widehat{E}_{11}$ is a transitive algebra on $\mathcal{K}$ containing $\widehat{E}_{11} \mathcal{A}^{(n)} \widehat{E}_{11}$. Thus, by definition, $\widehat{E}_{11} \mathcal{B} \widehat{E}_{11}$ is strongly dense in $B(\mathcal{K})$. By [**9**], Theorem 1.3, $\mathcal{B}$ is strongly dense in $B(\mathcal{K}^{(n)})$, which proves that $\mathcal{A}^{(n)}$ is catalytic on $\mathcal{K}^{(n)}$. □

COROLLARY 4.7. *Let $\mathcal{A}$ be a catalytic algebra acting on $\mathcal{K}$, and let $M_n(\mathcal{A})$ be the algebra of all $n \times n$ matrices with entries from $\mathcal{A}$. Then $M_n(\mathcal{A})$ is catalytic on $\mathcal{K}^{(n)}$.*

PROOF. $M_n(\mathcal{A})$ contains the catalytic algebra $\mathcal{A}^{(n)}$. Then the corollary follows from 1) of Proposition 4.6. □

Proposition 4.6 and Corollary 4.7 are due to Carl M. Pearcy.

We have the following theorem by combining Theorem 4.4 and Proposition 4.5.

THEOREM 4.8. *For the flat unitary bundle E over R, let $\mathcal{A}$ be a transitive subalgebra of $B(H^2_E(R))$ containing the algebra $\mathcal{J}_E(R)$. Then $\mathcal{A}$ is strongly dense in $B(H^2_E(R))$ or is a catalytic algebra.*

PROOF. Only one thing needs to be observed after noting that the bundles $R \times \mathbb{C}^n$ and E extend to a trivial and a flat unitary bundle over the closure of R and these extensions are similar [**3**]. If Φ is a bundle map from $clos(R) \times \mathbb{C}^n$ to the extension of E, establishing the similarity of $T_{\mathcal{H}}$ and T_E, then Φ induces a module isomorphism between $H^2_{\mathbb{C}^n}(R)$ and $H^2_E(R)$ conjugating $\mathcal{J}_E(R)$ and $\mathcal{J}_{\mathcal{R}\times\mathbb{C}^n}(R)$. Thus the similarity not only takes $T_{\mathcal{H}}$ to T_E, but also $\mathcal{J}_E(R) \otimes M_n(\mathbb{C})$ to $\mathcal{J}_E(R)$, which completes the proof. □

Note that for a separable Hilbert space $\mathcal{H}$, the proof goes through establishing that $\mathcal{J}(T_E)$ and $M_0(E)$ are catalytic, where $\Phi \in M_0(E)$ if $\Phi(z) \in K(E_z)$ for $z \in R$.

References

[1] M. B. Abrahamse, *Toeplitz operators in multiply connected regions*, Amer. J. Math. **96** (1974), 261–297. MR0361891 (50 #14333)

[2] M. B. Abrahamse and J. J. Bastian, *Bundle shifts and Ahlfors functions*, Proc. Amer. Math. Soc. **72** (1978), no. 1, 95–96, DOI 10.2307/2042541. MR503539 (80c:47029)

[3] M. B. Abrahamse and R. G. Douglas, *A class of subnormal operators related to multiply-connected domains*, Advances in Math. **19** (1976), no. 1, 106–148. MR0397468 (53 #1327)

[4] P. R. Ahern and Donald Sarason, *The H^p spaces of a class of function algebras*, Acta Math. **117** (1967), 123–163. MR0217600 (36 #689)

[5] Lars V. Ahlfors, *Bounded analytic functions*, Duke Math. J. **14** (1947), 1–11. MR0021108 (9,24a)

[6] William B. Arveson, *A density theorem for operator algebras*, Duke Math. J. **34** (1967), 635–647. MR0221293 (36 #4345)

[7] H. Bercovici, R. G. Douglas, C. Foias, and C. Pearcy, *Confluent operator algebras and the closability property*, J. Funct. Anal. **258** (2010), no. 12, 4122–4153, DOI 10.1016/j.jfa.2010.03.009. MR2609540 (2012b:47015)

[8] Guozheng Cheng, Kunyu Guo, and Kai Wang, *Transitive algebras and reductive algebras on reproducing analytic Hilbert spaces*, J. Funct. Anal. **258** (2010), no. 12, 4229–4250, DOI 10.1016/j.jfa.2010.01.021. MR2609544 (2011c:47169)

[9] Ronald G. Douglas and Carl Pearcy, *Hyperinvariant subspaces and transitive algebras*, Michigan Math. J. **258** (1972), 1–12. MR0295118 (45 #4186)

[10] Don Hadwin, Zhe Liu, and Eric Nordgren, *Closed densely defined operators commuting with multiplications in a multiplier pair*, Proc. Amer. Math. Soc. **141** (2013), no. 9, 3093–3105, DOI 10.1090/S0002-9939-2013-11753-3. MR3068963

[11] Theodore W. Gamelin, *Uniform algebras*, Prentice-Hall, Inc., Englewood Cliffs, N. J., 1969. MR0410387 (53 #14137)

[12] Kenneth Hoffman, *Analytic functions and logmodular Banach algebras*, Acta Math. **108** (1962), 271–317. MR0149330 (26 #6820)

[13] Kenneth Hoffman, *Banach spaces of analytic functions*, Dover Publications, Inc., New York, 1988. Reprint of the 1962 original. MR1102893 (92d:46066)

[14] Eric A. Nordgren, *Transitive operator algebras*, J. Math. Anal. Appl. **32** (1970), 639–643. MR0271746 (42 #6627)

[15] Stefan Richter, *Invariant subspaces of the Dirichlet shift*, J. Reine Angew. Math. **386** (1988), 205–220, DOI 10.1515/crll.1988.386.205. MR936999 (89e:47048)

[16] Michael Voichick, *Ideals and invariant subspaces of analytic functions*, Trans. Amer. Math. Soc. **111** (1964), 493–512. MR0160920 (28 #4129)

[17] Michael Voichick and Lawrence Zalcman, *Inner and outer functions on Riemann surfaces*, Proc. Amer. Math. Soc. **16** (1965), 1200–1204. MR0183883 (32 #1359)

[18] R. O. Wells, *Differential Analysis on Complex Manifolds*, Springer-Verlag, 1979.

Department of Mathematics, Texas A&M University, College Station, Texas 77843-3368

E-mail address: rgd@tamu.edu

School of Mathematical Sciences and Statistics, Chongqing University of Technology, Chongqing, China 400075

E-mail address: xuaj@cqut.edu.cn

Current address: Department of Mathematics, Texas A&M University, College Station, Texas 77843-3368

Contemporary Mathematics
Volume **638**, 2015
http://dx.doi.org/10.1090/conm/638/12816

Weak H^1, the real and complex case

Valentin V. Andreev and Joseph A. Cima

ABSTRACT. In this paper we present some basic (known) results on weak L^p spaces as well as some new results in the analytic case. The first section is devoted to giving some general background on the real harmonic analysis case. The second introduces the weak Hardy p spaces in R^n, especially with $p = 1$. We recall a decomposition theorem for distributions in such spaces. The third section develops more properties of the case of analytic functions on the unit disc, using for the most part the result of Aleksandrov (1981) on A-integrals for an interesting and natural subspace on the weak H^1 on the disc. The sections are essentially independent.

0. INTRODUCTION.

Several years ago Lorentz [**14, 15**] introduced a set of functions depending on two real parameters p and q $(0 < p < \infty,\ 0 < q < \infty)$ as well as $q = \infty$ for an important special case. His work was amplified and developed by several other authors [**6, 9, 12**]. With the development of the real Hardy space theory [**9**], the theory was introduced into that area and important developments were made in the area of harmonic analysis. Finally, the Soviet school [**1, 18–20**] made particular studies of the $q = \infty$ case for the Hardy space $H^p(\mathbb{T})$ on the unit disk. In our terminology the first two topics will be referred to as the real case and the study for the disk will be referred to as the complex or classical case.

Let us outline our approach: there will be three separate sections. The first will define a topology on the Lorentz space and characterize some of the topological duals. The second section will be concentrated on the introduction of the weak Hardy spaces for $\mathbb{R}^n$. In particular workers here [**9**] developed a form of atomic decomposition theory for functions in this weak Hardy space. From these results interesting distributional type inequalities follow. They also introduce a subspace of this weak H^1 and characterize the topological dual. In the third and final section we develop the theory of weak H^1 on the disk. Some of this development depends heavily on the A-integral results by Aleksandrov [**1, 2**]. In addition to the weak $H^1(\mathbb{T})$ we study a subspace (analogous to the one in the real weak $H^1(\mathbb{R}^n)$ space). We show this space is separable and introduce families of examples for inclusion (or exclusion) in the space.

In regards to the order of presentation of the sections let us say that (in our careers) we were first introduced to the classical $(H^p(\mathbb{T}))$ case first. After the seminal work of Fefferman and Stein [**8**] we tried to make comparisons or analogues of the two theories. Except for the VMO - BMO duality we found few results to compare. In our development beginning with the real theory and finally passing to the classical theory has proved helpful. In particular in the studies of the two subspaces (the real and then the complex one) the real case motivates ideas as to how to make progress in the classical case.

We do not give any substantial parts of proofs in the first two sections and refer the reader to the original papers. We have included a few calculations to illustrate or clarify a point in question. In short these sections (1 and 2) are a résumé of some of the strong results represented by the prominent authors listed in our references. We are sure there are other workers who have made significant contributions but either we are unaware of the works or found it not essential to our goals. There is an area, which has not impinged on our interests, where this machinery is a powerful tool in the theory of interpolating linear and sublinear operators on spaces. We omitted all this material on interpolation. Let us note the third section can be read independently of the first two sections.

1. The topology of the Lorentz $L(p,q)$ spaces

The spaces under consideration in this section were introduced by G. G. Lorentz [**14, 15**]. We have not gone back to these original papers but begin our study with a paper of R. A. Hunt [**12**].

The setup begins with a measure space $M = (X, \mathfrak{M}, m)$, where m is a positive measure. In section 2 and 3 X will be either $\mathbb{R}^n$ with the sigma finite measure dx or the unit circle $\mathbb{T}$ with measure $\frac{d\theta}{2\pi}$ (where $\zeta = e^{i\theta} \in \mathbb{T}$). It is easier to work in these spaces if m is non-atomic, as it is in all the cases we consider. We need the following well known definitions (see Folland [**10**]) of distribution functions. For $f \in M$ and $y > 0$ set

$$\lambda_f(y) = \lambda(y) = m(\{x \in X : \ |f(x)| > y\})$$

and for $t > 0$

$$f^*(t) = \inf\{y : \ \lambda(y) < t\}.$$

The functions λ and f^* are decreasing, non-negative and right continuous. Moreover the functions f and f^* have the same distribution functions. Lorentz used this latter fact in introducing spaces of functions $L(p,q)$ depending on $0 < p < \infty$ and $0 < q < \infty$. In addition the case $0 < p < \infty$ and $q = \infty$ is essential to the development. As motivation with p and q positive and finite and $f \in M$ one has

$$\begin{aligned} \|f\|_p &= \Big(\int_M |f(x)|^p dm(x)\Big)^{1/p} = \Big(\int_0^\infty [f^*(t)]^p \, dt\Big)^{1/p} \\ &= \Big(\frac{p}{p}\int_0^\infty \big[t^{1/p} f^*(t)\big]^p \, \frac{dt}{t}\Big)^{1/p}, \end{aligned}$$

Lorentz suggested replacing some of the p's by q's to arrive at a quantity

$$\|f\|^*_{p,q} = \Big(\frac{q}{p}\int_0^\infty \big[t^{1/p} f^*(t)\big]^q \, \frac{dt}{t}\Big)^{1/q}$$

so that when $p = q$ we arrive at the usual p norm. For these $f \in M$ for which $\|f\|^*_{p,q} < \infty$ we say that $f \in L(p,q)$. Of course if $q = \infty$ the above is not useful so the correct analogue for $q = \infty$ (with $0 < p < \infty$) is

$$\|f\|^*_{p,\infty} = \sup_{t>0}[t^{1/p} f^*(t)] = \sup_{y>0}[y(\lambda(y))^{1/p}].$$

If this is finite we say f is in weak L^p and use the notation $\|f\|^*_{p,\infty} = [f]_p$.

Of course there are many different cases to study here and we follow the presentation in Hunt [**12**]. First it is straight forward to check that $\|f\|^*_{p,q}$ is increasing as q decreases. Further, for $0 < q_1 \le p \le q_2 \le \infty$ the inclusions

$$L(p, q_1) \subset L^p \subset L(p, q_2) \subset L(p, \infty).$$

Hunt shows that $\| \cdot \|^*_{p,q}$ can be used to define a metric on M that makes it a topological vector space. In general it is not locally convex. It is a complete metric space. Other interesting properties are as follows.

(1) Simple functions are dense in $L(p, q)$ for $0 < p < \infty$ and $0 < q < \infty$.

(2) the (topological) dual

$$(L(p, 1))^* \simeq L(p', \infty)$$

for $1 < p$, where p and p' are conjugate Hölder indices ($\frac{1}{p} + \frac{1}{p'} = 1$).

(3) For $1 < p < \infty$ and $1 < q < \infty$,

$$(L(p, q))^* \simeq L(p', q')$$

where again the pairs (p, p') and (q, q') are conjugate Hölder indices.

(4) For $0 < p < 1$ and q finite

$$(L(p, q))^* = (0).$$

That is the zero linear functional is the only continuous linear functional on $L(p, q)$ for this range of p and q.

(5) If as mentioned above m is non-atomic

$$(L(1, q))^* \simeq (0)$$

for $1 < q < \infty$.

As a caveat let us point out two facts to show that one must be careful in considering containments between the $L(p, q)$ spaces. They are related only in special cases. For example if $m(X) < a$ then we conclude

$$L(p_2, q_2) \subset L(p_2, \infty) \subset L(p_1, q_1)$$

for $p_1 \le p_2$ and $q_1 \le q_2$, but not necessarily in general. Moreover the inclusions

$$L(p_1, q_1) \subset L(p_1, \infty) \subset L(p_2, q_2)$$

for $p_1 \le p_2$ only hold if m satisfies the stringent requirements that for every $E \in \mathfrak{M}$ with $m(E) > 0$ one must have $m(E) \ge 1$.

The inequality $L(p, q) \subseteq L(p, \infty)$ with $p = 1$, $q > 1$ is useful in showing the pathological nature of the topology on $L(1, \infty)$ in the following sense. A continuous linear functional Γ on $L(1, \infty)$ can be restricted to be a continuous linear functional when restricted to $L(1, q)$ (with q fixed in $(1, \infty)$). Characteristic functions of measurable sets are in $L(1, q)$ and since Γ restricted to $L(1, q)$ is continuous, and $(L(1, q))^* = (0)$, Γ must vanish on characteristic functions. This of course implies the characteristic functions are not dense in $L(1, \infty)$.

We return now to a paper of Cwikel and Sagher [**6**]. In this work they show the important fact that $L(1, \infty)$ contains non-zero bounded linear functionals (despite as noted above that they vanish on characteristic functions). The proof is constructive. This of course motivates the question here and in the following as to what is a characterization of $(L(1, \infty)^*)$. A partial answer to this question is given to an analogous interesting subspace of weak H^1 (Hardy) space.

Another reference to the Lorentz $L(p.q)$ spaces can be found in the book of Grafacos [**11**].

2. The real weak $H^1(\mathbb{R}^n)$space

In this section we narrow the focus of study to the real Hardy space $H^1(\mathbb{R}^n)$. The material here is coming from the paper of R. Fefferman and F. Soria [**9**]. We first recall the definition of $H^1(\mathbb{R}^n)$. Assume ϕ is a Schwartz function on $\mathbb{R}^n$ with $\int \phi \neq 0$. For $t > 0$ we set $\phi_t(x) = \frac{1}{t^n}\phi(\frac{x}{t})$. Define a maximal function for a tempered distribution f by setting

$$f^*(x) = \sup_{t>0} |f * \phi_t(x)|,$$

where $f * \phi_t$ is the usual distribution convolution.

DEFINITION 2.1. The distribution f is in the Hardy space $H^1(\mathbb{R}^n)$ if the function f^* is in $L^1(\mathbb{R}^n)$.

Note the notation (f^*) has been used earlier in Section 1 as the non-increasing rearrangement of f but in this section it appears only in the context used above.

DEFINITION 2.2. The tempered distribution f is in the weak Hardy space (denoted as $LH(1,\infty)$) if f^* belongs to the weak $L^1(\mathbb{R}^n)$ that is if

$$\sup_{y>0} y\, m(\{x : |f^*(x)| > y\}) \equiv [f]$$

is a real number.

It is known (and easy to check) that $L^1(\mathbb{R}^n) \subseteq LH(1,\infty)$. However, there are tempered distributions f in $LH(1,\infty)$ which are not in L^1. As an example consider $f(x) = \frac{1}{x}$ on $\mathbb{R}\backslash(0)$. Writing

$$f * \phi_t(1) = \lim_{\epsilon\to 0} \Big[\int_{\epsilon\le|1-y|\le 1} \frac{\phi_t(y)}{1-y}\, dy + \int_{1\le|1-y|} \frac{\phi_t(y)}{1-y}\, dy\Big].$$

We note that the second integral is bounded by $\int |\phi_t(y)|\, dy < \infty$. For the first we write

$$\begin{aligned} \Big| \int_{\epsilon\le|1-y|\le 1} \frac{\phi_t(y)}{1-y}\, dy \Big| &= \Big| \int_{\epsilon\le|1-y|\le 1} \frac{\phi_t(y)}{1-y}\, dy - \phi_t(1) \int_{\epsilon\le|1-y|\le 1} \frac{1}{1-y}\, dy \Big| \\ &= \Big| \int_{\epsilon\le|1-y|\le 1} \frac{\phi_t(y)-\phi_t(1)}{1-y}\, dy \Big|. \end{aligned}$$

Since ϕ has compact support for small $0 < t$ we have $\phi_t' = 0$. For large $t > 0$ it is easy to check that

$$|(\phi_t(y) - \phi_t(1))| = |\phi_t'(y^*)| \cdot |1-y|$$

is bounded. Hence

$$\int_{\epsilon\le|1-y|\le 1} \frac{c}{|1-y|}\, \|\phi'\|_\infty |1-y|\, dy = c\|\phi'\|_\infty.$$

This establishes that $f^*(1)$ is a finite number. By scaling for $x > 0$ one has

$$f * \phi_t(x) = \int \frac{1}{x-y}\,\phi_t(y)\,dy = \frac{1}{t}\frac{1}{x}\int \frac{1}{1-y/x}\,\phi(\frac{y}{t})\,dy$$
$$= \frac{1}{t}\frac{1}{x}\int \frac{1}{1-u}\,\phi(\frac{u}{(t/x)})\,du$$
$$= \frac{1}{x}\frac{1}{t/x}\int \frac{1}{1-u}\,\phi(\frac{u}{(t/x)})\,du.$$

Taking sup yields

$$f * \phi_t(x) = \frac{|f^*(1)|}{x}.$$

Treating the case $x < 0$ will finally yield

$$f^*(x) = f * \phi_t(x) = \frac{|f^*(1)|}{|x|}.$$

Hence, f^* satisfies the weak L^1 condition implying that $f(x) = \frac{1}{x} \in LH(1,\infty)$, with $[f] \leq |f^*(1)|$.

Another class of examples in $LH(1,\infty)$ which need not be in $H^1(\mathbb{R}^n)$ is given as follows.

Assume f is a nonnegative tempered distribution (i. e., $f(\phi) \geq 0$ for all nonnegative ϕ in the Schwartz class) in $LH(1,\infty)$. Then f is a finite, positive measure.

Although there are several significant results there are two principal ones that we discuss. The first is an atomic "decomposition" theorem for $LH(1,\infty)$. The second is the introduction of a very natural subspace of $LH(1,\infty)$ for which they characterize the dual in terms of oscillating functions. Using the above they obtain some distribution type inequalities which were previously known only for finite measures. The results on the dual space motivate natural questions for the Section 3 on classical Hardy space functions (on the unit disk).

We define the idea of finite overlap for cubes in $\mathbb{R}^n$.

DEFINITION 2.3. A family of cubes $\{Q_\alpha\}$ in $\mathbb{R}^n$ is said to have finite overlap if for every $x \in \mathbb{R}^n$,

$$\#\{\alpha : x \in Q_\alpha\} \leq c,$$

where c is an absolute constant.

THEOREM 2.4. *Given $f \in LH(1,\infty)$ there exist sequences of functions in $L^\infty(\mathbb{R}^n)$, $\{f_k\}_{-\infty}^{+\infty}$, and $\{g_{k,i}\}_{i=1}^{\infty}$ and a family of cubes $\{Q_{k,i}\}_{i=1}^{\infty}$ with finite overlap such that the following is true:*

(a) $(f - \sum_{|k|\leq N} f_k) \to 0$ in the sense of distributions.

(b) Each f_k may be decomposed as

$$f_k = \sum_{i=1}^{\infty} g_{k,i},$$

where the convergence is in L^1, and the $\{g_{k,i}\}$ satisfy the following:

(i) $g_{k,i}$ is supported in $Q_{k,i}$ for each k and i,

(ii) $\int_{Q_{k,i}} g_{k,i} = 0$,

(iii) $\|g_{k,i}\|_{L^\infty} \leq d\,2^k$ and

$$\sum_{i=1}^{\infty} m(Q_{k,i}) \leq \frac{C_1}{2^k} \leq \frac{C^*[f]}{2^k}.$$

Moreover, if f is a tempered distribution satisfying the above, with the slight modification $\sum_{i=1}^{\infty} m(Q_{k,i}) \leq \frac{C_1}{2^k}$ of the last inequality, then $f \in LH(1, \infty)$.

In the atomic condition for $H^1(\mathbb{R}^n)$ there is no overlap condition put on the cubes. Using this theorem they prove the following two propositions. Note that it need not be the case that the Fourier transform of an $f \in LH(1, \infty)$ need be bounded. But the following useful result is true.

PROPOSITION 2.5. *Let $B(0, R)$ denote the ball with center the origin and radius $R > 0$. For $f \in C_c^{\infty}(\mathbb{R}^n)$ the following a priori estimate holds:*

$$\frac{1}{m(B(0,R))} \int_{B(0,R)} \exp\Big(\frac{C|\hat{f}(\xi)|}{[f]}\Big)\, d\xi \leq D,$$

for some constants C and D depending only on the dimension n.

PROPOSITION 2.6. *Let $f \in C^{\infty}(\mathbb{T}^n)$ ($\mathbb{T}^n$ is the unit n torus). Then*

$$\frac{1}{N^n} \sum_{|m| \leq N} \exp\Big(\frac{C|\hat{f}(m)|}{[f]}\Big) \leq D,$$

where C and D depend on n (and $m = (m_1, m_2, \cdots, m_n)$ with $m_k \in \mathbb{Z}$).

It is known from classical Fourier analysis that if a continuous function f satisfies a Dini condition on its oscillation then the Hilbert transform $Hf(x)$ exists at every point and $Hf(x) \in L^{\infty}$. The authors use this fact as motivation to compute the topological dual of an interesting subspace of $LH(1, \infty)$. The subspace is defined as follows.

DEFINITION 2.7. Let $L^1(0) = \{f \in L^1 : \int f = 0\}$ and set $L(0) = \overline{\{L^1(0)\}}$, where the closure is taken in $LH(1, \infty)$.

In order to compute $(L(0))^*$ the authors introduce the following oscillation for functions. Let $\mathcal{O} \subseteq \mathbb{R}^n$ be an open set of finite measure and let $\{Q_k\}$ be a collection of open cubes, $Q_k \subseteq \mathcal{O}$, and the family $\{Q_k\}$ has finite overlap. For $\phi \in L^1_{\text{loc}}(\mathbb{R}^n)$ set

$$\phi_{Q_k} = \frac{1}{m(Q_k)} \int_{Q_k} \phi(x)\, dx$$

and

$$I_k(\phi) \equiv I(\phi, Q_k) = \int_{Q_k} |\phi(x) - \phi_{Q_k}|\, dx.$$

Define an oscillation as

$$\operatorname{Osc}(\phi, \mathcal{O}) = \sup \frac{1}{m(\mathcal{O})} \sum_{k=1}^{\infty} I_k(\phi),$$

where the sup is taken over all such families of cubes in the given set $\mathcal{O}$. Further, set

$$\omega(\delta) \equiv \sup \operatorname{Osc}(\phi, \mathcal{O}), \quad |\mathcal{O}| = \delta > 0.$$

Their theorem is as follows.

THEOREM 2.8. *The dual $(L(0))^*$ can be identified with the class of locally integrable functions ϕ ($\phi \in L^1_{\text{loc}}(\mathbb{R}^n)$) for which*

$$\|\phi\|_* \equiv \int_0^{\infty} \frac{\omega(\delta)}{\delta}\, d\delta < +\infty.$$

They prove more than the duality. The dual of $(L(0))^*$ is an algebra. This follows from the observation that for ϕ and ψ in $(L(0))^*$ and a cube Q one has

$$\begin{aligned}
&\int_Q |\phi(x)\psi(x) - (\phi\psi)_Q|\,dx \\
&\leq \int_Q |\phi(x)|\cdot|\psi(x) - (\psi)_Q|\,dx + \int_Q |\phi(x) - \phi_Q|\cdot|\psi_Q|\,dx + |\phi|\cdot|\psi_Q\phi_Q - (\psi\phi)_Q| \\
&\leq \|\phi\|_\infty \int_Q |\psi(x) - \psi_Q|\,dx + \int_Q |\phi(x) - \phi_Q|\,|\psi_Q|\,dx \\
&+ |\,|Q|\big(\frac{1}{|Q|}\int_Q \psi\big)\,\big(\frac{1}{|Q|}\int_Q \phi\big) - \frac{1}{|Q|}\int_Q \psi\phi| \\
&\leq \|\phi\|_\infty \int_Q |\psi(x) - \psi_Q|\,dx + \|\psi\|_\infty \int_Q |\phi(x) - \phi_Q|\,dx + |\int_Q \phi\psi_Q - \int_Q \psi\phi|.
\end{aligned}$$

This last expression is less than

$$\int_Q |\phi|\,|\psi_Q - \psi|\,dx.$$

So

$$\int_Q |\phi(x)\psi(x) - (\phi\psi)_Q|\,dx \leq 2\|\phi\|_\infty I(\psi, Q) + \|\psi\|_\infty I(\phi, Q).$$

Therefore

$$\operatorname{Osc}(\phi\psi, \mathcal{O}) \leq 2\|\phi\|_\infty \operatorname{Osc}(\psi, \mathcal{O}) + \|\psi\|_\infty \operatorname{Osc}(\phi, \mathcal{O})$$

and

$$\omega(\delta) \leq 2\|\phi\|_\infty \omega(\delta, \psi) + \|\psi\|_\infty \omega(\delta, \phi)$$

implying

$$\|\phi\psi\|_* \leq 2\|\phi\|_\infty \|\psi\|_* + \|\psi\|_\infty \|\phi\|_*.$$

3. The Complex Case

In this section we introduce the space weak $H^1(\mathbb{T})$, denoted by E. A particular efficacious tool in this regard is the A-integral studied early on by Titchmarsh [**17**], Kolmogorov [**13**], Ul'yanov [**19, 20**], and used by Aleksandrov [**1**] to establish a Cauchy integral formula for an interesting subspace of E, written as $E_0 \subset E$. Using the A-integral we prove that E_0 is separable. We give a large class of examples related to inclusion (and exclusion) in E and E_0. We obtain a large class of bounded linear functionals.

As usual weak $L^1(\mathbb{T})$ is the space of measurable functions on $\mathbb{T}$ for which

$$y\,\lambda_f(y) = y\,\lambda(y) = y\,m(\{\zeta \in \mathbb{T} : |f(\zeta)| > y\})$$

is bounded and m is the normalized measure $\frac{d\theta}{2\pi}$, $\zeta = e^{i\theta}$.

Consider the following definitions.

DEFINITION 3.1. $\mathbb{X}_1 \equiv$ weak $L^1(\mathbb{T}) \cap (\bigcap_{0<p<1} H^p)$.

DEFINITION 3.2. $\mathbb{X}_2 \equiv \{f$ analytic on $\mathbb{D} : [f_r] \equiv \sup_{y>0} y\lambda_{f_r}(y) \leq M,$ all $r < 1\}$.

DEFINITION 3.3. $\mathbb{X}_3 \equiv \{f$ analytic on $\mathbb{D} : Mf(\zeta) \equiv$ radial maximal function of $f \in$ weak $L^1(\mathbb{T})\}$.

In his paper Aleksandrov shows the spaces $\mathbb{X}_2$ and $\mathbb{X}_3$ are the same. We shall use the space $\mathbb{X}_1$ and show at the end of the paper that $\mathbb{X}_2 \subset \mathbb{X}_1$. In what follows we will denote $\mathbb{X}_1$ by E.

DEFINITION 3.4. $E_0 \equiv \{f \in E : \lim_{y\to\infty} y\lambda_f(y) = 0\}$ and weak $L_0^1 \equiv \{f$ measurable on $\mathbb{T}$ with $f \in$ weak L^1 and $\lim_{y\to\infty} y\lambda_f(y) = 0\}$.

We continue with earlier notation,

$$[f] = \sup_{y>0} y\, m(\{x : |f(x)| > y\}),$$

and for $0 < p < 1$ the (weak) metric on $L^p(\mathbb{T})$ is given by

$$[f]_p = \sup_{y>0} (y^p \lambda(y))^{1/p} < \infty.$$

Weak $H^p(\mathbb{T})$, $0 < p < 1$ is the space

$$(\text{weak}\, L^p(\mathbb{T})) \cap (\bigcap_{0<q<p} H^q(\mathbb{T})).$$

By Chebyshev's inequality

$$[f]_p \le \|f\|_p, \qquad 0 < p \le 1,$$

$$L^p(\mathbb{T}) \subseteq L(p,\infty;\mathbb{T}) \qquad (\text{weak } L^p(\mathbb{T})),$$

$$[f+g]_p \le 2([f]_p^p + [g]_p^p)^{1/p},$$

$$[\alpha f] = |\alpha|[f].$$

Hence, the weak Hardy spaces are topological vector spaces (which are not normable). In this situation (since m is a finite measure) $f \in L(p,\infty;\mathbb{T}) \Rightarrow f \in L^q(\mathbb{T})$, $q < p \le 1$.

To give the reader an idea where the weak Hardy spaces on $\mathbb{T}$ fit into the classical spaces we list the following.

(a) $H^p \subset$ weak $H^1 = E$ for $1 \le p$,

(b) the space C. T. of Cauchy transforms is a subspace E, e.g.

$$\frac{1}{1-z} = \int \frac{d\delta_1(\zeta)}{1-\bar{\zeta}z} \in E \backslash E_0$$

and if the measure $d\mu$ in

$$f(z) = \int \frac{d\mu(\zeta)}{1-\bar{\zeta}z}$$

is absolutely continuous with respect to Lebesgue measure, then $f \in E_0$.

Note. Cima and Nicolau [**5**] produce an infinite Blaschke product B on $\mathbb{D}$ for which $B' \in E$. In the proof in that paper the estimates on the Poisson kernel show $B' \notin E_0 \cap \frac{L^1}{H_0^1}$. We do not know if there is an infinite Blaschke product C with $C' \in E_0$.

(c) For $a > 0$ set

$$f_a(z) = \frac{1}{(1-z)(1-\log(1-z))^a}.$$

Also for $\theta > 0$ define the auxiliary

$$g_a(\theta) = \frac{1}{\theta(1-\ln\theta)^a}$$

$\theta \sim 0$. Computing

$$g_a'(\theta) = -\frac{(1-\ln\theta)^a + \theta a(1-\ln\theta)^{a-1}(-1/\theta)}{\theta^2(1-\ln\theta)^{2a}} = -\frac{1-a-\ln\theta}{\theta^2(1-\ln\theta)^{a+1}}.$$

For $|\theta| > 0$ sufficiently small we have $-\ln\theta > 0$ and $-\ln\theta > 1-a$, hence $g_a'(\theta) < 0$ for these values. So $g_a(\theta)$ is strictly decreasing for this range of θ's. With $g_a(\theta) = y$ or $g_a^{-1}(y) = \theta$ we have

$$ym(\{\theta : |f_a(\zeta)| > y\}) \simeq 2yg_a^{-1}(y).$$

We analyze the limit as $y \to \infty$ using L'Hospital's rule

$$\begin{aligned}\lim_{y\to\infty} \frac{2g_a^{-1}(y)}{\frac{1}{y}} &= 2\lim_{\theta\to 0} \frac{-(g_a(\theta))^2}{g_a'(\theta)} = 2\lim_{\theta\to 0} \frac{\frac{-1}{\theta^2(1-\ln\theta)^{2a}}}{\frac{-(1-a-\ln\theta)}{\theta^2(1-\ln\theta)^{a+1}}}. \\ &= 2\lim_{\theta\to 0} \frac{1-\ln\theta}{(1-a-\ln\theta)(1-\ln\theta)^a} = 0\end{aligned}$$

Hence, $f_a \in E_0$.

Note. For $0 < a < 1$ the primitive of g_a is $G_a(\theta) = -\frac{(1-\ln\theta)^{1-a}}{1-a}$, and for $a = 1$ the primitive of g_a is $G_a(\theta) = -\ln(1-\ln\theta)$, which are unbounded for $\theta \sim 0$. Hence $f_a \notin H^1$ for $0 < a \leq 1$.

(d) Further we consider the functions

$$F_a(z) = \frac{[1-\log(1-z)]^a}{1-z}, \quad z \in \mathbb{D}, \quad a > 0.$$

An analysis similar to the one above will show (using $G_a(\theta) = y$)

$$\lim_{y\to\infty} \frac{2G_a^{-1}(y)}{1/y} = 2\lim_{\theta\to 0} \frac{(1-\ln|\theta|)^{a+1}}{2-\ln|\theta|} = \infty.$$

Hence, $F_a \notin E$.

Note. For $a \leq 1$ the functions F_a in (d) above are in H^p, all $p < 1$.

PROPOSITION 3.5. *For $f \in E$, and $0 < p < 1$,*

$$[f]_p \leq \|f\|_p \leq C(p)[f].$$

PROOF. Since $f \in L^p$ the following integral formula holds

$$\|f\|_p = \big(p\int_0^\infty y^{p-1}\lambda(y)\,dy\big)^{1/p}.$$

Let $A > 0$ be a constant to be chosen later. Write the integral as

$$p\int_0^A y^{p-1}\lambda(y)\,dy + p\int_A^\infty y^{p-1}\lambda(y)\,dy.$$

An estimate on the first term yields

$$p\int_0^A y^{p-1}\lambda(y)\,dy \leq p\frac{A^p}{p} = A^p.$$

For the second integral write

$$p\int_A^\infty y^{p-1}\lambda(y)\,dy = p\int_A^\infty y^{p-2}(y\lambda(y))\,dy \le \frac{p[f]}{1-p}A^{p-1}.$$

It follows that choosing $A=[f]$ yields

$$[f]_p \le \|f\|_p \le (A^p + \frac{p}{1-p}A^p)^{1/p} \le [f]\cdot C(p).$$

□

The following proposition is an easy consequence of the previous proposition.

PROPOSITION 3.6. *If $\{f_\alpha\}\subseteq E$ and $[f_\alpha]\le M$ for all α, then $\{f_\alpha\}$ is a normal family.*

We now define the A-integral and mention some of its properties.

DEFINITION 3.7. For f measurable on $\mathbb{T}$,

$$\lim_{y\to\infty} y\, m(\{\zeta : |f(\zeta)|>y\}) = 0$$

define the A-integral for $\epsilon>0$ small, and $L>0$ large as,

$$\lim_{\epsilon\to 0, L\to\infty}\int_{\{\epsilon<|f|<L\}} f(\zeta)\,dm(\theta)$$

if it exists and is finite.

Notes. (a) The lower inequality on ϵ is not required, since our measure $dm(\theta)$ is finite.

(b) If the condition on $y\,m(\{|f|>y\})$ is not satisfied there are examples in $E\setminus E_0$ for which the A-integral may fail to exist. For example (see [**1**] or Remark 2.3.18(1) of [**4**]), the function $f(z)=i\,\frac{1+z}{1-z}$ is not in E_0 since $m(|f|\ge L) = \frac{2}{\pi}\arctan(\frac{1}{L})$ is not $o(1/L)$ as $L\to\infty$. Since $f(\zeta)\in\mathbb{R}$ for all $\zeta\in\mathbb{T}\setminus\{1\}$, the quantities $\int_{|f|\le L} f(\zeta)dm(\zeta)$ are real-valued for every L. However, $f(0)=i$ and thus the limit $f(0)=(A)\int_{\mathbb{T}} f(\zeta)dm(\zeta)$ in Theorem 3.5 fails.

(c) Titchmarsh [**17**] has proved the following. If f and g are measurable on $\mathbb{T}$ and if two of three A-integrals, $A(f)$, $A(g)$, $A(f+g)$ exist then $A(f+g) = A(f)+A(g)$. See Bary [**3**], page 130, for a proof. So if $\mathbb{X}$ is the set of functions f for which the A-integral exists then A is linear on $\mathbb{X}$.

Hence, the A-integral is linear on the subspace of weak $L^1(\mathbb{T})$ consisting of those functions $f\in$ weak $L^1(\mathbb{T})$ for which

$$\lim_{y\to\infty} y\,\lambda_f(y) = 0.$$

Write this subspace as $L(1, dm(\theta), \mathbb{T})$. We recall that Titchmarsh's theorem holds for vectors in this subspace. The space $E_0\subset L(1, dm(\theta),\mathbb{T})$. In fact the space we have labeled as $\mathbb{X}_1$ is the space E.

PROPOSITION 3.8. *The space E_0 is closed in E and hence complete.*

PROOF. Let $\{f_n\}_{n=1}^\infty$ be a sequence in E_0 such that $[f_n - f]\to_{n\to\infty} 0$ and $f\in E$. Define the sets $A(y)=\{\zeta\in\mathbb{T} : |f(\zeta)|>y\}$, $B_n(y)=\{\zeta\in\mathbb{T} : |(f_n-f)(\zeta)|>y\}$ and $C_n=\{\zeta\in\mathbb{T} : |f_n(\zeta)|>y\}$.

Let $\epsilon > 0$ and choose N so that (by assumption, for all y) the inequality $y\, m(B_n(y)) < \epsilon$ holds for all $n \geq N$. Since

$$A(y) \subset B_n(\frac{y}{2}) \cup C_n(\frac{y}{2}),$$

choose $n = N$ to obtain

$$y\, m(A(y)) \leq 2(\frac{y}{2})\, m(B_N(\frac{y}{2})) + 2(\frac{y}{2})\, m(C_N(\frac{y}{2})) < 2\epsilon + 2\epsilon = 4\epsilon$$

since $f_N \in E_0$. Hence, $f \in E_0$. □

The following theorem is Aleksandrov's [**1**] form of the "weak" Cauchy integral on E_0.

THEOREM 3.9. *If $f \in E_0$, and $z \in \mathbb{D}$, then*

$$Af(z) = \lim_{L\to\infty} \int_{\{|f|\leq L\}} \frac{f(\zeta)\, dm(\zeta)}{1 - \bar{\zeta} z} = f(z).$$

Note. The equality makes sense since f is defined on $\mathbb{D}$. It is only necessary to consider the case $z = 0$ as the general form follows in a straight forward way from this (see [**1**] or Section 2.3 of [**4**]). This type of integral was used in works by Tseretelli [**18**] and Uljanov [**19, 20**]. The result of Aleksandrov strengthens and improves some of these earlier results.

The following theorem follows by using the arguments (in modified form) of the proof of the above theorem.

THEOREM 3.10. *The space E_0 is separable.*

PROOF. Obviously it suffices to show that H^1 is dense in E_0. To this end let $f \in E_0$ and recall the symmetric decreasing rearrangement f^* of f is given by

$$f^*(x) = \inf\{y > 0 : \lambda_f(y) = \lambda(y) \leq x\}.$$

This function is non-negative, non-increasing and right continuous. Also, $f^*(x) = o(\frac{1}{x})$, $x \to 0^+$, and

$$xf^*(x) \leq y\lambda(y)$$

(recall f and f^* have the same distribution function). Hence, for $q \geq 0$, there is a map $h : \mathbb{T} \to [0, 1]$, satisfying

$$q = q^* \circ h(x).$$

Writing this in terms of f and f^* we have

$$f^* \circ h(\zeta) = |f(\zeta)|.$$

For $L > 0$, define a function on $\mathbb{T}$

$$g_L(\zeta) = \begin{cases} 0 & \text{if } h(\zeta) \geq \lambda(L) \\ \log \frac{h(\zeta)}{\lambda(L)} & \text{if } h(\zeta) \leq \lambda(L). \end{cases}$$

Since f^* is decreasing it follows that $h(\zeta) \to 0$ as $L \to \infty$ and $h(\zeta) \to 1$ as $L \to 0$. Also $g_L(\zeta) \leq 0$. Taking the Poisson kernel convolved with g_L yields a harmonic function on $\mathbb{D}$; with $g_L(\zeta) \leq 0$. Choose the conjugate harmonic function $\tilde{g}_L$ vanishing at 0, and set

$$F_L(z) = \exp(g_L(z) + i\tilde{g}_L(z)).$$

The integral
$$\int_0^{\lambda(L)} \log \frac{x}{\lambda(L)})\, dx = -\lambda(L)$$
tends to zero as $L \to \infty$. Hence, F_L is an outer function in the unit ball of H^∞ and by normality we have $\lim_{L\to\infty} F_L(z) = m(z)$ uniformly on compacta. But $F_L(0) \to e^0 = 1$ so $m(z) \equiv 1$ for all $z \in \mathbb{D}$ and a. e. on $\mathbb{T}$. Using this Aleksandrov shows $fF_L \in H^1(\mathbb{D})$ for all $L > 0$. Further, he proves
$$\int_{|f|\le L} |fF_L - f|\, dm(\theta) \le c\sqrt{\lambda(L)L},$$
which goes to zero if $f \in E_o$. Let $\epsilon > 0$ be given. For $y > 0$ note that
$$\{\zeta \in \mathbb{T} : |f(1-F_L)(\zeta)| > y\} \subseteq \{\zeta \in \mathbb{T} : |f(\zeta)| > y/2\}.$$
Then by assumption there is y_0 with
$$m(\{|f(1-F_L)| > y\}) \le 2\left(\frac{y}{2}\right) m(\{|f| > \frac{y}{2}\}) < \epsilon$$
if $y \ge y_0$. Note this inequality is independent of $L > 0$.

Considering $y_0 \ge y > 0$ we have
$$\begin{aligned} &y\, m(\{\zeta : |f(1-F_L)(\zeta)| > y\}) \\ &= y\, m(\{\zeta : |f(1-F_L)(\zeta)| > y_0\}) + y\, m(\{\zeta : y_0 \ge |f(1-F_L)(\zeta)| > y\}) \\ &\le y_0\, m(\{\zeta : |f(1-F_L)(\zeta)| > y_0\}) + y\, m(\{\zeta : y_0 \ge |f(1-F_L)(\zeta)| > y\}) \\ &< \epsilon + y\, m(\{\zeta : y_0 \ge |f(1-F_L)| > y\}). \end{aligned}$$
But for ζ in the last set
$$|f(\zeta)(1-F_L(\zeta))| \to 0 \quad \text{as} \quad L \to \infty$$
point wise and a. e. Hence, letting $L \to \infty$ we have
$$y\, m(\{y_0 \ge |f(1-F_L)| > y\}) \to 0$$
as $L \to \infty$. This completes the proof. □

PROPOSITION 3.11. *For $p \ge 1$, H^p is dense in E_0.*

Applying the A-integral we have the following

PROPOSITION 3.12. *Assume $f(z) = \sum_{n=0}^\infty a_n z^n$ is in E_o. Then*
$$a_n = (A)\int \bar{\zeta}^n f(\zeta)\, dm(\zeta)., \quad n \ge 0,$$
and $(A)\int \bar{\zeta}^n f(\zeta) dm(\zeta) = 0$ for all $n < 0$.

PROOF. First note the backward shift is bounded on E_0. That is if $f \in E_0$ then
$$f_1(z) \equiv \frac{f(z) - f(0)}{z} = a_1 + a_2 z + a_3 z^2 + \cdots \in E_0.$$
Moreover, $f_1 \in E_0$ since
$$\{y\, m(\{\zeta : |\frac{f(\zeta) - f(0)}{\zeta}| > y\})\} = \{y\, m(\{\zeta : |f(\zeta) - f(0)| > y\})\}$$
and it follows that
$$\lim_{y\to 0} y\, m(\{\zeta : |f(\zeta) - f(0)| > y\}) = \lim_{y\to 0} y\, m(\{\zeta : |f(\zeta)| > y - |f(0)|\}) = 0.$$

Hence,

$$f_1(z) = \frac{(A)\int \frac{f(\zeta)\,dm(\zeta)}{1-z\bar{\zeta}} - (A)\int f(\zeta)\,dm(\zeta)}{z}$$
$$= (A)\int \frac{f(\zeta) - f(\zeta)(1-\bar{\zeta}z)\,dm(\theta)}{z(1-\bar{\zeta}z)}$$
$$= (A)\int \frac{\bar{\zeta}f(\zeta)\,dm(\theta)}{1-\bar{\zeta}z}.$$

Evaluating at $z = 0$ yields

$$a_1 = f_1(0) = (A)\int \bar{\zeta}f(\zeta)\,dm(\theta).$$

Since $f_1 \in E_0$ we may apply $S_{\bar{z}}$ to f_1 to get $S_{\bar{z}}(f_1)(z) = a_2 + a_3z + a_4z^2 + \cdots = f_2(z)$ and $f_2(0) = a_2 = A\int \bar{\zeta}^2 f(\zeta)\,dm(\theta)$. Then by induction

$$a_n = (A)\int \bar{\zeta}^n f(\zeta)\,dm(\zeta).$$

Thus the nth Taylor coefficient of f is the nth A-distribution coefficient of f.

Finally for $n < 0$,

$$\int_{|\zeta^n f|\le L} \bar{\zeta}^n f(\zeta)dm(\zeta) = \int_{|f|\le L} \bar{\zeta}^n f(\zeta)dm(\zeta)$$

and by the proof of Aleksandrov's theorem

$$\int_{|f|\le L} \bar{\zeta}^n f dm = \int_{\mathbb{T}} \bar{\zeta}^n f F_L dm + \int_{|f|\le L} \bar{\zeta}^n f(1-F_L)dm - \int_{|f|>L} \bar{\zeta}^n f F_L dm.$$

The first integral on the right is equal to zero since $fF_L \in H^1$, while the other two integrals converge to zero as $L \to \infty$.

□

At this point we are trying to view E_0 as the analogue of the $L(0)$ subspace of $LH(1,\infty;\mathbb{R}^n)$. Although it appears to characterize the dual of E_0 we will have to follow the lead in Section 2 and develop an oscillation function for functions. We have been unable to do this at this writing. However we do prove that there is a large class of well known analytic functions that are in the dual of E_0. Further the action of these functionals agrees with some earlier results known for H^p $(p < 1)$ spaces.

First we recall the results of Duren, Romberg and Shields [**7**]. They prove the following: if $\lambda \in (H^p)^*$ then there exists a $g \in A_{\alpha(p)}$, $\alpha(p) = \frac{1}{p} - 1$ (i. e., g is an analytic Lipschitz function and $A_{\alpha(p)} \subset A$ (the disk algebra)) and

$$\lambda(f) = \lim_{r\uparrow 1} \int_{\mathbb{T}} f(r\zeta)\overline{g(\zeta)}dm(\zeta).$$

Also every $g \in A_{\alpha(p)}$ induces such a bounded linear functional on E_0.

This is proven as follows

Proposition 3.13. *For* $0 < p < 1$,

$$\cup_{0<p<1}(H^p(\mathbb{T}))^* \subseteq E_0^*.$$

PROOF. This follows easily from Proposition 3.2. Namely if $\lambda = \lambda_\phi$ is a functional from some $(H^p)^*$ (with $\phi \in A_{\alpha(p)}$) then for $f \in E_0 \subset H^p$

$$\begin{aligned}|\lambda_\phi(f)| &= |\lim_{r\uparrow 1} \int f(r\zeta)\overline{\phi(\zeta)}dm(\zeta)| \\ &\leq \|\lambda_\phi\| \cdot \|f\|_{H^p} \leq C(p)\|\lambda_\phi\| \cdot [f].\end{aligned}$$

□

We know from the closed graph theorem (e. g., Rudin [**16**]) that if Λ is a linear functional on E_0, f and f_n in E_0 are such that $[f_n - f] \to_{n\to\infty} 0$ and $\Lambda(f_n) = c_n \to_{n\to\infty} c_0$ and if $\Lambda(f) = c_0$ then Λ is a continuous linear functional on E_0. It suffices to show that $\lim_{n\to\infty} (A)\int f_n g = (A)\int fg = c_0$, hence, $\Lambda_g(f) = \Lambda(f) = (A)\int fg$, where $g \in H^\infty$.

From the following known results

$$H^\infty \subseteq BMOA = (H^1)^*$$

and from $[f] \leq \|f\|_{H^1}$ any bounded linear functional on H^1 restricts to a bounded linear functional on the subspace $H^1 \subset E_0$ with the E_0 topology. However, since the Hahn–Banach theorem does not necessarily hold in non-locally convex spaces this functional can not (in general) be extended from H^1 to E_0. But the following result will prove that vectors from H^∞ will induce bounded linear functionals on E_0.

The space we have labeled as E_0 is a subset of $\mathbb{X}_1(= E)$. Functions in $\mathbb{X}_2$ have well defined boundary values a.e. on the unit circle. If we take the set of those functions and consider their boundary values which satisfy the little "oh" growth condition on the $\lambda_f(y)$ as y goes to infinity we label this set as $\mathbb{X}_{2,0}$. Our proof on page 17 shows also that $X_{2,0} \subset E_0$.

A close study of the Aleksandrov proof of his result showes that he does not use the definitions involved in defining $\mathbb{X}_2$ but only the definitions from E_0. Hence, we are proving things for the larger set E_0.

One can check that for $f \in E_0$ and $g \in H^\infty$ the product $gf \in E_0$. By the Aleksandrov's theorem

$$\Lambda_g(f) = (A)\int fg$$

exists. It is linear and we shall show using the closed graph theorem that it is continuous on E_0.

THEOREM 3.14. *Λ_g as defined above is a continuous linear functional on E_0. That is H^∞ can be considered as a subspace of the dual space $(E_0)^*$.*

It suffices to prove the following theorem.

THEOREM 3.15. *If $f \in E_0$ and $f_n \in E_0$, $n = 1, 2, \ldots$, with $[f_n - f] \to_{n\to\infty} 0$ and if $g \in H^\infty$ then for every $\epsilon > 0$ there is an integer $N \geq 0$ such that*

$$|\int_{|f_n g|\leq L} f_n g - \int_{|fg|\leq L} fg| < \epsilon$$

for all $n \geq N$ and all $L > 0$.

PROOF. We define the following sets (they depend on L):

$$B(L) = B = \{\zeta \in \mathbb{T} : |(fg)(\zeta)| \leq L\},$$

$$Q_n(L) = Q_n = \{\zeta \in \mathbb{T} : |(f_n - f)(\zeta)| \geq L\},$$
$$A_n(L) = A_n = \{\zeta \in \mathbb{T} : |(f_n g)(\zeta)| \leq L\}.$$

We shall decompose the sets A_n and B as follows:

$$\begin{aligned} A_n &= (A_n \cap Q_n) \bigcup ((A_n \cap B) \setminus (A_n \cap B \cap Q_n)) \\ &\bigcup (A_n \setminus ((A_n \cap Q_n) \cup ((A_n \cap B) \setminus (A_n \cap B \cap Q_n))) \\ &=_{def} A_{1,n} \cup A_{2,n} \cup A_{3,n}, \end{aligned}$$

$$\begin{aligned} B &= (B \cap Q_n) \bigcup ((B \cap A_n) \setminus (B \cap A_n \cap Q_n)) \\ &\bigcup (B \setminus ((B \cap Q_n) \cup ((B \cap A_n) \setminus (B \cap A_n \cap Q_n))) \\ &=_{def} B_{1,n} \cup B_{2,n} \cup B_{3,n}. \end{aligned}$$

We observe the following. Assuming $[f_n - f] \to 0$ as $n \to \infty$ and let $\epsilon > 0$ be given, then

$$m\{\zeta : |f_n(\zeta) - f(\zeta)| > \epsilon\} \leq \frac{[f_n - f]}{\epsilon}.$$

Letting $n \to \infty$ we have $f_n \to f$ in measure. Also if $C_n = \{\zeta : |f_n(\zeta) - f(\zeta)| > \epsilon\}$ and $D_n = \mathbb{T} \setminus C_n$, then $1 = m(C_n) + m(D_n)$ and $m(D_n) = m\{\zeta : |f_n(\zeta) - f(\zeta)| \leq \epsilon\} = 1 - m(C_n)$. As shown above $m(C_n) \to 0$ as $n \to \infty$. Hence, $m(D_n) \to 1$ and this implies f_n tends poinwise a. e. to f.

Set

$$\int_{|f_n g| \leq L} f_n g = \int_{A_n(L)} f_n g = \int_{A_{1,n}} f_n g + \int_{A_{2,n}} f_n g + \int_{A_{3,n}} f_n g$$

and

$$\int_{|fg| \leq L} fg = \int_{B(L)} fg = \int_{B_{1,n}} fg + \int_{B_{2,n}} fg + \int_{B_{3,n}} fg.$$

It is sufficient to show

$$\int_{A_{1,n}} f_n g \to 0 \quad \text{and} \quad \int_{B_{1,n}} f_n g \to 0$$

and that

$$\int_{A_{2,n}} f_n g - \int_{B_{2,n}} f_n g \to 0 \text{ as } n \to \infty.$$

Finally, to conclude we show

$$\int_{A_{3,n}} f_n g \to 0 \quad \text{and} \quad \int_{B_{3,n}} f_n g \to 0 \text{ as } n \to \infty.$$

By assumption there exists N_1 such that if $n \geq N_1$ then

$$y\, m(Q_n(y)) = y\lambda_{([f - f_n])} = \epsilon(y) < \epsilon$$

for all $y > 0$, where N_1 is independent of y. Hence

$$|\int_{A_{1,n}} f_n g| \leq \int_{A_n \cap Q_n} |f_n|\,|g| \leq \int_{A_n \cap Q_n} L dm \leq L\, m(Q_n) \leq L(\frac{\epsilon}{L}) = \epsilon,$$

and

$$|\int_{B_{1,n}} fg| \leq \int_{B \cap Q_n} |f|\,|g| \leq \int_{B \cap Q_n} L dm \leq L\, m(Q_n) \leq L(\frac{\epsilon}{L}) = \epsilon.$$

Since $A_{2,n} = B_{2,n} = S_{2,n}$,

$$\int_{A_{2,n}} f_n g - \int_{B_{2,n}} fg = \int_{S_{2,n}} (f_n - f)g.$$

Notice that $(f_n - f)g \in L^1$ on the set $S_{2,n}$ since $|f_n - f|\,|g| \leq 2L$. By assumption $[f_n - f] \to 0$, thus for $0 < p_0 < 1$ one has

$$\|f_n - f\|_{p_0} \leq C(p_0)[f_n - f] \to 0.$$

Since, $|(f_n - f)|^{p_0}$ is in L^1 and $\|(f_n - f)^{p_0}\|_{L^1} \to 0$ then $(f_n - f)^{p_0}$ converges in measure, and so also $(f_n - f)$ converges in measure and point wise a. e. Hence for every $\epsilon > 0$ there is an integer $N_2 \geq 0$ such that

$$|\int_{S_{2,n}} (f_n - f)g| < \frac{\epsilon}{L}$$

for all $n \geq N_2 \geq N_1$.

It remains to consider the cases on the sets $A_{3,n}$ and $B_{3,n}$. First we want to rewrite these sets. For $A_{3,n}$ we proceed as follows. Set $X = (A \cap Q)$, $Y = (A \cap B) \setminus (A \cap B \cap Q)$, and $Z = A \setminus \{(A \cap Q) \cup [(A \cap B) \setminus (A \cap B \cap Q)]\}$. Then

$$\begin{aligned} Z &= A \setminus (X \cup Y) \\ &= A \cap (X \cup Y)^c \\ &= A \cap (X^c \cap Y^c) \\ &= A \cap ((A \cap Q)^c \cap Y^c) \\ &= A \cap ((A^c \cup Q^c) \cap Y^c) \\ &= A \cap [(A^c \cap Y^c) \cup (Q^c \cap Y^c)] \\ &= [A \cap (A^c \cap Y^c)] \cup [A \cap (Q^c \cap Y^c)] \\ &= A \cap Q^c \cap Y^c. \end{aligned}$$

Now

$$\begin{aligned} Y &= (A \cap B) \setminus (A \cap B \cap Q) \\ &= (A \cap B) \cap (A \cap B \cap Q)^c \\ &= (A \cap B) \cap (A^c \cup B^c \cup Q^c) \\ &= [(A \cap B) \cap A^c] \cup [(A \cap B) \cap (B^c \cup Q^c)] \\ &= [(A \cap B) \cap Q^c] \cup [(A \cap B) \cap B^c] \\ &= A \cap B \cap Q^c. \end{aligned}$$

Thus

$$\begin{aligned} Z &= (A \cap Q^c) \cap (A \cap B \cap Q^c)^c \\ &= (A \cap Q^c) \cap [B^c \cup (A \cap Q^c)^c] \\ &= [(A \cap Q^c) \cap B^c] \cup [(A \cap Q^c) \cap (A \cap Q^c)^c] \\ &= A \cap B^c \cap Q^c. \end{aligned}$$

Therefore

$$A_{3,n} = A_n \cap B^c \cap Q_n^c.$$

Thus

$$A_{3,n} = \{\zeta \in \mathbb{T} : |f_n(\zeta)g(\zeta)| \leq L \text{ and } |f(\zeta) - f_n(\zeta)| < L \text{ and } |f(\zeta)g(\zeta)| > L\}.$$

It remains to show that

$$\int_{A_{3,n}} f_n(\zeta)g(\zeta)\,dm(\theta) \to_{n\to\infty} 0$$

and that

$$\int_{B_{3,n}} f(\zeta)g(\zeta)\,dm(\theta) \to_{n\to\infty} 0$$

Let

$$M_n \equiv \bigcup_{k\geq n} A_{3,k},$$

and let

$$M = \bigcap_{n\geq 1} M_n.$$

That is M is the $\limsup_n A_{3,n}$. To reach a contradiction assume $m(M) = \delta > 0$. In this case since the M_n are nested we have for a fixed $\zeta \in M$ that ζ is in $A_{3,n(k)}$ for infinitely many $k \to \infty$. Then

$$L \geq \lim_{n\to\infty} |f_{n(k)}(\zeta)g(\zeta)|$$

and for which

$$\lim_{n\to\infty} |f_{n(k)}(\zeta)g(\zeta)| = |f(\zeta)g(\zeta)| > L.$$

This is a contradiction to $m(M) > 0$, and so $m(M) = 0$. In this case if there is a subsequence $n(k)$ for which $m(A_{3,n(k)}) \geq \delta > 0$ we have

$$0 \leq m(A_{3,n(k)}) \leq m(M_{n(k)}) \to 0$$

and again we have a contradiction. Hence, $m(A_{3,n})$ tends to zero on n.

Thus

$$\int_{A_{3,n}} |f_n(\zeta)g(\zeta)\,dm(\theta) \leq L\,m(A_{3,n}) \to 0$$

as $n \to 0$.

Similarly,

$$B_{3,n} = B \cap A_n^c \cap Q_n^c.$$

Thus

$$B_{3,n} = \{\zeta \in \mathbb{T} : |f(\zeta)g(\zeta)| \leq L \text{ and } |f(\zeta) - f_n(\zeta)| < L \text{ and } |f_n(\zeta)g(\zeta)| > L\}.$$

The same proof will show that $\int_{B_{3,n}} |f(\zeta)g(\zeta)\,dm(\theta) \to 0$. □

We will prove that if f satisfies $\mathbb{X}_2$ then f satisfies $\mathbb{X}_1$. Let us assume f satisfies $\mathbb{X}_2$ and show it satisfies $\mathbb{X}_1$. Let us assume f is analytic on $\mathbb{D}$ and there is $M > 0$ with $[f_r] \leq M$, all $r < 1$, and show $f \in E$. Since for $0 < p < 1$, we have

$$\|f_r\|_p \leq C(p)\,[f_r] \leq C(p)\,M,$$

we have $f \in H^p(\mathbb{T})$ for all $p < 1$. Then $f_r \to f$ in $L^p(\mathbb{T})$ and so $f_{r_{k(n)}}(\zeta) \to f(\zeta)$ p. w. a. e. (and some subsequence in measure).

Let $A_n(y) = \{|f(r_n\zeta)| > y\}$, $B = \{|f(\zeta)| = y\}$, and $A = \{|f(\zeta)| > y\}$. Let

$$\begin{aligned} S(y) &= \limsup\,(A_n(y)) = \cap_{k=1}^{\infty}(\cup_{n\geq k} A_n(y)) \\ &= \{\zeta : \zeta \in A_n \text{ for infinitely many } n\} \end{aligned}$$

and

$$\begin{aligned} I(y) &= \liminf (A_n(y)) = \cup_{k=1}^{\infty}(\cap_{n\geq k} A_n(y)) \\ &= \{\zeta : \zeta \in A_n \text{ for all but finitely many } n\}. \end{aligned}$$

Since $m(\mathbb{T}) = 1 < \infty$ and $A_n(y) \subseteq \mathbb{T}$ we have

$$m(I(y)) = m(\liminf A_n(y)) \leq \liminf m(A_n(y)).$$

Also,

$$m(S(y)) = m(\limsup A_n(y)) \geq \limsup m(A_n(y)).$$

Our assumption $[f_r] \leq M < \infty$ for all $r < 1$ implies if $r_n \uparrow 1$,

$$y\,\lambda_n(y) = y\,m(|f_{r_n}(\zeta)| > y) \leq M$$

all n and all $y > 0$.

Let $U_\alpha = \{\zeta \in \mathbb{T} : |f(\zeta)| = \alpha,\ \alpha > 0\}$ with $m(U_\alpha) > 0$. This set is countable, say $\{X_j\}_{j=1}^{\infty}$. If we show for every $y \notin X_j$, $j = 1, 2, \ldots$, that

$$y\,m(|f(\zeta) > y) \leq M$$

we have $f \in \text{weak } L^1(\mathbb{T})$ and we will be finished.

First claim $S(y) \subseteq A \cup B$. If $\zeta_0 \in S(y)$, there is a sequence $\{r_{k(n)}\}$, $r_{k(n)} \uparrow 1$ with $\zeta_0 \in A_{r_{k(n)}}(y)$. Then

$$|f(\zeta_0)| = \lim_{n\to\infty} |f(r_{k(n)}\zeta_0)| \geq y.$$

So $\zeta_0 \in A \cup B$.

Now claim $I(y) \subseteq A \cup B$. If $\zeta_0 \in I(y)$ there exists N such that $\zeta_0 \in A_n(y)$ for $n \geq N$. Hence, $|f_{r_n}(\zeta_0)| \mapsto |f(\zeta_0)| \geq y$.

Note if $\zeta_0 \in A_n(y)$ all $n \geq N$ then $m(A_n(y)) \leq M/y$ so $m(S(y)) = m(A(y)) = m(I(y))$ and $m(I(y)) \leq \liminf m(A_n(y)) \leq M/y$ implies $y\,m(A(y)) \leq M$, thus $f \in \text{weak } L^1(\mathbb{T})$.

As an unanswered question we ask for a proof that $\mathbb{X}_1$ implies $\mathbb{X}_2$ (or $\mathbb{X}_3$).

References

[1] A. B. Aleksandrov, *A-integrability of boundary values of harmonic functions* (Russian), Mat. Zametki **30** (1981), no. 1, 59–72, 154. MR627941 (83j:30039)

[2] A. B. Aleksandrov, *Essays on nonlocally convex Hardy classes*, Complex analysis and spectral theory (Leningrad, 1979/1980), Lecture Notes in Math., vol. 864, Springer, Berlin-New York, 1981, pp. 1–89. MR643380 (84h:46066)

[3] N. K. Bary, *A treatise on trigonometric series. Vols. I, II*, Authorized translation by Margaret F. Mullins. A Pergamon Press Book, The Macmillan Co., New York, 1964. MR0171116 (30 #1347)

[4] J. A. Cima, A. L. Matheson, and W. T. Ross, *The Cauchy transform*, Mathematical Surveys and Monographs, vol. 125, American Mathematical Society, Providence, RI, 2006. MR2215991 (2006m:30003)

[5] J. A. Cima and A. Nicolau, *Inner functions with derivatives in the weak Hardy spaces,* Proc. Amer. Math. Soc. (to appear).

[6] M. Cwikel and Y. Sagher, $L(p, \infty)$, Indiana Univ. Math. J. **21** (1971/72), 781–786. MR0295071 (45 #4139)

[7] P. L. Duren, B. W. Romberg, and A. L. Shields, *Linear functionals on H^p spaces with $0 < p < 1$*, J. Reine Angew. Math. **238** (1969), 32–60. MR0259579 (41 #4217)

[8] C. Fefferman and E. M. Stein, *H^p spaces of several variables*, Acta Math. **129** (1972), no. 3-4, 137–193. MR0447953 (56 #6263)

[9] R. Fefferman and F. Soria, *The space* Weak H^1, Studia Math. **85** (1986), no. 1, 1–16 (1987). MR879411 (88f:42042)

[10] G. B. Folland, *Real analysis*, 2nd ed., Pure and Applied Mathematics (New York), John Wiley & Sons, Inc., New York, 1999. Modern techniques and their applications; A Wiley-Interscience Publication. MR1681462 (2000c:00001)

[11] L. Grafakos, *Classical Fourier analysis*, 2nd ed., Graduate Texts in Mathematics, vol. 249, Springer, New York, 2008. MR2445437 (2011c:42001)

[12] R. A. Hunt, *On $L(p, q)$ spaces*, Enseignement Math. (2) **12** (1966), 249–276. MR0223874 (36 #6921)

[13] A. N. Kolmogorov, *Sur les functions harmoniques conjuquées et les series de Fourier,* Fund. Math. **7(1925), 23–28.**

[14] G. G. Lorentz, *Some new functional spaces*, Ann. of Math. (2) **51** (1950), 37–55. MR0033449 (11,442d)

[15] G. G. Lorentz, *On the theory of spaces* Λ, Pacific J. Math. **1** (1951), 411–429. MR0044740 (13,470c)

[16] W. Rudin, *Functional analysis*, 2nd ed., International Series in Pure and Applied Mathematics, McGraw-Hill, Inc., New York, 1991. MR1157815 (92k:46001)

[17] E. Titchmarsh, *On conjugate functions,* Proc. London Math. Soc. **29(1929), 49–80.**

[18] O. D. Cereteli, *Remarks on the theorems of Kolmogorov and of F. and M. Riesz* (Russian), Proceedings of the Symposium on Continuum Mechanics and Related Problems of Analysis (Tbilisi, 1971), Izdat. "Mecniereba", Tbilisi, 1973, pp. 241–254. MR0385444 (52 #6306)

[19] P. L. Ul′yanov, *The A-integral and conjugate functions* (Russian), Moskov. Gos. Univ. Uč. Zap. Mat. **181(8)** (1956), 139–157. MR0084627 (18,892d)

[20] P. L. Ul′yanov, *On the A-Cauchy integral. I* (Russian), Uspehi Mat. Nauk (N.S.) **11** (1956), no. 5(71), 223–229. MR0083552 (18,726a)

Department of Mathematics, Lamar University, Beaumont, Texas 77710
E-mail address: valentin.andreev@lamar.edu

Department Of Mathematics, University of North Carolina, Chapel Hill, North Carolina 27599
E-mail address: cima@email.unc.edu

Published Titles in This Subseries

638 **Javad Mashreghi, Emmanuel Fricain, and William Ross, Editors,** Invariant Subspaces of the Shift Operator, 2015

630 **Pierre Albin, Dmitry Jakobson, and Frédéric Rochon, Editors,** Geometric and Spectral Analysis, 2014

622 **S. Ejaz Ahmed, Editor,** Perspectives on Big Data Analysis, 2014

606 **Chantal David, Matilde Lalín, and Michelle Manes, Editors,** Women in Numbers 2, 2013

605 **Omid Amini, Matthew Baker, and Xander Faber, Editors,** Tropical and Non-Archimedean Geometry, 2013